RECENT ADVANCES IN AQUACULTURE

RECENT ADVANCES IN AQUACULTURE

Edited by
James F. Muir
& Ronald J. Roberts

CROOM HELM
London & Canberra

WESTVIEW PRESS
Boulder, Colorado

© 1982 James F. Muir and Ronald J. Roberts
Croom Helm Ltd, 2-10 St John's Road, London SW11

British Library Cataloguing in Publication Data

Recent advances in aquaculture.
   1. Aquaculture
   I. Muir, James F. II. Roberts, Ronald J.
   630'.9162   SH135 R43
   ISBN 0-7099-0303-0

First published in the United States of America in 1982 by
Westview Press, Inc., 5500 Central Avenue, Boulder, Colorado 80301
Frederick A. Praeger, President and Publisher

Library of Congress Catalog Card Number: 82-50692
ISBN 0-86531-464-0

Typeset by Elephant Productions, London SE19
Printed and bound in Great Britain

CONTENTS

PREFACE

Currently, as the main scientific foundations are being brought together with the essential organisational needs, aquaculture is becoming a more clearly defined area of study and development, as many successful sections of the aquaculture industry will testify. However, as in many multidisciplinary subjects, there is frequently a need to collect and integrate the many developments occurring in the individual component subjects, and to assess and express from the viewpoint of the wider discipline the significance of these developments.

In teaching and research in aquaculture, and in advising and assisting in aquaculture projects throughout the world, we have been particularly aware of the need to bring together some of the more recent developments and trends in a form not readily available in single scientific papers or in the more general textbooks in this field. We are also convinced of the need to treat aquaculture as an integrated subject and to present descriptions accordingly.

Thus we are aiming here to present a series of reviews, with the intention of providing up-to-date material in areas of aquaculture which are of particular current interest or development potential. While we hope this presentation will be of considerable interest and use to researchers, teachers and students in the field of aquaculture, who will find summaries on their particular field of study, we would also hope that the many planners, commercial investors, agency officials and managers associated with the current practical developments in aquaculture, will find the information they require for effective production and future expansion. For this reason particularly, a number of our first contributions specifically describe the economic implications of the biological and technical developments involved. While this does not exclude more specific reviews, we hope to maintain, as a priority, the practical aquaculturist's viewpoint.

Such is the state of current development in aquaculture, that a number of subjects are readily suitable for inclusion from the outset; it was felt that a cross-section of subjects spanning a range from habitat study, through straight-forward developments of culture techniques, to specific aquaculture technology, would best represent the breadth of interest and development in the field. For those familiar with the subjects chosen, their inclusion may be self-explanatory; it may be useful however to describe briefly the relevant background, and the involvement of the authors concerned.

The extensive mangrove areas, particularly of South East Asia, and also of Central and South America, have attracted considerable attention for their potential for aquaculture and their development for brackish-water fish production. While sometimes simplistically designated as unproductive areas, developers and planners are increasingly becoming aware of the vital ecological importance of mangroves, their importance as a source of local livelihood, and their significance in larval production of commercially important marine and brackish-water species. Dr Donald Macintosh's extensive experience in mangrove ecology at the University of Malaysia qualifies him particularly to explain the significance of mangroves to aquaculture in Chapter 1.

In the temperate developed economies great interest has centred on the production of the relatively high-valued crustacean species. In the context of the United Kingdom and European markets, Dr John Wickins of the MAFF Laboratory, Conwy, describes the selection of species suitable for intensive culture, the practical development to date of diets, husbandry methods and technologies, and the problems and the prospects for the future. As one of the principal and most effective researchers in this field, Dr Wickins' contribution is particularly valuable.

Tilapia has long been the mainstay of traditional pond culture, and it is perhaps unusual to consider tilapia as a candidate for the type of intensive culture normally typical of salmonids. Nevertheless, as John Balarin and René Haller point out, in appropriate situations the relative ease of culturing tilapia makes them very suitable for intensive production. Rene Haller is particularly well known for his pioneering developments in intensive culture of tilapia and their fry, at the Baobab Farm, near Mombasa in Kenya, and John Balarin, author of a widely popular manual on tilapia culture, has assisted in more recent developments as manager of the farm.

Dr Kim Jauncey's work on the nutrition of common carp has particular relevance to their culture in heated effluents and their growth in traditional pond cultures or polycultures. The science of fish nutrition is at a particularly interesting stage, as the development of diets for new species continues, and as the search for animal protein substitution grows more intense. As leading contenders for cheap food production and low feeding costs, carp are a particularly useful species for these developments.

Snakehead (*Ophiocephalus*, or more correctly, *Channa*) are among the cultured species possessing the highly desirable trait of air-breathing; thus the conventional limits to holding capacity do not apply and remarkable production rates can be obtained in simple pond systems. A very popular and highly priced species in South East Asia, their culture is currently facing problems in feed costs and husbandry-related diseases. Leong Wee, currently researching *Channa* nutrition at the University of Stirling, is attempting to tackle some of these problems, and is thus very well placed to describe current developments.

The final chapter, by one of the editors, J.F. Muir, concerns the technology of recirculated water systems for aquaculture. An area of considerable interest for some time, recirculated water systems have not gained significant acceptance in general use, in spite of the appeal of water economy and quality control. In addition to describing the practical aspects of recirculation system design, attention is given to the main limitations in the use of recirculated systems and to the identification of the most promising current and future applications. This is frequently a case of economic constraints, and this chapter serves to illustrate again the important interrelationship between biological, technical and economic factors in shaping the course of aquaculture.

James F. Muir,
Ronald J. Roberts.

1 FISHERIES AND AQUACULTURE SIGNIFICANCE OF MANGROVE SWAMPS, WITH SPECIAL REFERENCE TO THE INDO-WEST PACIFIC REGION

Donald J. Macintosh

1. Introduction

Mangrove swamps, those curious intertidal swampland communities of trees and animals variously adapted for semi-terrestrial existence, dominate sheltered coastlines throughout the tropics. As a unique protective margin between land and sea, mangrove swamps attract faunal components from adjoining terrestrial and aquatic ecosystems in addition to harbouring many indigenous animal species. Mammals, reptiles and birds exploit the landward mangrove periphery for food and shelter sites, while crabs, prawns and fish migrate into the mangrove zone with the tides for the same purposes. There is usually considerable tidal outflow of mangrove plant litter into coastal waters and a smaller, but comparable, inflow of freshwater-borne material from landward sources. The mangrove ecosystem is therefore an open one, interacting with other ecosystems, and extending in influence far beyond the intertidal zone. Mangrove prawns, for example, are known to migrate seawards to breeding grounds often many kilometres offshore, where they become additions to the marine community (Mohamed and Vedavyasa Rao, 1971; Chong, 1979).

Man has had a long association with mangrove communities. Historically, mangrove environments have been favoured sites for human settlement because of their sheltered coastal locations. The mangrove forest has provided native populations with a seemingly endless variety of derived products: timber, thatching, charcoal, tanning agents, resins, dyes, oils, medications, animal fodder, fish poisons (Watson, 1928, and de la Cruz, 1979, give detailed product inventories). The mangrove environment has also yielded an abundant supply of food: fish and prawns from its waterways, shellfish such as oysters and crabs from the shore zone, and bird's eggs, honey and edible fruits from forest areas.

This form of existence still survives in many parts of South East Asia. 'Sea-gypsies', probably descendants of the original aboriginal inhabitants of Malaya, can be found living in thatched huts on mangrove islands in the Malacca Straits, virtually on the doorstep of one of the busiest shipping lanes in the world (Figure 1.1).

Aquaculture activities in mangrove swamps date back about 500 years to the development of coastal milkfish culture in Indonesia during the fifteenth century. The following extract from Shao-Wen Ling's admirable book on the history of aquaculture in South East Asia (1977) suggests a somewhat ignominious origin to mangrove fish cultivation:

> It is not unreasonable to believe that many of the early pioneers in milkfish farming in Indonesia were prisoners or enemies of the government exiled to remote and venomous animal-infested mangrove swamps.

That milkfish culture was born of the ingenuity of desperate men supports the supposition that indigenous mangrove swamp dwellers were not predisposed towards fish farming because they could readily harvest the natural fish and shellfish populations. Even today, inestimable quantities of food are collected from mangrove areas by hand and by means of simple nets and traps. The fact that mangrove swamps in many countries have traditionally been under the jurisdiction of forestry directorates has also hampered the release of mangrove areas for aquaculture purposes. The mangroves of Peninsular Malaysia have been under forestry management since the 1900s (Watson, 1928), and in area less than 5 per cent (about 500 ha out of 120,000 ha) have so far been developed into culture sites.

Figure 1.1: Community of 'Sea-Gypsies' Living on a Mangrove Island (Pulau Ketam) in the Straits of Malacca. Huts are built entirely of mangrove timber and thatching made from the mangrove palm, *Nypa*.

Apart from milkfish farming the only other widespread culture practice of long standing in mangrove swamps is the familiar trapping and holding system of marine prawn cultivation. Based on the tidal inflow of juvenile prawns into mangrove ponds constructed in the intertidal zone, this culture method has been in existence for most of the present century in South East Asia (Delmendo and Rabanal, 1956; Tham, 1968; Teinsongrusmee, 1970; Cook and Rabanal, 1978). A similar form of prawn rearing, but in conjunction with paddy cultivation, is conducted in the coastal bheris of West Bengal and in the Cochin backwaters (described by Jhingran, 1975, and Kurian and Sebastian, 1976).

Coastal village communities excepted, the aquatic resources value of mangrove swamps has largely been disregarded in favour of their exploitation for wood products: building timber, poles, charcoal, firewood and woodchips. Because of a high resistance to decay in seawater, mangrove wood has been, and remains, one of the most extensively utilised building materials in coastal areas. It is less than 30 years ago that Schuster (1952) wrote, 'The greater part of down-town Singapore is built on *Rhizophora* piles driven into the mud of the harbour region'. Mangroves were the main cash crop in Puerto Rico in 1940, with an estimated 10 million mangrove poles being taken annually (Cerame-Vivas, 1977). Mangrove woodchip production in east Malaysia currently accounts for the annual loss of about 5,000 ha of forest (Ong *et al.*, 1979).

The scale of mangrove forest destruction has been alarming, even in countries such as Malaysia and Thailand where management systems of rotational tree-felling and regeneration or replanting are practised. Approximately 15 to 20 per cent of Thailand's mangrove forests have disappeared over the past decade (Vibulsreth *et al.*, 1976). The annual rate of mangrove deforestation in the Philippines is estimated to have been 24, 285 ha between 1967 and 1975 (Gatus and Martinez, 1977). Complete reclamation of mangrove environments for paddy cultivation and other forms of agriculture, and for solar salt production, has occurred traditionally (de la Cruz, 1979), and now, increasingly, mangrove reclamation is made to accommodate urban and industrial developments as human pressures on the coastal zone escalate.

Two developments over the past 20 years have moderated national opinions that mangrove swamps are expendable wastelands, towards an appreciation of their fisheries resource importance and aquaculture potential. First, ecological studies have shown that large quantities of energy, in the form of mangrove plant detritus, are exported from mangrove swamps into the coastal zone (Odum and Heald, 1975; Christensen, 1978; Ong *et al.*, 1979) and positive correlations have been established between the areal extent of mangroves and the size of fisheries yields from adjacent waters (Macnae, 1974; Martosubroto and Naamin, 1977). By inference, scientists argue that fish catches will decline in proportion to the degree of mangrove destruction. This view is highly credible in the case of the Malacca Straits, for example, where extensive mangrove reclamation for industrial purposes has been paralleled by a substantial drop in catches per unit of fishing effort (Khoo, 1976). Secondly, lucrative regional and international markets for high-quality seafoods have emerged — made possible by innovative technology in the packaging, storage and transportation of perishable foodstuffs. Many tropical countries now regard their coastal fish stocks as a valuable source of foreign exchange. Notable among these is India with an export of marine products totalling 77,946 tonnes in 1978, of which frozen prawns constituted 75.3 per cent (George, 1980).

It is axiomatic that countries like India are keen to develop coastal aquaculture as a means of increasing food production and reducing the country's dependence on capture fisheries. As the natural habitat of many edible species of oysters, prawns and crabs (shellfish with a premium export value), mangrove swamps are prominent among the coastal environments favoured for this purpose (e.g. Ajana, 1980).

Site preparation and design criteria for mangrove fishpond construction, i.e. the physical and engineering considerations involved in converting mangrove areas into culture sites, have been the subject of recent publications by Rabanal (1976), Tang (1976a, b), Gatus and Martinez (1977), Cook and Rabanal (1978), and others. Similarly, current management practices with respect to mangrove pond culture of fish and shellfish are outlined by Tham (1973), Cook (1976), Korringa (1976) and Cook and Rabanal (1978). Excellent earlier accounts include Schuster's (1952) monograph on mangrove tambak farming in Java and the description by Hall (1962) of a mangrove prawn pond in Singapore. However, it is more the objective here to assess the role of mangrove swamps in the ecology of tropical inshore waters and to review the limited biological information concerning aquatic organisms associated with mangrove environments that are of economic importance.

2. Distribution and Extent of Mangroves

Macnae (1968) defined mangroves as 'trees or bushes growing between the level of high water of spring tides and a level close to but above mean sea-level'. Very rarely mangroves may be found inland (see Stoddart *et al.*, 1973), but generally they are unable to compete with other forms of vegetation in non-tidal situations.

Mangroves occur throughout the tropics on sheltered shores bearing soft intertidal sediments. They also have a limited distribution within the sub-tropics. North of the equator, mangroves extend to approximately latitude 30° on the east coast of North America (Louisiana); to 30° on the Pacific Coast (north-west Mexico); to 32° in the Bermuda Islands; to 27° on the Red Sea coast of Africa (Gulf of Suez), and to 31° in Japan (Kyushu). In the southern hemisphere mangroves reach as far as latitude 32-33°S on the east coast of Africa (Kei River, South Africa). Their extreme southerly limit is Corner Inlet (38° 45'S) on the Victorian coast of Australia where *Avicennia marina* is the sole species (Wells, 1980). The absence of mangroves on certain coastlines within their latitudinal limits appears to be due largely to a lack of suitable soft sediments combined with an arid climate. Mangroves reach only 3° 40'S on the Pacific coast of South America, being absent from the coasts of Peru and Chile where these unfavourable conditions prevail.

Evaluation of the world's mangrove resources has been hampered severely by lack of competent regional surveys of the extent and types of mangrove environment. A comprehensive global assessment of mangrove lands, undertaken to overcome this deficiency, is in preparation by a SCOR/UNESCO Working Group (IMS Newsletter, 1980). Available estimates suggest that there is a total area of 9-10 million hectares of mangrove swamps in the Indo-West Pacific Region (Table 1.1). If adjoining tidal flats are included, the figure could well exceed 10 million hectares (Rabanal, 1976). Although mangrove vegetation can be found growing in the most adverse situations — even on coral reef flats and sand spits — development of extensive mangrove forest depends on the following conditions.

Temperature. The lowest average minimum monthly temperature must be above 20°C, with an annual range not exceeding 5°C (West, 1956; Van Steenis, 1962). On the South African coast, Macnae (1963) noted the presence of mangroves only in regions where the average and average minimum air temperatures were at least 19 and 13°C respectively. At their latitudinal limits, as in Louisiana, die-back of mangroves can occur during severe winters (Chapman, 1976).

Rainfall and Salinity. Mangrove development is greatest in regions of high and

Table 1.1: Areal Distribution of Mangrove Swamps in the Indo-West Pacific Region

Country	Estimated Area of Mangrove Swamps (ha)[a]	Source
Kenya	59,000	Macnae (1974)
Tanzania	50,000	Macnae (1974)
Mozambique	85,000	Macnae (1974)
Malagasy Republic	320,700	Macnae (1974)
South Africa	negligible	Macnae (1974)
Pakistan	249,500	Khan (1966)
India	256,000	Blasco (1976)
Sri Lanka	3,200	Seneviratne (1978)
Andaman Islands	100,000	Blasco (1976)
Bangladesh	600,000	Blasco (1976)
Burma	520,000	In Rabanal (1976)
Thailand	602,000	Vibulsresth *et al.* (1976)
Peninsular Malaysia	150,000	Unpublished estimate
Singapore	2,800	Ling (1973a)
Sabah	365,300	Liew (1977)
Sarawak	173,600	Chai (1977)
South Vietnam	286,400	Ross (1974)
Kampuchea	50,000	Ling (1973a)
Hong Kong	6,500	Ling (1973a)
Indonesia	3,600,000	In Snedaker (1980)
Philippines	401,000	Gatus & Martinez (1977)
Papua New Guinea	411,600	In Snedaker (1980)
Australia	1,161,700	Galloway (1979)
New Zealand	19,800	In Snedaker (1980)
Fiji	19,700	IUCN Report (1980)
Other Pacific Islands	10,000	IUCN Report (1980)
Total	9,503,800	

Note: a. To nearest hundred hectares. Estimates refer variously to mangrove forests, mangrove swamplands, mangroves and adjacent mudflats.

aseasonal rainfall. Although the majority of mangrove trees are euryhaline, with the most salt-tolerant species, *Avicennia marina*, surviving soil water salinities above 90 ppt, the species of *Avicennia, Bruguiera* and *Sonneratia* grow best where soil salinities do not exceed that of normal seawater (Macnae, 1966, 1967). This has also been shown experimentally for seedlings of these and other genera (reviewed by Walsh, 1974). Under arid conditions mangrove soils rapidly become hypersaline because of evaporative water losses. With progression into regions of increasingly low rainfall, mangrove vegetation becomes less diverse and noticeably more patchy and stunted (Macnae, 1966; Teas, 1979). Even in the humid tropics leaf production by *Rhizophora* mangrove is markedly reduced during dry seasons (Christensen and Wium-Anderson, 1977).

Substratum.   Mangroves flourish on fine alluvial muds composed predominantly of silt and clay particles. These soft substrata provide essential anchorage for young seedlings while their root systems are developing, and they retain moisture efficiently. Macintosh (unpublished) found that mangrove clay soils bordering

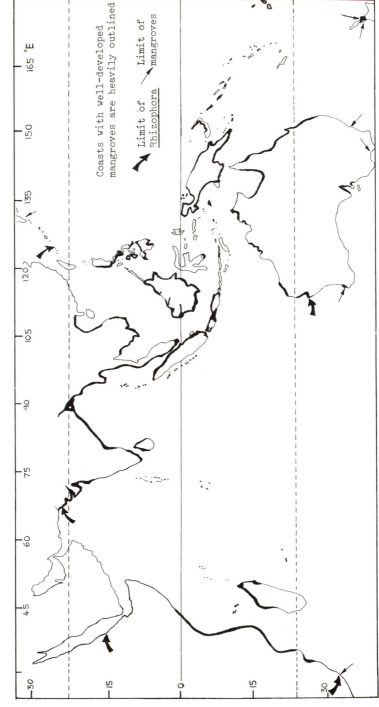

Figure 1.2: Distribution of Mangroves in the Indo-West Pacific Region.

Coasts with well-developed mangroves are heavily outlined

Limit of *Rhizophora*

Limit of mangroves

Source: Redrawn from Macnae (1974).

the Selangor estuary (Peninsular Malaysia) retained more than 31 per cent water by weight even in fully unshaded areas exposed to evaporation continuously for ten days.

Coastal Exposure. Shores protected from strong wave action and climatic forces encourage mangrove development by allowing alluvial sediment accumulation. Virtually continuous belts of mangrove border the Straits of Malacca which are sheltered from the open ocean by Sumatra, whereas the east coast of Peninsular Malaysia and the south-east coast of Sumatra are sea-facing, fully exposed to the monsoons, and are largely devoid of mangroves (Figure 1.2). Even mature mangrove trees can be toppled rapidly by eroding tides and similar damage can be done by flood conditions generated by monsoon rains. The catastrophic effects of hurricanes on mangrove forests are well documented and, in reviewing the subject, Lugo and Snedaker (1974) concluded that in addition to instances of complete devastation, mangroves in hurricane-prone regions such as Florida and Puerto Rico are probably limited in structure and scale.

The most luxuriant and extensive mangroves occur in the Indo-West Pacific region, and particularly in South East Asia (Figure 1.3). Here ideal conditions for mangrove development prevail: calm seas border much of the coastline, rainfall is high and fairly aseasonal, and rivers wash large quantities of silt into the coastal zone which accumulate to form emergent mud-banks and flats.

Approximately one-fifth of the world's mangroves border the shallow seas of the Sunda Shelf region enclosed by South Vietnam, the Gulf of Thailand, Malaysia, Sumatra, Java and Borneo. Elsewhere, mangroves are confined predominantly to estuaries and deltaic environments. The Irrawaddy Delta and Sunderbans region of the Gangenetic Delta together contain almost one million hectares. Mangroves in Nigeria, concentrated mainly within the Niger Delta, cover 710,000 ha (Moses, 1980). There are also vast areas of mangroves in eastern Africa, Papua New Guinea and eastern Australia (Table 1.1). The Caribbean region too has extensive mangroves, but estimates of their areal extent are not readily available.

Once established, mangroves promote the accretion of depositional shores by accelerating the rate of sedimentation. The pneumatophores of the pioneer *Avicennia* vegetation are especially effective in trapping sediments (Bird and Barson, 1977). Prodigious rates of mangrove shore progression have been documented, such as a rate of 125 m per year estimated for the south-east coast of Sumatra at Palembang (Macnae, 1968). However it is an overstatement to describe mangroves as 'land-builders'. This was recognised as early as 1928 by Watson who commented, 'In as much as silt deposits remain bare until they have attained a very definite condition of elevation and consistency, it is scarcely correct to regard mangroves as the causative agent in the silting process.' In short, mangroves augment rather than initiate sedimentation.

The mangrove-sediment relationship is of crucial importance when considering the suitability of sites for coastal aquaculture development. Clearly, eroding shores are to be avoided unless costs of establishing a mangrove 'buffer' zone can be accommodated.

Mangroves are likely to encroach on culture sites located on accreting shores and, through elevation of the substratum, the site may become gradually isolated from the tides (see Figure 1.4). The often stated view that vast areas of

Figure 1.3: Aerial View of a Mangrove Delta, West Coast of Southern Thailand.

mangroves are available for conversion to fishponds totally overlooks the vital role mangroves play in consolidating and developing sedimentary coastlines. As Macnae (1968) noted, 'erosion of a shore often, if not always, follows on the removal by man of a mangal'. Unfortunately, there are all too many examples to verify the truth of this statement.

Figure 1.4: An Accreting Mangrove Foreshore Bordering the Selangor River Estuary with a Pioneer Zone of *Avicennia* Saplings. The main forest is to the left of picture. The open foreshore area was the site of a mangrove cockle farm (*Anadara granosa*) until overgrown as the mangrove vegetation progressed seawards.

3. Ecological Features of Mangrove Swamps

3.1. Flora

Some 53 species of mangrove trees and shrubs are recognised (Chapman, 1970). Geographically there are two major groupings: the Old World mangroves of the Indo-Pacific region and those of the New World and West Africa. Although speciation appears to have occurred in both groups, the Old World mangroves, with 42 species, are by far the more diverse. The various specialisations of structure and physiology developed by mangroves for intertidal existence are extensively documented (e.g. Macnae, 1968; Walter, 1971; Walsh, 1974). They represent classical examples of biological adaptation that do not require elaboration here.

Mangrove vegetation is zoned in relation to shore level (which determines the frequency and duration of tidal inundation) and a number of moderating factors: notably the degree of water-logging of the soil and the soil-water salinity (Macnae, 1966; Clarke and Hannon, 1969, 1970). Zonation is best developed on shores with a tidal range of several metres, as on the west coast of Peninsular Malaysia where a classification of mangrove vegetation types in relation to the frequency of tidal inundation was established more than 50 years ago (Watson, 1928). Moreover, complete zonation to the landward limit of tidal penetration is confined to regions of high and non-seasonal rainfall. Elsewhere, as on some parts of the Queensland coast where annual rainfall is only 750-1000 mm, the arid upper shore may be completely devoid of mangroves (Macnae, 1966).

Because every mangrove shore has a unique topography and spatial pattern of microhabitats, it is possible only to generalise on the seral characteristics of the major vegetation types. Species of *Avicennia* and *Sonneratia* are consistently the pioneer mangroves on coastal deposition shores in the Indo-Pacific region. *Avicennia* seedlings are able to colonise soft, semi-fluid mud flats down to about mid-tide level. Typically a pioneer zone of *Avicennia* and *Sonneratia* extends from this seaward limit to around mean high water of neap tides (Figure 1.4); this zone is tidally inundated twice each day except during the 2-3 days of extreme neap tides each month. The pioneer vegetation consolidates and elevates the substratum and creates shade. These alterations in habitat generate conditions favourable to other mangrove types. The main forest zone, extending landwards as far as the limit of normal high tides, is usually dominated by stilt-rooted *Rhizophora* trees (Figure 1.5). *Bruguiera* species replace *Rhizophora* towards the landward margin reached only by exceptionally high spring tides. Where the substratum is well drained, other mangrove trees such as *Excoecaria* and *Xylocarpus* may be common. Species of *Ceriops, Lumnitzera* and *Aegiceras* are typical colonisers of open shore habitats.

In more estuarine localities *Rhizophora* usually replaces *Avicennia* and *Sonneratia* as the pioneer mangrove (see Figure 1.6). Upstream and towards other zones of freshwater influence, mangroves are replaced gradually by other vegetation communities dominated by nypah palm (South East Asia), *Heritiera* (Bay of Bengal) or *Barringtonia* (eastern Africa). The less complex mangrove communities of the New World and West Africa are characterised by species of *Avicennia, Rhizophora, Conocarpus* and *Laguncularia*.

A rich algal flora is also associated with mangrove environments. Mangrove algae are divisible into two main groupings: an epiphytic assemblage of macroscopic algae living on the stems, aerial roots and pneumatophores of mangrove trees, and an epiterrestrial community of predominantly micro-algae. Common genera of mangrove algae are listed in Table 1.2. It is these algal communities that provide the main food source for fish and prawns in mangrove culture ponds (see Section 5).

3.2 Fauna

The mangrove swamp community, or mangle, includes a complex faunal assemblage of resident, semi-resident and visiting species. The surface intertidal fauna is dominated by brachyuran crabs and gastropods, two groups that are physically and physiologically adapted to withstand exposure to high temperatures, desiccation and salinity fluctuations – conditions that intertidal mangrove animals face when their habitat is uncovered by the tides. They also show parental care of their eggs and time larval release to optimum phases of the

Figure 1.5: Dense Stilt Roots of *Rhizophora* Vegetation. Mangrove forest like this is difficult and expensive to clear for aquaculture development.

Figure 1.6: Illustration of the Typical Zonation Pattern of the Major Mangrove Vegetation Types in South East Asia. Abbreviated generic names are spelt out in the text.

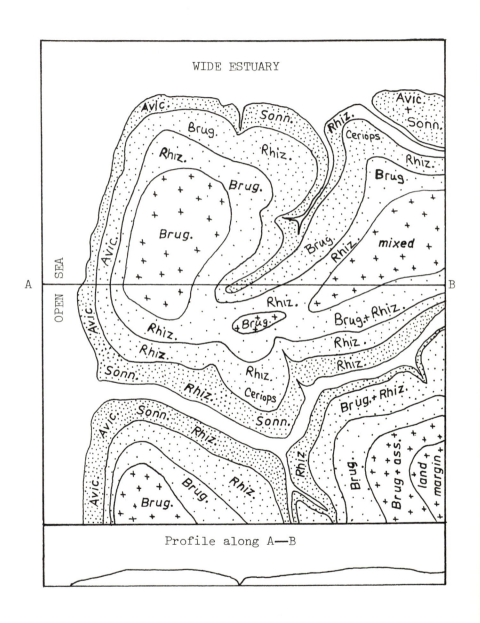

Source: Simplified from Watson (1928), p. 128.

Table 1.2: Common Genera of Mangrove Swamp Algae and Conditions Favouring Production of 'Lab-lab' and 'Lumut' Algal Communities in Mangrove Culture Ponds

	Blue-green Algae (Cyanophyta)	Red Algae (Rodophyta)
	Anabaena	*Bostrychia*
	Lyngbya	*Caloglossa*
	Microcoleus	*Catenella*
	Oscillatoria	*Gelidium*
	Phormidium	
	Diatoms (Bacillariophyceae)	Green Algae (Chlorophyta)
	Cyclotella	*Chaetomorpha*
	Cymbella	*Cladophora*
	Flageleria	*Enteromorpha*
	Navicula	*Vaucheria*
	Nitzschia	
Pond Conditions Favouring:	'lab-lab'	'lumut'
Salinity (ppt)	>25	<25
Pond water depth (cm)	<40	40-60
Type of pond bed	firm clay	soft mud

Sources: Various (e.g. Chan and Lee, 1979).

lunar tidal cycle. Details of these adaptations can be found in Macnae (1968), Berry (1972, 1975), Malley (1977) and Macintosh (1980).

The mangrove crab fauna is of major ecological and economic significance. It includes the much sought-after edible species *Scylla serrata*, a crab that lives both sub-tidally and in the lower intertidal zone. Crab burrows of characteristic shape and form permeate the mangrove floor: slanting burrows are made by *Metaplax* and *Ilyoplax* species; fiddler crabs (*Uca*) occupy simple vertical burrows; *Sesarma* species construct ramifying communal tunnel systems. Mounds of another burrowing crustacean, the mud lobster *Thalassina*, are a feature of many landward mangrove habitats (Figure 1.7). These burrows help aerate the substratum and provide refuge for many other animals, including mudskippers, frogs and sea slugs. However, *Thalassina* and *Sesarma* in particular are among the pests most feared by mangrove fish pond operators. By burrowing into the dykes of ponds they create holes that cause water leakage and facilitate the entry of pond predators. Crabs are themselves voracious predators of young fish and prawns. Cook and Rabanal (1978) state that prawn post-larvae will escape from ponds through crab holes.

The great majority of mangrove crabs are selective deposit feeders that extract particulate organic matter from the mangrove substratum. Recent studies in the Selangor mangroves of Peninsular Malaysia have shown that crabs are fed on by a great variety of animals: mudskippers, catfish, snakes, lizards, birds and even monkeys (Ong, 1978; Macintosh, 1979b; Sasekumar and Thong, 1980).

Figure 1.7: The Mangrove Mud Lobster *Thalassina anomala*, a Serious Pest in Many Mangrove Farms. Above: an adult specimen with massive chelae adapted for digging. Below: burrow mounds of *Thalassina*; larger mounds are about 1 m in height.

Crabs are therefore a link at the primary consumer level in food chains beginning with mangrove plant production and leading to higher level consumers of the adjoining terrestrial and aquatic communities. There is growing evidence to suggest that a similar ecological role is played by the mangrove gastropod fauna and the rich soil-dwelling populations of peanut worms (*Phascolosoma*) and nematodes (Prince Jeyaseelan and Krishnamurthy, 1980; Krishnamurthy *et al.*, 1980; Sasekumar and Thong, 1980; Lim and Sasekumar, 1979).

The mangrove tree fauna includes marine species in addition to birds and insects. Snails graze algae from the stems and leaves (*Littorina* and *Nerita* species are common); there are also predatory forms such as *Thais*. The lower trunks of trees support encrusting assemblages of barnacles, oysters and mussels. Being filter feeders, these animals are confined to levels below average high water and are usually only abundant on trees at the seaward forest edge. The culture potentials of the edible mangrove oyster (*Crassostrea spp.*) and the mussel *Mytilus smaragdinus* are discussed in Section 6.4. Natural mangrove populations of oysters and mussels are already exploited extensively for food. The edible mangrove mud snail *Cerithidea obtusa* is also eaten in small quantities. Snails of the genus *Cerithidea* are further notable as pests in fish ponds where they can accumulate in vast numbers and remove the benthic food supply. Schuster (1952) mentions heavy infestations of 750 *Cerithidea* per square yard, equivalent to 30,000 lbs of snails per acre (33,626 kg ha^{-1}) in milkfish ponds in eastern Java.

The great diversity of fish and prawn species that enter mangrove waterways attests to the significance of mangrove swamps as feeding and nursery sites for aquatic consumers. Studies in Florida (Odum and Heald, 1972), Malaysia (Ong, 1978), India (Prince Jeyaseelan and Krishnamurthy, 1980) and New Guinea (Collette and Trott, 1980) have, in each case, recorded more than 100 species of fish from mangrove waters. The Malaysian situation has been examined in particular detail. Among the species that ingress into mangroves along the west coast of Peninsular Malaysia are members of the families Clupeidae, Carangidae, Lutjanidae, Trichiuridae, Pomadasyidae, Sciaenidae, Tachysuridae and Serranidae. On a catch weight basis these families form a major part of the fisheries resources in the region (they comprise 49 per cent of the demersal fish resources in the southern Malacca Straits: Latiff *et al.*, 1976). Highly valued food and game fish that are indicated to have a close association with mangroves include mullets (*Liza, Mugil*), groupers (*Epinephelus*), snappers (*Lutjanus*), tarpons (*Megalops*), sea-perch (*Lates, Centropomus*) and catfish (e.g. *Arius, Tachysurus*).

Macnae (1974) concluded that mangrove swamps are utilised as nursery areas by the juvenile stages of many commercially important prawn species in the Indo-West Pacific region. These include nearly all the *Metapenaeus* group and several *Penaeus* species (*P. merguiensis, P. indicus, P. monodon* and others). Giant freshwater prawns (*Macrobrachium* spp.) make similar use of mangrove areas during their migrations between marine and freshwater areas. Immense mixed swarms of *Acetes* (Sergestidae) and mysid shrimps are present in mangrove waters on a more permanent basis. Table 1.3 gives details of the prawn landings in India in 1978 to illustrate the commercial importance of some of these mangrove-associated prawns.

The tidal runs of fish and prawns that take place in mangrove swamps are well

Table 1.3: Species Composition of Prawns Landed in India in 1978

Species[b]	Percentage of Catch
Acetes indicus[a]	21.7
Parapenaeopsis stylifera[a]	16.9
Metapenaeus dobsoni[a]	12.2
M. affinis[a]	11.5
Penaeus indicus[a]	10.1
M. monoceros[a]	9.1
Palaemon tenuipes	3.4
Penaeus semisulcatus[a]	3.3

Note: a. Species associated with mangrove swamps.
b. Only species forming more than two per cent of the catch are shown.

Source: Marine Fisheries Information Service, India (1979).

known to artisan fishermen who operate fixed nets and traps across the major waterways, or fish in mid-channel on the running tide with cast nets (Figure 1.8).

Figure 1.8: Fishing by Cast Net in a Mangrove Waterway, Pichavaram Mangroves, India; *Rhizophora* Trees in the background.

3.3 Soil and Water Conditions

Although mangroves are not restricted to fine particle substrata, it is the alluvial clay soils of estuarine and deltaic environments that are most suitable for fish pond construction. Coastal mangrove swamp substrata are generally much coarser in comparison (usually coming within the category of 'muddy sands') and consequently have less structural stability and a poorer nutrient and water-retaining capacity (see Kartawinata and Walujo, 1977, for comparison of nutrient levels in mangrove mud and sand soils in Java).

Mangrove clay soils from the west coast of Peninsular Malaysia have been analysed by Sasekumar (1970, 1974) and Macintosh (1979a). Silt and clay particles form up to 81 per cent by weight of the sediment (inorganic) fraction and as a result these soils are highly consolidated when semi-dry, and therefore excellent for the construction of pond embankments. Schuster (1952) gives a similar particle size composition for a representative soil from the tambak culture region of northern Java (Table 1.4).

Reducing conditions prevail in mangrove clay soils because of the accumulation of sulphides. These are produced during anaerobic decomposition of mangrove root debris in the soil by sulphate-respiring bacteria (Singh, 1980), oxygen penetration into the soil being insufficient to sustain aerobic breakdown of the organic material. High soil organic levels are particularly associated with *Rhizophora* mangroves because of their dense root systems (Figure 1.5). Kartawinata and Walujo (1977) recorded 62 per cent organic matter in soil supporting a *R. mucronata* community in Jakarta Bay. *Rhizophora* is necessarily extremely tolerant of sulphide soils. Diemont and van Wijngaarden (1975) found *R. apiculata* 'growing vigorously' where the soil hydrogen sulphide level was as high as 0.5 mmol l^{-1}.

Sulphides may combine with iron in the soil to form iron sulphide and subsequently the mineral pyrite (FeS_2). This produces a potentially acid sulphate condition, whereby drying of the soil may lead to a drastic drop in pH through oxidation of the pyrite to sulphuric acid. Potter (1977) cites low natural production, poor response to fertilisers and slow fish growth among the effects of acid sulphate soils utilised for fish ponds. Mass fish kills have resulted in severe cases when heavy rains have washed the soil acids into rivers (Dunn, 1965). Tang (1976a) suggests that 60 per cent or more of fish ponds in the Philippines are subject to acid sulphate effects. About 90 per cent of Malaysia's mangove swamps are formed of potentially acid soils. However, the extent to which mangrove clay soils acidify when drained has not been studied adequately. Being highly compacted they may resist oxidation. Sasekumar (1970) recorded only slightly acidic pH (range 6.0 to 7.1) in soils exposed to air for up to 20 days in the Port Kelang mangroves.

Turbid water conditions are another feature of mangrove estuaries and deltas. They result from the ease with which alluvial sediments are brought into suspension by tidal action. Mangrove waterways also carry large quantities of suspended organic matter in the form of plant debris washed out from the forest zone. Prince Jeyaseelan and Krishnamurthy (1980) recorded up to 0.97 g l^{-1} of particulate material in waters traversing a mangrove area of the Vellar-Coleroon estuarine system; light penetrated to a water depth of only 15 to 36 cm. Turbid waters show poor photosynthetic activity, with the result that aquatic primary production makes only a minor contribution to the total productivity of mangrove ecosystems. Ong *et al.* (1979) calculated that gross aquatic production in the Matang mangrove ecosystem in Perak, Malaysia, was only 2 per cent of the net production level of the mangrove forest. A similar

Table 1.4: Physical Composition of Mangrove Clay Soils Representative of Those Most Suitable for Fish Pond Construction

	Alluvial Mangrove Soil, Kuala Selangor, Malaysia (from Macintosh, 1979a)		Alluvial Mangrove Soil, Port Kelang, Malaysia (from Sasekumar, 1974)		Mangrove Tambak Soil, Java (from Schuster, 1952)	
	Particle size[a] (μm)	Percentage in sediment	Particle size (μm)	Percentage in sediment	Particle size (μm)	Percentage in sediment
	>63	0-4.7	>210	0-2.0	>100	2
	63-15.6	6.9-21.8	210-20	41.0-84.3	100-50	5
	15.6-3.9	15.9-81.0	20-2	12.0-54.3	50-10	30
	<3.9	9.4-73.6	<2	4.6-15.0	<10	63
Organic content		1.42-2.83[b]		1.5-5.2[c]		4.1[d]
Water holding capacity %		60.0-76.2		35.0-48.3		45.5

Notes: a. Particle size: >210 μm = sand, 210-63 μm = fine sand, 63-4 μm = silt, <4 μm = clay.
b. Particulate organic carbon only.
c. Total organic carbon.
d. Average humus content.

value of 3 per cent was given by Heald (1971) for the contribution of aquatic production to total primary production in a Florida mangrove estuary. These figures demonstrate why mangrove detritus, rather than phytoplankton production, is the energy source on which the main consumer food chains of the mangrove ecosystem are based (Odum and Heald, 1972).

Various studies have shown that there is a net export of organic matter from mangrove swamps into the nearshore aquatic zone (Snedaker and Lugo, 1973; Christensen, 1978; Golley *et al.*, 1962). Degradation of this material releases bound mineral elements that on subsequent re-entry into the mangrove zone are actively absorbed by the substratum where they become available for recombination into plant matter (Walsh, 1967; Aksornkoae and Khemnark, 1980; Nixon *et al.*, 1980). The networks of creeks and water channels that permeate mangrove swamps provide a large area of soil-water interface for nutrient exchange. Due appreciation of this function of water flow in mangrove environments is particularly applicable to the conversion of mangroves to fish ponds since high fish and prawn yields can be anticipated only from sites with nutrient-rich soils.

4. Productivity and Energy Flow in Mangrove Swamp Ecosystems

Lugo and Snedaker (1974) have reviewed our understanding of the physical and biological processes operating in mangrove ecosystems to sustain their high levels of productivity. Although the trophic position and functional importance of many faunal elements of the mangrove community remain to be elucidated, major energy linkages from mangroves to aquatic consumer organisms of fisheries and aquaculture significance (via mangrove detritus-based food chains) are demonstrable.

Total plant biomass production in mangrove forests can exceed 20 tonnes ha^{-1} year^{-1} (Christensen, 1978; Ong *et al.*, 1979). Part of this production, largely in the form of leaf and twig litter, is tidally exported into the adjacent aquatic zone. Estimates of mangrove leaf fall indicate an average annual release of 6-8 tonnes ha^{-1} of leaf material (Table 1.5). The flushing efficiency of the tides determines how much of this litter is exported directly. Christensen (1978) found very little retention of leaf litter in a mangrove study site in Thailand flooded by 65 per cent of high tides, whereas litter accumulation is commonplace in landward mangrove habitats above the range of average tides. Litter build-up at spring high-water level is further promoted because mangrove leaves can take many months to decompose *in situ* where conditions are dry (Odum and Heald, 1975). Based on studies in Florida (Snedaker and Lugo, 1973; Odum and Heald, 1975) and Thailand (Aksornkoae and Khemnark, 1980), it is reasonable to conclude that the annual net export of mangrove plant litter into coastal waters averages about 50 per cent of the total leaf fall, or approximately 3-4 tonnes ha^{-1}.

The initial stages of mangrove leaf decomposition generally commence before leaf-fall through attack by insects such as tailor ants (*Oecophylla* spp.). Although grazing insects remove little of the total leaf area (Christensen, 1978, estimated less than 10 per cent), the holes they create expose the leaf to secondary attack by micro-organisms. Decaying mangrove leaves become permeated by fungi, protozoans, micro-algae and bacteria (Odum and Heald, 1975).

Table 1.5: Estimates of Mangrove Leaf Litter Production

Locality	Type of Mangrove	Production ($t\ ha^{-1}\ year^{-1}$)	Source
Florida	*Rhizophora*	7.3	Heald (1971)
	Avicennia	4.85	Lugo & Snedaker (1974)
Puerto Rico	*Rhizophora*	3.65	Golley *et al.* (1962)
Mgeni estuary, South Africa	*Avicennia*	9.67	Steinke (1980)
	Bruguiera	9.71	
Chantaburi, Thailand	*Ceriops, Lumnitzera*	6.26	Aksornkoae &
	Rhizophora	7.52-8.31	Khemnark (1980)
	Avic., Brug., Xylocarpus	8.19-8.52	
Phuket, Thailand	*Rhizophora*	6.7	Christensen (1978)
Matang, Malaysia	*Rhizophora*	6.6-10.3	Ong *et al.* (1979)
Sydney region	*Avicennia*	4.58	Goulter & Allaway (1980)

The dynamics of microbial action on mangrove litter are unknown, but the major sequences leading to detritus formation appear comparable to those described for plant litter decomposition in other aquatic environments (see Lee, 1980, for recent coverage of the subject). Initially, there is a rapid loss of soluble and readily digested materials; then, as microbial colonisation proceeds, there is a rise in the relative protein content indicating a partial replacement of the plant constituents by microbial protein. Odum and Heald (1975) reported an increase from 6 to 20 per cent protein in *Rhizophora* leaf litter kept in seawater for six months.

The role of invertebrates in mangrove detritus formation continues to be important at all stages. Leaves that remain on the forest floor are fragmented by sesarmid crabs, some of which feed almost exclusively on leaf material (Malley, 1977). Amphipods are prominent among the animals that graze mangrove leaf litter in the aquatic zone (Odum and Heald, 1975; Boonruang, 1980). The mangrove soil meiofauna, particularly nematodes, are strongly implicated as regulators of the rate of microbial decomposition. There is evidence that, by selectively grazing litter micro-organisms, meiofauna stimulate microbial action (Lee, 1980). Deposit-feeding crabs and snails have a comparable effect because their feeding activities turn over the surface sediment layer, thereby exposing new litter surfaces to microbial colonisation. In a Malaysian mangrove Macintosh (1979a) found that mangrove fiddler crabs ingested only 10 per cent of the organic matter they scooped up, the remainder being returned re-sorted to the soil surface. Up to 41 g m^{-2} of soil was turned over daily by these animals.

Various techniques have been employed to measure rates of mangrove litter decomposition and a standardised method is urgently required to validate comparisons from different studies. Boonruang (1980) found that *Avicennia* and *Rhizophora* leaves submerged within nylon mesh bags in a mangrove channel lost

50 per cent of their dry weight in 20 and 40 days respectively. Under broadly comparable conditions these two leaf types have been found to decompose fully within 10 and 18 weeks (Odum and Heald, 1975; Loi and Sasekumar, 1980). Decomposition rates are considerably slower in sub-tropical climates: near Sydney *Avicennia* leaves showed 50 per cent decomposition after eight weeks (Goulter and Allaway, 1980), and six months was required for complete decay of *Avicennia* and *Bruguiera* leaves in a study by Steinke (1980) in South Africa.

An understanding of the dynamics of mangrove detritus formation would have direct practical value because coastal fish farm operators differ in their attitude towards the use of mangrove leaves as a manure cum food in their ponds. Experimentally, Vijayaraghavan and Ramadas (1980) have obtained an assimilation efficiency of 92 per cent for juvenile *Metapenaeus monoceros* fed a diet containing 60 per cent decomposed *Rhizophora* leaves.

The extent to which aquatic consumer organisms utilise mangrove food sources has emerged through analyses of gut contents from inshore fish and invertebrates. In addition to mangrove plant detritus, these sources include the benthic and epiphytic assemblages of mangrove algae, faecal matter, the intertidal fauna and larvae released by mangrove invertebrates. Information is available from two comprehensive studies: one conducted in a Florida mangrove estuary (North River) by Odum and Heald (1972, 1975), the other in the coastal mangroves of Selangor (Ong, 1978; Macintosh, 1979b; Sasekumar and Thong, 1980; Leh and Sasekumar, 1980).

Virtually all the commercially important species of mangrove-associated prawns are omnivorous. Differences in dietary preference are less evident between related species than between prawns of different size or stage of maturity. Many studies have shown that penaeid prawns consume a varied diet that can include zooplankton (including larval stages), diatoms, benthic algae, meiofauna, detritus, animal remains and inorganic particles. Post-larvae are strongly planktophagous, whereas benthic food sources are increasingly exploited by the juvenile and later prawn stages. The larger *Penaeus* species, such as *P. monodon*, *P. semisulcatus* and *P. merguiensis*, tend to be more carnivorous, taking whatever animals they can capture (Hall, 1962). Von Prahl (1978) has found that first-stage post-larvae of *P. stylirrostris* are able to digest wax from the surface of mangrove leaves, while the later post-larval and juvenile stages of this important neotropical species graze on the epiphytic micro-organisms that colonise decaying leaves.

Specimens of various inshore penaeid species collected from Selangor waters were found to have consumed 64 to 88 per cent of animal material (by volume) and 12 to 36 per cent of plant matter — of which 11 to 59 per cent was identified positively to be of mangrove origin (Leh and Sasekumar, 1980). Planktonic *Acetes* shrimp collected from the same waters had gut contents identified by Tan (1977) as mangrove detritus (32 to 42 per cent), diatoms and other micro-algae (11 to 31 per cent) and zooplankton (19 to 42.5 per cent). By virtue of their immense abundance *Acetes* must consume a significant proportion of the total detritus outflow from mangrove shores. Odum and Heald (1972) indicate that small palaemonid shrimps of the genus *Palaemonetes* occupy the equivalent niche to *Acetes* in the Florida mangrove ecosystem.

Omnivores and carnivores also predominate in the fish communities that ingress into mangrove waterways. Of 67 species from the Pichavaram mangrove ecosystem examined by Prince Jeyaseelan and Krishnamurthy (1980), 30 species were omnivorous, 32 species were carnivorous and only 5 species were entirely herbivorous. The edible mudskipper, *Boleophthalmus* (Figure 1.9) is one of the few common mangrove herbivores. This fish ingests surface fungi, diatoms and blue-green algae from the mangrove substratum (see Macintosh, 1979b). The data in Table 1.6, taken from the Selangor study, demonstrate that a diverse group of omnivorous fish species include mangrove detritus in their diets. Notable among these is *Liza subviridis*, one of the commercially important mullets in the region (Chan, 1977). The same data reveal that *Acetes* is a major food source for these fish, and confirm that this detritus-feeding shrimp is a key linkage organism in mangrove detritus food chains. Similarly, Chong (1979) has shown that adult white prawns (*Penaeus merguiensis*) feed heavily on *Acetes* and mysid shrimps.

Figure 1.9: The Edible Herbivorous Mudskipper, *Boleophthalmus*. Specimen shown is *B. boddaerti*, the common mangrove species in South East Asia.

Mangrove swamp invertebrates provide another detritus feeding link in the Selangor coastal ecosystem. Mud snails (*Cerithidea*), peanut worms (*Phascolosoma*) and crabs (*Metaplax, Sesarma*) have been identified in the guts of catfishes (*Arius sagor, Plotosus canius, Tachysurus maculatus* and other species), which implies that these predators scour over mangrove shores during high tide (Sasekumar and Thong, 1980). The predatory mudskipper, *Periophthalmodon schlosseri*, is a voracious consumer of *Sesarma* and *Uca* crabs;

Table 1.6: Gut Contents (Percentage Composition) of Representative Species of Fish from Selangor Waters that Exploit Mangrove Food Sources. (a) Fish caught by bag or drift net within mangrove waterways. (b) Fish caught by trawl net from coastal waters adjacent to mangroves. Only items forming 5 or more per cent of the diet are shown.

Species	Plant Detritus[a]	Conglomerates[b]	Acetes	Penaeidae Mysidae	Copepoda Amphipoda	Brachyura	Gastropoda	Teleostii	Others
(a)									
Johnius dussumieri	71.9			28.1					
Ambassis gymnocephalus	43.5	24.6		10.4		5.4	4.7		
Liza subviridis	37.6	21.3		23.3				7.2	
Trichiurus sp.	25.8	20.0	33.5	8.5					
Trissocles dussumieri	25.6	20.1	41.4	7.9					
Setipinna taty	15.4	37.8		41.4					
Pomadasys maculatus	9.1	12.7	69.9	8.3					
Arius sagor		6.5	15.2	12.0		46.0		17.1	
Arius sp.				7.5			92.5		
Plotosus canius						100			
Tachysurus maculatus						71.3		9.8	9.9 (sipunculids)
(b)									
Apogon ellioti	36.9	21.6	14.4					25.0	
Otolithus ruber	25.8	31.1	12.5	21.2				5.4	
Stolephorus sp.	23.3	12.5	19.2		38.1				
Ilisha elongata	18.4	24.9	40.5	9.1					
Johnius dussumieri	16.1	17.6	17.5	9.1	6.3		18.7		
Dussumieri hasselti	15.9	5.6	63.0		13.3				
Secutor ruconius	15.6	45.4		18.4		13.4 (zoeae only)			7.2 (diatoms)
Selar kalla	14.3		45.2	11.8					
Cynoglossus macrolepidotos	13.7		67.4				12.7	10.0	
Plotosus anguillaris			71.6	10.0					

Notes: a. Predominantly of mangrove origin.
b. Unidentified organic material.

the latter are also eaten by the semi-aquatic mangrove snake, *Cerberus rhychops*, and several terrestrial predators (Macintosh, 1979b). Edible mangrove crabs (*Scylla serrata*) feed mainly on benthic molluscs and other crab species (Hill, 1979; Natarajan and Radhakrishnan, unpublished).

5. Mangrove Pond Culture Practices

Mangrove pond culture depends for success on a controlled inflow and outflow of tidal water. The most desirable pond sites for fish culture have an average elevation around high-water level of neap tides but, depending on the tidal range, shore elevations between the levels of mid-tide and mean high water can be suitable (Figure 1.10). Where a modest tidal range prevails (1.0-1.2 m on average), a low elevation is preferable. For example, most milkfish ponds in the Pingtung region of Taiwan, where the normal tidal range is less than 1 m, have an elevation 10-30 cm above mid-tide level (Chen, 1976). Sites around mean high-water level may be acceptable where there is a large tidal range (2.0-3.0 m on average). Near Surabaya in eastern Java, where an extreme tidal range of 3.24 m occurs, about 80 per cent of a tambak area of 3,750 acres surveyed by Schuster (1952) was at an elevation of 0-30 cm below mean high water of spring tides. Ponds situated high in the intertidal zone can be tidally flushed only on the few days each month with extreme spring tides (Table 1.7). Where more frequent water exchange is necessary a lower elevation would be selected. The majority of mangrove prawn and crab farms are constructed at levels that allow ponds to be flushed on 15 to 20 days per month.

Table 1.7: Frequency of Tidal Inundation in Relation to Shore Elevation: Mangrove Prawn Culture Site, Sembawang, Straits of Johore

Shore Elevation (m above Chart Datum)		Average No. of Days per Month Reached by Tides[a]	Percentage of High Tides Reaching this Elevation[a]
2.5	(MHWN)	30	94-100
2.6		29	88.3
2.7		27	81.1
2.8		24	73.3
2.9	(MHWM)	22	62.6
3.0		18	52.9
3.1		15	42.7
3.2		12	33.4
3.3	(MHWS)	8	22.3
3.4		4	10.5
3.5		2	4.6
3.6	(EHWS)	0.5	1.1

Note: a. Calculated from tidal data for 1978.

Figure 1.10: Suitability of Potential Mangrove Fishpond Sites in Relation to Tidal Elevation. Applicable to conditions in the Philippines.

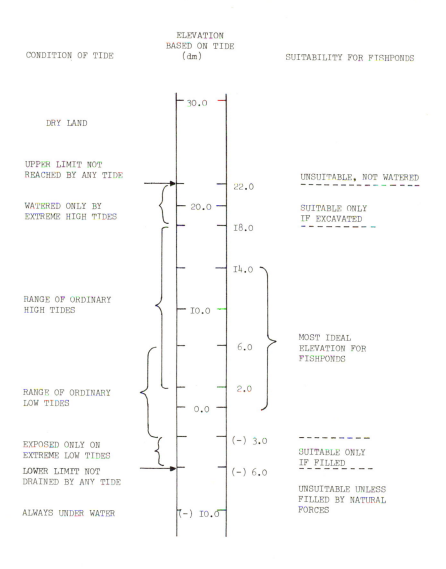

Source: Rabanal (1976).

Mangrove deltas and estuaries are favoured locations for pond culture operations. Swamplands bordering the bend of an estuary, or situated in the angle between two watercourses, are ideal because a throughflow water exchange system for the farm can be operated. A perimeter dyke, preferably rising to a height of 20-50 cm above the highest high-water level, is built to isolate the

Figure 1.11: Mangrove Prawn Culture Pond, Singapore. Above: view across the pond (area 25 ha) with perimeter dyke in the background and a protective mangrove zone extending to the Straits of Johore. Below: tidal water control gate with multiple sluices operated by a windlass system; a bag net and frame used to harvest prawns as they leave the pond is ready to be lowered into a sluice compartment.

ponds from the coastal zone. To prevent erosion it is important to leave a protective seaward margin of mangroves in front of this dyke (e.g. Figure 1.11a). A mangrove zone of 50-100 m is recommended by Tang (1976b) and Rabanal (1976). Similarly mangrove buffer zones at least 10 m wide should be left between the outer pond dykes and the water supply channels.

Mangrove ponds vary in size from 0.5 ha (some tambaks and crab culture ponds) to 100 ha and more in the case of some milkfish ponds in the Philippines. The common size, however, is 5-10 ha: this is a manageable unit and provides a reasonable water area in relation to the amount of earthwork construction and maintenance involved. To minimise labour costs the natural swamp topography is utilised to maximum advantage. It is not unusual for 20 per cent or more of the pond area to consist of unproductive high ground — the cost of removal being prohibitive. Natural depressions can be advantageous in prawn and crab culture ponds because they afford the animals refuge from temperature and salinity fluctuations in the surface water layer. A more even pond depth is desirable in fish ponds where the crop is harvested by drag net.

Soil excavated in preparing the pond bed is used for dyke-building. A puddle-trench and berm are usually constructed in conjunction with a dyke to reduce seepage and erosion respectively. Dyke stabilisation is promoted by planting *Avicennia* or *Rhizophora* mangrove trees, a traditional practice at tambak sites in Java that also provides a useful source of timber (Schuster, 1952). African star grass (*Cynadon plectostachus*) is also a good binding vegetation and is tolerant of acid sulphate conditions (Camacho, 1977).

Waterflow is controlled by sluice gates: main gates set in the perimeter dyke control inflow into the supply channels; secondary and tertiary gates regulate flow into and between pond units respectively. An equivalent system of discharge gates is operated at sites where water movements are on a flow-through basis. Experience has shown that one main gate is required per 10 ha of pond area (Rabanal, 1976). Large farms are supplied with a series of main gates or a single unit with multiple openings. A mangrove prawn pond in Singapore known to the author has an area of approximately 24 ha and is serviced by two inflow gates and a single discharge gate (Figure 1.11). Very elaborate water supply systems may be necessary for extensive pond complexes (see Tang, 1976b, for layout of a 1,600 ha mangrove fish farm and discussion of water-gate design). From a survey of 33,000 acres of tambak ponds in eastern Java, Schuster (1952) noted that 41 per cent of the culture area was tidally flushed directly from rivers and 50 per cent was serviced by water canals. The remaining 9 per cent was dependent on a system of secondary canals which did not provide for effective water exchange.

Salinity is the single most important ecological parameter in mangrove ponds. Good water quality and a stable pH can normally be maintained by regular tidal water exchanges, whereas sudden and severe decreases in salinity can occur as a consequence of heavy rainfall. Figure 1.12 shows the recorded effect of rainfall seasonality on water salinity in a coastal mangrove prawn pond in Singapore (from Hall, 1962). In estuarine localities seasonal variations in tidal water salinity of 15-25 ppt are commonplace, especially if there is a strong monsoon influence. For example, Gopalakrishnan (1971) recorded a monthly salinity range of 5.0-32.8 ppt in the lower reaches of the Hoogly estuary (part of the Sunderbans mangrove delta) where important coastal prawn culture operations exist. Korringa (1976) mentions 15-35 ppt as the salinity range in mangrove tambaks at a milkfish farm near Jakarta. Salinity conditions affect not only survival and growth of the cultivated species, but also influence the type of natural algal food source that develops in a pond.

Figure 1.12: Variations in Salinity of Water in a Mangrove Prawn Pond in Singapore.

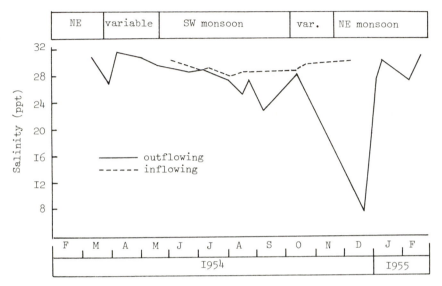

Source: Redrawn from Hall (1962).

Temperature variations in mangrove ponds can be moderated by planting trees along the dykes to provide shade. Over-heating can be a problem in shallow nursery ponds: an extreme of 38°C has been recorded in milkfish nurseries with a water depth of only 10 cm (Korringa, 1976). Hall (1962) observed a diurnal temperature range of 28.6-34.1°C in the prawn pond he studied in Singapore (the average water depth was 120 cm). In some cases prawn farmers reduce the danger of high temperature stress to their stock by cutting channels in the pond bed to provide refuges where water depths are 1.5-2.0 m approximately.

Mangrove farm operators encourage growth of natural food organisms by carefully preparing their ponds before stocking. Growth of one of two main food types can be promoted by manipulating environmental conditions appropriately. 'Lab-lab', a benthic association dominated by blue-green algae and diatoms, but including detritus, fungi, bacteria and meiofauna, develops in shallow water ponds where salinities are 25 ppt or higher (Table 1.2). By increasing the water depth and lowering salinity to below 25 ppt, growth of filamentous green algae ('lumut') is favoured. 'Lab-lab' is preferred for nursery ponds and most types of prawn culture. 'Lumut' is a suitable food for adult milkfish, mullet, tilapia and siganids (Cook and Rabanal, 1978). Dense growths of 'lumut' are hazardous to prawns which tend to become entangled in the algal filaments.

Figure 1.13: A Firm Mangrove Clay Soil Showing Surface Cracks after Several Days of Exposure to Drying. On re-wetting, pond soils in this condition give rise to rich 'lab-lab' algal pasture.

To prepare a pond for algal production, it is first drained and the bottom left to dry until surface cracks appear (Figure 1.13). Seven to ten days drying is normal for ponds prepared for 'lab-lab' production, compared to three days for ponds in which 'lumut' is to be the food source (Cook and Rabanal, 1978). Accumulations of organic debris and silt on the pond bed are removed at this time and necessary repairs made to the dykes. Drying causes organic matter in the soil to mineralise, thereby releasing nutrients for algal growth. The surface soil layer may be turned over to increase mineralisation, but this is not advisable in the case of potentially acid sulphate soils because pyrites may become oxidised, resulting in release of acids when the pond is refilled. Liming is recommended if soil pH is less than 6.5 (Cook and Rabanal, 1978); however, the cost of neutralising highly acid soils can be prohibitive. Kanapathy (1975) estimated that 12 tons of limestone per acre were needed to increase pH in the upper 30 cm of an acidic mangrove soil by one unit. Boyd (1976) gives a method for determining lime requirements in fishponds based on measurements of pond soil pH in water and in buffered solution. Low soil pH is less of a problem in mangrove ponds that can be flushed frequently with seawater to wash out and neutralise soil acids.

For 'lab-lab' production the pond is initially flooded to a depth of 3-5 cm. This is increased to 10-15 cm after one week and to 20-25 cm after two weeks, by which time a rich algal mat should have developed. To grow 'lumut', the pond bed is wetted and 'seeded' with young algal filaments of *Enteromorpha* or

Chaetomorpha. The pond is then flooded to a depth of 20 cm; this is increased to 40 cm after one week. It is a practice in the Philippines to arrange rows of twigs in 'lumut' ponds to trap algal filaments dislodged by water movements.

Benthic algal production in mangrove ponds is proportional to the amount of soil organic matter. The theory and practice of using fertilisers to improve algal growth is a subject of discussion by Cook and Rabanal (1978). Traditionally, only inexpensive organic fertilisers, such as rice bran, mangrove leaves and animal manure, have been utilised by mangrove fish farmers who have otherwise relied on the inherent high fertility of mangrove soils to sustain algal pasture production. Villaluz (1953), working in the Philippines, observed that algae developed abundantly in ponds where soil organic levels were above 9 per cent, a figure that is not exceptional for natural soils in estuarine mangroves (e.g. Diemont and von Wijngaarden, 1975; Chan and Lee, 1979).

Both organic and inorganic fertilisers are being used increasingly by coastal pond operators in South East Asia. Fertilisation is necessary particularly where large-scale conversion of mangrove land to culture ponds has resulted in a decline in fertility. For example, Fortes (1979) mentions that water flowing over undisturbed mangrove environments in the Philippines has been found to contain 59 and 26 μg 1^{-1} of nitrates and phosphates respectively, whereas at a milkfish pond site in Cebu where all mangrove trees had been removed, the levels of these nutrients were below 1 μg 1^{-1}. In a comparison of fish yields in earth ponds managed for 'lab-lab' production, Camacho (1977) found a combination of chicken manure and inorganic phosphate most effective, giving a yield of more than double that of an unfertilised pond (Table 1.8).

Table 1.8: Yields of Milkfish and Incidental Species from 500 m^2 Mangrove Ponds Managed for 'Lab-lab' Production in Relation to the Type of Fertilisation; Stocking Density of Milkfish Equivalent to 3000 ha^{-1}, Rearing Period 120 Days

Fertiliser Treatment	Yield (kg ha^{-1})
Urea (46-0-0)	623.0
Chicken manure	468.0
Phosphate (0-20-0)	382.5
Ammonium phosphate (16-20-0)	468.0
Chicken manure + urea	514.0
Chicken manure + phosphate	878.3
Chicken manure + ammonium phosphate	424.0
Unfertilised	346.5

Source: Camacho (1977).

Fertilisers are normally applied once the pond bed has been dried, and again after one week and two weeks during the period when a low pond water level is being maintained to encourage algal growth. Small additional quantities of organic fertiliser may be applied at about weekly intervals during the culture operation. Mandal (1962) showed that ammonium fertilisers are better suited for ponds managed for benthic algal production compared to nitrate fertilisers because a higher proportion of the nitrogen present becomes bound on to the pond soil surface. Nitrate fertilisers would be preferable if a phytoplankton food crop is desired. Similarly, fertilisers giving a high ratio of nitrogen to phosphorus will stimulate phytoplankton rather than benthic algae (Cook and Rabanal, 1978).

A fruitful line of research has been followed by Camacho (1977) who compared fish production in experimental mangrove ponds that received different levels of preparatory treatment. Based on a culture period of 120 days, data presented in Table 1.9 from Camacho's work show evidence of additive benefits derived from leaching and tilling operations (in this case the pond soil had an acid sulphate condition) followed by liming and fertilisation.

Table 1.9: Yields of Milkfish and Incidental Species in 500 m^2 Mangrove Ponds Prepared in Different Ways; Stocking Density of Milkfish Equivalent to 3000 ha^{-1}, Rearing Period 120 Days

Preparatory Treatment		Yield (kg ha^{-1})	
Lime (t ha^{-1})	Chicken Manure (t ha^{-1})	Before Tilling and Leaching	After Tilling and Leaching
0	2	336.5	419.3
8	2	402.7	565.6
8	8	412.2	805.0
4	2	478.5	555.4
4	8	689.0	929.8

Source: Camacho (1977).

6. Mangrove-Associated Organisms of Fisheries and Aquaculture Importance

6.1 Finfish

Milkfish. The milkfish *Chanos chanos* (Chanidae) is a large euryhaline clupeoid widely distributed in the Indian and Pacific Oceans. It is only of minor importance to open seas fisheries but, because juvenile milkfish enter coastal waters in vast numbers, it is the principal species reared in more than 400,000 ha of coastal mangrove ponds in South East Asia. Almost all the estimated 183,000 ha of tambaks in Indonesia, and more than 90 per cent of the 176,000 ha of fishponds in the Philippines, produce *Chanos* (Shang, 1976; Cruz and Laudencia, 1980). Intensive milkfish culture is practised in Taiwan, with reported yields of up to 3,000 kg ha^{-1} $year^{-1}$ in some ponds (Chen, 1976); the total Taiwanese production in 1978 was 30,153 tonnes (Taiwan Fisheries Bureau, 1979). Developments have also been reported towards establishing *Chanos* as a cultivated species in southern India (Silas *et al.*, 1980). The classical account of milkfish culture is given by Schuster (1952). Korringa (1976) summarises Schuster's description and includes details of the operation of a modern *Chanos* farm near Jakarta.

Milkfish breed only in fully marine conditions preferring, according to Schuster (1960), clear shallow waters overlying sandy or coral substrata. There have been reports that the species does mature in the hypersaline lagoons and other enclosed waters of Christmas Island — a phenomenon attributed to the presence of *Artemia* as a food source (Nash, 1978). With rare exceptions, only partial maturation of pond-reared *Chanos* has occurred. Spawning by milkfish kept in floating sea-cages has been reported recently (Lacanilao and Marte, 1980).

Milkfish fry appear seasonally in estuaries, mangrove areas and coastal shallows. On the north coast of Java, the major tambak region of Indonesia, peak numbers occur in October-November and April-May, at the beginning and end of the north-east monsoon. The largest incursion is from the east, probably from spawning grounds south of Sulawesi (Schuster, 1952). Fry appear somewhere along the Philippines coastline throughout the year with a peak in May-June (Smith *et al.*, 1978). The migration routes followed by *Chanos* are still being investigated (Schmitton, 1977; Kumagai and Bagarinao, 1979). Research in progress suggests that both active factors (behavioural responses) and passive factors (current movements) regulate the inshore flow of fry (Kumagai *et al.*, 1980).

Sweep nets, scoop nets and shore seines of various local designs are used to catch *Chanos* fry in coastal waters. Tidal filter nets are operated in estuarine areas, and light skimming nets are favoured in mangrove swamps where aerial roots are an impediment to the use of other gears. Kumagai *et al.* (1980) state that daily catch rates in the prime fry grounds of Panay (Philippines) vary from several hundred fry for small hand nets — often operated by children — to 20,000 for static shore gears. These authors also mention that some of the best fry waters border mangroves and comment on a belief of local fishermen that *Chanos* fry are positively attracted to mangrove environments.

Rearing ponds for milkfish vary considerably in size depending on location and topography, and intensity of the culture system. Some tambaks in west and central Java are only 0.5-2 ha in area; in east Java 3-10 ha is the common size (Schuster, 1952). Milkfish ponds as large as 100-500 ha are operated in the Philippines (Ling, 1977). Ponds of 1-6 ha have been adopted in Taiwan for highly intensive milkfish farming (Chen, 1976).

Milkfish are exclusively fine particle feeders. In mangrove ponds they graze the benthic organic complex of algae, diatoms, detritus and meiofauna. Experimentally, Lim *et al.* (1979) found that *Chanos* fry grew significantly faster on diets containing 40 per cent protein compared to their performance on 20 and 30 per cent protein levels. This finding suggests that the proteinaceous bacterial component of mangrove soil detritus may be important nutritionally in milkfish ponds. The natural food source is promoted by carefully preparing the pond bed before stocking as outlined earlier (p. 32). Milkfish show a strong preference for blue-green algae over filamentous green algae, although the latter are consumed if partially decayed (Schuster, 1952). Since a shallow water depth promotes development of blue-green algae (Table 1.2), water levels are kept low in *Chanos* ponds: about 12-15 cm in fry ponds, 20-30 cm in transition ponds, and 60-120 cm in rearing ponds. Some farmers in the Philippines do feed filamentous 'lumut' algae from the fingerling stage because they believe this improves the flavour of *Chanos*. Food conversion rates for 'lumut' algae are 20 kg: 1 kg milkfish compared to 15-16: 1 for blue-green algae of the 'lab-lab' type (Blanco, 1973).

Various forms of *Chanos* culture are practised in the Philippines and Indonesia according to the means and initiative of the operators. Subsistence

farmers owning small tambaks may stock only once annually — when fry prices
are lowest — and raise a single modest crop amounting to 300-400 kg ha^{-1} on
average (Ling, 1977). Increasingly, improved methods of culture are being
adopted in both countries which involve intensive use of fertilisers and
production of two or more crops per year. At the 'Gembira Bati' fish farm (near
Jakarta), described by Korringa (1976), animal manure or urea is used to fertilise
Chanos ponds before stocking. During the rearing periods, vegetable manure is
added in the form of *Avicennia* and *Rhizophora* leaves, *Pluchea* branches (a
common shrub in mangrove areas), and paddy straw (total 4,000-5,000 kg ha^{-1}
year^{-1}). Nursery ponds of 500-1,000 m^2 in area are stocked with *Chanos* fry
at a density of 30-50 m^{-2}. Fry attain a size of about 5 cm after only six to eight
weeks. They are then transferred to intermediate ponds at a stocking density of
15,000-20,000 ha^{-1}. A final transfer to the main rearing ponds is made after a
further ten weeks. Here they are grown at a density of 1,000-1,500 ha^{-1} for
four to six months until a marketable size of 200-350 g is reached. With this
type of operation two to three crops per year can be reared, giving a total yield
of 1,000 kg ha^{-1} approximately. Supplementary feeding, in the form of rice
bran, peanut cake and similar processed wastes, may be necessary to augment
the natural algal production in situations where over-grazing occurs.

The potential to increase milkfish production by intensification in favourable
areas is evidenced by Taiwan's milkfish industry, which produces an average of

Although culture methods are improving, average annual production of
milkfish in the Philippines and Indonesia is still only around 600 and 300 kg ha^{-1}
respectively (Cruz and Laudencia, 1980; Chong, 1980). The range in both
countries on a regional basis is 200-1,000 kg ha^{-1} or more. This high variation is
due, at least in part, to natural differences in soil fertility. Schuster recognised
the relationship between soil type and productivity in the case of *Chanos* ponds
in Java and Madura (Table 1.10). Alluvial soils are generally more productive
than indigenous coastal types, with those formed from volcanic and lateritic
materials being more fertile than alluvia of quartzite, marl or limestone origin
(Schuster, 1952; Macnae, 1968). This fact is often overlooked when impressive
production statistics from one locality are used to justify aquaculture developments elsewhere.

Table 1.10: Average Production of Milkfish in Tambaks of Different Soil Types
in Java and Madura

Soil Type	Production (lbs acre^{-1})	Equivalent Production (kg ha^{-1})
Juvenile volcanic soil	200-380	224-426
Colloidal clay	150-280	168-314
Juvenile lateritic soil	120-240	134-269
Calcareous clay	75-120	84-134
Senile lateritic soil	50-100	56-112
Rocky or sandy soil	40-75	45-84

Source: Schuster (1952).

The potential to increase milkfish production by intensification in favourable
areas is evidenced by Taiwan's milkfish industry, which produces an average of

more than 2,000 kg ha^{-1} by means of multiple selective stocking and harvesting (Shang, 1976). This is achieved in a growing season of only eight months. During winter, temperatures are below the 15°C lower thermal tolerance limit of *Chanos* (Korringa, 1976). Accounts of the Taiwanese method of milkfish culture are given by Ling (1977) and Chen (1973, 1976). During winter (November to March), rearing ponds are prepared for algal production as described earlier (p. 32), except that the process takes much longer because of the cold climate. Animal manure (500-1,000 kg ha^{-1}) or rice bran (300-500 kg ha^{-1}) is applied to pond soils deficient in natural organic matter. From April, multiple size stocking is practised using overwintered fingerlings and fry freshly caught from coastal waters (Table 1.11). Overwintered fish are either fry obtained in June to August of the previous year or small individuals (less than 150 g) selected out at harvesting at the end of the previous rearing season. Overwintered fish reach a size of 300-400 g in May to August. Fry stocked in April and May grow to a weight of 200-300 g by September. From late May several partial harvests are made to remove fish suitable for market (usually 250-300 g size). Ling (1977) has shown that milkfish of mixed size groups utilise the benthic algal crop more efficiently than fish of a single size class.

Table 1.11: Multiple-size Stocking Rates of Milkfish in Taiwan

Month of Stocking	Average Weight per Fish (g)	Stocking Rate (fish ha^{-1})	Marketable Weight (g) Attained in a Culture Period of		
			60 days	90 days	120 days
April	70	1,500	350	400	600
	30	2,000		350	400
	25	1,500			250
May	0.05	2,500	30	100	250
June	0.06	2,500			200
July	0.06	2,000			200
August and September	0.06	3,000		100	
	Total	15,000			

Source: Chen (1973).

The best-managed ponds produce a daily utilisable algal crop of 300-525 kg ha^{-1}. This is sufficient to feed a standing stock of about 700-800 kg ha^{-1} milkfish. Where necessary, supplementary feeding is given: rice bran 25 kg ha^{-1} day^{-1}) or soya bean and peanut meal (25 kg ha^{-1} day^{-1}), together with frequent applications of animal manure.

It is generally stated that shortage of *Chanos* fry is currently the most serious factor limiting milkfish production (e.g. Schmittou, 1977). This view has been challenged by Smith *et al.* (1978), who point out that fry requirements are usually based on a stocking rate of 10,000 ha^{-1}, whereas for the extensive systems of culture operated in the Philippines and Indonesia 4,000 ha^{-1} is a sufficient number. The present total catch of about 2.0 billion fry is therefore adequate (Table 1.12). However, annual quotas cannot be relied upon and a five-fold variation in catches has occurred in recent years. Moreover, seasonal fluctuations in fry availability result in price instability. Smith *et al.* give an example of this: in the Bulacan and Rizal region of Luzon, prices during 1976-7 varied from about 20 to 120 pesos per 1,000 fry.

Fry purchase accounts for 23 per cent of the production costs in *Chanos* culture in the Philippines and 40 per cent in Taiwan (Shang, 1976). Given these high figures, it is surprising that there are not more efforts made to reduce fry mortalities after capture. In experiments to determine the protein requirements of *Chanos* fry, Lim *et al.* (1979) concluded that the poor survival obtained (maximum 30 per cent over 30 days) was due to stress caused by conditions of high salinity (32-34 ppt) and low temperature (25-28°C). Taiwanese fry dealers rear their stocks in only 5-12 ppt salinity, and an increase to 25 ppt is considered detrimental to growth (Chen, 1976). Yet Schuster (1952) quotes 10-32 ppt salinity as suitable for rearing *Chanos* fry and peak ingressions of fry into Pulicat Lake Lagoon near Madras coincide with salinities of 34 ppt (Prabhakara Rao, 1971). Korringa (1976) implies that fry cope with decrease in salinity rather better than with change in the reverse direction. Mortalities of 30-50 per cent are usual during the nursery stage of milkfish culture. This, added to an estimated 16 per cent loss during fry collecting and shipping operations (Smith *et al.*, 1978), suggests that there is only 40 per cent survival to the fingerling stage.

Considerable effort is being made to produce *Chanos* fry artificially to overcome the difficulties created by present dependence on wild stocks. As coastal industrialisation proceeds this resource can only decline in the long term. Adoption of the Taiwanese system of intensive multiple stocking and cropping by milkfish farmers in the Philippines and Indonesia would increase the total annual fry requirement to about 5 billion; this level of demand could only be met through hatchery production.

Ripe female *Chanos* have been spawned after multiple injections of carp pituitary homogenate, or salmon pituitary powder, in combination with human chorionic gonadotropin (Vanstrone *et al.*, 1977; Kuo *et al.*, 1979; Liao *et al.*, 1979). Fish which have oocytes of at least 0.7 mm diameter are considered suitable for this technique, although success also depends on careful handling. Gradual acclimation of spawners to the salinity necessary for fertilisation (32-36 ppt) is advisable (Kuo *et al.*, 1979). Androgen injection has been employed to improve the flow of milt from *Chanos* males (Juario and Natividad, 1980). Liao *et al.* (1979) blamed the low volume of milt exuded by an untreated fish for the poor fertilisation rate they obtained (38 per cent). Scientists at the South East Asian Fisheries Development Center (SEAFDEC) in the Philippines have reared significant numbers of *Chanos* to sexual maturity in floating sea cages (Lacanilao and Marte, 1980). This achievement may herald the break-through that will enable regular natural spawning of milkfish in captivity.

Progress in culturing *Chanos* larvae has also been made at SEAFDEC. Liao *et al.* (1979) reared 2859 larvae to 21 days, obtaining a survival rate of 9-47 per cent (from the fertilised egg stage) under various rearing conditions. These larvae were fed initially on rotifers and fertilised oyster eggs; after 14 days the diet was

Table 1.12: Annual Numbers of Fry Required to Support the Present Milkfish (*Chanos chanos*) Culture Industry

Country	Area of Milkfish Ponds (ha)	Culture System	Fry Requirements		Source
			Per ha	Total (millions)	
Indonesia	183,000	extensive	4,000	840	Shang (1976)
Philippines	176,000	extensive	4,000	704	Smith *et al.* (1978)
Taiwan	16,000	intensive	10,000	160	Chen (1976)
			Total:	1704	
Mortalities in transport (16%: Smith *et al.*, 1978)				272	
			Grand total:	1976 million	

changed to a combination of copepods, *Artemia* nauplii, flour and prepared feed. A 'critical period' with high mortality occurred between days 3 and 6; at 13 days the larvae were comparable in size to fry caught in neighbouring coastal waters and by about 20 days they were of stockable size (mean length 14.5 mm).

Mullet. Grey mullet of the family Mugilidae are almost universally appreciated as food fishes. The great majority are catadromous, entering estuarine and brackish water environments when young, then returning to the sea when adult to spawn. Mullet contribute significantly to capture fisheries in many tropical countries, especially those with extensive inshore waters such as India and Nigeria.

Various forms of mullet farming are already practised. Mullet are the principal fish cultured in the 'valli' system, operated in bays and lagoons along the Adriatic coast of northern Italy (Ravagnan, 1978); in Israel and Taiwan they are reared in polyculture with carp and tilapia in freshwater ponds, while mullet and carp are farmed together in brackish-water ponds in Hong Kong. Accounts of mullet culture in Israel and Taiwan are given by Korringa (1976) and Chen (1976). Semi-culture of mullet, together with milkfish, sea perch, other fish species and prawns, has a long tradition in the brackish-water bheris of West Bengal. Several features of the biology of tropical mullet species point to their suitability for mangrove pond culture: they are extremely euryhaline; their natural feeding habit is to graze on benthic algae, plant detritus and the associated micro-flora and micro-fauna; and their juvenile stages enter mangrove waters in large numbers. These are characteristics shared in common with *Chanos*, and in fact Taiwanese fish pond operators avoid stocking mullet and milkfish together because they compete for the same food source: the algal pasture or 'lab-lab' (Chen, 1976).

Mullet fisheries in the tropics are based heavily on species of *Mugil*, *Liza* and *Valamugil*. Sivalingam (1975) states that the bulk of these mullet caught in Nigerian waters come from estuaries and brackish waters rather than from the open coast. Fishermen in India have long recognised that schools of mullet enter littoral areas with the rising tide and catch them using gill nets and various specialised nets of local design (Luther, 1973). The close association of tropical mullet with mangroves is well illustrated by fisheries statistics for Peninsular Malaysia which show that 96 per cent of the country's mullet landings came from the west coast (Annual Fisheries Statistics, 1968-75). Indeed reclamation of mangroves along this coastline is a factor blamed by Chan (1977) for the observed severe decline in mullet catches in Malaysia, from a peak of 2,859 tonnes in 1969 to 431 tonnes in 1975.

Mullet fry are typically 10-30 mm in size when they arrive in estuarine waters following their inshore migrations from spawning grounds (reviewed by De Silva, 1980). Surveys indicate that fry are present year-round in tropical estuaries, with each species showing seasonal peaks in abundance. In the Hoogly-Matlah estuary, for example, fry of *Liza parsia* occur from January to April, fry of *L. tade* from March to May and July to November, and fry of *Rhinomugil carsula* in October and November (Luther, 1973). Catch rates of up to 2,850 fry per man hour of operation have been recorded in exploratory fishing surveys in India (ICAR, 1978). Specific mangrove areas have not been surveyed in detail, but Sivalingam (1975) concluded that mullet fry resources in the Niger Delta could supply the stocking requirements of at least 10,000 ha of mangrove ponds. In brackish-water culture trials near Lagos the same author found that natural tidal entry of mullet fry (predominantly *L. falcipinnis*) into ponds reached levels equivalent to more than 3,000 ha^{-1}.

From the above evidence, there is potential to develop collection and distribution systems for tropical mullet fry — perhaps along the lines of the South East Asian milkfish fry industry — to support mangrove pond culture of mullet on a large scale. At present, mortalities of mullet fry during handling and transportation can be very high (Korringa, 1976; Durve, 1975) and comprehensive studies of the interactive effects of salinity, temperature, pH and dissolved oxygen concentration on fry survival are required.

Although most mullet species are extremely euryhaline, factors such as low temperature and low pH can adversely effect their salinity tolerance (Mires *et al.*, 1975). In Nigeria, Ezenwa (1973) found that fry of *L. falcipinnis* and other species could be transferred directly into freshwater provided the pH was at least 7.0. Almost complete mortality occurred in salinities below 4 ppt when the pH was less than 7.0. Thakurta and Pakrasi (1980) reported that conditioning of newly caught *L. parsia* fry for 24 hours in 400 l plastic tanks improved survival from 7.6 to 90 per cent when packed subsequently in polythene bags at a density in water of 250 fry per litre. In water of 23 ppt salinity oxygenated to give an initial dissolved oxygen concentration of 7.8 ppm, *L. parsia* fry could survive for 24 hours at a density of 400 l^{-1}. On the basis of studies with *L. tade*, Durve (1975) concluded that survival of mullet fry in transport could be greatly improved by using anaesthetics, of which chloral hydrate, tertiary amyl alcohol or MS-222 were found to be highly suitable. After a train journey in India of 400 miles, lasting 20 hrs, a mortality of 48 per cent was recorded in a group of 50 fingerlings packed in five litres of water, compared to only 2 per cent mortality in an equivalent group of fish anaesthetised with MS-222 (concentration 30 ppm).

Mullet have high rates of growth, an attribute which together with their wide environmental tolerances make them highly attractive for culture purposes. In Israel, where mullet are farmed in polyculture with carp and tilapia under virtually freshwater conditions (around 1.5 ppt salinity), young *M. cephalus* of 70-100 g size reach 500-800 g in a rearing period of 120-170 days (Korringa, 1976). Although smaller than the *Mugil* species, mullet of the *Liza* group are probably better candidates for culture in mangrove fish ponds because of their intimate association with mangrove environments. Chong (1977) showed that *Liza malinoptera* collected from mangrove waters along the Malaysian coast fed heavily on mangrove detritus and algae (44 and 17 per cent of stomach contents respectively). Thus there is every reason to believe that mullet will thrive in mangrove ponds prepared for algal production in the traditional way as described earlier. Up to 239 kg ha^{-1} of mullet were produced in unfertilised experimental mangrove ponds near Lagos as a result of natural tidal entry of mullet fry (Sivalingam, 1975). Tropical species of mullet also respond well to supplementary feeding. Several studies in India have demonstrated that survival and growth of *L. parsia* and *L. tade* fry is enhanced if both natural and supplementary food is available in nursery ponds (e.g. Chakrabati *et al.*, 1980; Roy and Chakrabati, 1980). In assessing the culture potential of *L. subviridis*, a common species of mullet in Malaysian waters, Chan (1977) estimated that fish of marketable size (10-12 cm) could be reared in about six months. This species occurs naturally in salinities ranging from 0.7 to 33.0 ppt and, experimentally, Chan acclimated fish to freshwater. The influence of salinity on growth of *Liza* species has not been investigated adequately, but appears to be slight. In brackish-water ponds of salinity 0.5-20 ppt Sivalingam (1975) reared mullet (mainly *L. falcipinnis*) up to a size of 300 g in a culture period of 411 days. This compares well with growth rates for *L. aurata* reported from salt-water ponds (salinity 42.7-46.5 ppt) on the Mediterranean coast of Israel: over two rearing seasons, representing 392 culture days, fish of 302 g mean weight were obtained (Chervinski, 1976).

Results of culture trials with *Liza* species in the Sunderbans mangrove region of West Bengal (ICAR, 1978) are indicative of the potential yields from intensive mangrove pond farming of mullet. In well prepared nursery ponds, fry of *L. parsia* stocked at densities equivalent to 100,000 ha^{-1} reached 6 cm — a size suitable for transfer to rearing ponds — in 90 days with a survival rate of 75 per cent. Fry of *L. tade* performed rather differently, with highest survival (50 per cent) resulting from a stocking density equivalent to 36,250 ha^{-1}. Fingerlings of *L. parsia* (5 cm) stocked at 30,000 ha^{-1} and reared for 180 days attained an average size of 13 cm with 70 per cent survival, giving a production equivalent to about 600 kg ha^{-1}. In growth trials with yearling *L. tade* (average size 196 g) stocked at a density of 6,000 ha^{-1}, a net production of 878 kg ha^{-1} was obtained in 100 days with repeated harvesting; over a culture period of 270 days, the figure was 1,775 kg ha^{-1}. In polyculture experiments with *Mugil curema, M. brasiliensis* and various species of Gerridae in mangrove ponds near Recife, Brazil, production figures of 400-1,500 kg ha^{-1} have been obtained (Cavalcanti *et al.*, 1978). As in the case of milkfish farming, it is likely that a system of multiple size stocking and selective harvesting would be the best management practice for mangrove culture of mullet.

That mullet fry are available in large numbers in mangrove waters is an observation of considerable relevance to the future of mullet farming in general because shortage of fry is a major factor limiting further development of existing mullet culture practices (Sebastian and Nair, 1975; Chen, 1976; Korringa, 1976). Moreover, a number of problems continue to impede mass-culture of mullet larvae despite consistent success in spawning fish artificially (reviewed by Nash and Kuo, 1975). Kuo and Nash (1975) found that partially purified salmon gonadotrophin (SG-G100) was the most effective agent for inducing *M. cephalus* to spawn. However, human chorionic gonadotropin (HCG), a more widely available and less expensive hormone, is also suitable. Female spawners of *M. cephalus* with oocytes of at least 600 microns diameter require a dose of approximately 60 i.u. of HCG per gram body weight, given in the form of a primary injection of about 20 i.u.g^{-1}, followed 24 hrs later by a main injection of 40 i.u.g^{-1} (Kuo *et al.*, 1973). The Indian species *M. macrolepis* has been induced to ovulate by injecting an extract from homogenised pituitaries of mature fish of the same species (Sebastian and Nair, 1975). Females receiving two injections, each of three or four pituitary glands, given six hours apart, were usually ready for stripping three to six hours after the second injection and some spawned spontaneously.

In the breeding season adult male mullet normally ooze milt freely and do not require hormone injection prior to stripping. Natural fertilisation in

M. cephalus can be obtained by placing two or three males with a hormone-treated female; each male may release milt when the female spawns (Nash *et al.*, 1974).

Mullet appear to have an annual breeding cycle throughout most of their range, with ripe adult fish suitable for induced spawning being caught during a confined period of the year even in tropical waters. An exception is *Liza subviridis*, which breeds year-round in mangrove estuaries along the coast of Peninsular Malaysia and in these waters adult spawners are available continuously (Chan, 1977). To some extent it may prove possible to manipulate the reproductive cycle of other mullet species to achieve a steady supply of ripe fish for commercial hatchery production of mullet fry. Kuo and Nash (1975) were able to reduce the post-spawning refractory period of female *M. cephalus* from 235 to 56 days by maintaining them in a constant light:dark photoperiod of 6:18 hrs at 21°C.

The major problems encountered in attempts to rear mullet artificially stem from the small size of their eggs and larvae: both stages are highly sensitive to environmental conditions, while size and density of food items offered are critical to larval survival (Nash and Kuo, 1975). Phytoplankton, oyster larvae, micro-crustaceans, other types of live food and various inert foods have been tried with only moderate success (Nash *et al.*, 1974; Sebastion and Nair, 1975). Two crucial periods, associated with sinking of mullet larvae, have been reported at 2.5 and 8 days after hatching (Kuo *et al.*, 1973). A hypothesis presented by Nash and Kuo (1975) is that the second descent is a consequence of osmotic imbalance between the larva and its culture medium, and that this could be overcome by giving foods with a higher freshwater content.

Siganids. Siganids, or rabbit fishes (Siganidae), are a small group of marine inshore herbivores. They are widely distributed in tropical and sub-tropical waters and support important local fisheries, particularly along the Red Sea coast and in Oceania. Only in the Philippines is there a tradition of siganid farming, which is done usually in conjunction with *Chanos* in coastal ponds (described by Pillay, 1962). However, there has been a co-ordinated research effort to develop siganid culture over the past decade, commencing with inauguration of a Siganid Mariculture Group in 1972.

Siganids are relatively small fishes; *S. canaliculatus*, which has received the greatest attention, attains a maximum length of 35 cm. Commercial catches of siganids typically contain mainly fish of less that 250 g size. Despite their small size, they are a favoured food fish in many countries. The Chinese belief that rabbit fishes symbolise good fortune results in their market price rising by up to 30 times during Chinese New Year in Malaysia (Chua and Teng, 1977). In a review of the biology of siganids, Lam (1974) provides data on the habitats of ten species. Several species are typical of both reef flats and mangrove waters. Siganids are schooling fishes and seem to migrate seasonally between these two habitats.

Available evidence suggests that several species could be cultured in mangrove environments. Lam (1974) claims that *S. canaliculatus* can be acclimated to withstand salinities down to 5 ppt and thrives in salinities in the region of 10-17 ppt. This species occurs naturally in waters that fluctuate from 17 to 37 ppt salinity and from 23 to 36°C. Wide salinity and temperature tolerances

are also indicated for *S. rivulatus* and *S. luridus*, two species that have entered
the Mediterranean from the Red Sea through the Suez Canal. Popper and
Gundermann (1975) found experimentally that fry of both species survived and
grew in salinities ranging from 20 to 50 ppt; fastest growth was recorded at
20 ppt. Similarly, fluctuating salinities do not deter members of the Indian
species *S. javus*, which ingress into the Vellar estuary in large numbers
(Balasubrahmanyan and Nataragan, 1980). Siganids have been stocked in saline
ponds (40-43 ppt) in Israel (Paperna *et al.*, 1977).

In nature, siganids feed predominantly on benthic algae, with interspecific
differences being shown in the types of algae preferred (reviewed by Lam,
1974). In captivity, they will accept a wide variety of alternative foods and
various studies have reported growth using diverse feeds including lettuce,
chopped fish, prawn heads, seaweed, rice bran, ground nut oil cake and pelleted
diets. This flexibility in food acceptance is obviously of significance in evaluating
the culture potential of siganids. Although data are lacking, it is reasonable to
conclude that siganids reared in mangrove ponds would utilise both naturally
occuring algae and artificial foods.

There is little information on the growth performance of siganids.
S. canaliculatus in Singapore waters reaches a size of about 150 g at the end of its
first year (Lam, 1974). This is approximately the growth rate achieved by the
same species in coastal ponds in the Philippines. Al-Aradi *et al.* (1980) report
that *S. canaliculatus* reared in a closed water system in Bahrain increased in
mean size from 4 to 40 g in 180 days when fed pelleted diets at a daily rate of
10 per cent body weight. In Israel, Kissil (cited by Lam, 1974) reared
S. rivulatus in cages to a maximum size of 185 g in 300 days. Bryan and
Madraisau (1977) comment that *S. lineatus* attains a larger size than
S. canaliculatus and may be the better species for culture.

Siganids seem to have a distinctly seasonal breeding cycle even in equatorial
regions. In Singapore waters for example, despite almost constant average
salinity and temperature conditions, spawning of *S. canaliculatus* is restricted to
February-April with fry appearing from February to May (Lam, 1974).
Spawning occurs in shallow waters — including mangrove areas in the case of
S. lineatus. The 'spawning runs' of siganids are better in some years than in
others (Popper and Gundermann, 1975). Although siganid juveniles come
inshore in large numbers to graze from sea grass beds (Lam, 1974; Popper and
Gundermann, 1975), their highly seasonal presence precludes major culture
practices based on the collection of wild fry. Fortunately, siganids have proved
relatively easy to spawn in captivity. Natural spawnings of captive *S. canaliculatus*
and *S. rivulatus* have been obtained (Popper *et al.*, 1973; Bryan and Madraisau,
1977). Lam (1974) suggests that spawning may be triggered by exposing ripe
fish to a drop in water level since spawning in nature is reported to coincide with
the receding tide.

Induced spawning and artificial fertilisation of *S. canaliculatus* using human
chorionic gonadrotropin (HCG), was achieved by Soh and Lam (1973) and
subsequently by Bryan *et al.* (1975) who developed this method into a routine
technique. Von Westerhagen and Rosenthal (1976) showed that multiple
stripping of female *S. canaliculatus* is possible and that 20.9-32.2 ppt is the
optimum salinity range for egg fertility. Popper *et al.* (1979) found HCG also
effective in spawning *S. argenteus* and *S. rivulatus*, two of the important species

in the Middle East. Fully ripe fish can be stripped, and fertilisation achieved, without recourse to hormones (Popper *et al.*, 1973, 1979).

Initial failures in attempts to rear siganid larvae have been overcome, at least on an experimental scale. *S. canaliculatus* and *S. lineatus* have been reared to the post-metamorphosis stage using successional feeding of mixed phytoplankton, rotifers (*Brachionis*), copepods (*Oithona*) and *Artemia* nauplii (May *et al.*, 1974; Bryan and Madraisau, 1977). Recently *S. luridus* and *S. rivulatus* larvae obtained by artificial fertilisation were reared using similar live foods and ongrown to a size of 9 g (Popper *et al.*, 1979).

Cichlids. Tilapia of the *Sarotherodon* and *Tilapia* groups, and the pearlspot, *Etroplus*, are potentially suitable cichlid fishes for mangrove pond culture in estuarine habitats. Although the Cichlidae have a fresh-water origin, the majority of *Sarotherodon* and *Etroplus* species are highly tolerant of brackish-water salinities and even full seawater conditions. Both tilapia and pearlspot are unselective grazers that consume a mixed diet of plankton, benthic algae, plant fragments and detritus (Spataru, 1976; Bowen, 1976; Jayaprakash and Padmanabham, 1980). Consequently, they are ideally adapted to exploit the rich natural algal and detrital food sources in mangrove ponds.

Tilapia are widely distributed in the tropics as a result of their introduction for culture purposes into Asia and America (beginning shortly after the Second World War) and their subsequent escape into natural waters (see Balarin, 1979, for review of tilapia distribution). A few species, especially *S. mossambicus*, are already familiar to mangrove pond operators in South East Asia. Schuster (1952) states that *S. mossambicus* has 'bred freely' in mangrove tambaks in Java since 1942 where it performs the valuable function of cropping excessive growth of green algae in milkfish ponds. Elsewhere, tilapia are generally regarded as ancillary fishes in brackish-water culture, or even pests because they breed prolifically and compete for food with more desirable species. This view is likely to change as modern techniques for production of fast-growing, all-male populations of tilapia become more widely adopted. Certain hybrid crosses, such as *S. niloticus* (female) x *S. hornorum* (male) and *S. niloticus* x *S. aureus* result in virtually all-male offspring (Pruginin *et al.*, 1975; Lovshin and Da Silva, 1975). Equally effective in mono-sex tilapia production is the method of hormonal sex reversal using androgens such as methyltestosterone (Guerrero, 1975, 1976; Shelton *et al.*, 1978).

An outstanding feature of tilapia is their ability to acclimate to high salinities. Unfortunately there is confusion in the early literature on salinity tolerance in *Sarotherodon* and *Tilapia*, but *S. mossambicus*, *S. aureus* and *T. zilli*, three species of major culture importance, can certainly tolerate full-strength seawater and populations have been reported thriving in salinities of 40 ppt or more (Chervinski and Yashouv, 1971; Chervinski and Hering, 1973). Al-Amoudi (1982) has demonstrated that *S. mossambicus* can reproduce in water of 36 ppt salinity, a finding that challenges the general view that salinities above 30 ppt inhibit spawning in this species (e.g. Chen, 1976).

Tilapia species vary in their food preferences but are generally omnivorous, taking plankton and plant fragments when young, and benthic algae and detritus when adult. Farmed stocks readily accept supplementary food such as rice bran, ground nut cake, waste maize and vegetables (see Balarin, 1979, for review).

Although production data for intensive brackish-water farming of tilapia are
lacking, it is reasonable to conclude that yields from well managed mangrove
ponds could be as high as those achieved in freshwater tilapia culture.
Canagaratnam (1966) found that growth of *S. mossambicus* was actually higher
in 50 per cent seawater than in freshwater. Using modern methods of
fertilisation, controlled stocking of monosex fish and supplementary feeding,
yields of 4-6 tonnes ha^{-1} year^{-1} or more are attainable from freshwater ponds if
two or three crops are grown annually, or if multiple selective cropping is
practised. For example, hybrid all-male tilapia, offspring of the cross *S. niloticus*
x *S. hornorum*, were stocked as fingerlings in earth ponds in Puerto Rico at rates
equivalent to 10,000 ha^{-1}. After a culture period of 120 days, during which
supplementary feeding was provided at a daily rate of 2-5 per cent of body
weight, a net production of 2,608 kg ha^{-1} was obtained and all fish harvested
were of marketable size (Fram and Pagan-Font, 1978). In Taiwan, yields of
6,500-7,800 kg ha^{-1} year^{-1} are achieved from tilapia ponds enriched by water
from sewage canals (Chen, 1976). Yields of around 20,000 kg ha^{-1} year^{-1} have
been reported for tilapia ponds in Israel where fish have been stocked at
densities of up to 100,000 ha^{-1} and have been fed pelleted diets (Sarig and
Arieli, 1980).

Pearlspot are estuarine-adapted cichlids native to India and Sri Lanka.
Jhingran (1975) lists three species: *Etroplus canarensis, E. maculatus* and
E. suratensis, among the commercially important fishes of India. Of these,
E. suratensis is the most highly esteemed — it is a highly priced delicacy in
south-west India, and has excellent culture potential. Mangrove estuaries, such as
the Vellar estuary, serve as natural nursery areas for *E. suratensis*, and juveniles
can be collected by drag net during the monsoon seasons for stocking in ponds
(ICAR, 1978; Jayabalan *et al.*, 1980). However, pearlspot will also breed readily
in brackish-water ponds provided there are stones or other hard substrata present
to which spawners can attach their eggs (Hora and Pillay, 1962).

In experimental culture trials at Cochin, a fertilised coastal pond of area
0.16 ha was stocked with pearlspot fingerlings (average weight 15 g) at a rate
equivalent to 10,000 ha^{-1}. After six months these fish had reached a size of 50 g
with a survival rate of 37 per cent. The production was equivalent to 316 kg ha^{-1}
(ICAR, 1978). Contrary to the situation with tilapia, these trials indicated that
pearlspot do not respond effectively to supplementary feeding. Although
pearlspot grow slowly in comparison to milkfish and tilapia, this disadvantage is
compensated for by their high market value.

Mudskippers. Mudskippers are ubiquitous inhabitats of mangrove environments.
These amphibious gobioid fishes are well known as biological curiosities, but in
Taiwan, Burma and some localities in South East Asia they are also a highly
regarded delicacy. Species of the genera *Periophthalmus, Scartelaos,
Periophthalmodon* and *Boleophthalmus* occur in the Indo-Pacific region
(Macnae, 1968; Macintosh, 1979b). The common edible forms are
Periophthalmodon schlosseri, a large carnivorous mudskipper reaching 25 cm in
length, and species of *Boleophthalmus* (Figure 1.9) which are herbivorous and
attain an average length of 10-15 cm. *Periophthalmodon* and *Boleophthalmus*
are readily collectable from mangrove foreshores using nets or traps.

There is considerable interest in farming mudskippers in Taiwan where the common species is *B. chinensis*. Chen (1976) states that it is the highest priced among local Taiwanese fishes and gives an account of the status of mudskipper culture in Taiwan. Mudskippers are reared in brackish-water ponds of size 0.1-1.0 ha, of which there was a total area of approximately 40 ha in 1974. New ponds are prepared with about 600 kg of animal manure and rice bran to encourage development of an algal food crop as in the case of milkfish ponds. Juvenile mudskippers, usually 1.5-3.0 cm in length, are purchased from collectors who obtain them from natural waters, mainly during June to August. These are stocked at rates equivalent to around 30,000 ha^{-1} and maximum 50,000 ha^{-1}. Initially, a pond water depth of about 15 cm is maintained to allow adequate light penetration for good algal growth. Channels cut in the pond bed provided deeper water areas where mudskippers can seek shelter when surface temperatures are high. Fencing is placed around the pond to prevent entry of predators and escape of fish: mudskippers are remarkably adept at walking and leaping on land by virtue of their modified skeletons and musculature (van Dijk, 1960; Harris, 1960). Once the young mudskippers are more than 5 cm in length, they are strong enough to burrow deeply into the pond bed (*Boleophthalmus* spp. simply force their way down into the substratum tail first). This makes it possible partially to drain the pond and add more fertiliser to promote a new crop of benthic algae.

Mudskippers cultured in Taiwan grow to a marketable size of at least 24 g in one to two years; the largest fish can reach 40 g. The rate of survival is 60 per cent, giving a production equivalent to around 600 kg ha^{-1} $year^{-1}$. Given the high demand for mudskippers in Taiwan, it is not unreasonable to suppose that they could be farmed profitably elsewhere for export to the Taiwanese market. A much shorter culture period probably would be possible in tropical mangrove ponds because Chen (1976) mentions that growth of *B. chinensis* is highest at temperatures above 28°C. Some progress towards rearing mudskippers artificially has been reported by Liao *et al.* (1973) who induced a female *B. chinensis* to spawn by injecting carp pituitary gland extract and synahorin. Eggs were fertilised by mixing with chopped testes from male mudskippers. Larvae hatched from these eggs survived for four days.

Predatory Fishes. Many marine predatory fishes utilise mangrove environments as nursery and feeding areas. In the Indo-Pacific region the most significant of these include groupers (*Epinephelus* spp.), snappers (*Lutjanus* spp.), sea-perch (*Lates calcarifer*), tarpons (*Megalops cyprinoides*), ten-pounders (*Elops* spp.) and catfishes (*Arius* and *Plotosus* spp.). Mangrove-associated species of all these groups have culture potential in addition to supporting important inshore fisheries. Farming efforts have focused mainly on sea-perch, groupers and snappers by virtue of their high market value, rapid rates of growth and abundance in coastal waters. However, all predatory fishes are of concern to mangrove pond owners because of the damage they can do to stocks of prawns and young fish. Every precaution may be taken to prevent predators entering culture ponds, but inevitably some pass in as juveniles even when tidal inflow gates are well screened. Small numbers of predators may be tolerated if they are a valuable species such as *Lates calcarifer*, although their worth rarely compensates in full for the quantities of prey species they consume. Predators can be eliminated completely only by draining a pond and allowing the bed to dry. Extraneous fish in prawn ponds can be killed by applying a selective poison, of which the most familiar is saponin, a substance present in seeds of *Camellia* (teaseed). Saponin is 50 times more toxic to fish than to prawns. (Cook and Rabanal, 1978, review methods of selective poisoning in mangrove ponds; see also Terazaki *et al.*, 1980).

Lates calcarifer has been cultivated traditionally in the brackish-water bheris of West Bengal where it is known locally as 'bhekti'. It is a voracious predator and can attain a size of around 500 g in twelve months (Hora and Pillay, 1962). Sea-perch will ingress into virtually fresh-water in search of food and therefore are not adversely affected by low salinities in culture ponds. Juveniles enter culture ponds naturally with tidal water or are collected for stocking from estuarine waters. In a survey of larval stocks of sea-perch in the Muriganga estuary, part of the Sunderbans mangrove region of West Bengal, peak availability of sea-perch larvae was recorded in May to June, when a maximum collection rate of 60 larvae per man hour was achieved (ICAR, 1978).

Encouraging results have been obtained from intensive pond culture trials with sea-perch in West Bengal by the Central Inland Fisheries Research Institute at Kakdwip (ICAR, 1978). Sea-perch fry were reared in experimental nursery ponds of 0.02 ha size. Fry of 20-25 mm were stocked at a rate equivalent to 10,000 ha^{-1} and reared for 90 days. In open nurseries flushed daily by tidal water, an average daily growth of 1.0 mm day^{-1} was achieved with a survival rate of 86 per cent. In static water ponds treated with cow manure and inorganic fertiliser at rates equivalent to 500 and 200 kg ha^{-1} respectively, the values for growth and survival were 1.6 mm day^{-1} and 90 per cent. Mysid shrimp were the main food organisms eaten by the fry.

Young sea-perch stocked in a pond water supply channel (area 0.12 ha) grew rapidly on a diet of fish and prawns (extraneous species collected from the tidal water supply). Over a twelve-month culture period sea-perch of average size 168 and 375 g attained a respective size of 800 and 1,900 g approximately. These trials indicated that, with a system of multiple stocking and harvesting, as much as 3,350 kg ha^{-1} of sea-perch could be reared.

The groupers and snappers of the Indo-Pacific region include a number of species that, like *L. calcarifer*, are highly adapted to estuarine conditions and are suitable for culture in inshore environments. The estuarine grouper, *Epinephelus tauvina*, is held in particular esteem throughout the region and grows well in confined waters. Studies in Penang have shown that estuarine groupers have an extreme salinity tolerance range from 2.5 to 45.5 ppt, with best growth occurring in waters of 15-26 ppt (Chua and Teng, 1980). Small-scale commercial culture of *E. tauvina* and related species is already practised in parts of South East Asia, either in coastal ponds or in fish cages. In Singapore marketable-sized grouper weighing one kati (605 g) have been reared in floating cages in six months on a diet of trash fish (Chen *et al.*, 1977). Chua and Teng (1980 and earlier studies) obtained a production figure of 90.45 kg m^{-3} of marketable fish when they reared estuarine groupers in sea cages for eight months, beginning with fingerlings stocked at 60 fish m^{-3}. Economic evaluation of grouper cage culture in Malaysia by these three authors indicated that a very favourable net income to total costs ratio of 32 to 115 per cent could be achieved depending on management practice. A better food conversion rate was obtained for grouper given pelleted foods (3,046 kcal food:1 kg fish) compared to a diet of trash fish (up to 3,347 kcal:1 kg) (Teng *et al.*, 1978).

With little modification the cage culture systems developed for groupers are also suitable for rearing sea-perch and snappers. Among the latter, the mangrove snapper *Lutjanus argentimaculatus* may be mentioned as a highly regarded, fast-growing species which is closely associated with mangroves. An evaluation of floating cage polyculture, involving *E. tauvina, Lutjanus argentimaculatus, Lates calcarifer* and other locally available species, is in progress in a mangrove-fringed lagoon in Trengganu, Malaysia (Chan *et al.*, 1978). At a target stocking rate of 10-15 fish m^{-3} (average size 150 g), a yield of 7.5-11.25 kg m^{-3} after a ten-month rearing period is anticipated (Lau and Cheng, 1978).

The extent to which mangrove environments are suitable locations for cage culture operations is a point of general importance that may be commented on here. Coastal and lagoonal mangrove areas, with sandy rather than muddy substrata, generally have the desirable feature of good clear-water circulation. In these situations mangroves have only a moderate physical impact on the habitat, but still act as a focus of biological activity where groupers and other cultivable fishes are likely to congregate. *Acetes* and mysid shrimps, which abound in mangrove waters, are the main food organisms consumed by juvenile groupers (Chua and Teng, 1980). These fish can serve as natural stock for the culture operation (see Chan *et al.*, 1978). Estuarine mangrove waters on the other hand are basically unsuitable sites for cage culture. In mangrove estuaries any form of cage mesh rapidly becomes clogged with silt and organic debris of mangrove origin. Heavy biological fouling, which is a general problem in tropical waters, can also be anticipated. Chua and Teng (1980) found that fish cages sited in the Straits of Penang became fouled rapidly by a varied community of tunicates, mussels, oysters, barnacles, serpulid worms, green algae and other organisms. Fouling caused water exchange rates through cage netting of 37.5 mm mesh diameter to drop by 60 per cent after an immersion period of two weeks and by 87 per cent after one month. For netting of 12.7 mm mesh diameter a 93 per cent reduction in water flow was recorded after three weeks of operation.

At present, juvenile groupers required for stocking in culture cages are collected from natural waters. Although mangrove environments are important nursery areas for groupers, they are not found there in sufficient numbers to support large-scale farming operations. In some cases it appears that the natural stocks of groupers have become depleted by over-fishing (Hussain and Higuchi, 1980). And, moreover, natural populations in coastal waters may contain fewer than 5 per cent male fish. This is because *Epinephelus* species show protogynous hermaphroditism: they mature as females and subsequently transform into males. In nature this sex reversal seems to occur only when fish are several years old. Tan and Tan (1974) examined gonads from 68 specimens of *E. tauvina* obtained from the South China Sea and found an intersex condition in fish of length 66-72 cm, while only individuals of more than 74 cm, and weighing more than 11 kg, were males with ripe testes.

To overcome these disadvantages considerable efforts are being made to rear grouper artificially using appropriate hatchery technology. Chen *et al.* (1977) were able to accelerate sex reversal in three-year-old female *E. tauvina* by incorporating the androgenic hormone methyltestosterone in their diet. Milt was obtained freely from fish that received methyltestosterone over a period of two months. Other fish were induced to ovulate by giving injections of human chorionic gonadotropin and pituitary gland extract from chum salmon or white snapper. In one case complete ovulation occurred 69 hours following a single injection of 5,000 i.u. HCG. A fertilisation rate of around 30-40 per cent was achieved for eggs produced by induced spawning. On hatching, larvae were reared to the juvenile stage (day 33) on a diet of cladocerans (days 10 to 30) and mysids (30-day stage onwards), but the rate of larval survival was only about 1 per cent.

Hussain and Higuchi (1980) reared grouper larvae produced from natural spawnings of *E. tauvina* in Kuwait. Larvae consumed rotifers from the first feeding stage (days 3 to 4), but as experienced by Chen *et al.* (1977), high mortality occurred at around the 5-day stage. Attempts to wean older larvae onto a diet of *Artemia* nauplii also resulted in high losses, apparently because this food source could not be digested. Large-scale hatchery production of grouper will not be realised until a successful alternative larval diet is found.

6.2 Prawns

Hall (1962), commenting on the commercial success of mangrove prawn ponds in Singapore, first raised the question of how widespread is the association between the Penaeidae and mangrove swamps. Since Hall's time several attempts have been made to relate marine prawn fisheries to the occurrence of mangroves. Macnae (1974) demonstrated that high prawn yields were derived from waters bordering mangrove coastlines in eastern Africa and South East Asia. Martosubroto and Naamin (1977) were able to correlate commercial prawn landings in Indonesian waters to the area of mangroves adjacent to the fishing grounds (Figure 1.14). Destruction of mangroves was blamed as a factor contributing to a decline in prawn catches in El Salvador (Daugherty, 1975). Nonetheless it is incorrect to generalise from these findings. Prawn landings are substantial in the state of Kerala (45,428 tonnes in 1978) where almost all mangrove land has been reclaimed for agricultural use; less than 1,000 ha of mangroves remain from an estimated total of about 70,000 ha (Blasco, 1975). It should be noted, however, that, in converting Kerala's coastal mangrove backwaters into paddy fields, the natural pattern of tidal water flow has largely been retained. These backwaters remain highly productive (Kurian and Sebastian, 1976) and it would be interesting to know to what extent this is due to organic enrichment from paddy straw debris.

Because of their high abundance and large size it is the species of *Penaeus* and *Metapenaeus* that are commercially most valuable. The current export price of *Penaeus* tails to Japan and USA is about US $15 per kilogram. Prawn species that are consistently prominent in mangrove waters of the Indo-Pacific region include *P. merguiensis, P. indicus, M. brevicornis, M. ensis* and *M. affinis* (Table 1.13). The first three named species account for over 80 per cent of Penaeidae landings from the mangrove coast of north-east Sabah (Simpson and Chin, 1978). *Penaeus monodon* and *P. semisulcatus* are usually also present among the mangrove prawn community, but in smaller numbers. However *P. semisulcatus* exhibits an inconsistent relationship with mangroves, for although found in mangrove waters throughout South East Asia, the major fisheries for this large species are in regions lacking mangroves: notably the south-east coast of India and the Arabian Gulf (Marine Fisheries Information

Figure 1.14: The Relationship between Shrimp Catches from Various Regions of Indonesia and the Area of Mangroves Adjacent to the Fishing Grounds — Data for 1973.

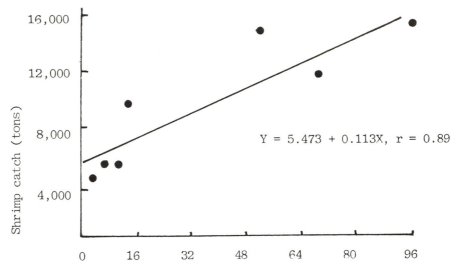

Source: Martosubroto and Naamin (1977).

Service, India, 1979; Al-Attar and Ikenoue, 1979). Several other mangrove-associated prawn species are of more localised commercial importance within the region. These include *M. stebbingi* (south-eastern coast of Africa), *M. dobsoni* (Indian waters), *P. latisulcatus* (Peninsular Malaysia) and *M. mastersii* (Singapore). In the Neotropics, mangrove prawns of economic significance include *P. aztecus, P. duorarum*, and *P. stylirrostris*.

The life cycles of the Penaeidae species mentioned above are broadly similar. Adult prawns spawn in fully marine conditions offshore. Their young move shorewards, arriving in vast numbers in estuaries and mangrove waters at a late mysis or post-larval stage. Some impression of the scale of this mangrove incursion is given by Hall (1962) who estimated that 166.4×10^9 mysis of *P. indicus* passed through the Singapore Strait (from breeding grounds bordering the island's east coast) in March 1956 during the annual breeding peak of this species. After a nursery period of up to eight months in mangrove waters (Table 1.14) prawns — now at the sub-adult stage — move progressively offshore and by maturity have returned to their spawning areas. Results from a study of *P. merguiensis* on the Selangor coast well illustrate how this behaviour pattern results in spatial separation of the different prawn stages. Chong (1979) found that only some 2.5 per cent of female prawns in a mangrove swamp population of this species were impregnated (i.e. in reproductive condition) compared to 75 per cent among those caught from spawning grounds 30 km offshore. A distinct cohort of sub-adult *P. merguiensis* was evident in nearshore waters.

In many tropical countries the bulk of commercial prawn catches are now being taken by mechanised trawlers operating in coastal waters with fine-meshed

Table 1.13: Species of Penaeidae Widely Distributed in the Indo-Pacific Region that Support Important Fisheries and are Closely Associated with Mangrove Environments

Species	English Name	Maximum Length (cm)	Average Number per kg
Penaeus monodon Fabricius	giant tiger prawn	35	5-6
P. merguiensis de Man	white prawn, banana prawn	25	7-8
P. indicus H. Milne-Edwards	white prawn, Indian prawn	22	10-12
P. semisulcatus de Haan	green tiger prawn	23	8-9
Metapenaeus ensis (de Haan)	greasy-back prawn	20	13-15
M. brevicornis (H. Milne-Edwards)	yellow prawn	15	45-50
M. affinis (H. Milne-Edwards)	pink prawn	17	20-30

Table 1.14: Periods of Residence and Growth of Species of Penaeidae in Mangrove/Estuarine Environments

Species	Nursery Site	Period of Residence (mth)	Length Reached (mm)	Source
Metapenaeus dobsoni	mangroves, Goa	5-6	50-60	Achathankutty & Sreekumaran Nair (1980)
M. monoceros	" "	7-8	70-80	
Penaeus merguiensis	" "	6-7	80-90	Chong (1979)
P. merguiensis	mangroves, Selangor	4-5	20[a]	
P. indicus	Cochin backwaters	6	80	Mohamed & Vedavyasa Rao (1971)
M. monoceros	" "	10	85	
M. dobsoni	" "	5	50	
M. affinis	" "	4	40	

Note: a. Carapace length; other measurements are total length.

trawl nets. But despite this recent development the traditional prawn fisheries operating in nearshore and estuarine waters are still of immense socio-economic importance. In Sabah, for example, a fleet of 200 mechanised trawlers exploit the rich prawn grounds bordering the north-east (Sandakan) coast, and yet some 20 per cent of Sabah's total catch is still provided by artisanal fishing methods (Simpson and Chin, 1978). A bewildering variety of trapping techniques is employed by local prawn fishermen who operate individually with cast nets and push nets, and in twos and threes using bag nets, lift nets and tidal barrier nets.

A comprehensive account of fishing methods adopted in India is given by Kurian and Sebastian (1976). Fisheries statistics for India in 1978 reveal that 70 per cent of the recorded prawn landings for that year (179,856 tonnes) was obtained by native fishing methods, mainly bag net trapping. About 4,000 bag nets are operated in the mangrove estuaries of West Bengal (Pillay and Ghosh, 1962; Kurian and Sebastian, 1976). Other examples of specific mangrove fisheries for prawns include the 'gamboa' (stake-net) fishery of Mozambique (see Macnae, 1974) and the 'gampang' (bag-net) fishery of the Malacca Straits. Prawn catches taken by these artisanal methods are probably considerably higher than official statistics indicate because they are mostly traded in local markets not covered by fisheries recorders. Bag nets and stake nets are most effective when set against the tides in mangrove channels at new and full moon. The swiftly-moving spring tides associated with these lunar phases carry migrating prawns through the mangrove zone in large quantities. It is this same migratory behaviour in the Penaeidae that makes possible trapping and holding forms of mangrove prawn pond farming, as exemplified by Singapore's prawn culture industry.

Immigrant Hokkien Chinese are accredited with introducing prawn farming into Singapore at the turn of this century (Le Mare, 1950). Prawn ponds on the island vary in size from 1 to 25 ha; a pond of average size can be operated by four men (Tham, 1973). At present a total area of about 465 ha is farmed. A gradual loss of pond sites has occurred over the past 20 years as mangroves have been reclaimed for industrial purposes. Prawn farmers in Singapore favour pond sites with an average tidal elevation below that of mean sea level so that prawns can be harvested on 20 or more days each month. This practice stems from a need to supply a daily market demand for fresh prawns. Elsewhere in South East Asia it is more common for mangrove pond operators to harvest only on days around new and full moon when maturing prawns have the greatest urge to migrate seawards (see Cook and Rabanal, 1978, for a general account of prawn culture practices in the region).

The day-to-day operation of a mangrove prawn pond is fairly straightforward. The sluice gates are opened when the flood tide reaches a height of about 60 cm above the base of the sluice. This ensures that tidal water will enter the pond with sufficient force to prevent exit of the prawn stock; at the same time, post-larvae and juvenile prawns are flushed into the pond. Once high water is reached, a wire screen of mesh diameter 1 cm is set across the sluices to prevent prawns escaping with the ebb tide. The sluice gates are closed again once the pond water depth has returned to about 60 cm. At evening high water, a filter net of rami netting is placed in the exit sluice gate to trap prawns as they leave the pond on the receding tide. The net is attached to a wooden frame which fits exactly into the sluice gate chamber (Figure 1.11b). At low water the net is raised and the catch is removed from the cod end and sorted for market. It is an interesting feature of the behaviour of penaeid prawns that adults show a strong urge to migrate seawards only at night. During daytime they are believed to remain close to the pond bottom.

The main disadvantages of tidal flow trapping and holding prawn culture practices are that stocking is uncontrollable and is dependent on the seasonal availability of prawn post-larvae. In species composition the stock may contain mainly prawns of low value and there is always the danger of undesirable predators entering ponds during the tidal trapping periods. Given these limitations, it is surprising that prawn farmers have not in the past made greater use of the natural supplies of penaeid post-larvae by collecting and sorting them before stocking in ponds. For many years 'sugpo' seed (*P. monodon*) have been gathered and sold to pond operators in the Philippines and Indonesia (Delmendo and Rabanal, 1956), but this enterprise developed only as a subsidiary to the more important *Chanos* fry industry in these countries.

Various studies recently have revealed that prawn post-larvae occur virtually year-round in mangrove waters, although numbers fluctuate seasonally and in relation to the lunar, diurnal and tidal cycles. In the estuaries and backwaters of west India, post-larvae are most numerous during the inter-monsoon period (October to May), while on the east coast there tends to be a pre-monsoon (August to November) and post-monsoon (January to April) peak in abundance (ICAR, 1978). The timing of highest recruitment varies with each species and reflects the spawning seasonality of the adult prawns. In the Cochin backwaters peak numbers of *M. dobsoni* post-larvae occur from August to January. December to May is the peak for *P. indicus*; for *P. semisulcatus* it is January to March, for *P. monodon* April to May, and for *M. affinis* and *M. monoceros* March to August (Suseelan and Kathirvel, 1980).

Breeding seasonality in the Penaeidae is not confined to regions with a strong monsoon influence. Chong (1979) found that peak spawning and larval recruitment of *P. merguiensis* on the Selangor coast occurs in October and November. This is the season of highest tides in Selangor waters and an increase in tidal amplitude may act as the 'cue' for breeding in this region where environmental conditions are otherwise almost constant. During periods of extreme tides post-larvae have a greater chance of being carried into mangrove nursery areas. Chong discovered that *P. merguiensis* post-larvae entered mangroves mainly with nocturnal flood tides and relatively few were exported with ebb tides. Three hundred kilometres south of Selangor, in Singapore waters, *P. indicus* shows a single peak in breeding intensity from February to April during the north-east monsoon season, while *M. ensis* and *M. mastersii* have two breeding peaks during the inter-monsoon periods of April-May and October-November (Hall, 1962). Evidently, therefore, there is still much to be learned about the factors influencing reproductive activity in mangrove species of the Penaeidae.

Prawn seed fisheries have expanded rapidly in the Indo-West Pacific region within the past few years with the adoption of improved methods of mangrove prawn culture involving controlled stocking of post-larvae of premium species such as *P. monodon* and *P. indicus*. Cast nets, scoop nets and tidal filter nets of various designs are operated in mangrove waters to catch migrating post-larvae (Motoh, 1980a). Bag nets set against the flood tide are widely employed in mangrove estuaries of the Sunderbans delta to trap prawn seed which are then sold to bheri owners. Seed collectors in Indonesia and the Philippines exploit the tendency of *P. monodon* juveniles to seek shelter under vegetation in shallow waters. Bunches of grass or mangrove twigs are suspended in the collecting areas to act as lures which are then fished with a scoop net (Cook and Rabanal, 1978). Fry rafts, consisting of a floating platform to which a cod end net and guiding funnel of bamboo are attached, are popular in the Philippines. These are anchored in mangrove channels to filter juvenile prawns migrating inshore (Motoh, 1980b).

Surveys of prawn seed resources in mangrove estuaries on the east coast of India have reported peak catch rates of 500-4,600 and 2,500-6,200 seed of *P. monodon* and *P. indicus* respectively per man hour of operation (ICAR, 1978). An important finding that has emerged from surveys of this nature is that the species composition of catches can be influenced considerably by the type of collecting gear used. In the Cochin backwaters it was found that plankton nets caught mainly *M. dobsoni* (88 per cent of the catch), whereas *P. indicus* was the predominant species in drag net collections (93 per cent). Approximately equal numbers of *M. affinis* and *M. dobsoni* (45 and 46 per cent of the catch) were trapped by miniature trawl nets (Suseelan and Kathirvel, 1980).

Despite their delicate appearance, penaeid post-larvae can be transported without undue mortality provided clean well-oxygenated water is used. By using chloral hydrate as a sedative, Singh *et al.* (1980) found that *P. monodon* post-larvae could survive transportation periods of more than 24 hours.

Since the fate of post-larvae released directly into prawn rearing ponds is difficult to determine, and reasonably it may be concluded that they suffer high predation mortality from older prawn and extraneous pond fishes, it is preferable to make use of nursery ponds or tanks for the first phase of culture. Delmendo and Rabanal (1956) have described a typical model for a prawn farm in the Philippines in which nursery ponds of 0.5 ha size are incorporated within rearing ponds of 5 ha size. In this type of facility, nurseries are stocked with prawn seed at rates equivalent to 300,000-500,000 ha^{-1}. These feed on 'lab-lab' algae produced by careful preparation of the pond as described on p. 32. Small quantities of rice bran may also be added. Prawns attain a length of 25 mm after four to six weeks and at this stage they are released into the rearing pond.

Prawn farmers are also able to rear post-larvae in small containers. For example, boxes made from marine plywood and measuring 2.4 x 1.2 x 1.2 m, are stocked with 10,000-40,000 post-larvae of 5-6 mm length. Aeration must be provided and it is beneficial to change 25-50 per cent of the water each day. Ground fish, mussel meat and pelleted fish foods are fed. A survival rate of 70-90 per cent to the 25 mm size can be achieved (Cook and Rabanal, 1978). More sophisticated tank rearing systems for post-larvae are gradually being adopted in South East Asia as prawn farmers become more practised in nursery stock management.

In semi-intensive forms of mangrove prawn culture rearing ponds are stocked with 10,000-50,000 juveniles per hectare (Delmendo and Rabanal, 1956; Chen, 1976; Sundararajan *et al.*, 1979). Stunting has been reported in *P. indicus* farmed at higher densities (CMFRI, 1978).

Production figures for various types of pond culture of penaeid prawns practised in the Indo-West Pacific region are compared in Table 1.15. With

Table 1.15: Yields from Coastal Pond Culture of Marine Prawns (Penaeidae) in the Indo-West Pacific Region

Locality	Principal Species	Culture System Practised	Yield (kg ha^{-1} year^{-1})	Source
Kerala	*Penaeus indicus* *Metapenaeus dobsoni*	tidal trapping in paddy fields	500-1,200 av. 900	CMFRI (1978) CMFRI (1978)
West Bengal	*P. monodon*	tidal trapping in bheris	100-300	CMFRI (1978)
Calcutta	*P. monodon*	controlled stocking in paddy fields	500-800	Macintosh (pers. obs., 1980)
Madras	*P. monodon*	controlled stocking in brackish-water ponds	1,000-1,500 (exptl scale)	Sundararajan *et al.* (1979)
Thailand	*P. merguiensis* *M. ensis*	tidal trapping in mangrove ponds	250-900 av. 440	Sribhibhadh (1973) Sribhibhadh (1973)
Thailand	*P. monodon*	controlled stocking in mangrove ponds	av. 885	Cook (1976)
P. Malaysia	*P. merguiensis* *P. indicus*	tidal trapping in mangrove ponds	730-1,220 (Johore) 320-1,110	Fish. Div., Malaysia (1973) Macintosh (pers. obs., 1976)
Singapore	*P. indicus* *M. ensis* *M. mastersii*	tidal trapping in mangrove ponds	800-1,200 av. 1,200	Tham (1973) Macintosh (pers. obs., 1976)
East Java	*P. indicus* *M. ensis*	tidal trapping in mangrove tambaks	av. 466	Schuster (1952)
Philippines (Iloilo)	*P. monodon*	controlled stocking in coastal ponds	av. 397 (traditional) 900-1,000 (improved)	Blanco (1973) Blanco (1973)
Taiwan	*P. monodon*	controlled stocking in coastal ponds	av. 1,400	Chen (1976)

controlled stocking of prawn seed, production rates of around 1,000-1,500 kg ha^{-1} year^{-1} are obtained, or approximately double those achieved by traditional trapping and holding methods. Although the prawn seed resources of mangrove waters are still far from being fully exploited, major efforts are being made to produce penaeid larvae artificially. Despite the high cost of prawn hatchery operations this is justified in the case of *P. monodon*, because this species is rarely as abundant in coastal waters as its relatives, and it is the fastest growing and highest priced prawn (because of its large size) in the region. In ponds, *P. monodon* juveniles of length 3 cm have been reared to 75-100 g size in five to six months (see Cook and Rabanal, 1978, for comparison of the merits of each mangrove penaeid species for cultivation).

Hatchery technology for the production of tropical penaeid larvae has progressed rapidly since 1977 when viable eggs were first obtained from *P. monodon* females induced to spawn using an eye-stalk ablation technique (Santiago, 1977). Although some females will spawn naturally in captivity given suitable conditions, their rate of maturation and overall fecundity are poor in comparison to ablated animals (Aquacop, 1977; Primavera, 1980). Unilateral eye-stalk ablation is usually practised. The procedure adopted in the Philippines at the South East Asian Fisheries Development Centre (SEAFDEC) is to make an incision into the eyeball, squeeze out the contents and then to crush the eye-stalk tissues which are the sites of production and storage of hormones inhibitory on ovarian maturation (Primavera and Yap, 1979). On average, *P. monodon* females spawn three to four weeks following this treatment (Primavera, 1978; Primavera and Borlongan, 1978). Multiple spawnings within a single intermoult period have been reported from ablated prawns. Using *P. monodon* of 70-100 g size, Beard and Wickins (1980) obtained more than 2 million nauplii from 34 batches of viable eggs released from March to May, 1978, by five females kept in laboratory conditions. Egg batches varied in size from 19,000 to 460,000. This compares well with an average of 120,000 eggs per batch reported for pond reared *P. monodon* in the Philippines (Primavera *et al.*, 1978).

The larval stages of *P. monodon* and other tropical penaeid species have been reared successfully on live planktonic foods. At SEAFDEC, diatoms of the genera *Skeletonema* and *Chaetoceros* are fed from the zoeal stage. Rotifers (*Brachionis*) and *Artemia* nauplii are given from the mysis stage. The flagellate, *Tetraselmis*, has also proved to be a suitable larval food. Post-larvae are offered finely ground fish or shellfish flesh (Table 1.16). With proper husbandry technique, survival rates of around 80 per cent from the nauplius to post-larval stage can be achieved (Aquacop, 1980; Beard and Wickins, 1980). Build-up of metabolites and fouling of water by waste food are the major causes of poorer survival (Silas and Muthu, 1977; Primavera and Yap, 1979).

Research in India, at Narakkal near Cochin where the Central Marine Fisheries Institute operates an experimental prawn hatchery, has shown that prawn larvae can be produced using simple and inexpensive methods. Adult spawners of the various local species of *Penaeus* and *Metapenaeus* are collected from the wild and placed individually in 50 1 basins containing aerated seawater. After spawning, the female is removed and the eggs are left undisturbed to hatch. Phytoplankton collected from brackish-water ponds is added daily from the protozoea larval stage at a concentration of 10-15 x 10^3 cells ml^{-1}. The

Table 1.16: Feeding Schedule for Penaeid Prawn Larvae Reared under Hatchery Conditions in the Philippines

Day:	0	1	2	3	4	5	6	7	8	9	10	11	12	13	14	15	16+
Stage:	Egg	Nauplius		$Zoea_1$		Z_2		Z_3		$Mysis_1$		M_2	M_3		Post-larva$_1$		$P_5{}^a$

Food Provided

1. Diatoms or *Tetraselmis*

2. *Brachionis*

3. Brine shrimp nauplii

4. Minced fish, clam or shrimp

Note: a. The hatchery phase of culture is usually terminated at the five day post-larval stage (P_5).

Source: Primavera and Yap (1979).

mysis and post-larval stages progress onto a diet of mainly tintinnids, rotifers and copepod nauplii. Using this method post-larvae of the mangrove-associated species *P. indicus, M. dobsoni, M. affinis* and *M. monoceros* have been reared with a survival rate from nauplius to post-larva of 1.0-18.2 per cent (Silas and Muthu, 1977). To reduce the demand for spawners from natural waters, the goal of this and other prawn larval culture programmes must be routinely to spawn pond-reared stocks.

In addition to penaeid prawns, mangrove environments support vast numbers of small shrimp — known collectively as 'djembret' in Indonesia — of which *Acetes* (Sergestidae) and fairy shrimps (Mysidae) are the most prominent. Despite their tiny size, the total catch of *Acetes* by countries in the Indo-West Pacific region (including Korea, China and Japan) is of the order of 170,000 tonnes annually (Omori, 1974). *Acetes* and mysid shrimp swarm through mangrove channels during spring high tides where they are caught by means of triangular push-nets (Figure 1.15). The shrimp are partially dried and made into a fermented paste that forms a key ingredient in South East Asian cooking. Pickled shrimp or 'chinchalok' is a traditional speciality of the Malacca district (Aziz, 1973).

Figure 1.15: 'Geragau' Net and Catch of *Acetes* being Partially Dried to Prepare Fermented Shrimp Paste, Local Name 'Belachan', Selangor Coast Mangroves, Malaysia.

Schuster (1952) noted that 'djembret' shrimp can be the main crop in new tambak ponds in Java before they are developed for milkfish culture. Shrimp enter these ponds during spring flood tides and are netted at the sluice gates as they try to leave on the ebb tide. Schuster observed brief peaks in abundance of mangrove shrimp when daily catches in some coastal tambaks were as high as 5.6 kg ha^{-1}. The average annual yield was 16.8 kg ha^{-1}.

6.3 Crabs

The portunid crab *Scylla serrata* is distributed from eastern Africa to the central Pacific. Known as the mangrove or mud crab, *Scylla* is a ubiquitous inhabitant of mangrove estuaries and coastal waters, and is sufficiently abundant to support significant local fisheries and semi-culture operations throughout the Indo-West Pacific region. Unlike most other portunid species, *S. serrata* is heavily built: large specimens (carapace width above 20 cm) can weigh more than 1 kg; an average adult crab weighs 250-400 g (CW 10-15 cm). Males bear magnificent chelae (these can weigh almost as much as the body) highly prized for their flesh. The ovaries of ripe females are a much sought-after delicacy.

The habits of *Scylla* are not well known. The major populations live sub-tidally in mangrove estuaries and creeks, although burrows of this crab can also be numerous intertidally on mangrove shores. Fishermen claim that burrowing crabs make regular migrations into the estuarine zone, returning to their mangrove refuges during high tides. Tagging experiments conducted by Hill (1975) in two South African estuaries revealed that *S. serrata* is a free-ranging species that does not establish territories. *Scylla* is a fairly opportunistic feeder on slow-moving benthic invertebrates and a preference for small crabs and bivalves is indicated from Hill's work. With their tremendously powerful chelae adults are able to break open mussel shells of at least 5 cm length (Williams, 1978). Dead or immobilised fish and prawns are also consumed.

Mangrove crabs can reach a marketable size of 10-15 cm (carapace width) in 7-8 months. Spawning is fairly aseasonal except where monsoon conditions result in severe dilution of estuarine waters, as on the south-west coast of India during May to August (Pillai and Nair, 1968). Hill (1974) found that *Scylla* stage I zoeae larvae could not tolerate salinities below 14 ppt. Larvae showed particular sensitivity to combinations of low salinity and high temperature, indicating a lack of physiological adaptation to estuarine conditions. It is also the case in some other mangrove crab species (*Uca and Metaplax* spp.) that the zoeae are less tolerant of low salinities than the post-larval stages (Macintosh, unpublished). Not surprisingly, therefore, female *S. serrata* migrate seawards after copulating (Hill, 1974). Up to two million eggs may be carried in a single brood (Arriola, 1940), although 300,000-600,000 is the average number. In tropical latitudes at least, females usually produce more than one brood per season. It is not difficult to rear small numbers of *Scylla* larvae, but attempts to mass-culture *Scylla* have been unsuccessful because the larval stages are extremely cannibalistic (Ling, 1973b).

The mangrove crab supports year-round fisheries in much of South East Asia. In India, where there is a strong monsoonal influence, the main landings are in the inter-monsoon seasons: from about July to September on the west coast and from April to September on the east coast. Two clever techniques are employed to catch *Scylla*. Burrow-dwelling crabs in mangrove forests are snared using a flexible pole bearing a metal hook to which a string noose is attached. The pole is carefully prodded inside the burrow, the noose catches around the crab's legs

Figure 1.16: A Mangrove Crab, *Scylla serrata*, Caught by Baited Trap in a Mangrove Channel.

or chelae and it is hauled to the surface. Crab trappers in Malaysia and Singapore make daily catches of up to 20 animals, representing an earning of around US $6.00. Small baited traps are set to catch members of the estuarine populations. These are made from fine netting suspended from a wooden cross-piece (Figure 1.16). Crabs attracted to bait (usually a morsel of fish or a prawn) hung from the cross-piece become entangled in the mesh. Traps are set in series within mangrove channels, or from the bank of an estuary. *Scylla* is also caught accidently in gill nets, stake nets and other fishing gears.

Few reliable fisheries statistics are available for *Scylla* because it supports an entirely artisanal fishery. Vedavyasa Rao *et al.* (1973) state that average yields of *Scylla* and other portunids from the lower reaches of the Godavari Estuary total 1.7 tonnes km^{-2} $year^{-1}$. This compares well with Hill's (1975) estimate of about 80 *Scylla* per hectare in the Kowie estuary in South Africa, which represents an annual production of 34 kg ha^{-1} (equivalent to 3.4 tonnes km^{-2}).

There are clear indications that *Scylla* is being overfished in many parts of its range. The average market size of crabs in Peninsular Malaysia has dropped from 200-300 g to 100-150 g over the past ten years; and similar under-sized crabs are prominent in catches sold in India and Singapore (personal observations). This situation has stimulated renewed interest in rearing *Scylla* and has boosted the market price for crabs produced by existing farming methods.

Mangrove pond culture of *Scylla* is practised in several South East Asian countries, usually in polyculture with milkfish or penaeid prawns. The extent to which *Scylla* enter culture ponds naturally is evident from Schuster's (1952) comment that a 29 acre area of coastal tambaks in eastern Java yielded approximately 3,200 crabs in one year: this indicates a production in the region of 55 kg ha^{-1}. 'Lab-lab' is a suitable natural food for young *Scylla* (Grino, 1977), which can be stocked therefore in mangrove ponds prepared in the same manner as for prawn culture. It is generally agreed that supplementary feeding, in the form of waste animal protein, should be given as the crabs grow to reduce cannibalism and promote maturation. Under-fed animals become increasingly aggressive and restless. Taiwanese crab farmers actually recognise three varieties of *S. serrata* of which the 'white' variety is favoured for pond culture partly because of its docility (Chen, 1976). The extent to which *S. serrata* is a polytypic species certainly merits further investigation.

Entrepreneurs in Singapore import *Scylla* by air from Thailand, Malaysia and Indonesia and 'fatten' them in mangrove ponds (Figure 1.17). The usual size of a crab pond is around 0.5 ha and this is stocked with 6,000-8,000 crabs. These are fed trash fish at a rate of 6 per cent of their body weight daily. After as little as 15 days the larger crabs are harvested by trapping; smaller individuals may be left for up to three months. Crabs fattened in this way fetch almost three times the value of those marketed straight from the wild. Recovery is about 80 per cent, losses being due to natural death, fighting and cannibalism, and escapes from the pond. Similar methods are used to 'fatten' *Scylla* in crab ponds in Thailand, Hong Kong and Taiwan (Chen, 1976; Ling, 1977). The prime interest of Taiwanese crab rearers is to produce females with ripe ovaries because these are regarded as a great delicacy. Snails are included in the crab diet to promote gonadal maturation. In southern Taiwan juvenile crabs of 1.5-3.0 cm or larger are also stocked in milkfish or *Gracilaria* culture ponds of 0.5-2.0 ha in area. Up to

10,000 crabs ha^{-1} are usually stocked and a marketable size of 12 cm (220 g) is reached in five to six months. Waste animal protein is fed and survival over the culture period is 50-70 per cent depending on the size of crabs stocked (Chen, 1976).

Figure 1.17: Mangrove Crab Ponds in Singapore Used to 'Fatten' *Scylla serrata*. Note the retention of mangrove trees to stabilise the earth dykes.

The total Taiwanese production of *Scylla* in 1978 was 445 tonnes (Taiwan Fisheries Bureau, 1979). Post-moult crabs — those with soft exoskeletons following ecdysis — face the highest risk of mortality in culture ponds. A moult cycle of 25-50 days is indicated for *Scylla* (Bensam, 1980), and therefore four to six moults are undergone by crabs during a typical rearing period. While the new shell is still uncalcified moult-stage animals are virtually defenceless against attack from other crabs. Provision of shelters on the pond bed for moulting crabs to hid in can serve to reduce this source of mortality. Experiments conducted in India suggest that problems of aggression and cannibalism in *Scylla* culture could be overcome by rearing crabs individually in wire-mesh cages. (Bensam (1980), using cages measuring 24 cm in diameter placed within a coastal pond, obtained an average growth increment of 17 g crab^{-1} month^{-1} based on trials with 165 animals reared for 3-15 months over a total period of two years; crabs were fed mainly trash fish and clam meat. Bensam concluded that the overall survival rate recorded (65.5 per cent) might have been improved had larger cages been used.

Hill (1975) has compared growth rates of captive *Scylla* reported from several other studies. Those of Ong (1966), Raphael (1970) and Du Plessis (1971) reveal that pond-reared crabs can be expected to grow from 4 to 12 cm in about twelve months (Figure 1.18). No obvious effects of salinity on growth rate has been shown. The degree to which *Scylla* is euryhaline is evident from an observation by Hill (1975) that crabs in the Kleinemond estuary, South Africa, experience a salinity range of 2-38 ppt.

Figure 1.18: Growth Rates of the Mangrove Crab, *Scylla serrata*.

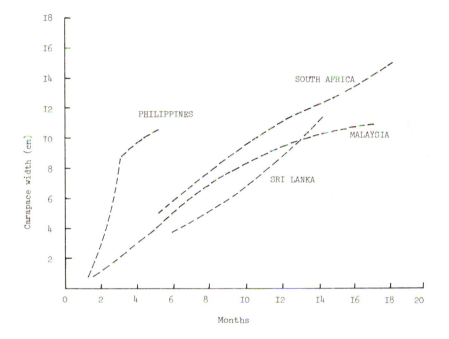

Source: Redrawn from Hill (1975). Original data for the Philippines from Arriola (1940), for South Africa from Du Plessis (1971), for Malaysia from Ong (1966) and for Sri Lanka from Raphael (1970).

In addition to *S. serrata*, there are minor fisheries for some of the larger species of mangrove sesarmid crabs in Burma and Thailand. These are caught by hand in mangrove forests and are usually sold pickled. One species of *Sesarma* (probably *S. meinerti*) is common in markets in Bangkok; the current price is 1 baht per crab (US $0.05).

6.4 Molluscs

Molluscs are numerous in almost every mangrove habitat and edible species of oysters, cockles and gastropods are collected extensively for local consumption, usually by the families of coastal fishermen. Mangrove molluscs are also fed to domestic animals; the discarded shells are collected for lime or concrete production. Rabanal *et al.* (1977) estimated that 29,500 tons of mussels, cockles and other shellfish were used in Thailand to feed ducks in 1976.

The mangrove cockle, *Anadara granosa*, has been farmed in Malaysia on a semi-culture basis since 1948. Cultivation of this species is also practised in Thailand, Kampuchea and Vietnam, but historically the natural stocks of edible mangrove molluscs have been more than sufficient to meet market demands. This situation has altered dramatically in many countries because of over-collection and loss of previously productive mollusc beds as a result of coastal industrialisation (Nikolic *et al.*, 1976a; Rabanal *et al.*, 1977; Saraya, 1980). Decline of the once important oyster beds in Kaohsiung Harbour, Taiwan, can be cited as a specific example of damage arising from industrial pollution (Chen, 1973). Mangrove mollusc culture is being viewed as a means of supplementing natural production to meet local requirements and also to supply the developing export markets for molluscs, particularly oysters.

Oysters. Of the many types of oysters that occur in mangrove areas the most abundant and valuable are members of the genus *Crassostrea*. Where environmental conditions are suitable, dense clusters of *Crassostrea* colonise the lower trunks, prop roots and pneumatophores of mangrove trees. Here they are usually in competition for space with barnacles, mussels and other encrusting organisms. As a result, individual oysters are typically stunted and distorted in shape. A single cluster of *Crassostrea* can weigh as much as 9 kg and consist of several hundred individuals ranging in size from 1 to 102 mm (Nikolic *et al.*, 1976a). Populations of large oysters do occur however, as in the Cowie Bay mangroves of Sabah, where Chin and Lim (1977) observed a 'fair abundance' of oysters greater than 150 mm.

In addition to mangrove forest populations of oysters some species of *Crassostrea* and *Saccostrea* inhabit mudflats and sub-tidal areas bordering the mangrove zone. Coastal villagers used to dive for oysters in Labuk Bay, Sabah as well as collecting them from intertidal areas (Chin and Lim, 1977). In the Pichavaram mangroves on the east coast of India large quantities of *C. madrasensis* shells are dredged from tidal channels for the purpose of extracting lime. Although *Crassostrea* species tolerate exposure to air, few oysters live above mid-tide level on mangrove shores. Berry (1975) recorded *C. cucullata* up to a height of 4.2 m above chart datum (MHWM) in a mangrove forest on the Selangor coast, but the main population was heavily concentrated in the seaward forest zone within the range of neap tides. Kamara *et al.* (1976) suggest that mangrove oyster spat settle broadly over the shore; thereafter survival is greater at lower tidal levels, no doubt as a consequence of better protection from temperature and desiccation stresses combined with longer feeding periods. Other factors known to influence distribution and survival of mangrove oysters include salinity, current velocity and dissolved oxygen concentration. *Crassostrea rhizophorae*, the most important mangrove species in Central and South America, is not found where salinities are outwith the range 12-45 ppt (Nikolic *et al.*, 1976a). High mortalities can result from sudden changes in salinity. Heavy losses of *C. gasar* occur in Lagos Lagoon during the wet season when salinities fall to as low as 0.5 ppt (Ajana, 1980).

Moderate tidal and current movements are necessary to supply oyster populations with a steady supply of food and oxygen, and to inhibit settlement of sediments. For *C. rhizophorae* water currents of 30 cm second^{-1} and dissolved oxygen levels of 2-5 parts per thousand are considered optimal (Nikolic *et al.*, 1976a).

The quantities of oysters gathered from mangrove swamps appear to be considerable, but declining. Annual harvests of *C. rhizophorae* in Cuba previously have reached 3,600 tons (Nikolic *et al.*, 1976a). Harvesting of *C. gasar* in Nigeria is now regulated by a licensing system and a closed season is enforced from February to April (Ajana, 1980). Mangrove oysters are usually collected from the shore at low tide (Figure 1.19), although boats may be used if the oyster zone is more accessible from the sea.

Figure 1.19: Women Collecting Mangrove Oysters, *Crassostrea* sp., in Southern Thailand — the Oysters are being Chipped off Pneumatophores of *Avicennia* Mangroves.

Inevitably, the natural oyster stocks will continue to decline as human pressures on mangrove coasts increase. This situation and the potentially lucrative export value of oysters, has prompted several tropical countries to investigate the feasibility of mangrove oyster culture using techniques modified from the well established systems of oyster culture operated in temperate latitudes. Notable progress has been made in the Caribbean region with *C. rhizophorae* and in Sabah with *C. belcheri*. Nikolic *et al.* (1976a, b) have described a low-cost culture system for *C. rhizophorae* developed in Cuba which utilises branches of the red mangrove (*Rhizophora mangle*) as spat collectors. These are suspended by ropes from 'stockades' constructed from mangrove tree trunks or other inexpensive woods. Collectors are positioned so that the majority of branches penetrate the water depths over which greatest natural spatfall occurs. At the oyster farms referred to by Nikolic *et al.* this is the lower 30-40 cm of the tidal range. The collectors are adjusted in height to accommodate seasonal changes in average tidal level: their correct placement with respect to the tidal regime is critical to successful spat settlement. The collectors are raised out of water for up to 24 hours once per month to eliminate fouling organisms such as sponges and tunicates.

Oysters are harvested gradually as they attain commercial size, beginning 5-6 months after the initial spat settlement. After eight months' growth the average yield per collector recorded by Nikolic *et al.* was 374 oysters, representing a total weight of 5.2 kg. Based on this level of production, a culture unit of 600-700 m^2 (containing about 7,500 collectors) could yield approximately 26 tonnes of oysters annually. Despite the heavy labour involved in establishing and maintaining collectors, Nikolic *et al.* (1976b) estimate that a unit of this size could be managed by one man.

There is a danger that wide-scale adoption of this culture technique could result in overfelling of *Rhizophora* trees through high demand for branch collectors (the average life of a collector is only about nine months). However Nikolic *et al.* mention that twigs (30 cm long) cut from mangrove shrubs and suspended vertically at 10 cm intervals along horizontal ropes, can be employed as substitute collectors. Since mangrove shrubs regenerate rapidly, even when cut down to ground level, large quantities of twig collectors could be obtained on a continuous yield basis.

A combination of the fixed rack and floating raft culture systems originally developed for European oysters has been adopted in Sabah for the South East Asian mangrove species, *C. belcheri*. At present there are only 4.1 acres under commercial oyster culture in Sabah, but enormous potential exists for expansion as Chin and Lim (1977) estimate that oysters occur naturally in 48,500 ha of mangroves around the Cowie Bay region alone. Moreover, spatfall occurs year-round on the Sabah coast, with an indication of two peaks of settlement around April-June and October-November (Chin and Lim, 1975).

Intensive research on oyster culture has been conducted since 1970 by the Sabah State Fisheries Department. After experimenting with various materials (see Table 1.17) corrugated asbestos sheeting was selected as the most suitable substrate for collecting oyster spat. To reduce costs, research is continuing to find a coating that will enable repeated use of asbestos collectors. The suitability of coconut shells as a culture material is also being investigated.

The rearing system established at the Sungei Mapan Oyster Culture Experimental Station, Taiwau, has been outlined by Chin and Lim (1975, 1977). Asbestos strips (10 x 22 cm) are arranged vertically in rows within wooden frames. After sufficient numbers of spat have settled, the frames are placed on racks held above the substratum on vertical supports. A spat density of 100-150 per 64 cm^2 is considered ideal: this can be achieved by setting out collectors when oyster larval densities in the surrounding waters range from 6 to 60 m^{-3}. From environmental observations Chin and Lim (1975) concluded that spawning was stimulated by heavy spatfall and the accompanying drop in temperature and salinity. Heavy rainfall (up to 8 cm^{-2}) followed two to three weeks after a spawning peak. Environmental monitoring has proved invaluable in avoiding heavy fouling of the oyster spat collectors by barnacles. Fairly turbid water conditions, with current velocities of about 22 cm s^{-1} were found to be unfavourable to barnacle settlement. Barnacles react to the same reproductive stimuli as oysters but fortunately their spawning response is slower so that, with care, oyster spat collection can be completed before serious barnacle infestation occurs.

Table 1.17: Comparison of the Suitability of Different Materials as Collectors for Mangrove Oysters, *Crassostrea belcheri*

	Type of Collector		
	Oyster shells	Cement slabs	Corrugated asbestos-cement sheets
Advantages			
Inexpensive	*		
Abundant supply			*
Unbreakable	*	*	
Easily handled and prepared			*
Both surfaces collect spat		*	*
Spat adhere well		*	*
High rate of spat growth	*		*
Spat easily protected		*	*
Resistant to fouling organisms	*	*	*
Remain free from silting		*	*

Source: Modified from Chin and Lim (1975).

After a rearing period of about three months, when the young oysters are 30 mm or more in length, the protective frames around the collectors are removed. Two months later the oysters are detached from their collectors and transferred to wire trays suspended from floating rafts. Following a period of twelve months in trays, 80-90 per cent of the oysters reared at the Sungei Mapam Station reached 140 mm in length, with an average meat weight of 14-21 g. Figure 1.20 shows a similar form of oyster culture in operation experimentally on the east coast of India.

The Sabah oyster culture system could yield approximately 18,000 kg per hectare of operation, over a total rearing period of 18 months. This figure is calculated from data given by Chin and Lim (1975) indicating that one asbestos collector can produce about 3.7 kg of oyster meats and that a one-acre nursery area with 6,000 collectors would provide sufficient stock for two acres of culture rafts.

The size and 'fatness' of oysters strongly influences their market value, so it is important that oyster growers are aware of environmental influences on growth and condition, and site their culture areas accordingly. A valuable study by Nascimento *et al.* (1980) has shown that water temperature, light intensity, food supply and tidal exposure level all influence the meat weight/shell size ratio of *C. rhizophorae*. It is customary to express this ratio in the form of a 'condition index' (CI) where

$$CI = \frac{\text{Dry weight of meat (g)} \times 1,000}{\text{Volume of shell cavity (cm}^3)}$$

Figure 1.20: Experimental Tray Culture of Mangrove Oysters, *Crassostrea madrasensis*, near Madras, India. Oyster meats of 30 g weight can be produced in 6-7 months.

Values of 60-120 are typical for populations of *Crassostrea*. The index usually fluctuates seasonally, increasing when oysters accumulate energy reserves or reproductive products and dropping sharply in the post-spawning period. Chin and Lim (1975) quote 50-150 as the CI range for cultivated *C. belcheri* in Sabah, with a high average value of 115-135 when weather conditions are good. Rojas and Ruiz (1972) give values of 50-75 and 70-105 for natural populations of *C. rhizophorae* in two localities in Venezuela. In Lagos lagoon, *C. gasar* varies in condition from 60 to 105 (Ajana, 1980). From studies in Brazil, Nascimento *et al.* (1980) recommend that *C. rhizophorae* should be exploited at a size of 4-6 cm because their meat weight does not increase substantially with shell growth above 6-7 cm. Oysters greater than 4 cm would have already reproduced, thus ensuring renewal of the population.

Cockles. Also known as the blood clam, *Anadara granosa* (Arcidae) occurs abundantly on mangrove foreshores, particularly soft mudflats, where it lives partially buried in the surface mud layer. Food in the form of benthic micro-algae and detritus is siphoned from the mud-water interface during periods of tidal cover. On good natural cockle beds in Malaysia annual production is in the 500-750 kg ha^{-1} (dry flesh weight) range (Broom, 1980). Ecologically *Anadara* is somewhat similar to the European edible cockle *Cardium*.

About 2,000 ha of mudflats are licenced to cockle farmers in the Malaysian states of Perak (1,600 ha) and Selangor (400 ha). The actual area under culture is much greater because licencees utilise far larger sites than they are licensed for. A farm of 16 ha requires one full-time watchman and the services of three or more casual workers; further assistance is needed during harvests (Fisheries Division, Malaysia, 1973). Seed cockles, usually of 6-12 mm size, are collected from natural nursery beds; in Perak seed are available in February-March. Culture beds are seeded at high tide from small boats. Approximately 4,500-5,500 1 of cockle seed are scattered per hectare. Subsequent thinning reduces their density to 400-600 m^2. The crop is harvested over a 2-3 month period commencing 8-9 months after sowing. Cockles mature in their first year at a size of 18-20 mm. Breeding is distinctly seasonal, with a peak from July to October. To ensure that cockles spawn before they are harvested, a minimum legal size of 31 mm has been adopted in Malaysia. The total yield of *Anadara* from both farmed and natural beds was 55,599 tonnes in 1978 (Annual Fisheries Statistics, Malaysia, 1979). The average yield per hectare is about 20.7 tonnes (Fisheries Division, Malaysia, 1973).

Further development of cockle culture in Peninsular Malaysia is being held back by lack of demand for the product. Several beds on the Selangor coast were abandoned during the mid 1970s because wholesale prices did not appreciate realistically with rising transportation costs. Notwithstanding these problems, total production has increased steadily over the past decade.

In Thailand about 500 ha of mangrove mudflats are utilised for cockle farming. A further 11,300 ha have been identified as potential culture sites (Rabanal *et al.*, 1977). Most of the present farms are in the north-west region of the Gulf of Thailand. Average production is 8.9 tonnes ha^{-1}. This is considerably below the Malaysian figure and may be explained by the fact that Thai cockle farmers sow only 250-600 litres of seed per hectare. Nevertheless, Rabanal *et al.* (1977) estimate that 80 per cent of the cockles marketed in

Thailand come from cultivated beds.

Approximately 120-180 tonnes of cockles are produced in Taiwan from a farmed area totalling 200 ha, mainly located on the Island's south-west coast (Chen, 1976). Contrary to the Malaysian situation, cockles are in high demand in Taiwan, but there are insufficient seed stocks to sustain greater production. Spawning occurs from January to April; seed cockles are available from March to September. Seed collected in March are reared in nursery beds 0.1-0.3 ha in area. Transfer to growing areas takes place in June and July. Cockles that have reached a size grade of 5,000 kg^{-1} at this stage require a further two-year rearing period before they attain commercial size (100-200 kg^{-1}). Seed obtained later in the season can be stocked directly onto the growing beds. Once a mudflat is seeded only minor maintenance is required until harvest. However losses from predators can be serious. Broom (1980a) calculated that two carnivorous snails, *Natica maculosa* and *Thais carinifera*, consumed some 53 per cent of the annual production of cockles in a rich spatfall area on the Selangor coast. Other predators of *Anadara* include crabs and seabirds.

Comparison of density and shore level effects on growth and survival of young cockles (initial size 7.5 mm) by Broom (1980b) indicated that mortality was significantly higher at low shore levels, but no density-dependent influence on mortality was evident over the range 625-2,500 cockles m^{-2}. Faster growth by low shore (1.5 m above chart datum) individuals does not compensate for their higher rate of mortality, and Broom concluded that sites at least 2.0 m above chart datum are preferable for culture operations. The relationship between cockle density and production at the site studied by Broom followed a declining exponential trend such that the maximum production attainable was 1,055 kg ha^{-1}.

Mussels. In addition to oysters and cockles, there is considerable scope to develop mussel culture in mangrove areas using species tolerant of muddy estuarine conditions, such as the green mussel *Mytilus smaragdinus*. This species spats naturally on the trunks of mangrove trees as well as on rocky substrata (Saraya, 1980). Green mussels are already cultivated on a moderate scale in Thailand and the Philippines and production rates have been most impressive. In fact *M. smaragdinus* has replaced oysters as the most important and more lucrative organism cultivated in Bacoor Bay on the south coast of Manila Bay (Tortell and Yap, 1976).

The normal culture practice in both countries utilises groups of wooden stakes as spat collectors. These are driven securely into coastal mud flats and muddy shallows with a water depth of up to 8 m. Bamboo poles of about 11 m length and spaced some 2 m apart are used in Bacoor Bay; up to 12.5 kg of mussels can develop on a single metre of bamboo. Stakes cut from mangrove date palms (*Phoenix paludosa*) are preferred in Thailand because they are inexpensive, readily available and durable — they last about three years which is more than double the life of a bamboo stake (Rabanal *et al.*, 1977). The Thai practice is to set the stakes very close together (about 5,000 are used per hectare) and to harvest after a culture period of about eight months. The average annual production of green mussel in Thailand is some 40 tons per hectare and the total culture area in 1976 was 2,626 ha. Two crops per year are possible in Bacoor Bay because a five-month growth period is sufficient for mussels to reach an average size of 3.5 cm which is acceptable to the Philippines market. Between 150 and 400 tonnes ha^{-1} are produced per crop depending on the water depth available for culture (Tortell and Yap, 1976). Chen (1977) gives the average annual production figure for mussels in the Philippines as 250 tonnes ha^{-1}.

Mangrove mudflats represent a vast potential habitat for tropical mussel culture. Rabanal *et al.* (1977), for example, estimate that a further 35,000 ha could be developed for culture of *M. smaragdinus* in Thailand. Of course it is by no means certain that adequate numbers of mussel spat will exist in every area. Available evidence suggests that spawning seasons for *Mytilus* vary between localities (Saraya, 1980) and this would be a further consideration in assessing prospective culture sites. The green mussel is obviously tolerant of estuarine conditions, but the influence of factors such as temperature, salinity and current velocity on survival and growth of this species require study. Such information is available for the related species, *M. viridis* (Tham *et al.*, 1973; Sivalingam, 1976), but the ecological requirements of *M. smaragdinus* and *M. viridis* must be very different as the latter is virtually restricted to clear water areas bordering rocky shores.

6.5 Seaweeds

Tropical seaweeds are an often overlooked resource, and yet the world volume of trade in seaweeds is increasing by some 20 per cent annually (Doty, 1977), due largely to high demand for seaweed colloids used in the manufacturing industries as stabilisers and emulsifiers. There are also important markets for edible seaweeds in Asia where they form the basis of many traditional recipes for soups, salads and desserts (the Japanese are thought to have discovered the method of extracting seaweed agar for culinary purposes in the seventeenth century).

The majority of commercially important tropical seaweeds, including *Eucheuma* (a source of carrageenan), are associated with coral reef environments where there is strong water circulation. However, *Caulerpa* (an edible green seaweed) and *Gracilaria* (a source of agar) occur naturally in some mangrove areas of South East Asia; both these seaweeds have excellent farming potential.

Caulerpa is regarded as a candidate for culture in abandoned mangrove fish ponds and other sites where mangroves have been clear-felled (Fortes, 1979). It is tolerant of poor water circulation and low nutrient levels. In seaweed farms in the Calawisan region of Cebu (the Philippines), *Caulerpa* yields of 35 tonnes fresh weight ha^{-1} or more can be produced in two months when conditions are most favourable. At present, *Caulerpa* farming is feasible only where there is local demand for the fresh product, as in parts of the Philippines. However substances of possible pharmaceutical value have been identified in *Caulerpa* (see Doty, 1977). Extraction of these on a commercial scale would create an export market for the dried seaweed. The usefulness of *Caulerpa* as an animal foodstuff or manure does not appear to have been investigated.

The red alga, *Gracilaria*, is also suitable for cultivation in disused mangrove fish ponds. This seaweed can be grown in estuarine salinities (8-25 ppt) and is intolerant of hypersaline conditions (Chen, 1976). Over the past ten years in Taiwan it has become popular to culture *Gracilaria* in ponds previously stocked with milkfish. Ponds of 1 ha in area are seeded in April with 3,000-5,000 kg of *Gracilaria* cuttings spread evenly over the pond bed. The water depth is kept at

20-30 cm during the first two months, then from July is increased to 60-80 cm as temperatures rise. Water changes are made every two to three days, or as necessary to prevent fluctuations in salinity. The ponds are fertilised with urea (3 kg ha^{-1} applied weekly) or fermented pig manure (120-180 kg ha^{-1} every two to three days). During the June to December growing season the seaweed is harvested at ten-day intervals, cleared of debris and sun dried. Average yields obtained are 7-12 tonnes of dried seaweed ha^{-1} year^{-1}. The crop is processed locally or is exported to Japan. The profit to the farmer is about five times that obtained from an equal area of milkfish culture (Shang, 1976).

The muddy foreshore areas bordering mangrove coastlines in South East Asia are considered adaptable for *Gracilaria* culture (Doty, 1977). In localities sheltered from strong water forces the seaweed could be seeded directly onto the soft substratum of the lower intertidal zone (below MLWN). Grazing of the crop by fish, and damage by crabs, are likely to be the main problems in this 'open' form of culture. This has been the case in south-east India where *Gracilaria* is grown on coconut fibre nets staked out in shallow waters (CMFRI, 1979).

Acknowledgements

Much of the material presented in this account was gathered during periods of study in mangrove swamp environments made possible through generous financial assistance provided by the University of Malaya, Marine Sciences Division, Unesco, Overseas Development Administration and the British Council. Help in various other ways was kindly provided by Dr A. Sasekumar, Dr M. Steyaert, Prof. E. Silas, Prof. R. Natarajan, Mr K. Singh, Mr G. Tay, Miss S. Bien, Mr R.B. Stewart and Mrs L. Cumming.

Finally I am indebted to Prof. R.J. Roberts who granted me extended leave to visit tropical Asia and awaited completion of the manuscript with great patience.

References

Achathankutty, C.T. and Sreekumaran Nair, S.R. (1980). Penaeid prawn population and fry resource in mangrove swamp of Goa. In: *Proc. Symp. on Coastal Aquaculture*, Cochin, January 1980

Ajana, A.M. (1980). Fishery of the mangrove oyster, *Crassostrea gasar*, Adanson (1757), in the Lagos area, Nigeria. *Aquaculture, 21*: 129-37

Aksornkoae, S. and Khemnark, C. (1980). Nutrient cycling in mangrove forest in Thailand. In: *Proc. Asian Symp. on Mangrove Environment, Research and Management*, Kuala Lumpur, August 1980

Al-Amoudi, M.M. (1982). Studies on the acclimation of commercially cultured *Sarotherodon* species to seawater. Unpubl. PhD thesis, University of Stirling, 183 pp.

Al-Aradi, J., Al-Bahrna, W. and Al-Alawi, Z. (1980). Feasibility of rearing *Siganus oramin* (= *Siganus canaliculatus*). In: *Proc. Symp. on Coastal Aquaculture*, Cochin, January 1980

Al-Attar, M.H. and Ikenoue, H. (1979). The production of juvenile shrimps (*Penaeus semisulcatus*) for release off the coast of Kuwait during 1975. *Kuwait Bull. Mar. Sci., 1*, 32 pp.

Aquacop (1977). Reproduction in captivity and growth of *Penaeus monodon* Fabricius in Polynesia. In: Avault, J.W. Jr (ed.), *Proc. 8th Annual Workshop World Maricult. Soc.*, San Jose, Costa Rica, January 1977. Louisiana State University Press, pp. 927-48

———— (1980). Reared broodstock of *Penaeus monodon*. In: *Proc. Symp. on Coastal*

Aquaculture, Cochin, January 1980

Arriola, F.J. (1940). A preliminary study of the life history of *Scylla serrata* (Forskal). *Phillip. J. Sci., 73:* 437-56

Aziz, A. (1973). Geragau fishing in Malacca. *Malay. Nat. J., 26:* 81-90

Balarin, J.D. (1979). *Tilapia: A Guide to their Biology and Culture in Africa.* University of Stirling Publication, 174 pp.

Balasubrahmanyan, K. and Natarajan, R. (1980). Food and feeding experiments in rabbit-fish *Siganus javus* (Linnaeus). In: *Proc. Symp. on Coastal Aquaculture*, Cochin, January 1980

Beard, T.W. and Wickins, J.F. (1980). Breeding of *Penaeus monodon* in laboratory recirculation systems. *Aquaculture, 20:* 79-90

Bensam, P. (1980). A culture experiment on the crab *Scylla serrata* (Forsskal) at Tuticorin to assess growth and production. In: *Proc. Symp. on Coastal Aquaculture*, Cochin, January 1980

Berry, A.J. (1972). The natural history of west Malaysian mangrove faunas. *Malay. Nat. J., 25:* 135-62

——— (1975). Molluscs colonising mangrove trees with observations on *Enigmonia rosea* (Anomiidae). *Proc. Malac. Soc. Lond., 41:* 589-600

Bird, E.C.F. and Barson, M.M. (1977). Measurement of physiographic changes on mangrove-fringed estuaries and coastlines. *Mar. Res. Indonesia, 18:* 73-80

Blanco, G.J. (1973). Status and problems of coastal aquaculture in the Philippines. In: Pillay, T.V.R. (ed.), *Coastal Aquaculture in the Indo-Pacific Region.* Fishing News (Books) Ltd, Farnham, England, pp. 60-7

Blasco, F. (1975). The mangroves of India. Institut Francais de Pondichery, Travaux de la section scientifique et Technique, No. 14, 175 pp.

——— (1976). Outlines of ecology, botany and forestry of the mangals of the Indian sub-continent. In: Chapman, V.J. (ed.), *Ecosystems of the World, I. Wet Coastal Ecosystems.* Elsevier, New York, pp. 241-60

Boonruang, P. (1980). The rate of degradation of mangrove leaves, *Rhizophora apiculata* and *Avicennia marina* at Phuket Island, western peninsula of Thailand. In: *Proc. Asian Symp. on Mangrove Environment, Research and Management*, Kuala Lumpur, August 1980

Bowen, S.H. (1976). Mechanism for digestion of detrital bacteria by the cichlid fish *Sarotherodon mossambica* (Peters). *Nature (Lond.), 260:* 137

Boyd, C.E. (1976). Lime requirements and application in fish ponds. In: Pillay, T.V.R. and Dill, W.A. (eds), *Advances in Aquaculture* (Based on FAO Technical Conference on Aquaculture, Kyoto, Japan, 1976). Fishing News (Books) Ltd, Farnham, England (published 1979)

Broom, M.J. (1980a). Size preferences of the gastropods *Natica maculosa* and *Thais carinifera* when feeding on *Anadara granosa* and the effect on mortality in an *A. granosa* population. In: *Proc. Asian Symp. on Mangrove Environment Research and Management*, Kuala Lumpur, August 1980

——— (1980b). The effect of exposure and density of the growth and mortality of *Anadara granosa* (L.) with an estimate of environmental carrying capacity. In: *Proc. Asian Symp. on Mangrove Environment, Research and Management*, Kuala Lumpur, August 1980

Bryan, P.G. and Madraisau, B.B. (1977). Larval rearing and development of *Siganus lineatus* (Pisces: Siganidae) from hatching through metamorphosis. *Aquaculture, 10:* 243-52

Bryan, P.G., Madraisau, B.B. and McVey, J.P. (1975). Hormone induced and natural spawning of captive *Siganus canaliculatus* (Pisces: Siganidae) year round. *Micronesia (J. Univ. Guam.), 11:* 199-204

Camacho, A.S. (1977). Implications of acid sulphate soils in tropical fish culture. In: *Joint SCSP/SEAFDEC Workshop on Aquaculture Engineering*, Vol. 2. Technical Report, SCS/GEN/77/15, pp. 97-102

Canagaratnam, P. (1966). Growth of *Tilapia mossambica* (Peters) at different salinities. *Bull. Fish. Res. Stn Ceylon, 19:* 47-50

Cavalcanti, L.B., Coelho, P.A., Leca, E.E., Luna, J.A.C., Macedo S.J. and Paranagua, M.N. (1978). Utilizacion de zonas de manglares en el Estado de Pernambuco (Brasil) para fines de acuicultura. In: *Memorias del Seminario Sobre el Estudio Cientifico e Impacto Humano en el Ecosistema de Manglares.* UNESCO, Montevideo, pp. 317-23

Cerame-Vivas, M.J. (1977). Management of the coastal zone with special reference to mangrove shores. In: Stewart, H.B. (ed.), *Proc. Symp. on Progress in Marine Research in the Caribbean and Adjacent Regions*, Caracas, July 1976. FAO Fisheries Rep. No. 200, FIR/R200 (E/Es), pp. 87-100

Chai, P.K. (1977). Mangrove forests of Sarawak. Contribution to the Workshop on Mangrove and Estuarine Vegetation, Serdang, Malaysia, December 1977 (pp. 1-6 in unpublished proceedings)

Chakrabarti, N.M., Karmakar, H.C. and Roy, A.D. (1980). Observations on the effect of supplementary feeds on growth and survival of grey mullet, *Liza parsia* (Hamilton), fry in brackish-water nursery ponds at Kakdwip. In: *Proc. Symp. on Coastal Aquaculture*, Cochin, January 1980

Chan, E.H. (1977). The declining mullet fishery in Peninsula Malaysia: the possible reasons and some solutions. In: *Current Research and Development in Marine Sciences in Malaysia*. Malaysian Society of Marine Sciences, Second Annual Seminar, Penang, October, 1977, pp. 27-32

Chan, K.Y. and Lee, S.W. (1979). Algal flora associated with Tai Po Kau mangrove, Hong Kong. *Int. J. Ecol. Environ. Sci., 5:* 23-8

Chan, W.L., Ho, J., Chiu, M., Saleh, I., Dusoh, R., Lam, L.Y., Chang, L.W. and Cook, H.L. (1978). Cage culture of marine fish in east coast Peninsular Malaysia. South China Sea Fisheries Development and Co-ordinating Programme, Manila, Philippines. Working Paper SCS/78/WP/69, 66pp.

Chapman, V.J. (1970). Mangrove phytosociology. *Trop. Ecol., 11:* 1-19

———— (ed.) (1976). *Ecosystems of the World, Vol. 1. Wet Coastal Ecosystems*. Elsevier, New York

Chen, F.Y. (1977). Preliminary observations of mussel culture in Singapore. First ASEAN Meeting Exp. Aquacult., Tech. Rep. ASEAN 77/FA. EgA/Rpt., 2: 73-80 (cited by Rabanal *et al.*, 1977)

Chen, F.Y., Chow, M., Chao, T.M. and Lim, R. (1977). Artificial spawning and larval rearing of the grouper, *Epinephelus tauvina* (Forskal) in Singapore. *Singapore J. Pri. Ind., 5:* 1-21

Chen, T.P. (1973). Status and problems of coastal aquaculture in Taiwan. In: Pillay, T.V.R. (ed.), *Coastal Aquaculture in the Indo-Pacific Region*. Fishing News (Books) Ltd, Farnham, England, pp. 68-73

———— (1976). *Aquaculture Practices in Taiwan*. Fishing News (Books) Ltd, Farnham, England, 162 pp.

Chervinski, J. (1976). Growth of the golden grey mullet (*Liza aurata*) (*Risso*) in salt water ponds during 1974. *Aquaculture, 7:* 51-7

Chervinski, J. and Hering, E. (1973). *Tilapia zilli* (Gervais) (Pisces, Cichlidae) and its adaptability to various saline conditions *Aquaculture, 2:* 23-9

Chervinski, J. and Yashouv, A. (1971). Preliminary experiments on the growth of *Tilapia aurea* (Steindachner) (Pisces, Cichlidae) in sea-water ponds. *Bamidgeh, 23:* 125-9

Chin, P.K. and Lim, A.L. (1975). Some aspects of oyster culture in Sabah. Fisheries Bulletin No. 5, Ministry of Agriculture and Rural Development, Malaysia, 13 pp.

———— (1977). Oyster culture development in Sabah. In: *Current Research and Development in Marine Sciences in Malaysia*. Malaysian Society of Marine Sciences, Second Annual Seminar, Penang, October 1977, pp. 48-54

Chong, K.C. (1980). Philippine milkfish production economics study underway. *ICLARM Newsletter, 3(i)*, pp. 6 and 13

Chong, V.C. (1977). Studies on the small grey mullet, *Liza malinoptera* (Valenciennes). *J. Fish. Biol., 11:* 293-308

———— (1979). Biology of *Penaeus merguiensis* in Pulau Angsa-Klang Strait Waters. MSc thesis, University of Malaya

Christensen, B. (1978). Biomass and primary production of *Rhizophora apiculata* BL. in a mangrove in southern Thailand. *Aquatic Botany, 4:* 43-52

Christensen, B. and Wium-Anderson, S. (1977). Seasonal growth of mangrove trees in Southern Thailand. I. The phenology of *Rhizophora apiculata* BL. *Aquatic Botany, 3:* 281-6

Chua, T.E. and Teng, S.K. (1977). Floating fishpens for rearing fishes in coastal waters, reservoirs and mining pools in Malaysia. Ministry of Agriculture, Malaysia, Fisheries Bulletin No. 20, 29 pp.

———— (1980). Economic production of estuary grouper, *Epinephelus salmoides* Maxwell, reared in floating net cages. *Aquaculture, 20:* 187-228

Clarke, L.D. and Hannon, N.J. (1969). The mangrove swamp and salt marsh communities of the Sydney district. II. The holocoenotic complex with particular reference to physiography. *J. Ecol., 57:* 213-34

——— (1970). The mangrove swamp and salt marsh communities of the Sydney district. III. Plant growth in relation to salinity and waterlogging. *J. Ecol., 58:* 351-69

CMFRI (1978). *Breeding and Rearing of Marine Prawns.* Central Marine Fisheries Research Institute, Cochin, Special Publ. No. 3, 128 pp.

——— (1979). *Coastal Aquaculture.* Proceedings of the first workshop on technology transfer, Cochin and Mandapam, July, 1979. Central Marine Fisheries Research Institute, Cochin, Special Publ. No. 6

Collette, B.B. and Trott, L. (1980). Mangrove fishes of New Guinea. Contribution to the 2nd Int. Symp. on Biology and Management of Mangroves and Tropical Shallow Water Communities, Papua New Guinea, July-August 1980

Cook, H.L. (1976). Problems in shrimp culture in the South China Sea region. South China Sea Fisheries Development and Co-ordinating Programme, Manila, 1976, Working Paper SCS/76/WP/40, 29 pp.

Cook, H.L. and Rabanal, H.R. (eds) (1978). *Manual on Pond Culture of Penaeid Shrimp.* ASEAN National Co-ordinating Agency of the Philippines, Ministry of Foreign Affairs, Manila, 132 pp.

Cruz, E.M. and Laudencia, I.R. (1980). Polyculture of milkfish (*Chanos chanos* Forskal), all-male nile tilapia (*Tilapia nilotica*) and snakehead in freshwater ponds with supplemental feeding. *Aquaculture, 20:* 231-7

Daugherty, H.E. (1975). Human impact on the mangrove forests of El Salvador. In: Walsh, G.E., Snedaker, S.C. and Teas, H.J. (eds), *Proc. Int. Symp. on Biology and Management of Mangroves,* 1974. University of Florida, Gainsesville, pp. 816-24

De la Cruz, A.A. (1979). The functions of mangroves. In: Srivastava, P.B.L. (ed.), *Proc. Symp. on Mangrove and Estuarine Vegetation in Southeast Asia,* Serdang, Malaysia, April 1978. Biotrop Special Publ. No. 10

Delmendo, M.N. and Rabanal, H.R. (1956). Cultivation of 'Sugpo' (Jumbo Tiger Shrimp), *Penaeus monodon* Fabricius, in the Philippines. *Proc. Indo-Pacif. Fish. Coun., 18:* 424-31

Denila, L. (1977). Improved methods of manual construction of brackish-water fishponds in the Philippines. In: *Joint SCSP/SEAFDEC Workshop on Aquaculture Engineering,* Vol. 2. Technical Report, SCS/GEN/77/15, pp. 233-59

De Silva, S.S. (1980). Biology of juvenile mullet: a short review. *Aquaculture, 19:* 21-36

Diemont, W.H. and van Wijngaarden, W. (1975). Nature conservation of mangroves in West-Malaysia. Agricultural University, Wageningen, Nature Conservation Dept., Report No. 293, 21 pp.

Dijk, D.E. van. (1960). Locomotion and attitudes of the mudskipper, *Periophthalmus,* a semi-terrestrial fish. *S. Afr. J. Sci., 56:* 158-62

Doty, M.S. (1977). Seaweed resources and their culture in the South China Sea region. South China Sea Fisheries Development and Co-ordinating Programme, Manila, Philippines, Working Paper SCS/77/WP/60, 19 pp.

Dunn, I.G. (1965). Notes on mass fish death following drought in Malaya. *Malay. Agric. J., 45:* 204-11

Du Plessis, A. (1971). A preliminary investigation into the morphological characteristics, feeding, growth, reproduction and larval rearing of *Scylla serrata* Forskal, held in captivity. Unpubl. Rep. of the Fisheries Development Corporation of South Africa, 24 pp.

Durve, V.S. (1975). Anaesthetics in the transport of mullet seed. *Aquaculture, 5:* 53-63

Ezenwa, B. (1973). Acclimation experiments with mullet fry for culture in fresh water ponds. *Ann. Rep. Fed. Dept. Fish Lagos, Nigeria* (cited by Sivalingam, 1975)

Fisheries Division, Malaysia (1973). Review of the status of coastal aquaculture in Malaysia. In: Pillay, T.V.R. (ed.), *Coastal Aquaculture in the Indo-Pacific Region.* Fishing News (Books) Ltd, Farnham, England, pp. 52-9

Fortes, M.D. (1979). Studies on farming the seaweed *Caulerpa* (Chlorophyta, Siphonales) in two mangrove areas in the Philippines. In: Srivastava, P.B.L. (ed.), *Proc. Symp. on Mangrove and Estuarine Vegetation in Southeast Asia,* Serdang, Malaysia, April 1978. Biotrop Special Publ. No. 10, pp. 125-38

Fram, M.J. and Pagan-Font, F.A. (1978). Monoculture yield trials of an all-male hybrid tilapia (♀ *Tilapia nilotica* × ♂ *T. hornorum*) in small farm ponds in Puerto Rico. In: Smitherman, R.O., Shelton, W.L. and Grover, J.H. (eds), *Proc. Symp. on Culture of Exotic Fishes*, Atlanta, 1978. American Fisheries Society Publ., pp. 55-64

Galloway, R.W. (1979). Distribution and patterns of Australian mangroves. Contribution to National Mangrove Workshop, AIMS, Townsville, April 1979

Gatus, A.R. and Martinez, E.S. (1977). Engineering considerations in the release of mangrove swamps for development into fishponds. In: *Joint SCSP/SEAFDEC Workshop on Aquaculture Engineering*, Vol. 2. Technical Report, SCS/GEN/77/15, pp. 85-95

George, M.J. (1980). Status of coastal aquaculture in India. In: *Present Status of Coastal Aquaculture in Countries Bordering the Indian Ocean*. Special Publ. of mar. biol. Assoc. India, Cochin, pp. 28-53

Golley, F., Odum, H.T. and Wilson, R.F. (1962). The structure and metabolism of a Puerto Rican red mangrove forest in May. *Ecology, 43:* 9-19

Gopalakrishnan, V. (1971). The biology of the Hooghly-Matlah estuarine system (West Bengal, India) with special reference to its fisheries. *J. mar. biol. Ass. India, 13:* 182-94

Goulter, P.F.E. and Allaway, W.G.(1980). Litter fall and decomposition in *Avicennia marina* stands in the Sydney region, Australia. Contribution to the *2nd Int. Symp. on Biology and Management of Mangroves and Tropical Shallow Water Communities*, Papua New Guinea, July-August, 1980 (in press)

Grino, E.G. (1977). Notes and observations on practises of culturing *Scylla serrata* in Western Visayas. In: *Readings in Aquaculture Practises*. SEAFDEC, pp. 154-9

Guerrero, R.D. III. (1975). Use of androgens for the production of all-male *Tilapia aurea* (Steindachner). *Trans. Am. Fish. Soc., 104:* 342-8

——— (1976). Culture of male *Tilapia mossambica* produced through artificial sex reversal. In: Pillay, T.V.R. and Dill, W.A. (eds), *Advances in Aquaculture* (Based on FAO Technical Conference on Aquaculture, Kyoto, Japan, 1976). Fishing News (Books) Ltd, Farnham, England, pp. 166-8 (published 1979)

Hall, D.N.F. (1962). *Observations on the Taxonomy and Biology of some Indo-West Pacific Penaeidae (Crustacea, Decapoda)*. Colonial Office Fishery Publications, No. 17. HMSO, London, 229 pp.

Harris, V.A. (1960). On the locomotion of the mudskipper *Periophthalmus koelreuteri* (Pallas): (Gobiidae). *Proc. Zool. Soc. Lond., 134:* 107-35

Heald, E.J. (1971). The production of detritus in a south Florida estuary. *Sea Grant Tech. Bull. (Univ. Miami). No. 6*, 110 pp.

Hill, B.J. (1974). Salinity and temperature tolerance of zoeae of the portunid crab *Scylla serrata. Mar. Biol., 25:* 21-4

——— (1975). Abundance, breeding and growth of the crab *Scylla serrata* in two South African estuaries. *Mar. Biol., 32:* 119-26

——— (1979). Aspects of the feeding strategy of the predatory crab *Scylla serrata. Mar. Biol., 55:* 209-14

Hora, S.L. and Pillay, T.V.R. (1962). *Handbook on Fish Culture in the Indo-Pacific Region*. FAO Fisheries Biology Technical Paper No. 14. FAO, Rome, 204 pp.

Hussain, N.A. and Higuchi, M. (1980). Larval rearing and development of the brown spotted grouper, *Epinephelus tauvina* (Forskal). *Aquaculture, 19:* 339-50

ICAR (1978). *Collected Reports from the Third Workshop of the All India Co-ordinated Research Project, Brackishwater Prawn and Fish Farming*. Indian Council for Agricultural Research Publication

IMS Newsletter No. 26 (1980). Article entitled: information sought on mangrove areas. Unesco, Paris, 4pp.

IUCN Report (1980). *The Global Status of Mangrove Ecosystems*. Circulated document by IUCN Working Group on Mangrove Ecosystems.

Jayabalan, N., Thangaraj, G.S. and Ramamoorthi, K. (1980). Finfish seed resources of Vellar estuary. In: *Proc. Symp. on Coastal Aquaculture*, Cochin, January 1980

Jayaprakash, V. and Padmanabham, K.G. (1980). Food and feeding habits of the pearlspot, *Etroplus suratensis* (Bloch). In: *Proc. Symp. on Coastal Aquaculture*, Cochin, January 1980

Jhingran, V.G. (1975). *Fish and Fisheries of India*. Hindustan Publ. Corp., India, 954 pp.

Juario, J.V. and Natividad, M. (1980). The induced spawning of captive milkfish. *Asian Aquaculture, 3, 8:* 3-5

Kamara, A.B., McNeil, K.B. and Quayle, D.B. (1976). Tropical mangrove oyster culture: problems and prospects. In: Pillay, T.V.R. and Dill, W.A. (eds), *Advances in Aquaculture* (Based on FAO Technical Conference on Aquaculture, Kyoto, Japan, 1976). Fishing News (Books) Ltd, Farnham, England, pp. 344-8 (published 1979)

Kanapathy, K. (1975). The reclamation and improvement of acid sulphate soils for agriculture. *Malaysian Agric. J., 50:* 264-90

Kartawinata, K. and Walujo, E.B. (1977). A preliminary study of the mangrove forest on Pulau Rambut, Jakarta Bay. *Mar. Res. Indonesia, 18:* 119-30

Khan, S.A. (1966). Working plan of the Coastal Zone Afforestation Division from 1963-4 to 1982-3. Govt. of West Pakistan, Agriculture Dept, Lahore

Khoo, K.H. (1976). Optimal utilisation and management of fisheries resources. In: L.J. Fredericks (ed.), *Proc. Seminar on the Development of the Fisheries Sector in Malaysia. J. Malaysian Econ. Ass., 13:* 40-50

Korringa, P. (1976). *Farming Marine Fishes and Shrimps, A Multi-disciplinary Treatise.* Developments in Aquaculture and Fisheries Science No. 4. Elsevier, Amsterdam, 208 pp.

Krishnamurthy, K., Sultan Ali, M.A. and Prince Jeyaseelan, M.J. (1980). Structure and dynamics of the aquatic food web community with special reference to nematodes in a mangrove ecosystem. In: *Proc. Asian Symp. on Mangrove Environment, Research and Management*, Kuala Lumpur, August 1980

Kumagai, S. and Bagarinao, T.U. (1979). Results of drift card experiments and considerations on the movement of milkfish eggs and larvae in the northern Sulu Sea. *Fish. Res. J. Philipp., 4:* 64-81

Kumagai, S., Bagarinao, T. and Uriggui, A. (1980). A study on the milkfish fry fishing gears in Panay Island, Philippines. Southeast Asian Fisheries Development Centre, Aquaculture Dept. Tech. Rep. No. 6., 34 pp.

Kuo, C.M. and Nash, C.E. (1975). Recent progress on the control of ovarian development and induced spawning of the grey mullet (*Mugil cephalus* L.). *Aquaculture, 5:* 19-29

Kuo, C.M., Nash, C.E. and Watanabe, W.O. (1979). Induced breeding experiments with milkfish, *Chanos chanos* Forskal, in Hawaii. *Aquaculture, 18:* 95-105

Kuo, C.M., Shehadeh, Z.H. and Nash, C.E. (1973). Induced spawning of captive grey mullet (*Mugil cephalus* L.) females by injection of human chorionic gonadotropin (HCG). *Aquaculture, 1:* 429-32

Kurian, C.V. and Sebastian, V.O. (1976). *Prawn and Prawn Fisheries of India.* Hindustan Publ. Corp., India, 280 pp.

Lacanilao, F. and Marte, C. (1980). Sexual maturation of milkfish in floating cages. *Asian Aquaculture, 3:* 4-6

Lam, T.J. (1974). Siganids: their biology and mariculture potential. *Aquaculture, 3:* 325-54

Latiff, A., Weber, W. and Liong, P.C. (1976). Demersal fish resources in Malaysian waters – 8. Ministry of Agriculture, Malaysia, Fisheries Bull. No. 10, 21 pp.

Lau, F. and Cheng, C.L. (1978). Recent innovations in the fish cage culture activity at the Kuala Besut small-scale fisheries pilot project, Malaysia. South China Sea Fisheries Development and Co-ordinating Programme, Working Paper SCS/78/WP/76, 15 pp.

Lee, J.J. (1980). A conceptual model of marine detrital decomposition and the organisms associated with the process. In: Droop, M.R. and Jannasch, H.W. (eds), *Advances in Aquatic Microbiology, Vol 2.*, Academic Press, London, pp. 257-91

Leh, C. and Sasekumar, A. (1980). Feeding ecology of prawns in shallow waters adjoining mangrove shores. In: *Proc. Asian Symp. on Mangrove Environment, Research and Management*, Kuala Lumpur, August 1980

Le Mare, D.W. (1950). The prawn pond industry of Singapore. *Rep. Fish. Dept. Malaya, 1949:* 121-9

Liao, I.C., Chao, N.H., Tseng, L.C. and Kuo, S.C. (1973). Studies on the artificial propagation of *Boleophtalmus chinesis* (Osbeck). I. Observations on Embryonic development of early larvae. JCRR Fisheries Series, No. 15 (in Chinese, cited by Chen, 1976)

Liao, I.C., Juario, J.V., Kumagai, S., Nakajima, H., Natividad, M. and Buri, P. (1979). On the induced spawning and larval rearing of milkfish, *Chanos chanos* (Forskal). *Aquaculture, 18:* 75-93

Liew, T.C. (1977). Mangrove forest of Sabah. Contribution to the Workshop on Mangrove and Estuarine Vegetation, Serdang, Malaysia, December 1977 (pp. 7-42 in unpublished proceedings)

Lim, B.H. and Sasekumar, A. (1979). A preliminary study on the feeding biology of mangrove forest primates, Kuala Selangor. *Malay. Nat. J., 33:* 105-12

Lim, C., Sukhawongs, S. and Pascual, F.P. (1979). A preliminary study on the protein requirements of *Chanos chanos* (Forskal) fry in a controlled environment. *Aquaculture, 17:* 195-201

Ling, S.W. (1973a). Status, potential and development of coastal aquaculture in the countries bordering the South China Sea. FAO/UNDP, SCS/DEV/74/5, 51 pp.

——— (1973b). A review of the status and problems of coastal aquaculture in the Indo-Pacific Region. In Pillay, T.V.R. (ed.), *Coastal Aquaculture in the Indo-Pacific Region.* Fishing News (Books) Ltd, Farnham, England, pp. 2-25

——— (1977). *Aquaculture in Southeast Asia: A Historical Overview.* A Washington sea grant publication. Univ. Washington Press, Seattle, 108 pp.

Loi, J.J. and Sasekumar, A. (1980). Litter production, accumulation and decomposition in the South Banjar mangrove forest reserve, Kuala Selangor. In: *Proc. Asian Symp. on Mangrove Environment, Research and Management,* Kuala Lumpur, August 1980

Lovshin, L.L. and Da Silva, A.B. (1975). Culture of monosex and hybrid tilapias. Contribution to the FAO/CIFA Symp. on Aquaculture in Africa, Accra, Ghana, CIFA/75/SR 9, 16 pp.

Lugo, A.E. and Snedaker, S.C. (1974). The ecology of mangroves. *Ann. Rev. Ecology and Systematics, 5:* 39-64

Luther, G. (1973). The grey mullet fishery resources of India. In: *Proc. Symp. on Living Resources of the Seas around India,* Cochin, 1968. CMFRI Special Publ., pp. 455-60

Macintosh, D.J. (1979a). The ecology and energetics of mangrove fiddler crabs (*Uca* spp.) on the west coast of the Malay Peninsula. PhD thesis, University of Malaya, 332 pp.

——— (1979b). Predation of fiddler crabs (*Uca* spp.) in estuarine mangroves. In: Srivastava, P.B.L. (ed.), *Proc. Symp. on Mangrove and Estuarine Vegetation in Southeast Asia,* Serdang, Malaysia, April 1978. Biotrop Special Publ. No. 10, pp. 101-10

——— (1980). Ecology and productivity of Malaysian mangrove crab populations (Decapoda: Brachyura). In: *Proc. Asian Symp. on Mangrove Environment, Research and Management,* Kuala Lumpur, August 1980

Macnae, W. (1963). Mangrove swamps in South Africa. *J. Ecol., 51:* 1-25

——— (1966). Mangroves in eastern and southern Australia. *Aust. J. Bot., 14:* 67-104

——— (1967). Zonation within mangroves associated with estuaries in North Queensland. In: G.H. Lauff (ed.), *Estuaries.* Publs. Am. Ass. Advmt Sci., No. 83, pp. 432-41

——— (1968). A general account of the fauna and flora of mangrove swamps and forests in the Indo-West Pacific region. *Adv. Mar. Biol., 6:* 73-270

——— (1974). Mangrove forests and fisheries. Indian Ocean Programme Publ. No. 34. Indian Ocean Fishery Commission, Rome, 1 OFC/DEV/74/34, 35 pp.

Malley, D.F. (1977). Adaptations of decapod crustaceans to life in mangrove swamps. *Mar. Res. Indonesia, 18:* 63-72

Mandal, L.N. (1962). Nitrogenous fertilizers for brackish-water ponds – ammonium or nitrate form? *Indian J. Fish. (A), 9:* 123-4

Marine Fisheries Information Service, India (1979). Synopsis of marine prawn fishery of India – 1978. Technical and Extension Series No. 10, 18 pp.

Martosubroto, P. and Naamin, N. (1977). Relationship between tidal forests (mangroves) and commercial shrimp production in Indonesia. *Mar. Res. Indonesia, 18:* 81-8

May, R.C., Popper, D. and McVey, J.P. (1974). Rearing and larval development of *Siganus canaliculatus* (Park) (Pisces: Siganidae). *Micronesia (J. Univ. Guam), 10:* 285-98

Mires, D., Shilo, S. and Shak, Y. (1975). Further observations on the effect of salinity and temperature changes on *Mugil cephalus* fry. *Aquaculture, 5:* 110

Mohamed, K.H. and Vedavyasa Rao, P. (1971). Estuarine phase in the life-history of the commercial prawns of the west coast of India. *J. Mar. Biol. Ass. India, 13:* 149-61

Moses, B.S. (1980). Mangrove swamp as a potential food source. Contribution to the Workshop on Mangrove Ecosystem, Port Harcourt, May 1980

Motoh, H. (1980a). Traditional devices and gear for collecting fry of 'sugpo' giant tiger prawn, *Penaeus monodon* in the Philippines. Aquaculture Dept, Southeast Asian Fisheries Development Centre, Tech. Rep. No. 4, 15 pp.

——— (1980b). Fishing gear for prawn and shrimp used in the Philippines today. Aquaculture Dept, Southeast Asian Fisheries Development Centre, Tech. Rep. No. 5, 43 pp.

Nascimento, I.A., Pereira, S.A. and Souza, R.C. (1980). Determination of the optimal commercial size for the mangrove oyster (*Crassostrea rhizophorae*) in Todos os Santos Bay, Brazil. *Aquaculture, 20:* 1-8

Nash, C.E. (1978). Milkfish at Christmas. *Fish Farming International, 5, 2:* 8-13

Nash, C.E. and Kuo, C.M. (1975). Hypotheses for problems impeding the mass propagation of grey mullet and other finfish. *Aquaculture, 5:* 119-34

Nash, C.E., Kuo, C.M. and McConnel, S.C. (1974). Operational procedures for rearing larvae of the grey mullet (*Mugil cephalus* L.). *Aquaculture, 3:* 15-24

Natarajan, P. and Radhakrishnan, K.V. (unpublished). A study of crabs of south Kanara coast.

Nikolic, M., Bosch, A. and Alfonso, S. (1976a). A system for farming the mangrove oyster (*Crassostrea rhizophorae* Guilding, 1828). *Aquaculture, 9:* 1-18

Nikolic, M., Bosch, A. and Vazquez, B. (1976b). Las experiencias en el Cultivo de ostiones del mangle (*Crassostrea rhizophorae*). In: Pillay, T.V.R. and Dill, W.A. (eds.). *Advances in Aquaculture* (Based on FAO Technical Conference on Aquaculture, Kyoto, Japan, 1976). Fishing News (Books) Ltd, Farnham, England, pp. 339-44 (published 1979)

Nixon, S.W., Furnas, C.N., Lee, V., Marshall, N., Ong, J.E., Wong, C.H. and Sasekumar, A. (1980). The role of mangroves in the carbon and nutrient dynamics of Malaysian estuaries. In: *Proc. Asian Symp. on Mangrove Environment, Research and Management*, Kuala Lumpur, August 1980

Odum, W.E. and Heald, E.J. (1972). Trophic analyses of an estuarine mangrove community. *Bull. Mar. Sci., 22:* 671-738

—— (1975). Mangrove forests and aquatic productivity. In: Haster, A.D. (ed.), *Coupling of Land and Water Systems*. Ecological Studies, Vol. 10. Springer-Verlag, New York, pp. 129-36

Omori, M. (1974). The biology of pelagic shrimps in the oceans. In: Russell, F.S. and Yonge, M. (eds), *Advances in Marine Biology, Vol. 12*. Academic Press, London, pp. 233-324

Ong, J.E., Gong, W.K. and Wong, C.H. (1979). Productivity of a managed mangrove forest in West Malaysia. Contribution to the Conference on Trends in Applied Biology in Southeast Asia, Penang, Malaysia, October, 1979, 9 pp.

Ong, K.S. (1966). Observations on the post-larval life history of *Scylla serrata* Forskal reared in the laboratory. *Malay. Agric. J., 45:* 429-43

Ong, T.L. (1978). Some aspects of trophic relationships of shallow water fishes (Selangor Coast). BSc honours thesis, University of Malaya

Paperna, I., Colorni, A., Gordin, H. and Kissel, G.W. (1977). Diseases of *Sparus aurata* in marine culture at Elat. *Aquaculture, 10:* 195-214

Pillai, K.K. and Nair, N.B. (1968). Observations on the reproductive cycles of some crabs from the south-west coast of India. *J. Mar. Biol. Ass. India, 10:* 1-2

Pillay, T.G. (1962). Fish farming methods in the Philippines, Indonesia and Hong Kong. *FAO Fish. Biol. Tech. Paper, 18:* 50-2

Pillay, T.V.R. and Ghosh, K.K. (1962). The bag-net fishery of the Hoogly-Matlah estuarine system (West Bengal). *Indian J. Fish. (A), 9:* 71-99

Popper, D., Gordin, H. and Kissil, G.W. (1973). Fertilization and hatching of rabbit fish *Siganus rivulatus. Aquaculture, 2:* 37-44

Popper, D. and Gundermann, N. (1975). Some ecological and behavioural aspects of siganid populations in the Red Sea and Mediterranean coasts of Israel in relation to their suitability for aquaculture. *Aquaculture, 6:* 127-41

Popper, D., Pitt, R. and Zohar, Y. (1979). Experiments on the propagation of Red Sea Siganids and some notes on their reproduction in nature. *Aquaculture, 16:* 177-81

Potter, T. (1977). The problems to fish culture associated with acid sulphate soils and methods for their improvement. Contribution to the Joint FAO-UNDP/SCSP and SEAFDEC Regional Workshop on Aquaculture Engineering, Tigbauan, Iloilo, Philippines, 1977, SCSP/SFDC/77/AEn/BP22, 13 pp.

Prabhakara Rao, A.V. (1971). Observations on the larval ingress of the milkfish *Chanos chanos* (Forskal) into the Pulicat Lake. *J. Mar. Biol. Ass. India, 13:* 249-57

Primavera, J.H. (1978). Induced maturation and spawning in five month old *Penaeus monodon* Fabricius by eyestalk ablation. *Aquaculture, 13:* 355-9

—— (1980). Studies on broodstock of sugpo *Penaeus monodon* Fabricius and other penaeids at the SEAFDEC Aquaculture Department. In: *Proc. Symp. on Coastal*

Aquaculture, Cochin, January 1980

Primavera, J.H. and Borlongan, E. (1978). Ovarian rematuration of ablated sugpo prawn *Penaeus monodon* Fabricius. *Ann. Biol. Anim. Bioch. Biophy., 18:* 1067-72

Primavera, J.H., Borlongan, E. and Posadus, R.A. (1978). Mass production in concrete tanks of sugpo *Penaeus monodon* Fabricius spawners by eyestalk ablation. *Fish. Res. J. Philippines, 3:* 1-12

Primavera, J.H. and Yap, W.G. (1979). Status and problems of broodstock and hatchery management of sugpo (*Penaeus monodon*) and other penaeids. Southeast Asian Fisheries Development Center, Aquaculture Department Contr. No. 47, 21 pp.

Prince Jeyaseelan, M.J. and Krishnamurthy, K. (1980). Role of mangrove forest of Pichavaram as fish nursery. *Proc. Indian Nat. Sci. Acad. B, 46:* 2-6

Pruginin, Y., Rothbard, S., Wohlfarth, A., Haley, A., Moav, R. and Hulata, G. (1975). All-male broods of *Tilapia nilotica* and *T. aurea* hybrids. *Aquaculture, 6:* 11-21

Rabanal, H.R. (1976). Mangroves and their utilisation for aquaculture. Contribution to the National Workshop on Mangrove Ecology, Phuket, Thailand, January, 1976

Rabanal, H.R., Pongsuwana, U., Saraya, A. and Poochareon, W. (1977). Shellfisheries of Thailand: background and proposal for development. South China Sea Fisheries Development and Co-ordinating Programme, Manila, Philippines, Working Paper SCS/77/WP/61, 48 pp.

Raphael, Y.I. (1970). A preliminary report on the brackish-water pond culture of *Scylla serrata* in Ceylon. *Proc. Indo-Pacif. Fish. Con., 14:* 1-10

Ravagnan, G. (1978). *Vallicultura Moderna.* Edagricole, Bologna, 283 pp.

Rojas, A.V. and Ruiz, J.B. (1972). Variacion estacional del engorda des ostion *C. rhizophorae* da Baia de Machima y Laguna Grande. *Bol. Inst. Ocean. Univ. Orient, II:* 39-43 (cited by Nascimento *et al.*, 1980)

Ross, P. (1974). The effects of herbicides on the mangrove of South Vietnam. Working Paper in: *The Effects of Herbicides in South Vietnam.* Nat. Acad. Sci., Nat. Res. Counc., 33 pp.

Roy, A.K. and Chakrabarti, N.M. (1980). Studies on the effect of supplementary feeds and fertilisation on the growth and survival of grey mullet, *Liza tade* (Forskal) fry in brackish-water nursery ponds. In: *Proc. Symp. on Coastal Aquaculture*, Cochin, January 1980

Santiago, A.C. Jr (1977). Successful spawning of cultured *Penaeus monodon* Fabricius after eyestalk ablation. *Aquaculture, 11:* 185-96

Saraya, A. (1980). Physico-chemical properties of mussel farm at Samaekhae, Chachoengsao Province. In: *Proc. Asian Symp. on Mangrove Environment, Research and Management*, Kuala Lumpur, August 1980

Sarig, S. and Arieli, Y. (1980). Growth capacity of tilapia in intensive culture. *Bamidgeh, 32:* 57-65

Sasekumar, A. (1970). Aspects of the ecology of a Malayan mangrove fauna. MSc thesis, University of Malaya, 111 pp.

—— (1974). Distribution of macrofauna on a Malayan mangrove shore. *J. Anim. Ecol., 43:* 51-69

Sasekumar, A. and Thong, K.L. (1980). Predation of mangrove fauna by marine fishes at high tide. In: *Proc. Asian Symp. on Mangrove Environment, Research and Management*, Kuala Lumpur, August 1980

Schmittou, H.R. (1977). A study to determine the spawning grounds of milkfish and the environmental conditions that influence fry abundance and collection along the Antique Coast of Panay Island, Philippines. In: Avault, J.W. Jr (ed.), *Proc. 8th Ann. Meeting World Maric. Soc.*, San Jose, Costa Rica, January 1977. Louisiana State University, pp. 91-106

Schuster, W.H. (1952). *Fish-culture in Brackish-water Ponds of Java.* Indo-Pacific Fisheries Council, Special Publications No. 1, 140 pp.

—— (1960). Synopsis of biological data on milkfish *Chanos chanos* Forskal, 1775. FAO Fish Synop., No. 4, 67 pp.

Sebastian, M.J. and Nair, A. (1975). The induced spawning of the grey mullet, *Mugil macrolepis* (Aguas) Smith and the large scale rearing of its larvae. *Aquaculture, 5:* 41-52

Seneviratne, E.W. (1978). The Sri Lanka mangroves. Cited by Snedaker, S.C. (1980). The mangroves of Asia and Oceania: status and research planning. In: *Proc. Asian Symp. on Mangrove Environment, Research and Management*, Kuala Lumpur, August 1980 (in press)

Shang, Y.C. (1976). Economics of various management techniques for pond culture of finfish. South China Sea Fisheries Development and Co-ordinating Programme, Manila, Philippines, Working Paper SCS/76/WP/37, pp. 32.

Shelton, W.L., Hopkins, K.D. and Jensen, G.L. (1978). Use of hormones to produce monosex tilapia for aquaculture. In: Smitherman, R.O., Shelton, W.L. and Grover, J.H. (eds), *Proc. Symp. on Culture of Exotic Fishes*, Atlanta, 1978. American Fisheries Society Publ., pp. 10-33

Silas, E.G., Mohanraj, G., Gandhi, V. and Thirunavukkarasu, A.R. (1980). Spawning grounds of the milkfish and seasonal abundance of the fry along the east and southwest coasts of India. In: *Proc. Symp. on Coastal Aquaculture*, Cochin, January 1980

Silas, E.G. and Muthu, M.S. (1977). Hatchery production of penaeid prawn larvae for large scale coastal aquaculture. In: *Proc. Symp. on Warm Water Zooplankton*. UNESCO/NIO Special Publication, pp. 613-18

Simpson, A.C. and Chin, P.K. (1978). The prawn fisheries of Sabah, Malaysia. Ministry of Agriculture, Malaysia, Fisheries Bulletin No. 22, 24 pp.

Singh, H., Chowdhury, A.R. and Pakrase, B.B. (1980). Observations on the transport of the postlarvae of tiger prawn *Penaeus monodon* (Fabricius). In: *Proc. Symp. on Coastal Aquaculture*, Cochin, January 1980

Singh, V.P. (1980). Management of fish ponds with acid sulphate soils. Part I. *Asian Aquaculture, 3, 4:* 4-6

Sivalingam, P.M. (1976). Aquaculture of molluscs as a means of overcoming the economic problems of the coastal fishing community. In: Fredericks, L.J. (ed.) *Proc. Seminar on the Development of the Fisheries Sector in Malaysia. J. Mal. Econ. Ass., 13:* 219-28

Sivalingam, S. (1975). On the grey mullets of the Nigerian coast, prospects of their culture and results of trials. *Aquaculture, 5:* 345-57

Smith, I.R., Cas, F.C., Gibe, P. and Romillo, L.M. (1978). Preliminary analysis of the performance of the fry industry of the milkfish (*Chanos chanos* Forskal) in the Philippines. *Aquaculture, 14:* 199-219

Snedaker, S. and Lugo, A. (1973). The role of mangrove ecosystems in the maintenance of environmental quality and a high productivity of desirable fisheries. US Bureau of Sports, Fisheries and Wildlife, NTIS distribution

Snedaker, S.C. (1980). The mangroves of Asia and Oceania: status and research planning. In: *Proc. Asian Symp. on Mangrove Environment, Research and Management*, Kuala Lumpur, August 1980

Soh, C.L. and Lam, T.J. (1973). Induced breeding and early development of the rabbitfish, *Siganus oramin* (Schneider) (= *S. canaliculatus*). In: *Proc. Symp. Biol. Res. Natl. Dev. Singapore*, pp. 49-56 (cited by Lam, 1974)

Spataru, P. (1976). The feeding habits of *Tilapia galilaea* in Lake Kinneret, Israel. *Aquaculture, 9:* 47-59

Sribhibhadh, A. (1973). Status and problems of coastal aquaculture in Thailand. In: Pillay, T.V.R. (ed.), *Coastal Aquaculture in the Indo-Pacific Region*. Fishing News (Books) Ltd, Farnham, England, pp. 74-83

Steinke, T.D. (1980). Degradation of mangrove leaf and stem tissues *in situ* in Mgeni Estuary, South Africa. Contribution to the 2nd Int. Symp. on Biology and Management of Mangroves and Tropical Shallow Water Communities, Papua New Guinea, July-August, 1980

Stoddart, D.R., Bryan, G.W. and Gibbs, P.E. (1973). Inland mangroves and water chemistry, Barbada, West Indies. *J. Nat. Hist., 7:* 33-46

Sundararajan, D., Chandra Bose, S.V. and Venkatesan, V. (1979). Monoculture of tiger prawn, *Penaeus monodon* Fabricius, in a brackish-water pond at Madras, India. *Aquaculture, 16:* 73-5

Suseelan, C. and Kathirvel, M. (1980). Prawn seed calendars of Cochin backwater. In: *Proc. Symp. on Coastal Aquaculture*, Cochin, January 1980

Taiwan Fisheries Bureau (1979). Cited by Pullin, R.S.V., 1980. Aquaculture in Taiwan. *ICLARM Newsletter, 3:* 10-12

Tan, K.F. (1977). Some aspects on the biology of the *Acetes erythraeus* in the Sungei Sementa Besar, Malaysia. BSc honours thesis, University of Malaya, 37 pp.

Tan, S.M. and Tan, K.S. (1974). Biology of tropical grouper, *Epinephelus tauvina* (Forskal) I. A preliminary study on hermaphroditism in *E. tauvina. Singapore J. Pri. Ind., 2:* 123-33

Tang, Y.A. (1976a). Physical problems in fish farm construction. In: Pillay, T.V.R. and Dill, W.A. (eds), *Advances in Aquaculture* (Based on FAO Technical Conference on Aquaculture, Kyoto, Japan, 1976). Fishing News (Books) Ltd, Farnham, England, pp. 99-104 (published 1979)

———— (1976b). Planning, design and construction of a coastal milkfish farm. In: Pillay, T.V.R. and Dill, W.A. (eds), *Advances in Aquaculture* (Based on FAO Technical Conference on Aquaculture, Kyoto, Japan, 1976). Fishing News (Books) Ltd, Farnham, England, pp. 104-7 (published 1979)

Teas, H.J. (1979). Silviculture with saline water. In: Hohaender, A. (ed.), *The Biosaline Concept*. Plenum, New York, pp. 117-61

Teinsongrusmee, B. (1970). A present status of shrimp farming in Thailand. Contr. No. 18, Invertebrate Fisheries Investigations Dept. of Fisheries, Bangkok, 34 pp.

Teng, S.K., Chua, T.E. and Lim, P.E. (1978). Preliminary observation on the dietary protein requirement of estuary grouper, *Epinephelus salmoides* Maxwell, reared in floating net-cages. *Aquaculture, 15:* 257-72

Terazaki, M., Tharnbuppa, P. and Nakayama, Y. (1980). Eradication of predatory fishes in shrimp farms by utilisation of Thai teaseed. *Aquaculture, 19:* 235-42

Thakurta, S.C. and Pakrasi, B.B. (1980). Experiments on the transport of mullet *Liza parsia* Hamilton fry. In: *Proc. Symp. on Coastal Aquaculture*, Cochin, January 1980

Tham, A.K. (1968). Prawn culture in Singapore. Contribution to the FAO World Scientific Conf. on the Biology and Culture of Shrimps and Prawns at New Mexico, June 1967. *FAO Fish. Rep., 57, 2:* 85-93

———— (1973). Fish and prawn ponds. In: Chuang, S.H. (ed.), *Animal Life and Nature in Singapore*. Singapore University Press, pp. 260-8

Tham, A.K., Lam, S.L. and Tan, W.H. (1973). Experiments in coastal aquaculture in Singapore. In: Pillay, T.V.R. (ed.), *Coastal Aquaculture in the Indo-Pacific Region*. Fishing News (Books) Ltd, Farnham, England, pp. 375-83

Tortell, P. and Yap, W.G. (1976). Mussel culture gathers momentum in the Philippines. *Fish Farming International, 3:* 26-8

Van Steenis, C.G.G.J. (1962). The distribution of mangrove plant genera and its significance for paleogeography. *Kon. Neder. Akad. van Wetensch., 65:* 164-9

Vanstrone, W.E., Tiro, L.B.Jr., Villaluz, A.C., Ramsingh, D.C., Kumagai, S., Doldoco, S., Barnes, P.J. and Duenas, C.E. (1977). Breeding and larval rearing of the milkfish, *Chanos chanos* (Pisces: Chanidae). Southeast Asian Fisheries Development Center, Aquaculture Dept Tech. Rep., No. 3, pp. 3-17

Vedavyasa Rao, P., Thomas, M.M. and Sudhakara Rao, G. (1973). The crab fishery resources of India. In: *Proc. Symp. on Living Resources of the Seas around India*, 1968. CMFRI Special Publ., pp. 581-91

Vibulsreth, S., Ketruangrote, C. and Sriplung, N. (1976). Distribution of mangrove forest as revealed by earth technology satellite (ERTS–1) imagery. Contribution to the National Workshop on Mangrove Ecology, Phuket, Thailand, January 1976

Vijayaraghavan, S. and Ramadas, V. (1980). Food conversion in the shrimp. *Metapenaeus monoceros* (Fabricius) fed on mangrove leaves. In: *Proc. Symp. on Coastal Aquaculture*, Cochin, January 1980

Villaluz, D.K. (1953). *Fish Farming in the Philippines*. Bookman, Manila, 336 pp.

Von Prahl, H. (1978). Imporancia de manglar en la biologia de los camarones Peneidos (English Summary). In: *Memorias del Seminario Sobre el Estudio Cientifico e Impacto Humano en el Ecosistema de Manglares*. Unesco, Montevideo, pp. 341-3

Von Westernhagen, H. and Rosenthal, H. (1976). Induced multiple spawning of reared *Siganus oramin* (Schneider) (= *S. canaliculatus* Park). *Aquaculture, 7:* 193-6

Walsh, G.E. (1967). An ecological study of a Hawaiian mangrove swamp. In: Lauff, G.H. (ed.), *Estuaries. Publs. Amer. Ass. Advmt Sci., 83:* 420-31

———— (1974). Mangroves: a review. In: Reimhold, R. and Queen, W. (eds), *Ecology of Halophytes*. Academic Press, New York, pp. 51-174

Walter, H. (1971). *Ecology of Tropical and Subtropical Vegetation*. Oliver and Boyd, Edinburgh, 539 pp.

Watson, J.G. (1928). *Mangrove Forests of the Malay Peninsula*. Malayan Forest Records No. 6, 275 pp.

Wells, A.G. (1980). Distribution of mangrove species in Australia. Contribution to the 2nd Int. Symp. on Biology and Management of Mangroves and Tropical Shallow Water

Communities, Papua New Guinea, July-August, 1980

West, R.C. (1956). Mangrove swamps of the Pacific coast of Colombia. *Ann. Assoc. Am. Geogr., 46:* 98-121

Williams, M.J. (1978). Opening of bivalve shells by the mud crab *Scylla serrata* Forskal. *Aust. J. Mar. Freshwat. Res., 29:* 699-702

2 OPPORTUNITIES FOR FARMING CRUSTACEANS IN WESTERN TEMPERATE REGIONS

J.F. Wickins

Figure 2.1: The Experimental Culture of Tropical Prawns in a Controlled Environment. To the left of the two layers of raceway tanks are four percolating biological filters.

Source: MAFF, Conwy (1973).

1. Introduction

There is an almost insatiable demand for Crustacea in North America and Europe. In the USA the average annual landing of 174,000 tonnes of shrimps is augmented by imports of 91,000 tonnes. In addition about 30,000 tonnes of lobsters (USA plus Canada) and 4,000 tonnes of marine crawfish (USA) are landed. In Europe nearly 2,000 tonnes of lobsters, 38,000 tonnes of scampi (*Nephrops*), and 230 tonnes of marine crawfish (= spiny lobster) were landed in 1977. In 1975 some £4.5 million worth of frozen shrimps and prawns (largely penaeids) were imported into Great Britain. As examples of the value of Crustacea, at Billingsgate market (London) in early 1979, lobsters averaged £6.70 per kg, scampi (shelled and frozen) and marine crawfish were £6.60 per kg, freshwater crayfish were £3.96 per kg and prawns (*Pandalus borealis*) averaged £1.80 per kg. A survey of several importers in April 1977 suggested that the following prices were being paid for frozen penaeid prawn tails according to their size (Wickins and Beard, 1978): 12 g – £1.70 per kg; 18g – £2.60 per kg; 30 g – £3.00 per kg.

Although there is no established commercial culture of Crustacea in the temperate zones of Europe or North America, crustaceans have been cultured for many years in the Indo-Pacific region. The most advanced nation in this field is undoubtedly Japan where the estimated production of cultured prawns increased from 200 tonnes in 1965 to 1,300 tonnes in 1974 (Hirasawa and Walford, 1976). The keen demand for crustacean flesh and its high value prompted many organisations to study the possibility of culturing crustaceans in other parts of the world. Indeed, the extent of interest is such that some 46 species of prawn have been cultured experimentally in over 40 countries (Wickins, 1976a).

This chapter examines the factors governing the choice of species for cultivation, and some of the problems which need to be overcome before crustacean culture becomes an economically viable proposition in the developed countries. North America and Europe lack the natural advantages – cheap labour and warm water – which have enabled aquaculture to develop so widely in the tropics; hence special techniques must be developed to overcome these handicaps. Some of these techniques are now well-established and are described below; others are less well-developed and call for further research. In the concluding paragraphs progress is summarised, and a synthesis of what appear to be the most successful emergent practices is made in an attempt to highlight the most promising ways forward.

There is every reason to be optimistic about the future success of crustacean cultivation in temperate countries. At this point, however, it is necessary to sound a warning. Although advances in culture technology have substantially improved the quality of life for traditional farmers who produce, at the village level, crustaceans as a cash crop, crustacean culture as considered in this chapter is not for the philanthropist who views world protein shortages with alarm; our

89

Table 2.1: Classification of Commercially Important Crustacea

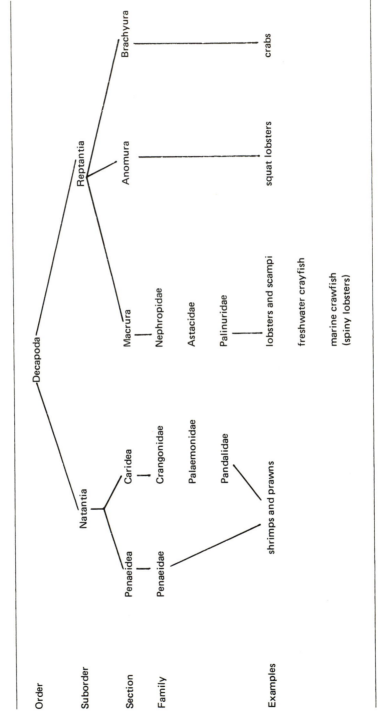

Order	Suborder	Section	Family	Examples
Decapoda	Natantia	Penaeidea	Penaeidae	shrimps and prawns
		Caridea	Crangonidae	
			Palaemonidae	
			Pandalidae	
	Reptantia	Macrura	Nephropidae	lobsters and scampi
			Astacidae	freshwater crayfish
			Palinuridae	marine crawfish (spiny lobsters)
		Anomura		squat lobsters
		Brachyura		crabs

visionary would be better employed redistributing existing feedstuffs or cultivating mussels. Likewise, crustacean culture is not for those hoping to 'get rich quick'. Press reports and popular scientific articles abound with euphoric predictions of yields extrapolated from small-scale trials, but it is worth pausing to ask how many organisations outside Japan and middle America are making a worthwhile profit growing Crustacea? Our speculator would perhaps be best advised to invest in real estate or in the leisure industry. Crustacean culture, in the Western world at least, is only approaching adolescence. Much development work has already been done but the time is fast approaching for its commercial potential to be put to the test. This is only likely to be achieved through adequately funded and properly managed pilot projects on a scale sufficient to provide costings that would satisfy investors. In the present world economic climate such programmes might appear rather more strategic than immediately applicable. Funding might therefore only be forthcoming from governments, very large companies or business consortia.

2. Biology

2.1 Classification

Most commercially important crustaceans are classified in the order Decapoda of the class Crustacea. Characteristically decapods have five pairs of walking legs, the first often bearing substantial chelae, as, for example, *Macrobrachium, Homarus* and *Austropotamobius*. The order is further sub-divided as shown in Table 2.1 into suborders, sections and families. Some specific examples of these that might be suitable for cultivation are listed in Table 2.2, together with their common names and countries where they are native.

2.2 Life History

In most decapod species the sexes are separate. A few species, notably among the pandalid prawns, change sex at some time during their lives. A good example is the spot prawn, *Pandalus platyceros*, which first matures and functions as a male; after 1.5-2.5 years it matures and functions as a female. During cultivation this habit can complicate the management of breeding stock.

The penaeid prawns spawn a large number ($1-2 \times 10^6$) of small eggs (200 μ diameter) directly into the sea. The eggs hatch after a few hours and each larva is left to fend for itself as it develops through some twelve stages of nauplius, protozoea and mysis before metamorphosing to the post-larva (Figure 2.2). In caridean prawns and reptantians the females care for their eggs which are attached to the pleopods during an incubation period which may last for several months. Fewer, larger eggs are laid by females in these groups and they contain sufficient yolk to allow some of the early larval stages to be 'by-passed' in the egg. These larvae, therefore, hatch at a more advanced stage of development and often have fewer moults to complete before metamorphosing into the adult form. In the most extreme cases, exemplified by the freshwater crayfish (Astacidae) there is no free-living larval phase. The eggs hatch to release post-larvae which cling to their mother until their first or second post-larval moult, after which they are able to begin foraging for food. There is a tendency for brood size to be smaller in many species where larval development is abbreviated or the eggs hatch to release post-larvae (Figure 2.3).

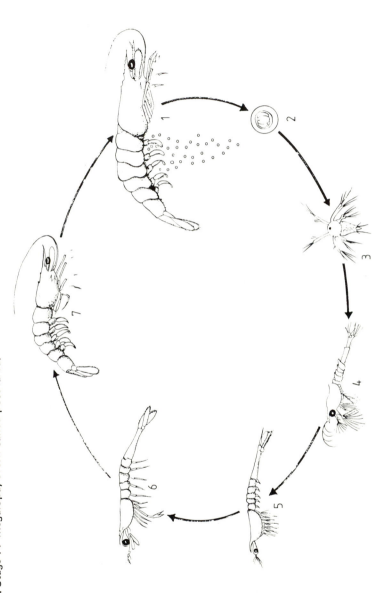

Figure 2.2: A. The Life Cycle of a Penaeid Prawn. 1. Adult spawning female; 2. Egg; 3. Nauplius — 5 or 6 moults; 4. Protozoea — 3 moults; 5. Mysis — 3 moults; 6. Post larva; 7. Juvenile. B. The Life Cycle of the Lobster *Homarus* sp. 1. Adult; 2-4. Stages I-III zoea larvae; 5. Stage IV megalopa, often called post-larva.

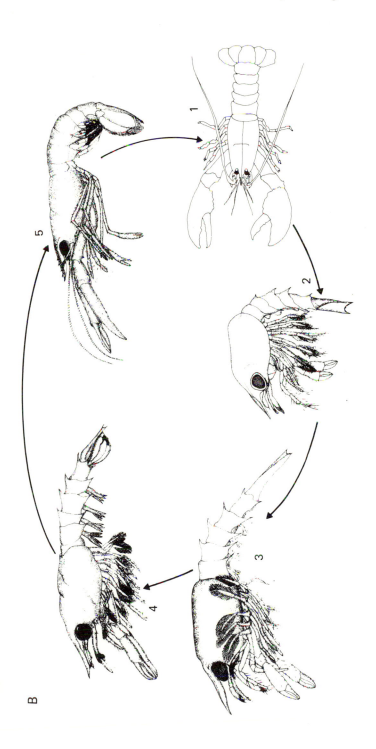

Source: B (2-4), from Nichols and Lawton (1978).

B

Table 2.2: The Origin and Common Names of Some Commercially Important Species of Crustacea

SPECIFIC NAMES	COUNTRY	COMMON NAMES
Natantia — shrimps and prawns		
Penaeidea		
Penaeidae		
Penaeus japonicus	Japan	Kuruma ebi
P. merguiensis	Australia	Banana prawn
P. monodon	Taiwan	Grass shrimp
	Philippines	Jumbo tiger shrimp or sugpo
P. stylirostris	Central/S. America	Blue shrimp
Hymenopenaeus mülleri	Argentina	Langostino
Caridea		
Crangonidae		
Crangon crangon	Germany	Nordsee-garnele
	France	Crevette grise
	Netherlands	Garnaal
	Great Britain	Common, brown or sand shrimp
Palaemonidae		
Macrobrachium rosenbergii	Malaysia	Udang galah
	India	Giant freshwater prawn
Palaemon serratus	France	Crevette rose Bouquet
	Great Britain	Common prawn
Pandalidae		
Pandalus platyceros	USA	Pink shrimp
	Canada	Spot prawn
Reptantia		
Macrura — lobsters crayfish and crawfish		
Nephropidae		
Homarus americanus	USA and Canada	American lobster
H. gammarus	Great Britain	European lobster
Nephrops norvegicus	Great Britain	Norway lobster Dublin Bay prawn Scampi

Table 2.2 continued

Astacidae

Austropotamobius pallipes	Great Britain	Crayfish
Astacus astacus	Sweden	Noble crayfish
A. leptodactylus	Turkey	East European crayfish
Orconectes limosus	USA	Striped crayfish
Pacifastacus leniusculus	USA	Signal crayfish
Procambarus clarkii	USA	Red swamp crayfish
Cherax tenuimanus	Australia	Marron
C. destructor	Australia	Yabbie

Palinuridae

Palinurus elephas	Great Britain	Crawfish
Panulirus argus	USA	Spiny lobster

Features of the life cycles of species often considered for cultivation are compared and contrasted in Table 2.3. The information was selected as being most pertinent to culturists and, where available, was taken from studies in which the animals were grown under good culture conditions. Otherwise, data from wild populations were used.

Many reptantians have low reproductive rates (long incubation period and low fecundity) compared with natantians. In this respect *Panulirus* is especially notable, because of its long larval life.

2.3 Food and Feeding

Larvae. Newly-hatched nauplii of penaeids and the first stage of some crayfish do not feed and are nourished by internal yolk reserves. Penaeid nauplii moult three or four times in two days before moulting to the protozoea stage when they feed on unicellular algae. They then moult a further two or three times in some four or five days before reaching the mysis stage. Mysis larvae are conveniently fed on nauplii of the brine shrimp, *Artemia*. The transition involves changes in the morphology, digestive enzymes (Van Wormhoudt, 1973) and dietary requirements of the larvae. In cultures, the protozoea to mysis transition is frequently associated with heavy mortalities, often more than at metamorphosis, which follows in 5 or 6 days after a further 2-3 moults. Most caridean and reptantian larvae feed voraciously on zooplankton, and in cultures *Artemia* nauplii or grown *Artemia* are widely used (New, 1976; Hanson and Goodwin, 1977). The availability of *Artemia* eggs and a method for their decapsulation are described by Sorgeloos (1979) and Sorgeloos *et al.* (1977). Research on artificial diets for larvae has been stimulated by the cost and variability in quality of live foods (e.g. *Artemia*; Wickins, 1972a, c). Bound and microencapsulated diets have been studied using *Penaeus japonicus* (Hirata *et al.*, 1975; Jones *et al.*, 1979), *Macrobrachium rosenbergii* (Jones *et al.*, 1975) and *Penaeus merguiensis* (Möller *et al.*, 1979). It is likely to be some time yet before such prepared foods beome commercially acceptable alternatives to live foods.

Figure 2.3: The Relation between the Number of Larval Stages and Fecundity in Selected Decapods.

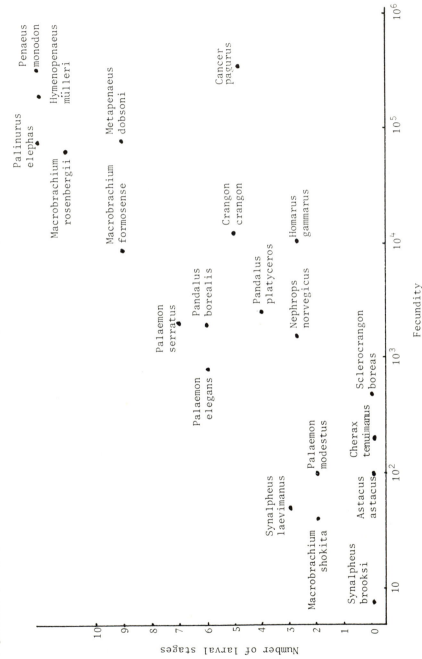

Source: Modified from Wickins (1976a).

Table 2.3: Aspects of the Life History of Selected Crustacea

Species	Incubation Period	Fecundity	Length of Larval Life (days)	Age at First Maturity	Size at First Maturity (female)	Interval Between Spawning	Suitable Culture Temperature (°C)
Penaeus monodon	none	5×10^5 - 1×10^6	12-14	8-10 mo	60-80 g[a]	3-4 days (induced)	28
Hymenopenaeus mülleri	none	5-7×10^5	19-23	?	?	?	19-24
Macrobrachium rosenbergii	19-21 days	8×10^4	34-44	7-8 mo	25-30 g	3-4 mo	28
Palaemon serratus	30-40 days	1-2×10^3	18-35	8 mo	3-3.5 g	4 mo	22
Pandalus platyceros	4-5 mo	2.5-4.5×10^3	15-24	Protanderous hemaphrodite 24-36 mo	25-35 g	12 mo	16-18
Homarus gammarus	4-9½ mo	5×10^3 - 1×10^5	10-14	3-6 yrs	400 g	1-2 yrs	20
H. americanus	4-7½ mo	5×10^3 - 5×10^4	9-14	3-4 yrs	400 g	10 mo-2 yrs	22
Austropotamobius pallipes	6-8 mo	60-150	0	3-4 yrs	24 mm CL[b]	12 mo	15-18
Pacifastacus leniusculus	4-5 mo	70-260	0	2-3 yrs	100 mm TL[c]	12 mo	22
Astacus astacus	6 mo	70-130	0	3-4 yrs	90 mm TL	12 mo	20
Cherax tenuimanus	4-8 wks	200-600	0	1-2 yrs	30-50 mm CL	12 mo	20-25
Nephrops norvegicus	3-4 mo	900-6×10^3	14-21	2-3 yrs	20-23 mm CL	12 mo	11
Panulirus argus	8-10 mo	8×10^4 - 1.6×10^5	150-210	?	100 mm CL	?	10-12
Cancer pagurus	5-7 mo	5×10^4 - 1×10^5	30	4-6 yrs	110 mm CW[d]	?	10-12

Notes: a. Live weight; b. carapace length; c. total length; d. carapace width.

Juveniles and Adults. Most post-larval Crustacea considered here are omnivorous scavengers, but the majority of species seem to prefer animal rather than vegetable tissue. Many prawns browse for food among detritus but when larger pieces of food are available these are readily taken. Large fragments are manipulated and broken up by the chelae and mouthparts outside the body. The process is particularly well defined in caridean prawns where the gastric mill is poorly developed in comparison with penaeids and reptantians (see Patwardhan, 1935, a-e; Reddy, 1935, 1938; Barker and Gibson, 1977). In cultures where food is provided by the farmer this process can be wasteful, since food particles are scattered, carried away by water currents coming from the gills, and may cause fouling in the tanks.

The mid-gut gland, or hepatopancreas, is the main, perhaps only, source of digestive enzymes (see Van Weel, 1970). Discrete phases of secretion have been observed in some species: two in *Palaemon serratus* (Van Wormhoudt and Ceccaldi, 1975) and *Astacus leptodactylus* (Hirsch and Jacobs, 1930; Hirsch and Buchmann, 1930, both quoted in Barker and Gibson, 1977), and three in *Homarus gammarus* (Barker and Gibson, 1977). In *H. gammarus* the phases occur between 0-15 min, 1-2 hrs and 3.5-5 hrs after a meal. After the third period of secretion the cells are gradually replenished and about 12 hrs later are ready for the next meal. It was shown that the speed at which food was cleared from the gut in an estuarine palaemonid prawn, *Palaemonetes varians*, varied according to its composition. A detritus or algal meal took 4-6 hrs, while a meal of pelleted shrimp took 27 hrs to digest and absorb (Snow, 1969). Although differences do occur between species, penaeids in general seem to be more leisurely feeders than the more predatory lobsters, some crayfish and palaemonid prawns. It has been suggested (Conklin, 1980) that small juvenile *Homarus americanus* may feed continuously.

In trials made with *Penaeus merguiensis*, Sedgwick (1979a) observed that the protracted feeding behaviour allowed food given once daily to deteriorate in the tanks. When fed every 6 hrs the diet was more efficiently utilised and gave an improved growth rate, lower conversion rate and a more consistent feeding response. Further studies (Sedgwick, 1979b) showed that in *P. merguiensis* the amount of food consumed daily was proportional to its caloric value and that the prawns were feeding to satisfy an energy requirement. The potential capacity for ingesting food of low energy content was extremely large as would be expected with a browsing form of feeding behaviour.

The daily amount of food consumed by *Homarus gammarus* varied through the moult cycle, intake being greatest at the beginning and at the end of the intermoult period (Richards and Wickins, 1979). Periodic starvation as a method for improving food utilisation has been investigated with juvenile *H. gammarus* (Richards, 1981; Richards and Wickins, 1979) and *H. americanus* (Schleser, 1974). Maximum growth of juvenile lobsters was only obtained when food was continuously available, but the weight of *H. gammarus* fed every two weeks was only 9 per cent less after twelve weeks than those fed daily. It was suggested that a small reduction in growth may in some circumstances be offset by improved food utilisation and labour costs.

In the case of the penaeids, red prawn (presumably *Penaeus penicillatus*) (Liao and Lee, 1972), *P. monodon* (Lee *et al.*, 1974) and *P. japonicus* (Lee *et al.*, 1976), efficiency of food utilisation, growth rate and survival were improved by periodic starvation. In the red prawn alternate days of feeding and starvation gave good results in juveniles up to 1.2 g weight, while larger prawns of 6.8-15.1 g benefited from one day's feeding, followed by three days' starvation. In *P. japonicus* good results were claimed when periods of starvation increased as the prawns increased in size. A similar result was not obtained with *P. monodon*, possibly because the latter is less carnivorous and needs to feed more frequently than *P. japonicus* (Liao and Huang, 1972). It has been suggested (Wickins, 1976a) that the predominantly browsing penaeids and some more predatory carideans and reptantians may show similar differences in the response to chemical attractants. The importance of such differences would become apparent in controlled culture situations where water quality is greatly influenced by feeding frequency (see pp. 149-54). In these circumstances not only dietary formulation but also the husbandry of feeding must be in sympathy with the animal's nutritional, behavioural and physiological requirements, otherwise food is wasted and fouling of the water may occur. The influence of environmental factors on feeding behaviour and digestion, for example presence of a substrate (Möller and Jones, 1975) and light (Van Wormhoudt and Malcoste, 1976), may be used with advantage by the culturist to stimulate consumption and thereby improve food conversion ratio and reduce fouling of the water by uneaten food.

Studies on the composition of prepared diets for prawns have been reviewed in the last four years (New, 1976; Hanson and Goodwin, 1977) and commercial formulations can be purchased in the USA and Japan. Not all species are expected to do equally well on these diets and there is considerable scope for further research. No satisfactory diet has yet been developed for lobster (Conklin *et al.*, 1977 and Conklin, 1980) and there are few studies on diets for crayfish (Nose, 1964; Huner *et al.*, 1975). Most dietary studies for the reptantians have been limited to the first 3-6 months of post-larval life, and an economically attractive diet that will support good growth for the whole culture period has yet to be developed. Despite this, promising results have been achieved with penaeids when pelleted diets were supplemented once a week with fresh foods (Wickins, unpublished results).

2.4 Moulting and Growth

All Crustacea have an external skeleton or shell which serves for both muscular attachment and protection: this is capable of only limited expansion. Although true growth (the addition of new protoplasm and cell division) is probably continuous, the increase in linear dimensions occurs only at moulting. The rate of growth is a function of the frequency of moulting and the increase in size at each moult. The main sequence of events in the moult cycle are:

(1) accumulation of mineral and organic reserves;
(2) removal of material from the old shell and formation of the new exoskeleton;
(3) ecdysis, accompanied by an uptake of water;
(4) molecular strengthening of the exoskeleton by rearrangement of organic matrices and deposition of inorganic salts; and
(5) replacement of fluid by tissue growth.

Figure 2.4: The Relation between Intermoult Period and Size of Prawns and Lobster. 1. *Homarus gammarus*, 20° C; 2. *Macrobrachium rosenbergii*, 28° C; 3. *Penaeus monodon*, spawning females, 28° C; 4. *Pandalus platyceros* larvae, 15° C; 5. *Penaeus monodon* juveniles, 28° C.

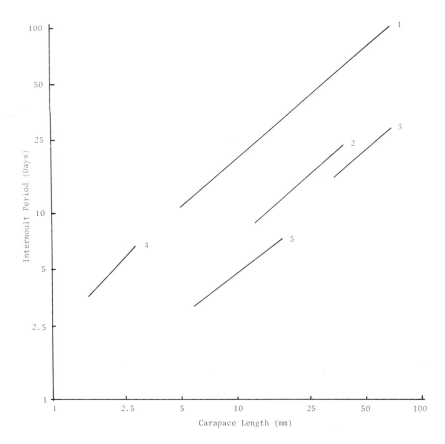

The intermoult period varies naturally between species, within species with size and, in normally growing individuals, with age. Young penaeid larvae may moult two or three times in a day, lobster larvae every two or three days, while adult lobsters and crayfish may moult only once or twice each year, large lobsters even less frequently (Figure 2.4). Variation in the rate of growth can be caused by either a change in the frequency of moulting or a change in the size increment. Both moult frequency and length increment may be affected by changes in environmental factors such as photoperiod, temperature, pollutants, diet and confinement (see Wickins, 1976a). Moult frequency usually changes more readily than size increment, but in confined lobsters reduced increment at moulting was judged to be predominantly responsible for the observed reduction in growth (Van Olst and Carlberg, 1978). Indeed, under severe conditions, for example with inadequate feeding, the animal may continue to moult but not grow (Reeve, 1969; Lee, 1971). Crustaceans will often eat cast shells (exuvia);

a convenient source of minerals which might otherwise be lost. Under crowded conditions, during intensive culture, for example, newly moulted animals may be attacked, killed and eaten by their companions. While soft, they are also unable to tolerate high current speeds. Moulting animals are particularly sensitive to stress, and mortalities among immediate pre- and post-moult Crustacea are good indicators of adverse environmental or nutritional conditions in cultured populations.

A comparison of the rates of growth, under good culture conditions, of selected commercially important species is shown in Figure 2.5. By far the quickest growth is that shown by *Penaeus monodon* which, even at high stocking densities, can reach marketable sizes of 20 and 35 g in 3-6 months, and therefore holds the promise for the commercial culturist of a more rapid cash flow than most other cultured fish and shellfish. Worthy of note is the growth of *Cherax tenuimanus* at 18°C (Morrissy, 1979). This species is known to survive at temperatures over 25°C and its maximum growth rate is judged to be considerably faster than that shown in Figure 2.5. It is unlikely that the growth rates and temperatures in Figure 2.5 for the crayfish species represent the best that could be obtained under good culture conditions.

3. Choice of Species for Cultivation

3.1 Culture of Larvae

The advantages to a culturist of a species with a short larval life, e.g. penaeids when contrasted with *Macrobrachium*, include reduced operational risks and costs of juvenile production. Reduced costs are often due to the shorter period of time for which specialised, often live, larval food is prepared. Similar advantages for species with abbreviated larval development have been proposed by Williamson (1968). In the extreme case of freshwater crayfish, free swimming post-larvae can be stocked almost immediately into the on-growing facility. A major disadvantage for the commercial culturist is the low fecundity usually associated with abbreviated larval development, which demands the maintenance of a large brood stock. In some extensive cultures, however, where young or brood stock females are obtained from the crop, low fecundity is not so important.

In vitro culture of eggs has been practised with many carideans and reptantians as a means of reducing the space and effort required to hold the large brood stocks necessary for commercial ventures: *Palaemon serratus* – see Figure 2.6 – (Phillips, 1971), *Homarus americanus* (Jamieson *et al.*, 1976), *Astacus astacus* (Strempel, 1973), *Pacifastacus leniusculus* (Cabantous, 1975; Mason, 1977). With the possible exception of small crayfish farms *in vitro* incubation and hatching is judged not to have great commercial potential. In lobsters there is evidence that the method results in abnormal (Ennis, 1973) or pre-larval stages (Brown and Aiken, 1976).

The techniques used for the mass culture of the larval and early juvenile stages up until the time they are ready for stocking has been, for most species, well documented, viz., Natantia: Western Atlantic penaeids (Tabb *et al.*, 1972; Yang, 1975); *Penaeus japonicus* (Shigueno, 1975); comparison of American and Japanese methods (Mock and Neal, 1974); *P. monodon* (Platon, 1978); general accounts (Heinen, 1976; Wickins, 1976a); *Macrobrachium rosenbergii*,

Figure 2.5: The Growth Rates of Cultured Crustacea. 1. *Penaeus monodon*, 28° C; 2. *Macrobrachium rosenbergii*, 28° C; 3. *Cherax tenuimanus*, 18° C; 4. *Pandalus platyceros*, 18° C; 5. *Homarus gammarus*, 20° C; 6. *Palaemon serratus*, 22° C; 7. *Pacifastacus leniusculus*, 20° C; 8. *Astacus leptodactylus*, 22° C; 9. *Astacus astacus*, ambient; 10. *Austropotamobius pallipes*, ambient.

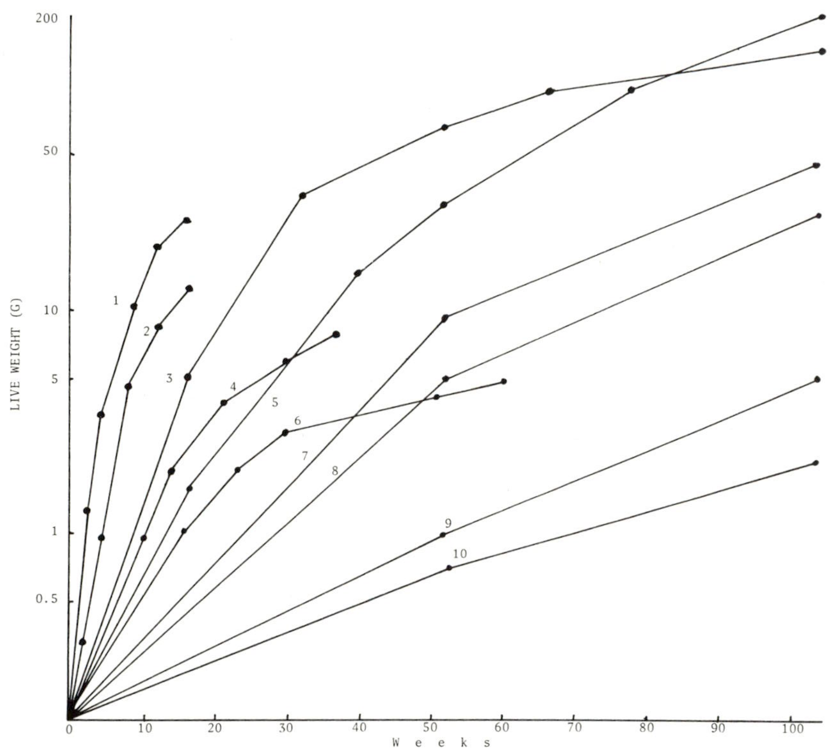

engineering analysis (Wang and Kuwabara, 1975); design of continuous
production system (Wang and Williamson, 1976), controlled culture in clear
water (AQUACOP, 1977a), closed circulation nursery culture (Sandifer and
Smith, 1977); Reptantia: *Homarus americanus*, culture vessels and system
(Hughes *et al.*, 1974; Serfling *et al.*, 1974; Schuur *et al.*, 1976), adaptation of
American methods to *H. gammarus* (Richards and Wickins, 1979), *Astacus
astacus* and *Pacifastacus leniusculus*, hatching and parental care (Strcmpel, 1976;
Hofmann, 1980); *P. leniusculus* reproductive efficiency (Mason, 1977).

Figure 2.6: Illuminated Larval Culture Vessels and, Front Right, Conical Tanks
for the Incubation of Eggs Removed from Ovigerous *Palaemon serratus* Females

Source: Photograph courtesy of Dr G.C. Phillips, Unilever Research (1968).

Figure 2.7: Body Shape and Proportions of Twelve Species of Decapod.
1. *Penaeus aztecus;* 2. *Crangon crangon;* 3. *Palaemon serratus;* 4. *Macrobrachium rosenbergii;* 5. *Pandalus platyceros;* 6. *Homarus gammarus;* 7. *Nephrops norvegicus;* 8. *Austropotamobius pallipes;* 9. *Cherax tenuimanus;* 10. *Palinurus elephas;* 11. *Mya squinado;* 12. *Cancer pagurus.*

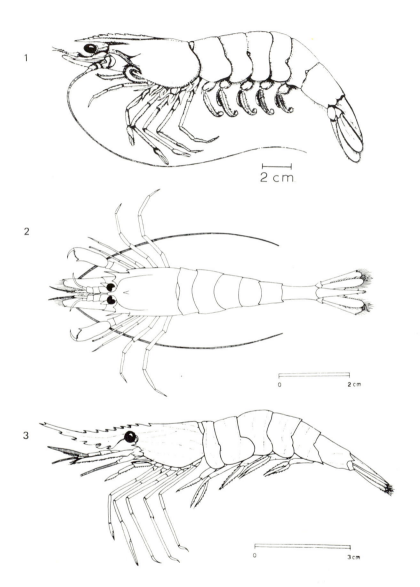

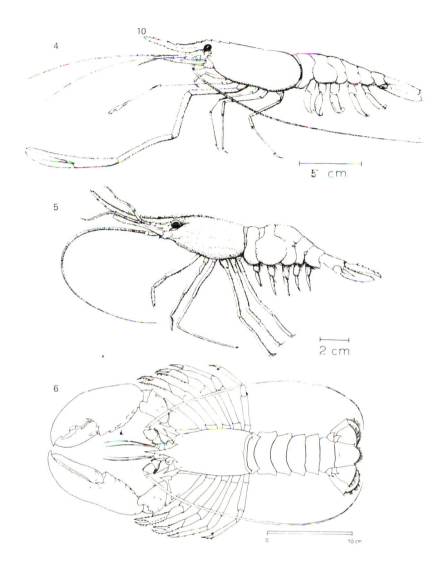

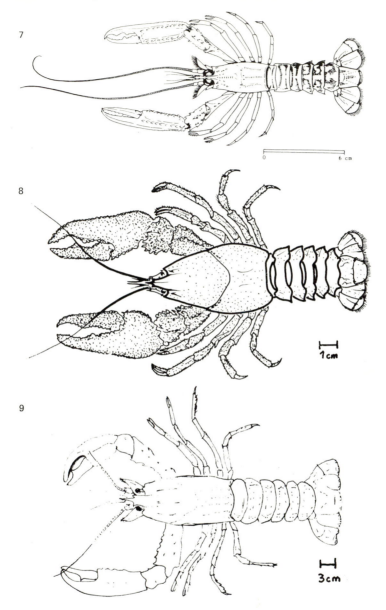

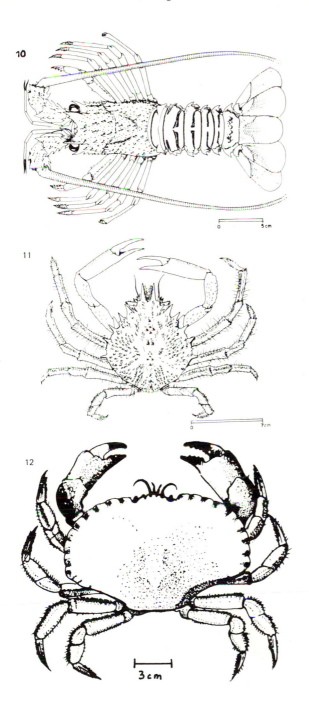

Source: 1, 4, 5 from Forster and Wickins (1972); 2, 3, 6, 7, 10, 11 from Fischer (1973); 8 from Allen (1967); 9, 12 original.

3.2 Morphology

The gross morphological characteristics of twelve representative species are shown in Figure 2.7. Two features are significant for the culturist: the presence of large chelae, typically indicative of a territorial or predatory species, and the proportion of the body that comprises the edible 'tail' portion. *Macrobrachium, Homarus, Nephrops* and the crayfish are naturally territorial or solitary, and when crowded together are likely to become unduly aggressive. Of the Natantia, the penaeid prawns, represented here by *Penaeus aztecus*, have the highest edible flesh to total weight ratio, about 60 per cent compared with 40 per cent for carideans.

3.3 Marketing

Selection, however, must not be based entirely upon biological considerations since a fusion of biological and economic judgements are prerequisites for a sound choice. On the one hand the animal must grow quickly, survive well in culture, tolerate crowding, breed in captivity, be resistant to disease and efficient in its conversion of food to body flesh. On the other hand, it must have good market acceptability at a price which will not collapse as production increases, and be able to sell at a premium over existing fished or imported produce.

Since some Crustacea e.g. lobster and crawfish, are very highly priced because of their scarcity and the costs of catching, preparing and transporting them to market, a moderate increase in supplies could cause prices to fall sharply. Markets for more moderately priced species (e.g. medium-sized prawns) may well be able to absorb changes in volume without causing significant fluctuations in price. To sell at a premium in established markets a cultured animal might be available out of the normal season, have better gustatory or visual appeal, or be more consistent in overall quality and availability.

One of the most challenging prospects for farmers would be to consider the problem of creating markets for unfamiliar products. Examples for consideration in Britain might include the potential for:

(1) large Pacific King prawns (35 g live weight) sold as individually quick frozen (IQF) tails (25 tails/lb count) and cooked, hand-peeled IQF tails (30/lb);

(2) giant freshwater prawns (*Macrobrachium rosenbergii*) sold live to Asiatic caterers at 30-40 g fresh weight;

(3) regular year-round supplies of live, plate-sized lobsters (350-450 g; 0.75-1 lb) for the catering and hotel trades where lobsters larger than 1.5-2 lbs are less convenient to use;

(4) native and exotic species of freshwater crayfish.

The value or sheer volume of a species in an established market may, therefore, not be the sole best economic indicator of whether or not it is worth farming. It is also likely that a farmer wishing to produce relatively small quantities of Crustacea (10-20 tonnes per year) may be forced to consider very different markets and produce a size of animal different from the company considering production of 100-200 tonnes per year, incidentally avoiding direct competition. Useful annotated bibliographies for economic evaluations in aquaculture are

Johnston and Collinsworth (1973) and Vondruska (1976).

Biological considerations can rule out (at least at the present state of knowledge) some superficially attractive candidates. First of these must be 'scampi' (*Nephrops norvegicus*), which is unsuitable for culture because of its preference for offshore environments, pronounced burrowing habit and whose larvae have proved difficult to rear in large numbers (Farmer, 1974).

Among the natantians, small (4-6 g) commercial species — the sand shrimp (*Crangon crangon*) and pink shrimp (*Pandalus montagui*) — fetch far too low a price to be worth cultivating. The larger (10-15 g) deep-water prawn (*Pandalus borealis*), being an offshore species, does not adapt well to land-based or coastal culture conditions and in Britain, because prawn prices are predominantly related to the size rather than species of prawn, this too is not considered to be a worthwhile candidate.

Marine crawfish (*Palinuridae*), also known as spiny or rock lobsters, are represented in Britain by *Palinurus elephas* and typically have a complicated, often protracted, larval life (Dexter, 1972). Although their culture has not yet been mastered, the benthic pueruli stages may sometimes be located in such numbers and with sufficient regularity to support 'capture and fatten' cultures. It would seem, however, that such cultures are unlikely to be economic (Ting, 1973).

Economic considerations at present rule out the two species of crab that are fished in Britain — the edible crab (*Cancer pagurus*), which grows to a marketable size at a similar rate to lobster but has a much lower market value, and the spider crab (*Mya squinado*). Spider crabs were regarded as a by-catch until recently when they became the basis of a new fishery and export trade to Europe.

There is a market for freshwater crayfish in Europe but it is small by comparison with the markets for many marine Crustacea. Two species, *Orconectes limosus* and *Austropotamobius torrentium*, sometimes found on the markets, can probably be ruled out on grounds of unpalatability and being difficult to catch (*O. limosus*) or having stringent environmental requirements and slower growth (*A. torrentium*). Of the warm-water species that might be considered for culture in thermal effluent or other heated waters two, *Procambarus clarkii* and *Cherax destructor*, are notorious for their destructive burrowing habits and the problems caused by their introduction to a number of countries have been well documented (Bardach *et al.*, 1972; Carstairs, 1975; Huner, 1977; Lowery and Mendes, 1977). These then are the more well-known crustacean species that are currently considered unsuitable for cultivation in most Western temperate regions on biological or economic grounds.

3.4 Diseases

Crustacean aquaculture in Western temperate regions is, with the exception of crayfish culture, a new business. It is generally accepted that animals kept in artificial (culture) conditions experience stress at some time or another and become more susceptible to disease. In research and pilot-scale cultures stressful periods may occur while operators gain experience with the husbandry of their charges and with the management of water quality. It is to be expected, therefore, that some diseases will be encountered which may differ from those experienced during commercial operation. Indeed, the true cost and significance of disease will probably not be known until large-scale cultures are operational.

Publications on the diseases of crayfish, lobsters and prawns that are considered to be useful to the culturist as well as some of these describing outbreaks of disease in culture trials include: Delves-Broughton and Poupard (1976), disease outbreaks in British prawn culture trials; Lightner (1973), normal post-mortem changes in shrimp; Wickins (1976a), disease problems of prawns in recirculation systems — brief review; Fisher *et al.* (1975, 1978), diagnosis of disease in cultured *Homarus americanus*; Richards and Wickins (1979), brief mention of organisms associated with mortality in cultured *H. gammarus*; Unestam (1973), significance of disease in crayfish; Unestam (1975b), susceptibility of Australasian crayfish to plague fungus; Unestam and Weiss (1970), responses of *Astacus astacus* and *Pacifastacus leniusculus* to plague fungus; Vey and Vago (1973), observations of diseased crayfish in France; Cossins (1973), histological study of *Thelohania*; Sindermann (1971), internal defences of Crustacea; and Sindermann (1977), descriptive account of crustacean diseases. It is worth noting that among the Natantia, *Macrobrachium rosenbergii* is outstanding in its ability to resist infection (Hanson and Goodwin, 1977).

4. Culture Options

Before going on to discuss the choice of species in further detail, a consideration of the environments and culture options available in Western temperate regions, particularly Great Britain, is necessary. In most of the strategies discussed here the provision of seed or juveniles for stocking will be made from hatcheries (see p. 101). Since hatchery procedures are well documented and the cost of juvenile production is likely to be less than 5-10 per cent of the total production costs (Wickins and Beard, 1978; Richards and Wickins, 1979), emphasis is placed on methods for fattening and on-growing the juveniles to marketable size.

Methods of cultivation may conveniently be categorised under three headings: marine, freshwater and controlled environment cultures. In the first two categories practices may range from extensive to intensive, while the third will invariably be intensive. These terms were defined arbitrarily for prawn cultures by Wickins (1976a) on the basis of annual yields. The same definitions are used here but it is emphasised that such categories cannot be rigid and some overlap may be expected.

(1) Extensive: range from stocked and managed stretches of rivers and waterways to ponds and sea enclosures; some natural food is available but generally substantial supplementation with prepared foods; yields up to 1,000 kg/ha/year.

(2) Semi-intensive: ponds, raceways or other land-based facilities, marinas, sea cages carefully managed; mainly compounded or prepared foods used; yields from 1,000 to 10,000 kg/ha/year.

(3) Intensive: in controlled environment land-based facilities, sophisticated management and feeding solely with compounded feeds: yields above 1 kg/m^2/harvest, equivalent to over 10,000 kg/ha generally with continuous year-round production.

The options open to crustacean farmers in Britain and climatically similar regions are:

(1) temperate or warm-water marine species grown extensively or semi-intensively in sea enclosures, marinas, cages or land-based facilities with or without a supply of heated water, e.g. industrially heated effluent;

(2) temperate or warm-water freshwater species cultivated extensively or semi-intensively in streams, rivers or ponds with or without a heated water supply, as above;

(3) marine or freshwater tropical or sub-tropical species grown intensively in a heated, controlled and covered environment.

In both (1) and (2) economy of water and heat may be achieved by treatment, e.g. oxygenation, filtration, and recirculation of water. In (3) it is expected that most, but not necessarily all, cultures would use recirculated water.

In the pages that follow the choice of species and methods of culture are considered in the light of each available option. Examples have been made of specific existing and defunct aquaculture projects in order to illustrate certain approaches from which lessons for the future may be learned. No attempt has been made to include all possibilities or combination of options.

4.1   *Marine Species — Prawns*

Background Information. Over the past three decades perhaps the majority of the literature on crustacean farming has been concerned with the culture of prawns and shrimps in warm-water regions of the Far East (IPFC, 1956; Hall, 1962; Mistakidis, 1968, 1969, 1970; Pillay, 1972; Shigueno, 1975; Wickins, 1976a; Hanson and Goodwin, 1977; Pillay and Dill, 1979). In India and Asia a number of fish farms traditionally produced an incidental by-catch of prawns which were sold locally and it was only with the advent of refrigeration and improved transportation that access to the higher-priced city and international markets was possible (Figure 2.8). This so increased the value of the by-catch that many farmers encouraged the trapping of naturally occurring juvenile prawns in their ponds or rice fields, and ultimately set aside areas specifically for prawn culture from which fish and predators were eliminated. One of the major constraints to this kind of prawn culture was the unreliable nature of natural supplies of juveniles. After the pioneering Japanese studies of Hudinaga (≡ Fujinaga) (1942), who first reared juvenile Kuruma prawn (*Penaeus japonicus*) from ripe females caught from the sea, Government laboratories and hatcheries were built in many Far East countries to provide R and D support and, in some cases, supplies of juveniles for the developing industry. General accounts of the problems and present state of the art are published and the reader is referred to Cook (1976), and Cook and Rabanal (1978). More recently, hatcheries run by fishermen's co-operatives, private industry and government have released many millions of cultured post-larvae or juvenile prawns into selected nursery areas in attempts to enhance a natural fishery. In one area at least, the Seto Sea of Japan, this practice is considered to be profitable (Kurata and Shigueno, 1976; Hanamura, 1976).

Other countries, including the USA, showed interest in prawn farming and in the mid-1950s scientists at the Bears Bluff laboratories, South Carolina, began

Figure 2.8: Packing Live Cultured 'Kuruma ebi' (*Penaeus japonicus*) for the Tokyo Tempura Market.

Source: Photograph courtesy of Dr J.W. Lee, Oriental Brewery Co., Korea (1974).

studies on the growth of naturally occurring penaeid post-larvae in ponds. Shortly afterwards in 1963 the US Bureau of Commercial Fisheries at Galveston, Texas, collaborated with Dr Fujinaga and reared the larvae of *Penaeus setiferus* and *P. aztecus*. In 1964-8 the State of Louisiana was constructing prawn culture ponds at Grand Terre Island and the National Marine Fisheries Service Tropical Biological laboratory at Maine were rearing *P. duorarum*. American industry began to play an active role in the late 1960s, and thereafter interest and research effort rapidly expanded (Hanson and Goodwin, 1977). Also in the 1960s Ling (1969) successfully reared the young of a large freshwater prawn (*Macrobrachium rosenbergii*) and widened the choice of a suitable species for culture in Malaysia, Thailand and India, where *M. rosenbergii* is endemic, and also in Hawaii and Mauritius, where it was introduced. Throughout the 1960s

and 1970s publicly funded laboratories made significant advances and
contributions to the understanding of prawn biology and culture methods.
Notable in this field, outside Japan, were: the Tungkang Marine Laboratory,
Taiwan; Gulf Coast Fisheries Center, Galveston, USA; Centre Oceanologique du
Pacifique, Tahiti; and recently the South East Asian Fisheries Development
Center, Aquaculture Department, Philippines.

By early 1970 the eastern technology for shrimp and prawn culture had been
exported to the USA, Middle and South America and the Middle East.
Commercial pressures and perhaps a failure to assess properly the different
marketing situations for penaeids in America and Japan, and possibly a certain
initial inflexibility in the translation of Japanese methods to the species and
conditions of the other countries, led to a number of expensive failures. By
the late 1970s the relocation of sites and adaptation of new ideas had produced
a number of promising pilot operations in Asia, Middle America and the islands
of Hawaii and Mauritius. The upsurge of interest in Great Britain coincided with
the boom in demand for 'scampi' (*Nephrops norvegicus*) which, until the
mid-1950s, had been largely regarded as a by-catch by commercial fishermen.
European prawn culture research began in earnest during the period 1964-7 in
Great Britain and France, with Spain, Italy, Portugal and Eire being involved
a little later. Early studies were made with the native common prawn or
'Bouquet' (*Palaemon serratus*) (Forster and Wickins, 1967; Reeve, 1969; Puff,
1971) mainly because juveniles and ovigerous females were readily available.
In Great Britain early research was government-sponsored as a result of
widespread commercial interest and the enthusiasm of Dr H.A. Cole, who was
then Director of Fisheries Research. At the Fisheries Experiment Station,
Conwy, experiments began in 1964 and prawn culture trials continue to the
present. Commercial trials began shortly afterwards at the Unilever Research
laboratory, Aberdeen. In France the initial trials made in 1965 were conducted
privately by M. William Herter at Kerhoustin, near Quiberon in Brittany. Later
in 1968 the Département des Traveaux Recherches et Exploitations Oceaniques
of the Compagnie Général Transatlantique took over and, with backing from
CNEXO, extended the work to include the species *Penaeus japonicus*,
P. kerathurus and *Homarus gammarus*. The company used a variety of sites for
the work, including l'Ile des Embiez, near Toulon, and on the Ivory coast,
North Africa (Puff, 1971). These and the other European studies that were
made mainly at ambient temperatures in warmer waters around the Mediterranean
fall outside the scope of this chapter (see San Feliu, 1973; Lumare *et al.*, 1973).

A few years later at Quiberon a new company, Ferme Aquicale du Morbihan
(FAEM), of the Compagnie Générale Transatlantique (now Compagnie Générale
Maritime) was reported to be investigating the culture of *Penaeus kerathurus*,
P. japonicus, Palaemon serratus and *Homarus vulgaris*. Cultures were planned in
such a way as to take advantage of the different spawning seasons of these
species to maximise the yield from the culture facility (San Feliu *et al.*, 1973).
By 1974, however, the project was closed in favour of activities elsewhere
(de Bondy, 1974).

Culture. Attempts to culture only two temperate marine species of prawn
have been made in Europe and North America, the European common prawn,
Palaemon serratus, and the Pacific spot prawn, *Pandalus platyceros*. There are

other temperate-water species which it has been suggested could be domesticated in Europe, for example the Langostino of Argentina (*Hymenopenaeus mülleri*) (Forster and Wickins, 1972). The biology and aquaculture potential of this and another penaeid *Artemesia longinaris* have been studied in Argentina for at least a decade (Boschi and Scelzo, 1976).

Experience has shown that four factors must be considered if one of these or a similar species is to be chosen for culture in Britain:

(1) Ability to mature and breed in captivity. In this respect only *Palaemon serratus* breeds readily (Wickins, 1972a). In *Pandalus platyceros* mating has been observed between captured females and laboratory reared males (Rensel and Prentice, 1977). In Argentina, however, reliance is still on the supply of impregnated mature penaeid females from the wild.

(2) Fecundity. Cultures based on species with low fecundity necessarily involve the expense of maintenance of large numbers of brood stock or *in vitro* incubation and hatching, e.g. *P. serratus, P. platyceros* (see Phillips, 1971).

(3) The weight of edible meat that is produced. This is particularly important where prawns are to be sold as tails. *Hymenopenaeus mülleri* has up to 20 per cent more meat than *Pandalus* or *Palaemon* and slightly more than *A. longinaris*.

(4) Growth rate and temperature. The species grow well between 16° and 24°C and *P. platyceros* does best in the range 15°-19°C. Recorded growth rates under good conditions are shown in Table 2.4.

Table 2.4: The Growth Rates of Four Temperate-water Prawns

Species	Growth in 150 Days from Metamorphosis (g)	Temperature (°C)	Author
Hymenopenaeus mülleri	2.7	17-24	) Boschi &
Artemesia longinaris	1.4	16-23	) Scelzo (1976)
Palaemon serratus	1.2 (male) 1.5 (female)	20	Campillo (1975)
Pandalus platyceros	3.8	15-19	Wickins (1972b)

Experiments performed in Britain (Reeve, 1969; Forster, 1970; Wickins, 1972a) and France (Campillo, 1975) with *Palaemon serratus* showed that while large-scale production was biologically and technically possible, the commercial feasibility was very much in doubt. The main reasons were:

(1) *P. serratus* did not survive for more than a few days at temperatures below 4°C and therefore if prawns were to be overwintered out of doors supplemental heating would be required.

Figure 2.9: Tanks for the Experimental Culture of *Palaemon serratus* and *Pandalus platyceros* at Ambient Temperatures.

Source: MAFF, Conwy (1970).

(2) Only females reached a reasonable commercial size (6-8 g); males rarely grew larger than 5 g and would not therefore fetch maximum prices.
(3) Males matured sexually at 2.0 g live weight and females at 3.0-3.5 g. After maturation much of the growth effort will go towards the development of the gonads, which under good conditions may ripen three times a year. Following maturation, therefore, growth slows and food conversion efficiency will be lower, thereby increasing costs.
(4) The growth rate is slow. Under near optimum conditions it takes at least a year to reach 4-6 g, but in outdoor tanks and ponds it could take considerably longer (Figure 2.9).

Recent Culture Trials. Attempts to culture *Palaemon serratus* and *Pandalus platyceros* (as well as four warm-water penaeids, *Penaeus japonicus, P. setiferus, P. merguiensis* and *P. monodon*) were also made in Britain in the warm-water effluent of Hinkley Point Nuclear Power Station by Marine Farm Ltd, Somerset, during the 1970s. Temperatures were about 8°C above ambient throughout the year and ranged from 14°C to 30°C in the experimental tanks. Temperatures higher than 25°C, lethal to *Pandalus platyceros* but the minimum for good growth in the penaeids, were limited to periods within the five summer months. The best results showed that the number of crops obtainable per year were limited to one for *Palaemon serratus* (at 3-5 g mean live weight) and *Pandalus platyceros* (6-8 g), and two for the penaeids at 10-15 g due to their faster growth rate (M. Ingram, 1977, pers. comm.). Yields obtained for the two carideans under conditions of controlled temperatures in the laboratory are shown in Table 2.5. Of the four temperate species for which a reasonable amount of data is available, the most promising are *P. platyceros* and *Hymenopenaeus mülleri*.

Table 2.5: The Yields of *Palaemon serratus* and *Pandalus platyceros* Achieved in Laboratory Trials at Conwy, Great Britain

Species	Initial		Final		Duration (weeks)
	Density (no/m^2)	Mean Weight (g)	Yield (g/m^2)	Mean Weight (g)	
Palaemon serratus	347	0.1	315	1.5	20
Pandalus platyceros	185	0.4	492	5	23

It seems unlikely that either could be grown outdoors in Britain as commercially viable monocultures. Indeed, there is as yet no commercially viable outdoor culture of prawns for human consumption in western temperate regions. If, however, a suitable market could be found or created for small prawns (10-20 g), their culture in warm-water effluent might warrant further attention, particularly if they could be grown in conjunction with other valuable species such as Pacific oyster (*Crassostrea gigas*) or abalone (*Haliotis* sp.) — see Kelly *et al.* (1977).

An interesting study was made on the Delaware River, New Jersey, USA, in which alternate crops of rainbow trout and *Macrobrachium rosenbergii* were grown in power station effluent. The alternation of crops was attractive because

of the wide natural temperature variation that occurred annually at the site —
0° to 29° C — (Guerra *et al.*, 1975). The main cash crop was trout and was
economically attractive in its own right. If worthwhile yields of prawns could
also be achieved during the summer months, even greater returns would accrue
(Godfriaux *et al.*, 1977). It was judged, however, that better use would be made
of the depth of water available in the culture raceways by growing bass or eels
instead of the prawns.

One of the more recent attempts known to the author to culture temperate
marine prawns with fish is that of Rensel and Prentice (1979). These researchers
grew coho salmon (20 g weight) at 8.2 salmon/m^3 of water in a net pen
(10.8 m^2 of submerged substrate surface area) with 100 *Pandalus platyceros*
(0.6 g weight). This gave a prawn density of 9.3/m^2 of substrate. After 206 days
93 per cent of the prawns survived to reach a mean weight of 8.6 g. The prawns
were not fed but scavenged what they could from the salmon feed, faeces and
net-fouling organisms. No adverse interactions between prawns and salmon were
noted, although the prawn density used was rather low. A major problem was
the need to stock prawns large enough to be retained by the mesh used for
salmon. It was suggested that the reduction in net fouling achieved at the
experimental site by the feeding habits of the prawns would aid the salmon
culture by allowing improved water exchange with the pens. Whether this was a
significant advantage in salmon cage cultures and, if so, whether it would be best
achieved commercially by prawns or the use of antifouling treatments or
equipment, was not discussed. Further studies on the potential of temperate
prawns for cultivation in conjunction with other species should be preceded by
extensive economic background studies showing that worthwhile yields of a
valuable 'by-product' could be obtained under commercial conditions. There is
often the danger that under nutritive stress one cultured species might become
food for the other(s). It is also possible that where Crustacea are cultured with
fish, some undesirable parasites may be able to complete their life cycle.

4.2 Temperate Marine Species – Lobsters

Background Information. Interest in augmenting the marine fisheries for
lobsters, *Homarus gammarus, H. americanus*, in Europe and North America was
widespread in the late nineteenth century. Experimental work on the hatching
and rearing of lobsters began in Europe between 1860 and 1865 and in North
America in 1885. European research was conducted in Norway, Denmark,
Sweden, Germany, France and Great Britain, while marine hatcheries were set
up in southern Norway, on the Isle of Man (UK) and later in France. In Canada
and the United States some 20 hatcheries or rearing stations were operational
over the turn of the century, but by 1917 most of these had closed. Some of
the units released millions of newly hatched larvae into the sea on the
assumption that this would bypass the heavy mortalities which normally
occurred in nature during incubation and hatching. Others, notably in the State
of Rhode Island, raised the larvae and released the post-larvae hoping to
overcome much of the natural mortality of larvae in the sea. In 1951 the
Massachusetts State lobster hatchery was set up at Martha's Vineyard,
Massachusetts, and between 1951 and 1970 reared and released over 7 million in
attempts to assist the State lobster fishery — see Figure 2.10 — (Kensler, 1970).

Canadian studies on the feasibility of transplanting lobsters to the Pacific coast in order to establish a breeding population began on Vancouver Island, British Columbia, in 1965. Considerable numbers of *Homarus americanus* have been shipped to the west coast of North America and to Europe for commercial resale and attempts at aquaculture.

Figure 2.10: Vessels Used in the Mass Culture of Lobster Larvae

Source: MAFF/Worshipful Company of Fishmongers, Conwy (1978).

With the transplanted lobsters came the risk of disease transfers, one example being Gaffkemia (*Aerococcus viridans*), a bacterial infection of the blood. Gaffkemia is also known to be pathogenic to other commercially important shellfish, including *Pandalus platyceros, Cancer magister, C. innoratus, Panulirus interruptus* (Steenbergen and Shapiro, 1974) and *Homarus gammarus* (Egidius, 1978). In addition Krekorian *et al.* (1974) have suggested that commercial pressure to transplant large numbers of *H. americanus* to California could result in serious dislocation in resident populations of *Panulirus interruptus.* In Europe a transplantation was made by Audouin *et al.* (1969) who imported adult South African Cape crawfish, *Jasus lalandei,* to France, where they were found to be hardy enough to survive in the waters of the English Channel.

A number of North American laboratories became involved with research into the biology and culture of *Homarus americanus* between 1965 and 1975, including units in Maine and California, which investigated the effects of warm-water effluent on the growth and aquaculture potential of lobsters. In Brittany, France, in 1973, a hatchery was jointly set up by a fishermen's co-operative and the government to produce and release late stage juveniles (three months old) into sanctuary areas of the sea (ICES, 1978). In 1977 three hatcheries released 5th-7th stage juveniles – these were at: (1) l'Ile de Yeu, 132,000; (2) l'Ile de Houat, 35,000; (3) l'Ile de Sein, 5,000 (Tiews, 1978).

It is interesting to note the similar attitudes taken by French and Japanese authorities, whose legislation and customs allow fishing in such areas to be restricted to the members of the co-operatives who contribute to the cost of juvenile production. The position contrasts with that in the USA and Britain where such restrictions would be difficult to legislate for. Legislation in Britain does exist, however, to protect cultivators of molluscs.

Culture. At this point it is worth considering some of the problems of marine fishery enhancement programmes or culture-based marine fisheries since, although these exercises have been tried, sometimes successfully, with other species elsewhere, for example with fish and prawns in Japan (Hanamura, 1976), attempts with Crustacea in Europe and North America have been limited to lobster.

In the broadest sense culture-based fisheries include:

(1) Closure and management of sea areas to conserve a breeding stock (see Audouin *et al.*, 1971). Major difficulties include enforcement of the closure because the prospect of high catches per unit of effort attracts poachers.

(2) Release of hatchery-reared larvae or juveniles into a fishery (USA) or into sanctuary areas (France). The problem here is trying to evaluate the usefulness of the approach since five to six years must pass before attempts can be made to demonstrate an increase in landings. Even so, such a demonstration would be difficult, due largely to natural fluctuation in catches. In addition the habitat and behaviour of the released young is poorly understood – one Canadian study (ICES, 1978) showed almost 100 per cent mortality in 10 minutes after releasing 5th-7th stage juvenile lobsters onto the sea bed in daylight. Growth, survival and movement have also proved difficult to monitor but recent studies with hybrids and on genetic tagging in USA and France using, for example, rare coloured sports or biochemically identifiable traits show promise (Hedgecock *et al.*, 1976).

(3) Habitat improvement (sometimes called ranching) by the provision of hides (artificial reefs) to increase habitat availability in areas where lobsters naturally occur, or in areas where they are not found but to which they are transferred. Cost effectiveness is generally considered to be a major problem. Where fishing intensity is high and stocks low, habitat availability is not likely to be a limiting factor. Other less valuable species – crabs, eels – could compete with lobster for any habitats provided. Areas suitable for reef construction could include thermally polluted areas, although the present costs of reefs and of culturing the juveniles with which to stock them are thought to make investment unattractive (ICES, 1978). Accounts of North American work include Scarratt (1973) and Sheehy (1976). Results from recent trials in France are not yet available.

The practice of holding Crustacea prior to marketing or shipment is widespread. Commercial lobster, crawfish and crab fishermen customarily hold their catches in keep-pots or boxes at sea, or in land-based pond or tank facilities. This is done in order to accumulate sufficient animals to make up a worthwhile consignment or, for longer periods, to take advantage of seasonal fluctuations in market price or availability. Apart from keep-pots several types of storage facility are used in Great Britain and France and have been described by Ayres and Wood (1977); the following is taken from their account. Shore-based units are often embayments, ponds or tanks, where the water is exchanged tidally or by pumping. Some tank systems operate by continuously pumping sea water, while others have the facility to recirculate the water, for example at periods of low tide. Inland units, often built near to the main markets, are frequently used for short-term storage and may incorporate recirculation and the use of artificial sea water made from mixtures of five chemical salts and tap water. The use of artificial sea water in short-term lobster storage is described by Wood and Ayres (1977). Examples of an inland storage system and an artificial sea water formulation are given in Figure 2.11 and Table 2.6 respectively.

An initial step towards cultivation is the possibility of holding small (but legally landed) lobsters in modified storage pounds until they moult to a larger size and thus are worth proportionately more. One attempt was made some ten years ago to hold lobsters in sea-bed cages in Scotland by the Kinlochbervie Shellfish Company. A major difficulty experienced by the company was in obtaining a sufficient number of small lobsters either from their own fishermen, or by buying from elsewhere, for the exercise to proceed on a profitable basis. Labour costs were high since the cages were serviced by divers and the project was abandoned soon after it was started. More recently Bishop and Castell (1978) have reported a similar project at Clarks Harbour, Nova Scotia, Canada, run jointly by the Federal Fisheries Department and a private company, Island Lobsters Ltd. In this trial over 3,000 male lobsters (*H. americanus*), each weighing 400-600 g, were held for four months in individual containers in the company's lobster pounds. Bilateral eye-stalk ablation was performed on 90 per cent of the lobsters to induce premature moulting. Throughout the trial water was pumped from the sea and run through the tanks to waste and the lobsters were fed on a compounded diet of crab (40 per cent), squid (20 per cent),

Figure 2.11: An Example of an Inland Lobster Storage System Using Recirculated Water. 1. Stacked trays; 2. Removable trays; 3. Sand filter; 4. Sump; 5. Cooling coils; 6. Pump; 7. Refrigeration unit. Arrows denote direction of water flow.

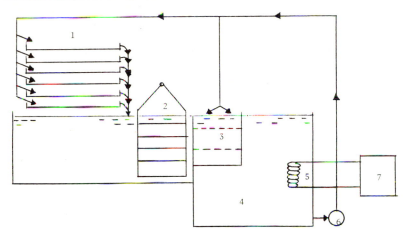

Source: Redrawn from Ayres and Wood (1977).

Table 2.6: Composition of Artificial Sea Water for Short-term Lobster Storage

Salt	kg/1,000 l
NaCl	23.7
$MgSO_4.7H_2O$	5.7
$MgCl_2.6H_2O$	4.6
$CaCl_2.2H_2O$	1.2
KCl	0.6
Salinity 30^0/oo	

Source: Ayres and Wood (1977).

lobster flesh (10 per cent), the remainder being wheat middlings, white fish-meal, Carageenan and 'alginate'. The project was adversely affected by water supply failures in the fourth and final month which caused heavy losses. The authors concluded that their bilateral eye-stalk ablation technique caused no undue mortality and that a holding period of nine months would be required for a successful operation. This time would allow the lobsters to be brought in, checked for weakness and disease, and operated on; after moulting they would then require four to six months of careful husbandry and feeding before the flesh was fully ready for marketing. It was estimated that a 30,000 lobster unit could be profitable and would require six workers for feeding and maintenance and tables of the estimated cash flows were appended to the report. This approach, although seasonal and subject to the vagaries of natural supply, has the advantage of a moderately fast cash turnover.

A German programme to culture lobsters in sea bed enclosures early in the last decade was reported by Bardach *et al.* (1972) but the idea was apparently not taken up commercially.

Attempts to use the sea have been made in a recent Norwegian project (Anon, 1978) which is believed to involve the culture of lobsters from the egg to late-stage juveniles in a hatchery prior to stocking them in sea-bed enclosures. The latter project is an example of co-operation in aquaculture between a university industrial research group (SINTEF) and private industry, but results have not yet been made public.

Other offshore ventures have been considered in the USA and are described by Van Olst *et al.* (1977). They include groups of floating individual containers, sea-bed cages serviced by divers, and a fixed sea-bed-to-surface cage system serviced at the surface by an access pier. The principal advantage of offshore systems would be lower capital, water treatment and pumping costs. The disadvantages would be lack of temperature control, difficulty of inspection, feeding, harvesting and predator and disease control.

A proposed new venture which received some advance publicity in Britain involved the possible use of the large circular concrete cassions (100 m^2 floor area) which form the outer wall of Brighton Marina for the containment of hatchery-reared fish and shellfish, including lobsters (Chamberlain, 1978). The sea water in the cassions is subject to tidal exchange and can be up to 2°C warmer in winter than the sea water outside. It is believed that each of the 70 available cassions could produce 100-150 marketable lobsters in 4-5 years when stocked with 300-400 stage 5-6 juveniles. Although now abandoned, the success of the operation would have been dependent upon the achievement of 30 per cent survival by the provision of hides with a minimum input of labour, and concurrent production of other valuable fish and shellfish species.

The use of heated industrial effluent for lobster culture has been considered in North America (Ford *et al.*, 1975; Dorband *et al.*, 1976; Muench, 1976) and a summary of commercial and research projects is given by TVA (1978). Problems associated with water supply when the effluent is used directly are chemical contamination of the effluent and temperature cycling of the generating units with load and periodic shut-down. The latter would also be a problem with indirect use of the heated effluent as, for example, when heat pumps or exchangers are used.

By far the most important aspect of lobster culture in heated water is the economic requirement for high yields of lobsters per unit area of culture tank. Various attempts have been made to mass-rear lobsters at high densities and unless shelters are provided, mortality is high (Shleser and Tchobanoglous, 1974). The design of such shelters also appears to be important. Lobsters of 11-90 mm carapace length were observed to choose shelters with low profile openings (height = ½ width) in preference to square (height = width) openings, and spent more time in shelters when the tanks were illuminated (3 and 117 lux) than when they were only dimly lit — below 1 lux (Cobb, 1971). Survival in mass cultures has been increased through abnormal light regimes, 18 hrs light-6 hrs dark, causing lobsters to remain in their shelters longer (Carlberg *et al.*, 1978). Estimates of possible stocking densities on various types of substrate were made by Van Olst *et al.* (1975). Using stage 5-12 lobsters, oyster shell, rock and PVC tubes were similar and allowed 30 lobsters/m^2; sand, however, only supported 6/m^2. In nature densities of 1/m^2 or less are usual. These and similar studies lead to the conclusion that only in the first six to seven months of life might losses due to cannibalism in mass culture be tolerated. For most of the culture period of two to three years, the lobsters must be held individually. Individual containment and the use of shelters for individually held lobsters will be discussed later under controlled environment systems.

4.3 Temperate Freshwater Species — Crayfish

Background Information. Probably the only form of crustacean farming traditionally practised in Europe was the cultivation of freshwater crayfish — Astacidae. In effect this may have been little more than primitive attempts to enhance production of a natural fishery or to fatten a few individuals prior to consumption.

Interest was greatly stimulated in the late-nineteenth and early-twentieth centuries by the widespread decline of natural fisheries. The decline was a result of increasing industrialisation, utilisation and alteration of water resources which polluted and altered crayfish habitats. Concurrently, stocks of most mainland European crayfish were being seriously affected by the 'crayfish plague'. A fungus, *Aphanomyces astaci*, Schikora, is commonly held to be responsible and attacks the soft areas of the exoskeleton, particularly the joints and abdominal segments (Unestam, 1973). The first outbreak was recorded in northern Italy in 1860, and before 1900 it had penetrated deep into Russia and north to Finland (1893). By 1907 it had spread to Sweden and in Europe outbreaks in old and new areas occur almost every year. Rapid dissemination of the disease may have been aided by the extensive commercial exploitation of crayfish and the associated transport of infected consignments and equipment. European crayfish susceptible to the plague seemed unable to become re-established in chronically infected localities. In addition, a number of uninfected habitats have been lost each year through change wrought by man on bodies of water (Kossakowski, 1973; Hobbs and Hall, 1974).

One of the earliest attempts to cultivate crayfish documented since the outbreak of plague was that made by the Marquess de Selve during 1864-70, who stocked the 'channels' of his castle at Villiers-sur-Essonne with the Noble crayfish, *Astacus astacus* (Laurent, 1973). In 1890 one hundred American crayfish, *Orconectes limosus*, were introduced into Germany where they thrived, and with further introductions spread to become widespread in Poland, Austria and France. The success of *O. limosus* over the native species in Europe is no doubt related to its greater resistance to the plague and tolerance to polluted waters. North American species of crayfish are notably more resistant to the plague, probably because the disease there is endemic.

As early as the turn of the century the environmental requirements and growth rate of *Pacifastacus leniusculus*, the American signal crayfish, were studied and found to be similar to those of *Astacus astacus* (Andrews, 1907, quoted in Abrahamsson, 1971). It was not until 1960, however, that *Pacifastacus leniusculus* was imported to Sweden, and initial trials commenced on its suitability for stocking in European waters. Unfortunately, their introduction from three widely separated water bodies in California resulted in outbreaks of plague in *Astacus astacus* in Sweden and Finland when the imported animals made contact with native populations (Abrahamsson, 1971). Further importation was prohibited. In 1968 a Swedish hatchery was set up and in a

short period of time developed into an alternative source of supply of
Pacifastacus leniusculus for restocking programmes.

Introductions of crayfish, particularly *Pacifastacus leniusculus* and
Procambarus clarkii, have been made for commercial reasons to Europe, Japan,
Central Africa and Hawaii, too often with disastrous results. Quite apart from
the introduction of the plague (Goldman, 1973; Unestam, 1975a), *P. clarkii*
became a serious pest by virtue of its burrowing habits (Bardach *et al.*, 1972).
In Japan it became so abundant (up to 1,200 kg/ha) that it caused substantial
losses by eating rice shoots; in Central Africa it consumed and destroyed
peripheral weeds and fish eggs to cause a decline in lake fisheries; in Hawaii and
Japan its burrowing habits caused damage to dykes and affected taro and rice
production.

In 1976 the first introduction of 1,000 hatchery-reared juvenile *Pacifastacus
leniusculus* to Great Britain was made. In 1977 and 1978 some 6,500 and
148,000 juveniles respectively were imported from Sweden and released into
ponds and rivers for commercial trials (Richards and Fuke, 1977; Anon, 1979;
Harvey, 1979). As yet no results have been published and in some, but not all,
cases, breeding individuals have been found. There are no reports of any profits
being made. In North America crayfish farming has expanded to considerable
proportions during the past thirty years. In 1949 a Louisiana rice farmer, Voorhies
Trahan, found that by flooding and stocking his field after the autumn rice
harvest he could obtain a crop of crayfish in the spring, and thus rotate crops of
crayfish and rice. Crayfish farming grew from an estimated 40 acres (16 ha) under
production in 1949 to 2,000 acres (810 ha) in 1960, and thence to 18,000 ha in
1973. The majority of crayfish farmed were the red swamp crayfish (*Procambarus
clarkii*), and in 1974 the annual production was estimated at some 5,000 tonnes
(Gary, 1974). Present production is around 11,000 tonnes from 25,000 ha.

European Crayfish Culture. Altogether there are about 500 species of crayfish,
of which over 250 are North American. They are widely distributed from polar
to tropical regions, though not endemic in Africa. In Europe the major culture
studies are directed towards the supply of juveniles for restocking programmes
or hatchery-based fisheries (e.g. Westman, 1973; Abrahamsson, 1973; Brinck,
1977). The species of greatest interest in Europe are:

(1) The Noble crayfish, *Astacus astacus*, native throughout Scandanavia and
    Europe but which today only has a scattered distribution. This species is
    most highly acclaimed by gourmets.
(2) *A. leptodactylus*, originally an Asian species but now found in eastern
    Europe, particularly Turkey. It has been introduced to western Europe.
(3) The Signal crayfish, *Pacifasticus leniusculus*. Originally imported from
    North America, it is presently cultured and distributed largely by the
    Swedes. It is reputedly comparable in taste and market value to *Astacus
    astacus* and may be sold at double the price of *A. leptodactylus* and
    Orconectes limosus (Laurent, 1973; Karlsson, 1977; Behrendt, 1979).
(4) *Austropotamobius pallipes*, found in south-west Europe, France and the
    British Isles.

No controlled comparative trials on the palatability and market acceptability of these species have been reported.

A number of private hatcheries, for example in France (Cabantous, 1975; Arrignon, 1975) and Germany (Hofman, 1980) operate to supply ponds with *Astacus astacus, A. leptodactylus, Austropotamobius pallipes* and, more recently, *Pacifastacus leniusculus*. Doubtless many other farms operate throughout Europe and it would be expected that the majority are primarily fish farms where the crayfish provide an incidental crop for sale and consumption locally. Information on yields from private crayfish monocultures is limited. More data exist from research and experimental studies. In one trial post-larval *Astacus leptodactylus* stocked at a weight of 35 mg in 1,300 m^2 ponds grew to 3.34 g mean weight with 85-90 per cent survival in one year, but considerable size variation was recorded, the largest specimen being 12 g (Tcherkashina, 1977). In the second year the mean weight was 27-32 g. Yields were not clearly specified, but could have been around 150 g/m^2 (1,500 kg/ha/yr). In other trials with the same species, exceptional yields of 1 kg/m^2 (equivalent to 10,000 kg/ha) of 50 g crayfish were reported in France in tanks (up to 3,000 m^2 bottom area) with supplemental feeding (Arrignon, 1975). Localised production in natural populations of crayfish has been reported to range from about 200 to 3,000 kg/ha/yr at population densities of 0.1-77 crayfish/m^2 (Laurent, 1973; Mason, 1975; Kossakowski and Orzechowski, 1975; Brown and Bowler, 1977). For comparison, production of *P. clarkii* in Louisiana crayfish farms ranges from 131 kg/ha/yr (Gary, 1974) to 898 kg/ha/yr (Avault, 1973). With supplemental feeding in small laboratory tanks (273 cm dia.) *P. clarkii* cultures produced about 119 g/m^2, equivalent to 1,186 kg/ha (Clark *et al.*, 1975). In Australia extensive cultures of Yabbie, *Cherax destructor*, yield 100-200 kg/ha/yr (Anon, 1977a) and with Marron, *C. tenuimanus*, 2,100-3,175 kg/ha/yr are obtained (Morrissy, 1979).

Stocking rates vary widely (3-10/m^2) according to the type of substrate and number of hides provided. Clark *et al.* (1975) concluded that for *Procambarus clarkii* numbers greater than 6/m^2 were needed for optimal yields. In traditional tropical prawn farms, yields of 1,000 kg/ha are not easily obtained unless the post-larvae are stocked at at least 20/m^2, despite the use of sophisticated diets. In these circumstances oxygen depletion limits production to about 2,000-3,000 kg/ha (Wickins, 1976a). A similar problem is reported for cultures of *P. clarkii* in Louisiana (Avault *et al.*, 1975) and *Cherax tenuimanus* in Australia (Morrissy, 1979). Caution should be exercised in comparisons of published yields since 'area' is used in different ways; for example Brown and Bowler (1977) used production/m^2 of aquaduct lining area, Clark *et al.* (1975) production/m^2 of tank floor, and Niemi (1977) reported numbers of individuals both/m^2 of river bed and /m length of shore line.

There is no established fishery for *Austropotamobius pallipes* in the British Isles and only small quantities are caught and from time to time appear in local markets. This is surprising since crayfish are widely distributed, particularly in southern England (Jay and Holdich, 1981), and market prices are good (see Sherry, 1977). The abundance of the species is not known, however, and without commercial interest little incentive has existed in the past to support research into British crayfish populations. Similarly few attempts have been made to farm *A. pallipes*. Yet, during the past four years interest in crayfish culture has blossomed because of promises of good market prices (up to £12/kg retail in Sweden) in the trade press (Anon., 1979), and the availability of imported juvenile *Pacifastacus leniusculus*. In addition, personal experience indicates that many potential crayfish farmers believe that, since *P. leniusculus* 'can be farmed in lakes, rivers and ponds, and with very little managing, they provide an extra income for any water area owner' (Karlsson, 1977), they will be able to obtain yields of 560 kg/ha (Anon., 1979) with the minimum of effort. Indeed, it is suggested that 'once colonies become established the ponds or tanks can be expected to yield up to 1 kg/m^2/yr' equivalent to 10,000 kg/ha/yr (Anon., 1977b). Yields of this magnitude are judged to be extremely unlikely without careful, full-time attention to habitat and water management and the provision of prepared foods (see Arrignon, 1975). Given reasonable water quality it would be unwise to say that yields of 500-1,000 kg/ha/yr could not be obtained in Britain as they are on the Continent. Time and experience alone will reveal the effort required and the risks and costs involved in extensive culture.

The importation of *Pacifastacus leniusculus* and sometimes *Astacus leptodactylus* to the British Isles has caused much concern (Holdich *et al.*, 1978). Several problems have been identified. Commercially the most important is the possibility of the introduction of fish viruses in the water used to transport the crayfish. Many juveniles are currently being stocked in fish farms or waters linked to natural streams and rivers (Harvey, 1979). Quite separately it has been argued that continued imports increase the risk of the introduction of the crayfish plague fungus to Britain which could destroy the native population of *A. pallipes* (see Unestam *et al.*, 1977; Söderhäll *et al.*, 1977). It is also possible that an exotic species threatens the range or abundance of native species by competition for food habitat or perhaps by interbreeding (Jay and Holdich, 1981). Attempts to cross *Pacifastacus leniusculus* and *Astacus astacus* have resulted in eggs that developed to the eyed stage before dying, and it is not therefore known if the hybrids were potentially viable or fertile (Strempel, 1975). It has also been stated that *Pacifastacus leniusculus* will interbreed with *Austropotamobius pallipes* (Richards and Fuke, 1977), but this has not been confirmed in the scientific literature.

The choice of crayfish to cultivate in the British Isles should thus be restricted to the native species *A. pallipes* on social, ecological and environmental grounds (Holdich *et al.*, 1978; Unestam, 1975a). It is a fact of life, however, that in business the value and size of established markets abroad and the availability of juvenile *Pacifastacus leniusculus* have had an overriding influence on the choices currently made. For these reasons it is worth considering the factors governing the choice of species to farm in order to assess better the culture potential of *Austropotamobius pallipes*. A logical choice of crayfish species for cultivation in the British Isles is dependent on:

(1) whether the chosen markets are species-oriented (i.e. based on the taste, texture and appearance of the whole animal), or are primarily influenced by size of the individual, and hence weight of meat produced per animal;

(2) the risks of plague fungus to native and European species and the risk of the introduced species proving susceptible to diseases endemic to the new environment;

(3) performance in relation to temperatures in terms of both growth rates

and survival;
(4) fecundity and ease of propagation;
(5) the potential for creating markets for *A. pallipes* at home, and the value of *A. pallipes* in Europe.

Each of the points may be considered in greater detail.

(1) The majority of crayfish are sold whole by weight and therefore small change in the meat yield between species is normally of less importance to the producers than if the animal were sold as tails as, for example, 'King prawns'. Meat yield in animals sold as tails is particularly important if expensive food is used in their culture. Usually then, species with a very small tail weight to total weight ratio will not be popular; an example is the world's largest crayfish *Astacopsis gouldi* (Frost, 1975). Morphometric studies and estimates of the meat yields obtainable from crayfish chelae and tail are therefore eminently desirable (see Rhodes and Holdich, 1979). Examples of the approximate meat yields (including chelae) of European and North American crayfish are 17-25 per cent for *Astacus astacus* (Dabrowski *et al.*, 1968; Linqvist and Louekari, 1975); 24 per cent for *Orconectes limosus* (Dabrowski *et al.*, 1966a), and 13-17 per cent for *Astacus leptodactylus* (Dabrowski *et al.*, 1966b). There are no similar data published for *Pacifastacus leniusculus* and *Austropotamobius pallipes*, although yield data from *A. pallipes* is being prepared (C. Rhodes, 1979, personal communication). Considerable variation is reported for some species, e.g. *Astacus astacus*, due to locality and the difference between the sexes; males develop larger chelae than females at maturity.

(2) The greater resistance to plague of North American species compared to European and Australasian crayfish (Unestam, 1975b; Unestam *et al.*, 1977) makes the American species more promising candidates for cultivation on mainland Europe. Yet some organisations in France (see Arrignon, 1975) consider the disease risks small enough to invest in native species, *Austropotamobius pallipes* and *Astacus leptodactylus*.

(3) Very little information has been published on the effect of temperature on survival and growth rates of crayfish grown under conditions that would enable realistic comparisons between species to be made (see Figure 2.5). Species growing in ambient conditions in favourable areas of the British Isles are likely to reach market size in 2-3 years. Supplemental heating, for example from freshwater power station effluent, could conceivably reduce this to two years. Culture strategies based on annual or biannual cropping of an established population which is continuously supplemented by hatchery-reared juveniles may not be so greatly influenced by growth rate differences of a year as when a complete stock is annually harvested. This means that provided the stocking and harvesting rates are sensibly balanced and on a sufficiently large scale, slower growing species with other advantages could be considered. In particular those with a specialist market appeal or low social and ecological impact could compete commercially with faster growing forms. The major drawback would be the slower rate of return on capital at the beginning of the project and the slower rate of recovery if the stock were lost.

(4) The slightly greater fecundity and reduced incubation period in *Pacifastacus leniusculus* (Table 2.3) are advantageous, but it would seem that all four crayfish, *Austropotamobius*, *Pacifastacus*, *Astacus* and *Cherax* may be

propagated with equal facility.

(5) Judging by the current interest in *P. leniusculus* as a replacement for *Astacus astacus* in both public and private sectors throughout Europe, little market education will be needed for its successful acceptance. Other exotic species, e.g. Marron (*Cherax tenuimanus*), that have hitherto not been seriously considered for cultivation in western temperate regions could well require the creation of new markets, particularly outside Great Britain, where crayfish-eating traditions are strong. In Britain relatively few crayfish are eaten. Approximately 50 tonnes/yr of *A. leptodactylus* are imported through Billingsgate where prices in 1979 ranged up to £3.96/kg (G. Watkin, 1979, pers. comm.). *Austropotamobius pallipes* was reported retailing at £3.3/kg in a Hampshire market (Goddard and Holdich, 1979).

It seems likely that on the basis of size and meat yield alone modest quantities of any palatable species could be caught and marketed direct to the public or to catering outlets. Unfortunately, little information is available to the author for prices paid in Europe specifically for *A. pallipes* though there are television reports of this species retailing at £1 each.

4.4 Controlled Environment Culture

The high prices paid for crustacean flesh in nearly every industrialised nation has provided the stimulus for private and public investment in the research and development of a technologically advanced system of controlled environment culture. The new method differs radically from those already described for the enhancement of 'natural' fisheries or for the improvement of traditional, largely extensive, aquacultural practices. The approach is attractive to investors because of three premises:

(1) Cost-effective processing, distribution and marketing operations rely upon predictable and regular supplies of the product in order to keep personnel and plant operating continuously and at full capacity.
(2) The product is consistent in its quality, particularly in respect of its texture, taste, appearance and size.
(3) Appearance and hence market acceptability may be influenced by culture techniques. For example, the colour of some crustaceans may be improved by the diet and colour of the culture tank; starvation for a day or so before harvesting results in a gut empty of unsightly contents; and gonad development in some species may be suppressed, allowing more effort to be put into profitable somatic growth.

Compared with other forms of aquaculture in western temperate regions, however, controlled environment culture is very much in its infancy, particularly in terms of commercial commitment. This is because it is a novel, sophisticated and high-risk process. Even so, it is attractive to a number of industrialists because of its similarity to a manufacturing process.

Tropical Marine and Freshwater Prawns. Among the earliest European attempts to culture prawns in controlled environment culture was that made privately by Mr Keir Campbell in Yorkshire, Great Britain, with imported specimens of the giant freshwater prawn, *Macrobrachium rosenbergii*. Government scientists at the Fisheries Research Station, Conwy, became involved in 1969 when asked by Mr Campbell to assist with the culture of larvae (Wickins and Beard, 1974). Since that time three or four private concerns have reared and studied *M. rosenbergii* in Britain. Of the largest, Unilever Ltd, who had previously tested *Palaemon serratus* (Figure 2.12) concluded that their best approach was to grow the prawns, *Macrobrachium rosenbergii* and *Penaeus monodon*, in ponds overseas (G.C. Phillips, 1979, pers. comm.). Rank Hovis MacDougal made substantial investigations from 1970 to 1975 and developed and operated a pilot plant operation on the inland site of Mr Campbell's original studies. The work demonstrated that regular and reliable supplies of juvenile *Macrobrachium rosenbergii* for stocking could be obtained from captive broodstock. By continuously culling marketable animals and replacing them with juveniles, yields in excess of 9 kg/m^2/yr of live prawns were actually achieved. The further development of the project, however, was postponed, largely because of rapidly rising heat costs during the early 1970s. The unpredictable heating costs prevented a satisfactory assessment of profitability margins. Unfortunately, species more suitable for mass culture (penaeid prawns) could not at that time be bred in captivity.

Figure 2.12: Experimental Tank Room used to Evaluate the Commercial Potential of *Palaemon serratus* and *Macrobrachium rosenbergii*.

Source: Photograph courtesy of Dr G.C. Phillips, Unilever Research (1968).

Similar investigations with *Macrobrachium rosenbergii* were conducted around the same period by American industrialists such as Syntex Aquacultural Services and Solar Aquafarms, both of California. Some attempts were also made to develop individual rearing containers for this species (Van Olst *et al.*, 1977).

For a summary of American studies and commercial trials with *M. rosenbergii* the work of Hanson and Goodwin (1977) is recommended. The experiences reported indicate that in America it is not yet practical to plan for an intensive controlled environment system for the culture of *M. rosenbergii*. Among the major constraints were the uncertainty of the market and the relationships between yield and cost of the culture tanks.

Government research in Britain into the suitability of penaeid prawns for controlled environment culture began at Conwy in 1970 and the results were summarised by Wickins and Beard (1978). Before 1978, commercial involvement with penaeids was not encouraged because it was uncertain whether penaeids could be bred in captivity with any degree of reliability. It was necessary, therefore, first to select a species that would fit into a controlled culture environment (Figures 2.13 and 2.14) and then to determine conditions needed to ensure successful reproduction. In the selection studies no attempt was made to tailor comprehensively an environment to each of the species investigated. In particular many penaeids require a sand or mud substrate in which they bury during daylight hours. If deprived of this substrate many like *Penaeus japonicus* grow and survive markedly less well (see Table 2.7). In addition, therefore, to looking for a species that would survive and grow well in the bare tanks (thought to be essential in large-scale cultures), other criteria could be specified:

(1) good market qualities, appearance, texture and flavour;
(2) ability to grow quickly;
(3) ability to tolerate crowded conditions, recirculated water and disturbance;
(4) acceptance and good conversion of compounded feeds;
(5) resistance to disease;
(6) ability to breed in captivity.

Ten species of penaeid and one caridean prawn were studied (Forster and Beard, 1974; Wickins and Beard, 1978).

Table 2.7: The Yields of Eleven Species of Prawn Obtained in Bare Culture Tanks for Three and Four Months When Initially Stocked at 162 Prawns/m^2

Species	Yield (g/m^2)		Heavy Mortalities During Weeks:
	12 weeks	16 weeks	
Penaeus monodon	1,337	1,908	
P. orientalis	1,155	971	12-16
P. occidentalis	753	889	
P. aztecus	564	612	
P. merguiensis	449	585	
M. rosenbergii	478	486	
P. indicus	343	432	
P. setiferus	347	366	1-2 and 13-15
P. semisulcatus	480	345	
P. japonicus	387	262	
P. schmitti	77	—	8-12

Figure 2.13: A Dense Culture of *Penaeus monodon*. Circular pellets of food and irregular faecal pellets can also be seen.

Source: MAFF, Conwy (1979).

Figure 2.14: An Experiment with *Penaeus monodon* to Study the Effects of Tank Size and Stocking Density.

Source: MAFF, Conwy (1973).

The species were initially selected for study because fast growth rates had been reported for them in natural and pond culture conditions; also, they could be obtained from abroad as laboratory reared or wild-caught juveniles. Market acceptability was implied by the fact that they all supported commercial fisheries and were often included in the many species already exported to Britain. The prawns were cultured at two stocking densities, 25 and 162 prawns/m^2 of tank floor in grey fibreglass tanks (86 x 72 x 20 cm depth of water). Because the conditions were the same for all species tested they were doubtless not optimal for some. Thus, a poor result did not necessarily mean that a species was totally unsuitable since better yields might have occurred under different conditions. A good result, however, clearly indicated that a species was capable of giving high yields in very artificial conditions and that further improvements in performance might be possible by paying attention to the specific environmental requirements of the chosen species. The results are shown in Figure 2.15 and Table 2.7.

Figure 2.15: The Growth and Survival of Ten Species of Penaeid Prawn and *Macrobrachium rosenbergii* at Two Stocking Densities. The trial with *Penaeus schmitti* was curtailed after eight and twelve weeks at the low and high densities respectively.

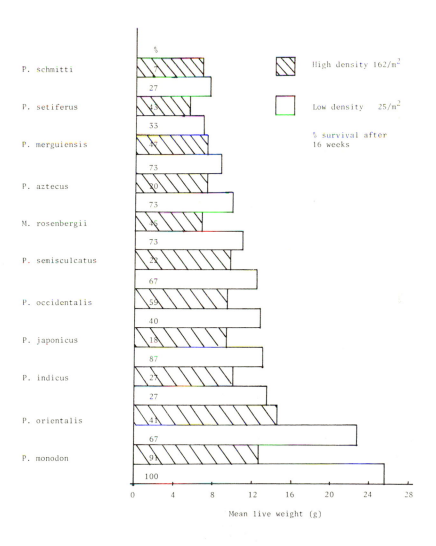

Source: from Wickins and Beard (1978).

Penaeus monodon showed exceptionally good growth and survival suggesting that this species was likely to be more tolerant than most of the culture environment. Further growth trials lasting four months were conducted with *P. monodon* at a series of stocking densities from 25 to 162/m^2 in tanks ranging in size from 0.6 to 3.4 m^2. The results are shown in Table 2.8.

Table 2.8: The Increase in Mean Weight of *Penaeus monodon* Reared at Four Stocking Densities in Laboratory Recirculation Systems

Weeks	Density/m^2							
	25	25	47	146	146	146	162	162
0	0.15	0.20	0.13	0.29	0.29	0.29	0.15	0.30
2	1.27	1.00	1.04	0.95	0.85	0.87	0.86	0.70
4	3.43	2.10	2.93	1.73	1.61	1.57	1.87	1.30
8	10.00	6.70	8.16	4.57	4.41	4.36	4.73	3.30
12	18.47	15.50	14.43	6.84	6.89	7.06	9.00	7.90
16	25.43	20.90	21.50	10.66	11.46	11.99	12.93	10.30
Survival %[b]	100	73	90	73[a]	57[a]	60[a]	91	68

Notes: a. Survival values adjusted because of sudden mortality during the last four weeks of culture.
b. Mean % is 76.5.

There is some indication from the measured growth rates that some factor such as water quality or tank size limited growth towards the end of the culture period by comparison with published growth rates for pond-reared *P. monodon* (Norfolk *et al.*, in prep.). Even so, these results show that two to four harvests per year could be obtained. The variation would depend on the size of prawns produced. In turn, this would be governed by the value and stability of the market price for the crop. Yields of 1.4-2.0 kg/m^2 of tank floor are to be expected in 15-20 cm water depth under these conditions. Elsewhere, in through-flow rather than recirculated waters even higher yields are obtained in commercial and large-scale pilot projects. Japanese farms based on the designs of Shigueno (1975) in which *P. japonicus* are cultured in 1 m deep circular tanks fitted with a false bottom and sand substrate produce 2.5-3.0 kg/m^2 per harvest. In a project set up jointly by the Coca Cola Company and University of Arizona at Puerto Penasco, Mexico, *P. stylirostris* was grown in long plastic covered raceways (Figure 2.16) in about 30-60 cm depth of water (see Salser *et al.*, 1978). Yields of 3 kg/m^2 of 20-25 g prawns have been obtained in 6-8 months at ambient water temperatures (N. Stamp, 1979, pers. comm.).

In western temperate regions continuous production, without regard to season would be required for commercial viability. A study of costs (Wickins and Beard, 1978) made on the basis of trials at Conwy showed that the major costs would be tanks (capital cost) and heating (running cost) — see Figure 2.17. These were also of greatest variability in the estimates. Heating cost estimates varied according to the amount of water reuse that could be achieved. Tank cost

Figure 2.16: The Covered
Raceway Culture of *Penaeus
stylirostris* in Mexico.

Source: Photograph courtesy of
P.O. Söderstrom, Coca Cola Co., and
C.N. Hodges, University of Arizona
(1979).

Figure 2.17: The Proportionate Costs (%) for Selected Items in the Culture of Penaeid Prawns from 0.2 g to 35 g in 21-26 Weeks.

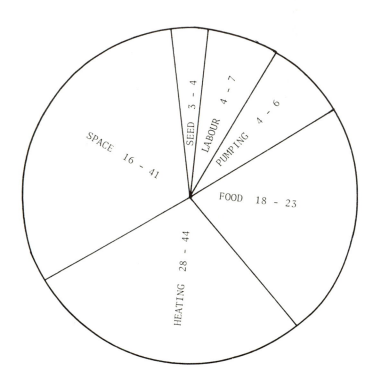

Source: from Wickins and Beard (1978).

estimates depended on whether the prawns could be grown in shallow raceways or whether deep tanks fitted with shelves would be required. The crucial issue for successful prawn culture in temperate regions is (as with lobster culture) the efficient utilisation of farm floor area. Where land is at a premium and a building is to be provided, maximum use of height must be obtained either by stacked shallow tanks or in deep tanks fitted with shelves. Development of the latter is likely to require R & D input from both biologists and engineers to solve the problem of husbandry and cleaning (Wickins and Beard, 1978).

A major constraint on the early development of British commercial interest in penaeids and therefore publicly funded research programmes was the inability to control reproduction (see HMSO, 1978, p. 14). It was not until 1975 (*P. merguiensis*) and 1978 (*P. monodon*) that penaeids matured and spawned viable offspring in Great Britain (Beard *et al.*, 1976; Beard and Wickins, 1980). It is well-known that most (but not all) captive female *P. monodon* do not naturally develop ripe ovaries and spawn either in ponds in the tropics or in laboratory systems in the west (Wickins, 1976a). Until knowledge of the natural trigger for ovarian maturation becomes available researchers must circumvent the problem by unilateral eyestalk ablation (Figure 2.18). This technique was used with success with *P. duorarum* (Idyll, 1971; Caillouet, 1972) and has proved reliable with *P. monodon* in a number of research establishments in Britain (Arnstein and Beard, 1975), Indonesia (Alikunhi *et al.*, 1975), Tahiti (AQUACOP, 1977b) and the Philippines (Santiago, 1977).

Figure 2.18: The Effect of Unilateral Eyestalk Ablation on Ovarian Development in *Penaeus orientalis*. Upper, intact prawn, no ovarian development visible; lower, one eyestalk removed, ovary clearly visible as a dark area extending the entire length of the prawn.

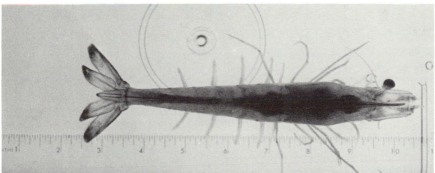

Source: Photograph courtesy of D.R. Arnstein, Conwy (1973).

Control of maturation in intact *P. japonicus* has also been achieved by photoperiod manipulation (Laubier-Bonichon, 1978). The process has been patented in France where two groups (DEVA Sud and Les Compagnons de Maguelonne) utilise the technique to produce post-larvae in time for the summer growing season. The prawns are stocked in ponds by the Mediterranean and since 1976 annual yields of 2-3 tonnes of *P. japonicus* have been produced.

At Conwy, however, another problem arose (Beard and Wickins, 1980). Although ablated females regularly matured, spawned and rematured, their eggs were not fertilised. Examination of thelyca from the cast exoskeleton of females suggested that copulation was not taking place. The males, which weighed over 40 g, showed signs of full maturity, including the presence of spermatophores. Sperm were confirmed in the spermatophores but no indication of their viability could be assessed visually as penaeid sperm are non-motile. It is possible to speculate on the reasons for the reluctance of *P. monodon* to copulate in the Conwy recirculation systems. During the four years of 1974-8 numerous factors that could conceivably affect courtship and mating, such as sex ratio, depth of water, tank size, photo-period, disturbance, diet, pH, salinity and temperature were tested without success. Throughout 1978, however, a batch of *P. monodon* females matured, mated and produced viable larvae. Unlike previous batches this group had been cultured at low density and not subjected to adverse experimental conditions prior to attaining maturity. With hindsight and the benefit of discussions with research workers from the tropics, particularly Mrs J. Primavera of SEAFDEC, Philippines, it was possible to list a number of factors that together could have significantly influenced the prospects for successful reproduction. These were:

(1) broodstock should be grown from metamorphosis to maturity in about ten months under the best possible conditions and at low densities;
(2) males should be over 50 g, females over 70 g live weight;
(3) tanks where courtship and mating are to occur should be at least 3 m^3 capacity;
(4) sex ratio to be in favour of males;
(5) at least ten animals should consitute a mating group;
(6) pH should not fall below pH 8.0-8.2;
(7) mature prawns should not experience salinities below 30‰.

The problem with mating among cultured *P. monodon* has been encountered elsewhere, but the nature of the parameters listed above support the hypothesis that at Conwy at least normal mating behaviour and successful fertilisation tend only to occur in prawns that are in prime condition. Growth rate is a good indicator of condition; thus males that have grown quickly to 50-90 g live weight, and females to 70-100 g, in 8-12 months, are most likely to breed.

Lobsters. There is almost a natural progression from the culture practices where lobsters are held in hatcheries prior to release and the development of culture facilities where juveniles are grown until they reach marketable size. It is difficult, therefore, precisely to trace the beginnings of this idea, but certainly in the last eight to ten years the strategy has attracted considerable speculation.

The initial impetus seems to have come from the pioneering studies of Hughes *et al.* (1972) who showed that *Homarus americanus* could be grown to a marketable size of 450 g in two years instead of the five to eight normally taken in the wild. In Great Britain Mr D. Gott of Lincoln was perhaps the first to study privately the mass production of juvenile *H. gammarus* in a recirculation system at his small, inland hatchery (Figure 2.19). By 1978 he had achieved a survival rate of 70 per cent to the 4th stage using artificial sea water. It was not until

1973 that government studies to determine the potential of *H. gammarus* for controlled environment culture began. Here the impetus was provided by the Worshipful Company of Fishmongers who sponsored a six-year programme of research with the Ministry of Agriculture, Fisheries and Food at Conwy (Richards and Wickins, 1979).

Figure 2.19: Experimental Containers Used in the Culture of Juvenile Lobsters.

Source: Photograph courtesy of D. Gott, Lincoln (1979).

Current research in North America and Europe indicates that three or four steps might be appropriate for intensive lobster culture. In the first place egg-bearing adults would be purchased from fishermen or obtained from specially cultured broodstock and held until their larvae hatched. In the second phase larvae would be cultured *en masse* (Serfling *et al.*, 1974). The third phase might be the mass culture of juveniles 2-26 weeks of age stocked in raceways fitted with plastic honeycomb compartments as suggested by Van Olst *et al.* (1977), although survival rates with this method will be largely unpredictable (but see Sastry and Zeitlin-Hale, 1977). In the final stage, losses due to cannibalism in mass culture would be too costly and the lobsters would need to be held individually. Many authors believe that all the post-larval stages will require individual containment. This necessity for individual confinement is crucial to the economic concept of indoor lobster culture. It is, therefore, worth reviewing a number of the studies relating to the survival and growth of lobsters in individual containers.

The shape of the rearing container, whether circular, square or rectangular, does not seem to influence the growth of *H. americanus* over the first 7-10 months of post-larval life (Shleser, 1974). In studies made with post-larval *H. gammarus* (Richards, 1981) growth and survival in circular containers of different floor areas were measured. Survival was over 95 per cent in all but the smallest containers (7 cm^2) where only 50 per cent survived after three months. A minimum acceptable size of 100-135 cm^2 was determined which gave between 85 and 90 per cent of the maximum yield obtained in 'unlimited' space (544 cm^2). The relationship between floor area of the container and total length of the lobster for no more than 10-15 per cent reduction in maximum yield after three months was: area (cm^2) = (2 x total length)^2. This agreed with the findings of Van Olst *et al.* (1976) who determined that for unrestricted growth in *H. americanus* the short side of a rectangular container should be 60 per cent larger than the total length (TL) of the lobster, i.e. area (cm^2) = 4 x (TL)^2. Richards (1981) found that the smaller final size of lobster grown in small containers was brought about by reduced growth rate from an early stage rather than by a sudden limitation after a period of normal growth. This was an important finding since it indicated that over-zealous space-saving could be a false economy. By extrapolation, larger lobsters of 25 cm total length would require an area of 0.25 m^2, and clearly some reduction in growth must be tolerated if space-saving is to be significant (Van Olst and Carlberg, 1978). In further studies with *H. americanus*, which lasted 18 months (Van Olst *et al.*, 1977), it was suggested that to avoid significant reductions in growth the width of a rectangular container should be at least as large as the total length of the lobster. In other trials Van Olst and Carlberg (1978) reported that survival was affected by confinement before reduction in growth could be detected and, using 560 stage four *H. americanus*, the relation between carapace length (mm) and container width (cm) at 10 per cent mortality was CL = 12 + 1.7 x width. The time (t) in months for the 10 per cent mortality level to be reached was t = 5.9 x 1.05 x width. No significant differences in growth were noticed between lobsters grown in containers with transparent walls and those grown in containers with opaque walls.

The interactions of shelter, light intensity, wavelength and photoperiod have received scant attention. Growing lobsters exhibit negative phototaxis and positive thigmotaxis. Using wild-caught lobsters (60-70 mm CL), Cobb (1971) determined that negative phototaxis was more important in the lobster's choice of shelter than positive thigmotaxis. Similar results were obtained with small (7th stage) *H. gammarus* (Howard and Bennett, 1976), but these authors also thought that positive thigmotaxis could predominate when the lobsters first experienced transparent shelters when in dark conditions prior to illumination. In bare, well-illuminated culture containers where neither response is satisfied the situation must be considered stressful to the lobster. In growth trials with *H. gammarus* made under normal laboratory lighting conditions (Richards 1981) better growth was achieved when shelter was provided than when it was not. The provision of shelters in systems where lobsters are held separately because of cannibalism complicates husbandry and therefore does not make economic sense. It is arguable that culture in reduced light intensities could alleviate some of the stress and the need for shelters. Juveniles grown for three months under conditions of constant illumination and 12 hrs light:12 hrs dark grew more slowly than juveniles kept in complete darkness, though the differences were small. It was thought that the effect could have been due to differences in the activity of the juveniles.

The effect of photoperiod on locomotor activity in *H. americanus* has revealed weak rhythmic activity in total darkness similar to the stronger pattern shown in a light:dark regime (12 hrs:12 hrs) (Cobb, 1976). A similar pattern was found for juvenile *H. gammarus* (Richards, 1981). Groups of individuals were exposed to three photoperiod regimes (24, 18 and 0 hrs light) with three intensities of illumination (0, 5 and 220 lux) in the presence or absence of shelter. Growth was best in constant low-level illumination 6 hrs L:18 hrs D, dim light and constant darkness (Figure 2.20). Under these regimes growth of lobsters without shelter was more variable and tended to be greater than those with shelters (Richards, 1981). Similarly, good growth has been noted in lobsters held in such abnormal regimes as 23 hrs dark:1 hr light (Zeitlin-Hale and Sastry, 1978). It is clear then that while reduced light intensity or even darkness can promote faster growth in individually held lobsters without shelters, the causes are by no means easily understood. More work would be most advisable before parameters for the design of the individual containers for economical operation are defined.

The most pressing problem that faces intensive lobster culture (and the one that will probably decide its future commercial existence) is the design of a suitable configuration and arrangement of containers for large-scale use. It is vital that any design provides for adequate water exchange and cleaning facilities. Current thinking is that the containers will be designed in such a way as to save space, and in a limited range of sizes to reduce manufacturing costs. The design must also ensure that lobsters can be moved, or the cells expanded as infrequently as possible during the culture period. This would reduce labour costs and allow the best compromise between good growth and economy of space. Systems tested or in use include:

(1) Simple troughs divided into compartments by wooden or plastic slats and mesh screens through which water continuously flows (as originally used in the Massachussetts State lobster hatchery.

(2) Rectangular tanks or troughs containing blocks of compartments fitted with perforated floors. Water is distributed by specialised flushing systems or tidal exchange. Examples include: USA — San Diego State University aquaculture project and a commercial pilot plant in Monterey; Canada — Environment Canada, St Andrews; UK — Fisheries Experimental Station, Conwy.

(3) Various experimental systems tested in N. America:
  (a) Deep tanks containing stacks of tubes or corrugated sheets of plastic.
  (b) Deep tanks containing vertical or horizontal stacks of perforated trays. These are serviced by sequential removal of the stacks by an overhead gantry or by continuous slow rotation of trays.
  (c) 'Care o cell' which is a shallow, round tank containing a revolving group of floating mesh-bottomed containers. Water is jetted upwards into the containers as they revolve. Further details and illustrations of these and other, privately developed systems may be found in Van Olst *et al.* (1977).

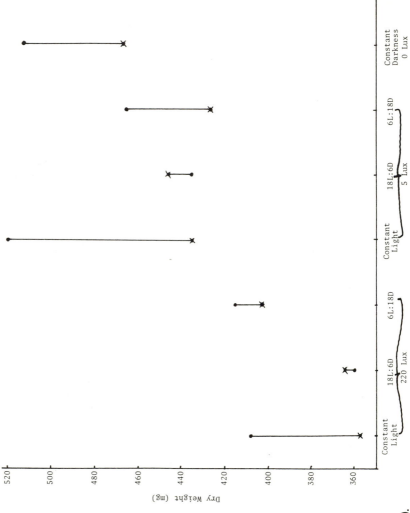

Figure 2.20: The Growth of Individually Held Juvenile Lobsters, *Homarus gammarus*, Reared for Twelve Weeks under Different Photoperiod Regimes. Closed circles — lobsters without shelters. Crosses — lobsters provided with shelters.

Source: Redrawn from Richards (1981).

Of all these systems, those using shallow water ((2) above) are the most widely used in research establishments. Since the rectangular tanks or troughs are easily stacked in layers (Figure 2.21) this seems to be the most readily constructed, easily serviced and least expensive (Schuur *et al.*, 1974). Yet Van Olst *et al.* (1977) favour deep tanks containing vertical stacks of trays. Despite difficulties of feeding and harvesting they claim that such problems 'can be compensated for, at least in theory, by incorporating several recent technological developments'. Their estimates suggest that yields of up to 73 kg/m^2 of farm floor area are possible from pits 3 m wide, 3 m deep. Wisely they stress that further evaluation of such systems is necessary before production larger than laboratory scale is attempted. No published data seem, however, to be available on the effect of container (ceiling) height and its effect on growth or moulting of individually held lobsters.

Figure 2.21: Controlled Environment Culture of Individually Held Lobsters.

Source: MAFF/Worshipful Company of Fishmongers, Conwy (1978).

If controlled lobster culture is to succeed in Britain the ability to produce regular supplies of marketable animals irrespective of season is likely to be of fundamental importance. This aspect was studied in a laboratory-scale production unit in which lobsters (*H. gammarus*) were hatched and reared through the larval stages and stocked, 50 at a time, at intervals of three months for two years, 1977-9 (Richards and Wickins, 1979; Richards *et al.*, in prep.). Conventional stacked rectangular tanks fitted with individual containers were used and at the end of the trial the unit held 400 lobsters of eight different age groups. Lobsters were reared in containers of 25, 100, 400 and 1058 cm^2 for 1, 3, 8 and 12-18 months respectively. The normal hatching season for *H. gammarus* is May to September with a peak in June and July. In order to obtain larvae for stocking in January, April, July and October, female lobsters bearing eggs were selected from a commercial storage pound according to the stage of development the eggs had attained. The berried females were then taken to the laboratory and held in individual tanks in a recirculation system where the temperature was controlled at up to 12°C above ambient. The time of hatching was advanced as required by maintaining an elevated water temperature. A relationship between water temperature and the weekly increase in the size of the developing lobster eye was established by Perkins (1972) for *H. americanus*, and enabled the time of hatching to be predicted. The same formula was found to be of practical use for *H. gammarus:*

$$\text{Time to hatch (weeks)} = \frac{\text{Eye index at hatching} - \text{Initial eye index}}{\text{Increase in eye index each week at the appropriate temperature}}$$

The eye index is defined as half the sum of the greatest length and breadth of the eye measured in microns. The incubation of many broods of *H. gammarus* at the Conwy laboratory provided sufficient data to determine the time to hatching from the initial eye index at 13°-15°C (Table 2.9). For example, when the eye index was 250-300, hatching could be induced in ten weeks by holding the female for that period in water of 13-15°C. Larvae produced in October and January tended to survive less well, probably as a result of warming. It was felt that delay of hatching by a reduction in temperature would have given better results at these times. Loss of larval vigour following incubation at elevated temperatures is well-known in the caridean prawns *Pandalus platyceros* (Price and Chew, 1969; Wickins, 1972b), and *Palaemon serratus* (Wickins, 1972a; Wear, 1974).

Table 2.9: The Relation between Eye Index (see text) and Incubation Period for *Homarus gammarus* Eggs Held at 13-15°C

Eye Index (μm)	Time to Hatch (weeks ± 2 weeks)
50-100	15
100-150	14
150-200	13
200-250	12
250-300	10
300-350	8
350-400	7
400-450	5
450-500	4
500-550	2
600-620 (Hatching time)	

Source: Richards and Wickins (1979).

Table 2.10: The Growth and Survival of *Homarus gammarus* at Conwy

Weeks	16			40			52			78			104			130		
Batch	wt	v	%	wt	v	%	wt	v	%	wt	v	%	wt	v	%	wt	v	%
1	1.25	29	98	9.20	33	98	18.34	29	91	80.90	32	87	194.00	39	60	285.60	29	24
2	1.42	21	80	10.85	25	78	23.70	23	76	75.26	22	44	166.02	34	22	339.50	33	7
3	1.56	18	98	13.67	26	98	30.16	22	95	95.45	29	63	208.60	24	33			
4	1.66	20	100	15.61	27	100	29.52	32	94	110.73	30	72	255.40	38	47			
5	1.75	18	98	13.59	31	98	27.43	34	98	85.32	35	73						
6	1.35	17	100	13.97	23	98	29.04	24	96	122.40	25	62						
7	1.48	22	100	17.67	23	100	33.85	25	93									
8	1.47	20	100	15.87	24	84	36.65	25	58									
9	0.90	14	91															
Target weight for 2½ year grow-out (g)	1.43			13.22			26.89			85.90			186.50			358.70		

Key: wt = mean weight (g); v = coefficient of variation of weight (%); % = survival.

The first batch of lobster post-larvae were stocked in April 1977 and further batches added at three-monthly intervals. The growth and survival of the nine batches are shown in Table 2.10 (Richards *et al.*, in prep.). Survival was good (above 78 per cent) for the first twelve months when mortalities began, chiefly among the older lobsters. Speculative reasons for this include (1) the possibility that the earlier batches were exposed to elevated levels of ammonia and nitrite during the establishment of nitrification in the biological filter and had become permanently weakened (see TVA, 1978) and (2) during 1979 the condition of the laboratory sea-water supply was unusual, as indicated by the fact that normal coastal algal blooms did not occur for the first time in ten years.

Each batch of lobsters was made up of progeny from a single parent and the coefficient of variation within a batch of their weights increased from about 20 to 40 per cent during the culture period. Variation of this magnitude possibly indicated a stressful environment although the component due to the lobsters' inherent variation was unknown.

Marketable size (350 g) was reached in 30 months by 10 per cent of the first two batches of lobsters (see Figure .2.22). In the event production targets were not achieved but the potential for year-round culture was successfully demonstrated and the 0.75 lb lobsters produced made excellent eating.

An analysis of the costs (Richards *et al.*, in prep.) demonstrated that manual feeding and servicing of each compartment could not be economic even in a small commercial operation and that the development of an optimal configuration of container was vital. A temperature of $20°C$ would be required in a production unit but this is low enough to allow economic use of high-volume, low-grade industrial heat sources, e.g. power station effluents. The usefulness of some power station effluents for *H. americanus* in America is now established (TVA, 1978), but it is judged that their use in America and Great Britain would be more appropriate to less-controlled semi-intensive culture because of the relative values of the energy generating and fish farming industries. The advantage of sophisticated indoor culture is control over production and product quality and should not be lost by using a seasonally variable heat supply whose availability is determined by other priorities. Under these circumstances it would be wise to incorporate conventional heating systems as a reserve.

Warm-water Crayfish. One of the few serious attempts to assess the commercial potential of freshwater crayfish in controlled environment cultures was that made privately by Mr K. Campbell in the mid-1970s in Yorkshire, Great Britain (see also Hofmann, 1980). The species studied was *Procambarus clarkii*, chosen because of its rapid growth rate and because supplies of juveniles were readily forthcoming from abroad. Initial growth rates were encouraging. Specimens grew to 40 g in six months at $21-23°C$. They were fed with compounded foods and stocked at 15 crayfish/m^2 in tanks in which hides were provided. Damage that was sustained during fighting reduced the marketability of some specimens and there was some evidence that second and third generations reared from the original stock grew less well than the first. The project was eventually abandoned to give priority to other commitments. These findings perhaps endorse the opinion that the small European and North American crayfish are unlikely to be worthwhile candidates for intensive indoor culture. Possibly the giant Australian Marron (*Cherax tenuimanus*) might well repay investigation. It is a large, rapidly growing crayfish (Figure 2.5) which, even in Australian pond cultures will grow faster at 15-20°C than *H. gammarus* does at the same temperature. In addition, its eating qualities are excellent and cooked meat yield, including claws is 40 per cent (Morrissy, 1976, 1978). The technology for its culture under controlled environment conditions would probably be very similar to that currently envisaged for lobsters and, being a freshwater species, would be ideally suited to cultivation inland.

Figure 2.22: Weighing a Two-year-old Lobster.

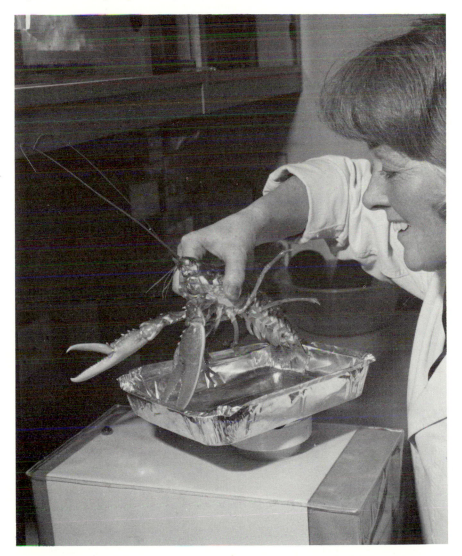

Source: MAFF/Worshipful Company of Fishmongers, Conwy (1979). Crown copyright, Central Office of Information.

Four possible handicaps to its culture in Europe or North America may be discussed at this stage:

(1) There is every possibility that it could be susceptible to the plague fungus or carry other pathogens if imported. Similar precautions to those currently taken in existing fish and shellfish hatcheries could presumably be taken prior to its culture in controlled indoor environments.

(2) The tolerance of *C. tenuimanus* to recirculated water is not yet known, although it is grown under a wide range of conditions in Australia (Morrissy, 1976, 1979). At least one attempt to introduce *C. tenuimanus* to Louisiana, USA, is known (Shireman, 1973) but survival was poor in both aquaria and ponds.

(3) Marron is an unfamiliar product outside Australia since it was a protected species until 1976. Attempts to export processed Marron to Sweden have been encouraging (Morrissy, 1980), even though the product was sold in competition with *Astacus astacus* on a specialised market. The value of *C. tenuimanus* in Britain is unknown and because of this its potential for intensive culture in controlled environments in temperate regions is only likely to be realised in countries where shellfish marketing is adaptable and progressive.

(4) The exportation of live Marron *was* prohibited until 1981 when the ban was lifted. Some juveniles are believed already to have been sent to France and Japan for aquaculture trials (N.M. Morrissy, 1981, personal communication).

5. The Reuse of Water in Controlled Environment Cultures

There are three basic water supply strategies suitable for crustacean culture in controlled environments:

(1) Single pass, where natural water is pumped, filtered, warmed, used and discarded.

(2) Partial recirculation – supply as (1) above but with limited treatment during use within the farm, e.g. oxygenation, mechanical filtration, biological filtration.

(3) Predominantly recirculated – supply as (1) above but with extensive reliance upon sophisticated water treatment plant and only 5-10 per cent per day input of new water to the system.

The choice will in most cases be determined by the availability to the farmer of water, heat and land. For example, a manufacturing industry wishing to utilise its 'waste' heat might be able to operate a single pass system. Heat pumps or exchangers would be used if the heat was in an inappropriate medium for aquatic life. In contrast a company with only small amounts of heat or good water would be obliged to consider a recirculation system, if necessary using artificial sea water.

It is expected that most crustacean aquaculture ventures in Britain will opt for indoor or covered cultures, possibly using a high degree of recirculation since these have a number of demonstrable advantages over uncovered and through flow systems:

(1) optimal conditions for growth can be permanently maintained;
(2) stock maintenance is greatly simplified and close control can be exerted over feeding, harvesting and disease;
(3) there is less reliance on large quantities of water from natural sources which are always vulnerable to contamination;
(4) the system is particularly suitable for automation and the culture of a number of species;
(5) there are no predator or poaching problems.

A flow diagram of a generalised recirculation system modified from those used at Conwy for the culture of lobsters and prawns is shown in Figure 2.23. A water treatment plant suitable for intensive culture of Crustacea must achieve the stabilisation of chemical changes that occur during culture by (1) replacement of depleted substances (e.g. oxygen), (2) removal or detoxification of substances which accumulate (e.g. ammonia and dissolved organic matter), (3) control of levels of suspended particles.

Plant design will include consideration of three main factors:

(1) quantitative prediction of the changes that occur;
(2) estimation of the tolerance of the cultured species to these changes;
(3) methods of water treatment.

5.1 Changes that Affect the Quality of Recirculated Water

During culture, dissolved oxygen, some anions (for example carbonate and bicarbonate) and cations (magnesium, calcium) become depleted in recirculated water, while ammonia, nitrite, nitrate, phosphate, carbon dioxide, dissolved organic compounds and particulates accumulate.

Data on the approximate oxygen consumption and excretion of ammonia nitrogen by some Crustacea under laboratory conditions are given in Table 2.11. Smaller individuals consume more oxygen and excrete more nitrogen per unit weight than larger animals and, using prawns as an example, the oxygen consumed by forty 20 g prawns is approximately $800 \times 0.3 = 240$ mg O_2/hr, while 800 g of small prawns (5 g) would consume $800 \times 2 = 800$ mg O_2/hr. The production of carbon dioxide probably also differs with size of animal, and in shallow water cultures (15 cm) an excess is easily removed by moderate agitation (see below). Not all the changes are due entirely to the cultured animals. Some, notably the production of nitrite, nitrate and hydrogen ions which cause the loss of bicarbonate and carbonate (buffering capacity) are generated by the activities of autotrophic nitrifying bacteria living in the biological filter. Comprehensive reviews of nitrification and nitrifying bacteria have been published: Painter (1970); McCarty and Haug (1971); Sharma and Ahlert (1977).

A proportion of the oxygen consumed and the ammonia and carbon dioxide released will be as a result of the activities of heterotrophic microorganisms (bacteria, protozoa, small metazoa) in suspension and attached to surfaces throughout the culture system. Under some circumstances, particularly after feeding, the contribution from these sources can represent as much as 60 per cent of some chemical changes (e.g. oxygen consumed) produced in the water — more than that caused by the cultured organism (see Shigueno, 1975). Normally, lower levels would be expected.

Figure 2.23: A Flow Diagram of a Generalised Marine Recirculation System Based on Those Used at Conwy for the Culture of Prawns and Lobsters.

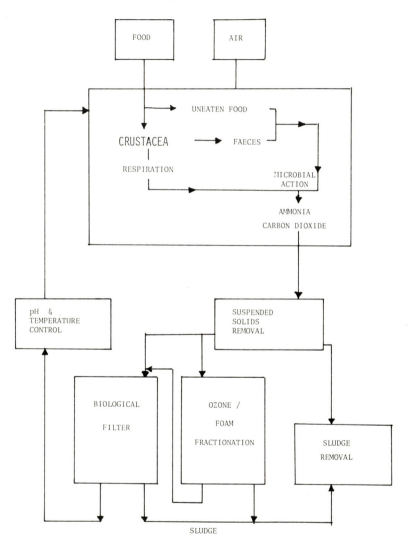

Table 2.11: Approximate Rates of (a) Oxygen Consumption (mg O_2/g wet weight/hr) and (b) Nitrogen — Mainly as Ammonia — Excretion (mg $NH_4^+ + NH_3$–N/g wet weight/day) by Crustacea. Published data have been recalculated where necessary for ease of comparison. Where data were expressed in terms of dry weight of animal a water content of 75 per cent has been assumed.

(a) Oxygen Consumption

Species	Temp. (°C)	Live Weight (g)	Consumption (mg O_2/g/hr)	Author
Natantia				
Macrobrachium rosenbergii	28	0.04-2.40	0.3-0.4	Nelson *et al.* (1977a)
Penaeus spp.	28	0.1-5.0	1-2	from Wickins (1976a)
Reptantia — lobsters				
Homarus gammarus	18	above 450	0.11	Thomas (1954)
H. gammarus	15	above 600	0.02-0.06	Spoek (1974)
H. americanus	15	below 100	0.06-0.09	McLeese (1964)
Reptantia — crayfish				
Pacifastacus leniusculus	20	3	0.4-0.5	Moshiri &
P. leniusculus	20	84	0.04	Goldman (1969)
Orconectes limosus	18	6-10 mm CL	0.23	Kossakowski &
		46-50 mm CL	0.15	Orzechowski (1975)

(b) Nitrogen Excretion

Species	Temp (°C)	Live Weight (g)	Excretion (mg N/g/day)	Author
Natantia				
Macrobrachium rosenbergii	28	2-27	0.85-0.25	from Wickins (1976a)
M. rosenbergii	28	4	0.2	Nelson *et al.*
M. rosenbergii	28	12	0.05	(1977b)
Penaeus spp.	28	0.01-5.00	1.0	from Wickins
		10-20	0.5	(1976b)
Reptantia — lobsters				
Homarus americanus	22	0.04	1.2	Delistraty *et al.* (1977)
H. americanus	22	0.7	0.3	Logan & Epifanio (1978)
H. americanus	—	0.7-400	0.24	Sastry & Barton (1975)
Reptantia — crayfish				
Orconectes rusticus	22-25	7-10	0.4	Sharma (1966)

The amount of ammonia nitrogen it was necessary to oxidise in a small-scale lobster culture unit holding 50 kg of lobsters was calculated by Richards and Wickins (1979): if lobsters excrete 0.4 g ammonia nitrogen/kg/day then 50 kg of lobsters excrete 50 x 0.4 = 20 g/N/day. Trials showed that when food (mussel meat containing 16 g N/kg wet weight) was given at a rate of 10 per cent of the lobsters' body weight/day, on average half of this (equivalent to 40 g N) was eaten. Of the remaining half, 80 per cent could be removed manually after 24 hrs and 20 per cent was lost within the culture system. Assuming that all the nitrogen in the 'lost' food (mussel tissue) was converted to ammonia, then some additional 8 g of ammonia nitrogen (28 per cent of the total) would be produced. This proportion was similar to the 30.8 per cent found by Gerhardt (1978) during the laboratory culture of juvenile *Penaeus indicus*. It is therefore of paramount importance to ensure that the design of the treatment plant is sufficient to cope with, and is in sympathy with, the changes in water condition caused by necessary husbandry practices — feeding, cleaning and disease prevention.

Perhaps the most vital single component for the stabilisation of these changes is an efficient means of continuously removing suspended solids, for example by mechanical filtration or foam floatation. There is, of course, no doubt that efficient feeding practices with properly prepared and bound diets are prerequisites for economical plant operation. Examples of the amounts of suspended solids, dissolved organic matter and bacteria produced in laboratory scale intensive prawn cultures at Conwy are shown in Table 2.12.

Table 2.12: Levels of Particulates and Dissolved Organic Matter in Laboratory Recirculation Systems in which 44-2,400 g of Penaeid Prawns were Grown at 28°C for Periods of 28-196 Days

	Mean	Range	No. of Trials
Load [a] (g/l)	0.096	0.012-0.166	5
Total wet wt of food given (kg)	42.9	1.4-73.8	5
Weight of particulates above 1 μm (mg/l)	6.56	2.25-10.55	5
Weight of sub-micron particulates (mg/l)	14.8	3.33-38.90	4
Dissolved organic carbon (mg/l)	9.84	6.37-14.60	4
Bacteria/ml x 10^3	3,484	1,251-7,527	3

Note: a. Load is calculated from the weight of animals held, divided by the total volume of water used throughout the culture period.

The dissolved organic carbon content increased from 4 to 16 mg C/l with increasing load in nine marine recirculation systems at 28°C according to the equation

$$\log_e C = \log_e 0.0039 + 0.2748 L$$
$$r = 0.8118 \, df = 10$$

where C = dissolved organic carbon mg/l, and L = animal load (see Table 2.12 caption for definition).

Figure 2.24: The Relation between the Total Dry Weight of Food Given Per Day and the Concentration of Filterable Solids (above 0.22 μ) Produced in Marine Recirculation Systems Containing Tropical Prawns.

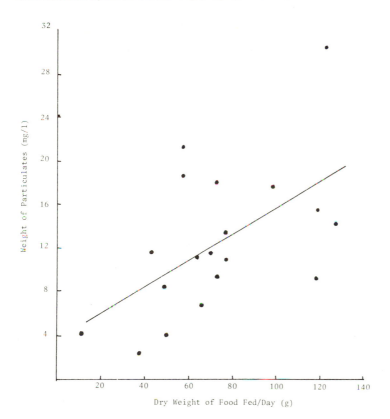

Natural sea water at Conwy normally contains 1-5 mg C/1 with a maximum of about 20 mg C/1 (Wickins and Helm, 1981). Concentrations of particulate matter varied widely but were proportional to the dry weight of food fed to the prawns in the systems (Figure 2.24). In these trials the prawns were fed for convenience once each day with fresh foods or a pelleted diet supplemented with fresh food twice each week. Feeding was therefore wasteful and gave rise to periodic contamination of the water.

5.2 Tolerance

Estimates of the tolerance of Crustacea to changes in water chemistry are essential to determine the maximum and minimum levels that will permit economic operations. Such levels are called 'acceptable' levels and are usually estimated from conventional acute and chronic toxicity tests. These tests may measure lethal or sub-lethal effects. In acute lethal tests the animals are exposed for periods of 48-96 hours and survivors placed in clean water for a further five days to check for delayed mortality. Calculation of acceptable levels from acute lethal tests involves the use of an arbitrary application factor which is assigned on the basis of judgment of scientists (Sprague, 1971). The judgment is often based on existing data for other chemicals. Such factors have no sound technical basis (Perkins, 1976) and can be misleading if used to formulate water quality standards. They can be used, however, in initial calculations of maximum or minimum acceptable levels in order to estimate the margin of safety in any given situation (Lloyd, 1979).

In chronic sub-lethal tests physiological and behavioural responses are measured — food consumption, growth rate, activity. Chronic tests lasting several months can give a good practical guide to 'safe' levels conveniently expressed in terms of the calculated concentration that affects only 1 per cent of the population in a specified time. In aquaculture the specified period would ideally be equal to the grow-out time. Single values of maximum or minimum acceptable concentrations might, however, be inappropriate as criteria, particularly in long-term culture operations (e.g. lobsters), and it might be better to express acceptable levels as a minimum percentile distribution over a known period, six months or one year (Alabaster, 1974). Extensive reviews of toxicity tests and their application are given by Sprague (1969, 1970, 1971), Perkins (1976) and Lloyd (1979).

Most of the published water quality standards have been determined in order to maintain fisheries and an ecological balance in the natural environment. In aquaculture situations practical experience with fish shows that many of the published standards are not only expensive to achieve, but are unnecessary in that production and product quality can be maintained at much higher levels of contamination. In addition, the characteristics considered by aquaculturists are generally fewer and sometimes different (e.g. NO_2, Ca levels) from those given priority by fishery biologists and environmentalists (Wickins, 1981). Most commonly considered characteristics are oxygen, ammonia, nitrite, nitrate, carbon dioxide, pH, alkalinity, suspended solids, dissolved organic and inorganic substances. In controlled environment cultures temperature and salinity are assumed to be controlled by the operator and will not be considered further; contamination by heavy metals and pesticides is likely to be site-specific and, likewise, is not considered here. Recommended levels for seven factors apparently satisfactory for Crustacea are shown in Table 2.13. There is a shortage of data on minimum and maximum acceptable levels based on long-term growth studies.

Table 2.13: Some Examples of the Levels of Various Water Quality Parameters for Cultured Crustacea

Parameter	Desirable Level or Range	Species	Author
Oxygen	above 4.65 mg/l @ 33°C	*Macrobrachium rosenbergii*	TVA (1978)
Oxygen	above 85% saturation	*Penaeus japonicus*	Liao (1969)
Oxygen	above 70% saturation	*Cherax tenuimanus*	Morrissy (1979)
Calcium	100 mg/l, approx 200 mg/l	*Metapenaeus bennettae*	Dall (1965)
Ca CO$_3$	65-200 mg/l	*Macrobrachium rosenbergii*	Hanson & Goodwin (1977)
Total hardness	50-200 mg/l	*Procambarus clarkii*	de la Bretonne *et al.* (1969)
pH	6.7-7.0	*P. clarkii*	de la Bretonne & Avault (1971)
pH	7.0-8.0	*Austropotamobius pallipes*	Jay & Holdich (1977)
Unionised ammonia	below 0.14 mg/l[a]	*Homarus americanus larvae*	Delistraty *et al.* (1977)
Unionised ammonia	0.09-0.11	*Penaeus* spp.	Wickins (1976b)
Nitrite	below 1.4	*Macrobrachium rosenbergii* larvae	Armstrong *et al.* (1976)
Nitrate	below 100-200	*Penaeus monodon*	Wickins (1976b)

Note: a. from 8 days' tests only.

For long-term toxicity tests with Crustacea to be meaningful it is judged that experiments should be long enough to allow the animals to pass through at least five moults. Clearly this would be impracticable with large decapods moulting once every six or twelve months. As in the natural environment, the effects of each water quality characteristic in a culture system are likely to be influenced by the other characteristics, sometimes favourably, sometimes unfavourably, for the culturist. In controlled environment cultures a considerable amount of control over water quality is implied and deficiences in one factor may be made good to some extent by optimisation of another. A simple example is given by the operator of a recirculation system who, on observing a rise in total ammonia concentration, causes the pH to fall, thus ensuring that a greater proportion of the total ammonia is ionised (NH_4^+) and therefore non-toxic (see Sousa *et al.*, 1974). Obviously this procedure treats only the symptom — the rise in ammonia — not the cause, but it can allow time for remedial action to be taken to prevent harm to the stock.

5.3 Water Treatment

Most important in crustacean cultures is an adequate removal of suspended solids, which is necessary to prevent an excessive increase in the biochemical oxygen demand of the water (BOD); an increase in BOD increases the amount of ammonia that has to be oxidised by the biological filter and of carbon dioxide that has to be lost to the air. In controlled environment cultures it is expected that settlement tanks and compact sedimentation devices may be used as they are in some fish farms. It is likely that more energy-intensive methods — mechanical filtration, foaming — will also be considered. A number of methods for particle removal in marine and freshwater systems have been reviewed by Kinne (1976), Muir (1978) and Wickins and Helm (1981).

Maintenance of levels of dissolved oxygen is vital for good growth and survival in Crustacea. In many fish farms oxygen levels are maintained by an adequate exchange rate of well-aerated water. The culture of Crustacea in shallow waters (up to 30 cm depth) maximises the benefits of gaseous exchange at the water surface and with moderate agitation can reduce the requirement for high water exchange rates. In deeper water (more than 100 cm depth) tank design must be such that dissolved oxygen levels can be maintained by a balance between water exchange rate and aeration or oxygenation. It is important to avoid high current speeds because newly moulted individuals are readily swept away and damaged.

The amount and rate of ammonia oxidation in a nitrifying filter is dependent on:

(1) input concentration of ammonia nitrogen;
(2) specific surface area of the filter matrix;
(3) filter volume;
(4) hydraulic load;
(5) pH, temperature, salinity and oxygen levels.

For more detailed consideration of the operation of nitrifying filters in aquaculture the reader is referred to Wickins and Helm (1981). Estimates of the rates of nitrification that can be expected from marine biological filters (see Figure 2.25) at 20° and 28°C are shown in Table 2.14. Rates for gravel and plastic media range from 0.03 to 0.84 g N oxidised/m^2 of filter surface area/day. Under routine operating conditions levels of ammonia and nitrite remain low while phosphate and nitrate increase. Examples of the levels that occurred in recirculation systems at Conwy are shown in Tables 2.15 and 2.16. Mean levels were: 0.02-0.39 mg total NH_4-N/1; 0.01-0.44 mg NO_2-N/1; 3.07-54.76 mg NO_3-N/1 and 0.21-4.04 mg PO_4-P/1. In all these systems the volume of the water treatment plant (reservoirs plus percolating filter) ranged from 22 to 85 per cent of the total water volume in the system. It was necessary in some of the systems to employ high rate mechanical filters for several hours each day to reduce levels of particulate matter. In addition some 30-50 per cent of the water in the systems was renewed once or twice each week; calculated on a daily basis this was 4-14 per cent renewed/day.

Higher nitrification rates and thus economy of filter size may be achieved with elevated input concentrations of ammonia. Concurrently, it would be necessary to minimise the toxic effect of similarly elevated levels of unionised ammonia

Figure 2.25: Model Recirculation Systems Used to Study Changes in Sea Water Chemistry Caused by Biological Filtration.

Source: MAFF, Conwy (1977).

Table 2.14: Examples of Rates of Nitrification in Marine Biological Filters

Input Concentration (mg N/l)	Hydraulic Load (m^3/m^3/day)	Temp. (°C)	Media and Specific Surface Area (m^2/m^3)	Nitrification Rate (g N Oxidised /m^2/day)	Author
Model systems (no animals)					
10	276	28	Plastic 164	0.28	Original
50 (initial dose)	276	27	Plastic 164	0.49	Original
1	80	26	Gravel 350	0.32	Forster (1974)
1	246	26	Gravel 350	0.62	Forster (1974)
Systems containing Crustacea					
0.1	95	20	Plastic 164	0.22	Richards & Wickins (1979)
approx. 0.1-0.3	114	20	Plastic 164	0.14	Richards et al. (in prep.)
approx. 0.2	182	28	Plastic 164	0.03-0.38	Original
approx. 0.28	153	28	Plastic 164	0.08-0.39	Original
approx. 0.39	26	28	Gravel 200	0.03-0.10	Original
?	360	20	Gravel 210	0.84	Goldizen (1970)

Table 2.15: The Levels and Ranges of Concentration of Total Ammonia Nitrogen in Marine Recirculation Systems Used for Nursery and Maintenance of Lobsters and for Grow-out Trials. Temperate-water system (20° C) holding 290-400 g of juvenile lobsters for periods of 44-186 days.

	Nursery (Load Normally Constant)		Grow-out (Load Increasing)		
System Number	7a	7b	8a	8b	9
Load[a]	0.011	0.103	0.036	0.100	0.026
pH	8.16	8.02	8.10	8.02	8.09
NH_4 –N (mg/l) mean	0.07	0.02	0.07	0.03	0.05
Range	0.01-0.21	0.01-0.05	0.01-0.18	0.03-0.04	0.03-0.10
NO_2 –N (mg/l) mean	0.06	0.01	0.06	0.02	0.03
Range	0.01-0.60	0.01-0.02	0.01-0.36	0.02-0.03	0.01-0.11

Note: a. Calculated as described in Table 2.12.

by maintaining depressed pH levels in the water. Since buffering capacity tends to be reduced in recirculation systems (see below) precise control of pH could be difficult. Equipment for the reliable control of low pH levels under commercial aquaculture conditions is likely to be expensive.

In heavily loaded marine recirculation systems the tendency for inorganic carbon (buffering capacity) to be lost is due to a change in the bicarbonate equilibrium caused by the production of H^+ ions during nitrification. The reactions are summarised (McCarty and Haug, 1971):

$$55\,NH_4{}^+ + 5\,CO_2 + 76\,O_2 \xrightarrow{Nitrosomonas} C_5H_7O_2N + 54\,NO_2{}^- + 52\,H_2O + 109\,H^+$$

$$400\,NO_2{}^- + 5\,CO_2 + NH_4{}^+ + 195\,O_2 + 2\,H_2O \xrightarrow{Nitrobacter} C_5H_7O_2N + 400\,NO_3{}^- + H^+$$

$$4\,H^+ + 2\,CO_3{}^{2-} \longrightarrow 2\,H^+ + 2\,HCO_3{}^- \longrightarrow 2\,H_2CO_3 \longrightarrow 2\,H_2O + CO_2\uparrow$$

Restoration of the buffering capacity of marine systems may be made by the addition of sodium hydroxide (indirect restoration) or sodium carbonate and bicarbonate (direct restoration). Examples of the amounts added to maintain reasonable levels of pH in warm sea-water recirculation systems at Conwy are shown in Table 2.17. The uses of limestone and shell as carbonate sources do not give predictable results in heavily loaded marine systems because they tend to become covered with organic matter or an insoluble coating of apatite (Siddall, 1974).

Table 2.16: The Levels and Ranges of Concentration of Total Ammonia, Nitrite and Nitrate Nitrogen and Inorganic Phosphate Phosphorus in Marine Recirculation Systems Used for Maintenance and for Grow-out Trials. Warm-water systems (28° C) holding 400 to 2,600 g of juvenile and adult prawns for periods of 90-352 days.

System Number	Stock Maintenance (Load Normally Constant)					Grow-out (Load Increasing)		
	1a	1b	1c	2	3	5	6	Beard & Forster (1973)
Load[a]	0.112	0.192	0.116	0.029	0.026	0.166	0.103	0.102
pH	7.64	8.41	7.55	7.98	8.04	7.90	8.06	6.80
NH_4-N (mg/l) mean	0.22	0.13	0.23	0.21	0.22	0.28	0.13	0.39
Range	0.12-0.29	0.03-0.37	0.06-0.58	0.11-0.78	0.09-0.81	0.11-0.46	0.04-0.34	0.15-0.61
NO_2-N (mg/l) mean	0.10	0.03	0.14	0.05	0.07	0.27	0.44	0.32
Range	0.02-0.51	0.01-0.13	0.00 (4) 0.33	0.02-0.07	0.03-0.13	0.09-0.64	0.05-3.52	0.08-0.60
NO_3-N (mg/l) mean	15.23	42.61	10.05	3.16	3.07	—	—	54.76
Range	0.20-32.38	0.24-111.06	4.98-18.36	2.16-4.47	2.05-4.31	—	—	10.90-94.60
PO_4-P (mg/l) mean	1.46	1.18[b]	4.04	0.22	0.21	3.71	—	—
Range	0.36-2.93	0.20-2.28	2.31-5.48	0.14-0.34	0.10-0.37	1.45-8.19	—	—

Notes: a. Calculated as described in Table 2.12.
b. Analysis did not cover entire period.

Table 2.17: Some Examples of the Weights of Sodium Hydroxide, Carbonate and Bicarbonate Added to Six Marine Recirculation Systems in Order to Maintain Reasonable Levels of pH. The systems were initially stocked with between 6 and 2,600 g of tropical prawns (size range 0.02-50 g). Temperature and salinity were 28 ± 2° C and 30 ± 2°/oo respectively.

	Stock Maintenance (Load Normally Constant)						Grow-out (Load Increasing)	
System number	1a	1b	1c	2	3	4	5	6
Duration (days)	90	211	352	125	125	231	112	196
Load [a] (g/l)	0.112	0.192	0.116	0.029	0.026	0.018	0.166	0.103
System volume (l)	10,425	10,425	10,425	2,784	2,784	8,909	1,253	885
Filter media Specific surface area approx. 190 m^2/m^3	1:1 Mixture silica gravel and dolomite chips						Plastic rings	
Mean pH	7.64	8.41	7.55	7.98	8.04	8.07	7.90	8.06
Weight of sodium hydroxide used (g)	—	3,566	—	108	103	7,656	1,581	—
Weight of carbon used (sodium carbonate and bicarbonate) (g)	—	964	768	9	12	—	30	321

Note: a. Calculated as described in Table 2.12.

The potential of denitrification plants for the control of nitrate levels and stabilisation of pH in closed systems has been demonstrated in brackish water fish cultures by Otte and Rosenthal (1978) and in freshwater by Meade (1974). Sophisticated control of the process, which is anaerobic and requires a carefully metered supply of carbon (e.g. methyl alcohol) is crucial. pH is restored by the production of hydroxyl ions (McCarty and Haug, 1971).

$$NO_3^- + \frac{5}{6} CH_3OH \longrightarrow \frac{1}{2} N_2 + \frac{5}{6} CO_2 + \frac{7}{6} H_2O + OH^-$$

Unless water supply was severely limited and nitrate accumulation became a problem, simpler chemical methods of pH control would seem more appropiate in a commercial setting (Westers and Pratt, 1977).

Short-term instability of pH was noticed in a number of laboratory cultures, particularly in small systems (200 1 capacity) and those heavily stocked with rapidly growing juvenile prawns. Part of the instability may have been due to the periodic excessive production of carbon dioxide following feeding. Some of this was readily removed by vigorous aeration (Table 2.18) which resulted in higher and slightly less variable pH values.

Table 2.18: The Effect of Aeration on the pH of Sea Water in Four Marine Recirculation Systems Each of 200 l Capacity. The experiment lasted 8 weeks; temperature $29 \pm 2^\circ$ C, salinity $30 \pm 2\%$.

System Number	1	2	3	4
Aeration (l/min)	14.1	13.8	4.5	4.4
pH mean	8.18	8.11	8.00	7.97
Interval of measured pH range	0.13	0.32	0.36	0.39
95% confidence limits of mean	8.17-8.19	8.08-8.15	7.98-8.03	7.94-8.01

Perhaps the most significant technological advance for water treatment in aquaculture has been the successful combination of ozone with foam fractionation columns. Foam fractionation is sometimes referred to as dispersed air floatation, air stripping, foaming or protein skimming (see Spotte, 1970). Small air bubbles are introduced into a column of water and as they rise a skin of surface-active micro-particulate and dissolved organic substances form at the air-water interface. At the surface the skin forms a foam which is discharged to waste. When used in conjunction with ozone the treatment is effective in:

(1) oxidising refractory organic substances and toxic metabolites, particularly nitrite;
(2) stabilising fluctuations in C, N and pH levels;
(3) controlling the growth of microorganisms.

There is some evidence (see Kinne, 1976; Rosenthal, 1974) that trace elements, e.g. manganese and heavy metals, may be removed from solution and that potentially toxic intermediary, primary ozonides may be formed in sea water. Hypobromous acid and other bromine compounds may also be formed. Useful diagrams of foam separation equipment are provided by Sander and Rosenthal (1975) and Kinne (1976).

The successful operation of ozone-foam fractionation in experimental fish culture units in fresh, brackish and sea-water systems has recently been demonstrated (Honn and Chavin, 1975, 1976; Otte and Rosenthal, 1978; Rosenthal *et al.*, 1978a, b; Rosenthal and Otte, 1978), while studies made with Crustacea (*H. americanus*) are beginning at San Diego State University, USA (J.M. Carlberg and J.C. Van Olst, 1979, pers. comm.).

Ozone-foam fractionation has exciting possibilities as a water treatment process but it might not be the panacea for all aquaculture treatment problems. It is expensive to operate and requires sophisticated control gear to prevent the escape of free ozone and toxic by-products to the culture tanks. It has not, as far as Crustacea are concerned, been tested in sea water systems.

Treatment methods now available enable significant economies of water to be made in small laboratory systems where precise control can be maintained over husbandry, feeding and water chemistry. Some considerable R & D effort by hydraulic and chemical engineers in the design and operation of scaled-up treatment units will be required before similar efficiencies of water reuse can be approached in commercial crustacean cultures. This is largely because of the present high capital and running costs of water treatment plant currently available. Indeed, most equipment was not designed for use under aquaculture conditions.

6. Conclusions

A number of Crustacea are unsuitable for cultivation in Great Britain at the present time for biological or economic reasons. These include small shrimps and prawns, scampi, crawfish, some freshwater crayfish and crabs.

Species currently under investigation for culture include some exotic inshore and brackish water prawns, lobsters and some crayfish. Methods of culture that are being considered for possible use in Great Britain include culture-based fisheries or sea ranching, extensive and semi-intensive marine and freshwater cultures and intensive controlled environment cultures.

Only lobsters are likely to be suitable for sea ranching. On the basis of past and current work, however, it is concluded that any further attempts to promote programmes for the release of juveniles into enclosed or protected sea areas should be preceded by a demonstration to the funding agency that a return of data would be continuously forthcoming about the released juveniles. How such monitoring might be done is not at all clear. The use of divers and possibly the enclosure of large areas with protective fences (Davies *et al.*, 1980) might be feasible but would be expensive.

In Britain, at least, it seems most unlikely that extensive or semi-intensive outdoor culture methods operating at ambient temperatures could be profitable without the support of a very specialised and high-priced market for the crustacean product. The majority of farmed fish occupy the whole water volume and their commercial culture at ambient temperatures succeeds because they can be grown in water deep enough or shallow waters flowing fast enough to prevent intolerable fluctuations in temperature. Decapod crustaceans, particularly soft, newly-moulted individuals, do not withstand the high current speeds that would probably be required in shallow water to maintain satisfactory environmental conditions and deep waters (more than 0.5-1.0 m) would be more efficiently employed for fish rather than crustacean culture.

There may be some scope for polyculture where significant quantities of Crustacea (prawns, lobsters or native crayfish) could be reared from the same facility as another species (e.g. salmonids) provided little or no extra labour costs or risks to the main fish crop were incurred.

Assuming fish/crayfish polyculture was practical and profitable, there are a number of reasons why the native crayfish, *Austropotamobius pallipes*, should be considered in preference to imported species such as *Pacifastacus leniusculus* or *Astacus leptodactylus*, particularly for projects operated on a continous 'stock and cull' basis. Not least is that it is well adapted to the British environment. The only reservation might be the market value of the crop in Scandinavia if British markets for crayfish did not develop. Certainly the risks attached to importing and releasing in Britain species that are in many ways ecological homologues of *Austropotamobius pallipes* have not been demonstrably justified in terms of profitability. No one seems to have even compared the culture performance of these species under defined conditions. Yet, until this is done or prohibitive legislation appears, there remains a risk of antagonism between businessmen and environmentalists (see Karlsson, 1978 and Bowler, 1978).

Semi-intensive or intensive culture of Crustacea in heated effluent would seem more attractive provided the problems of continuity of heat supply, reduced growth and the higher risk of mortality in winter could be overcome. In this context the approach adopted in the Clark's Harbour project (Bishop and Castell, 1978) would seem promising. Again, culture with or adjacent to an established culture practice, though not necessarily aquaculture, would seem a promising way forward.

Controlled environment cultures are, in part, analogous to a manufacturing process but differ because of the higher risks involved in an aquaculture enterprise. On paper the prospects for control over production seem very attractive; there is also considerable scope for patenting novel processes and tank designs. Yet the capital costs of setting up a pilot plant to make a commercial assessment on a scale large enough to convince investors of its potential seems prohibitive. British economic appraisals seem traditionally to concentrate on aiming production at an existing market with little thought of innovation or consideration of alternative markets including exports. As a result it is felt that crop values consistently tend to be underestimated. This argument applies more to the larger companies who would have to operate on a considerable scale (perhaps 200 t/yr) to cover overheads. In contrast smaller concerns aiming at 10 t/yr production, more often than not, over-estimate crop value. Another point is the realisation that many of the early pilot plants and even those being proposed use conventional tank designs, thus postponing the consideration of a crucial issue, namely economy of space and capital. It would seem likely that, in contrast to fish cultures, the development of a novel tank/management system will confer a major economic and commercial advantage on its designers. Progress in this area will depend upon a multi-disciplinary approach where husbandry, science and engineering skills are brought together, probably under the financial initiative of industry.

The question of which species to farm and which would be the best way to farm them are influenced in many cases by the facilities available to each potential culturist. These range from a source of waste heat, unwanted, potential food ingredients, land, buildings, water, past experience or expertise to a marketing network. Because of such considerations no simple answers can be given. It is my belief that in time controlled environment culture using a variety of species (the most promising is *Penaeus monodon – Figure 2.26*) will become commercially viable.

Figure 2.26: A Year-old Tiger Prawn (*Penaeus monodon*) Grown in a Recirculation System at Conwy, Weight About 90 g.

Source: Photograph courtesy of D.R. Arnstein, Conwy (1974).

Although aquaculture is unlikely to make a significant contribution to Britain's food supply for many years yet, it can enlarge the range of consumer choice and could stimulate new sources of wealth as older industries decline and trade in traditional manufacturing industries become more competitive (McAnuff, 1979). Expectations of early profitability will probably not be realised and repeated disappointments may result in the loss of public and private financial funding vital to the development of aquaculture (Shell, 1978). Indeed, the potential of controlled environment systems will not be realised if progress in technology is not matched by vigorous commercial exploitation. One study concluded that at present in Britain the fastest growth in the aquaculture industry was the sale of technical knowledge, consultancy and insurance rather than the expansion of productive operations (Pullin, 1977).

A number of areas where further research seems worthwhile can be defined in the light of the results reported here. These are:

(1) Tank and container design.
(2) Dietary requirements and the husbandry of feeding.
(3) Reduction of fluctuations in water quality in recirculation systems — applicability of ozone and denitrification treatments to marine systems.
(4) Hybridisation of species for controlled environment cultures, particularly lobsters (see Hedgecock *et al.*, 1977).
(5) Acclimatisation of *Penaeus monodon* to fresh water as recently claimed in the Philippines (Pantastico, 1979). This would immediately open up a large number of opportunities for intensive prawn culture at inland sites in Britain using waste heat from manufacturing processes.
(6) The use of artificial sea water in controlled environment cultures.
(7) Monosex cultures, which would maximise on changes of morphological characters at sexual maturity and differences in growth rate between sexes.
(8) Determination of minimum water quality standards required by each species under commercial culture conditions and associated pathological changes.
(9) Disease prevention and control.
(10) The creation of markets for unfamiliar products.

Acknowledgements

I am particularly grateful to specialists and colleagues at home and overseas who supplied me with information and photographs. I regret that not all the material received could be included. It is a pleasure to thank my colleagues within the Directorate of Fisheries Research (Ministry of Agriculture, Fisheries and Food) for advice, criticism and help in the preparation of the drafts. I am grateful to Dr D. Holdich, Nottingham University, for help with the crayfish sections.

Although much of the material in this chapter derives from my employment by the Ministry of Agriculture, Fisheries and Food at the Fisheries Experiment Station, Conwy, and is published with the consent of the Ministry, the compilation and presentation of it here and the views expressed are entirely the author's responsibility.

References

A number of items of limited availability have been included (e.g. theses, technical reports) since photocopy technology and the system of inter-library loans mean that few are completely out of reach.

Abrahamsson, S. (1971). Population ecology and relation to environmental factors of *Astacus astacus* Linné and *Pacifastacus leniusculus* Dana. Summary of dissertation, University of Lund, Sweden, 11 pp. (mimeo)

———— (1973). Methods for restoration of crayfish waters in Europe. The development of an industry for production of young of *Pacifastacus leniusculus* Dana. In: Abrahamsson, S. (ed.), *Freshwater Crayfish, 1.* Studentlitteratur, Lund, pp. 203-10

Alabaster, J.S. (1974). The development of water quality criteria for marine fisheries. *Ocean Management, 2:* 101-15

Alikunhi, K.H., Poernomo, A., Adisukresno, S., Budiono, M. and Busman, S. (1975). Preliminary observations on induction of maturity and spawning in *Penaeus monodon* Fabricius and *Penaeus merguiensis* de Man by eye-stalk extirpation. *Bull. Shrimp Cult. Res. Cent., I (1):* 1-11

Allen, J.A. (1967). *The Fauna of the Clyde Sea Area, Crustacea: Euphausiacea and Decapoda.* Scott. Mar. Biol. Ass., Millport, 116 pp.

Andrews, E.A. (1907). The young of the crayfishes *Astacus* and *Cambarus.* Smithsonian Institution, Washington, DC, 88 pp. (cited in Abrahamsson, 1971)

Anon. (1977a). Biology and farming of the Yabbie (*Cherax destructor*). Dept of Agriculture and Fisheries, Adelaide, South Australia, Fisheries Division Pamphlet, 8 pp.

——— (1977b). The profitable scavenger. *Fish Farmer, 1:* 20-1

——— (1978). Norway lobster project. *Fishing News Int., 17 (1):* 6

——— (1979). Signal crayfish are set for take-off in the south. *Fish Farmer, 2 (4):* 26-7

AQUACOP (1977a). *Macrobrachium rosenbergii* (de Man) culture in Polynesia: progress in developing a mass intensive larval rearing technique in clear water. Proc. 8th Ann. Wkshop Wld. Maricult. Soc., San José, Costa Rica, 9-13 January, pp. 311-26

——— (1977b). Reproduction in captivity and growth of *Penaeus monodon* Fabricius in Polynesia. Proc. 8th Ann. Wkshop Wld Maricult. Soc., San José, Costa Rica, 9-13 January, pp. 927-45

Armstrong, D.A., Stephenson, M.J. and Knight, A.W. (1976). Acute toxicity of nitrite to larvae of the Giant Malaysian prawn, *Macrobrachium rosenbergii. Aquaculture, 9:* 39-46

Arnstein, D.R. and Beard, T.W. (1975). Induced maturation of the prawn *Penaeus orientalis* Kishinouye in the laboratory by means of eyestalk removal. *Aquaculture, 5:* 411-12

Arrignon, J. (1975). Crayfish farming in France. In: Avault, J.W. Jr. (ed.), *Freshwater Crayfish, 2.* Louisiana State University Press, pp. 105-16

Audouin, J., Campillo, A., Fouillard, R. and Gueguen, T. (1969). Expériences sur l'acclimatation de la langouste du cap *(Jasus lalandei* Milne-Edwards 1837) dans les eaux de la Côte Atlantique Française. *Rev. Trav. Inst. Pêches Marit., 33 (2):* 213-21

Audouin, J., Campillo, A. and Legise, M. (1971). Les cantonnements a crustaces des Cotes Françaises de l'Atlantique et de la Manche. *Sciences et Pêche. Bull. Inst. Pêches Marit., 205:* 91 pp.

Avault, J.W. Jr. (1973). Crayfish farming in the United States. In: Abrahamsson, S. (ed.), *Freshwater Crayfish, 1.* Studentlitteratur, Lund, pp. 239-50

Avault, J.W. Jr., de la Bretonne, L.W. and Huner, J.V. (1975). Two major problems in culturing crayfish in ponds: oxygen depletion and overcrowding. In: Avault, J.W. Jr. (ed.), *Freshwater Crayfish, 2.* Louisiana State University Press, pp. 139-44

Ayres, P.A. and Wood, P.C. (1977). The live storage of lobsters. Lab. Leafl., *MAFF Direct. Fish. Res., Lowestoft, 37:* 9 pp.

Bardach, J.E., Rhyther, J.H. and McLarney, W.O. (1972). *Aquaculture – The Farming and Husbandry of Freshwater and Marine Organisms.* Wiley-Interscience, New York, 868 pp.

Barker, P.L. and Gibson, R. (1977). Observations on the feeding mechanism, structure of the gut, and digestive physiology of the European lobster *Homarus gammarus* (L.) (Decapoda: Nephropidae). *J. Exp. Mar. Biol. Ecol., 26:* 297-324

Beard, T.W. and Forster, J.R.M. (1973). A growth experiment with *Penaeus monodon* Fab. in a closed system. ICES Shellfish and Benthos Committee CM 1973, Doc. K 39, 6 pp (mimeo)

Beard, T.W. and Wickins, J.F. (1980). The breeding of *Penaeus monodon* Fabricius in laboratory recirculation systems. *Aquaculture, 20,* 79-89

Beard, T.W., Wickins, J.F. and Arnstein, D.R. (1976). The breeding and growth of *Penaeus merguiensis* de Man in laboratory recirculation systems. *Aquaculture, 10:* 275-89

Behrendt, A. (1979). Export push needed to open market for crayfish. *Fish Farmer, 2 (2):* 44-6

Bishop, F.J. and Castell, J.D. (1978). Commercial lobster culture feasibility study, Clark's Harbour, Nova Scotia. *Fish. Mar. Serv. Industry Rept., 102:* 18 pp.

de Bondy, E. (1974). Etat des travaux de la generale d'aquiculture. *CNEXO Série: Actes de Colloques, No. 1.* Brest, France, 22-4 October 1973, pp. 419-29

Boschi, E.E. and Scelzo, M.A. (1976). El cultivo de camarones comerciales peneidos en la Argentina y la posibilidad de su produccion en mayor escala. *FAO Tech. Conf. on Aquaculture, Kyoto, Japan,* 26 May-2 June 1976. FIR: AQ/Conf/76 Doc E 40, 3 pp. (mimeo)

Bowler, K. (1978). Letters to the editor. *Fish Farmer, 2 (1):* 34-5

de la Bretonne, L., Jr. and Avault, J.W., Jr. (1971). Liming increases crawfish production. *Louisiana Agriculture, 15 (1):* 10-13

de la Bretonne, L. Jr., Avault, J.W. Jr. and Smitherman, R.O. (1969). Effects of soil and water hardness on survival and growth of the red swamp crawfish, *Procambarus clarkii,* in plastic pools. *Proc. 23rd Ann. Conf. Southeast Ass. Game and Fish Commissioners,* pp. 626-33

Brinck, P. (1977). Developing crayfish populations. In: Lindqvist, O.V. (ed.), *Freshwater Crayfish, 3.* University of Kuopio, Finland, pp. 211-28

Brown, J.H. and Aiken, D.E. (1976). Maintaining and hatching eggs of *Homarus* isolated from maternal care. 1. Evaluation of hatching techniques. Aquaculture Research Report No. 7. Dept of Environment, Biological Station, St Andrews, New Brunswick, Canada, 13 pp.

Brown, D.J. and Bowler, K. (1977). A population study of the British freshwater crayfish *Austropotamobius pallipes* (Lereboullet). In: Lindqvist, O.V. (ed.), *Freshwater Crayfish, 3.* University of Kuopio, Finland, pp. 33-49

Cabantous, M.A. (1975). Introduction and rearing of *Pacifastacus* at the research center of les Clouzioux 18450 Brinon S/Sauldre France. In: Avault, J.W. Jr. (ed.), *Freshwater Crayfish, 2.* Louisiana State University Press, pp. 49-55

Caillouet, C.W. Jr. (1972). Ovarian maturation induced by eyestalk ablation in pink shrimp, *Penaeus duorarum* Burkenroad. *Proc. 3rd Ann. Wkshop Wld Maricult. Soc.,* St Petersburg, Florida, USA, 26-8 January, pp. 205-25

Campillo, A. (1975). Contribution a l'étude de l'elevage de la crevette rose *Palaemon serratus* (Pennant) en captivite. *Rev. Trav. Inst. Pêches Marit., 39 (4):* 381-93

Carlberg, J.M., Van Olst, J.C. and Ford, R.F. (1978). Pilot-scale systems for the culture of lobsters in thermal effluent. *Power Plant Waste Heat Utilization in Aquaculture, Workshop 2.* New Brunswick, New Jersey, USA, 29-31 March 1978, 13 pp. (mimeo)

Carstairs, I.L. (1975). The dangers of exporting live 'Yabbies' (freshwater crayfish) (*Cherax destructor*). *Australian Society for Limnology Newsletter, 13 (1):* 1-5

Chamberlain, D. (1978). Marina with a big shellfish potential. *Fish Farmer, 1 (5):* 18-19

Clark, D.F., Avault, J.W. Jr. and Meyers, S.P. (1975). Effects of feeding, fertilization and vegetation on production of red swamp crayfish *Procambarus clarkii.* In: Avault, J.W. Jr. (ed.), *Freshwater Crayfish, 2.* Louisiana State University Press, pp. 125-38

Cobb, J.S. (1971). The shelter related behaviour of the lobster, *Homarus americanus. Ecology, 52:* 108-15

——— (1976). The American lobsters: the biology of *Homarus americanus. University of Rhode Island, Marine Technical Report 49:* 32 pp.

Conklin, D.E., Devers, K. and Bordner, C. (1977). Development of artificial diets for the lobster, *Homarus americanus. Proc. 8th Ann. Wkshop Wld Maricult. Soc.,* San José, Costa Rica, 9-13 January, pp. 841-52

Conklin, D.E. (1980). Nutrition. In: Cobb, J.S. and Philipps, B.F. (eds), *The Biology and Management of Lobsters.* Academic Press, London, vol. 1, pp. 277-300

Cook, H.L. (1976). Problems in shrimp culture in the South China sea region. FAO South China Sea Fisheries Development and Coordinating Programme. SCS/76/WP/40, Working paper No. 40, 29 pp.

Cook, H.L. and Rabanal, H.R. (1978). Manual on pond culture of penaeid shrimp. Report of Association of Southeast Asian Nations Seminar/Workshop on shrimp culture, Iloilo City, Philippines, 15-23 November 1976. ASEAN/77/SHR/CUL 3, 132 pp.

Cossins, A.R. (1973). *Thelohania contejeani henneguy,* microsporidian parasite of *Austropotamobius pallipes* Lereboullet. An histological and ultrastructural study. In: Abrahammson, S. (ed.), *Freshwater Crayfish, 1.* Studentlitteratur, Lund, pp. 151-64

Dabrowski, T., Kolakowski, E. and Burzynski, J. (1968). Studies on the nitrogen components composition of crayfish (*Astacus astacus* L.) meat as related to its nutritive value. *Pol. Arch. Hydrobiol., 15 (28), 2:* 145-52

Dabrowski, T., Kolakowski, E. and Sokolowski, E. (1966b). Zusammensetzung und nährwert des Krebsfleisches von *Astacus leptodactylus. Z. Lebensm. Untersuch. Forsch.,*

129: 337-44

Dabrowski, T., Kolakowski, E. and Wareszuk, H. (1966a). Studies on chemical composition of American crayfish (*Orconectes limosus*) meat as related to its nutritive value. *J. Fish. Res. Bd Can., 23 (11):* 1653-62

Dall, W. (1965). Notes on the physiology of a shrimp, *Metapenaeus* sp. (Crustacea: Decapoda: Penaeidae). V. Calcium metabolism. *Aust. J. Mar. Freshw. Res., 16:* 181-203

Davies, G., Dare, P.J. and Edwards, D.B. (1980). Fenced enclosures for the protection of seed mussels (*Mytilus edulis* L.) from predation by shore crabs (*Carcinus maenas* (L.)). Fish. Res. Tech. Rep., MAFF Direct. Fish. Res., Lowestoft, (56), 14 pp.

Delistraty, D.A., Carlberg, J.M., Van Olst, J.C. and Ford, R.F. (1977). Ammonia toxicity in cultured larvae of the American lobster (*Homarus americanus*). *Proc. 8th Ann. Wkshop Wld Maricult. Soc.,* San José, Costa Rica, 9-13 January, pp. 647-72

Delves-Broughton, J. and Poupard, C.W. (1976). Disease problems of prawns in recirculation systems in the UK. *Aquaculture, 7:* 201-17

Dexter, D.M. (1972). Moulting and growth in laboratory reared phyllosomes of the California spiny lobster *Panulirus interruptus. Calif. Fish and Game, 58 (2):* 107-15

Dorband, W.R., Van Olst, J.C., Carlberg, J.M. and Ford, R.F. (1976). Effects of chemicals in thermal effluent on *Homarus americanus* maintained in aquaculture system. *Proc. 7th Ann. Wkshop Wld Maricult. Soc.,* San Diego, California, 25-29 January, pp. 391-413

Egidius, E. (1978). Lobster import: two outbreaks of Gaffkemia in Norway. ICES Shellfish Committee CM 1978, Doc. K 17, 5 pp. (mimeo)

Ennis, G.P. (1973). Endogenous rhythmicity associated with larval hatching in the lobster *Homarus gammarus. J. Mar. Biol. Assoc. UK, 53:* 531-8

Farmer, A.S.D. (1974). Reproduction in *Nephrops norvegicus* (Decapoda: Nephropidae). *J. Zool., Lond., 174:* 161-83

Fischer, W. (ed.) (1973). FAO species identification sheets for fishery purposes, Mediterranean and Black Sea (Fishing Area 37) Vol. 1 and 2. Rome: *FAO,* Pag. Var.

Fisher, W.S., Nilson, E.H. and Shleser, R.A. (1975). Diagnostic procedures for diseases found in eggs, larval and juvenile cultured American lobsters (*Homarus americanus*). *Proc. 6th Ann. Wkshop Wld Maricult. Soc.,* Seattle, Washington, USA, 27-31 January, pp. 323-33

Fisher, W.S., Nilson, E.H., Steenbergen, J.F. and Lightner, D.V. (1978). Microbial diseases of cultured lobsters: a review. *Aquaculture, 14:* 115-40

Ford, R.F., Van Olst, J.C., Carlberg, J.M., Dorband, W.R. and Johnson, R.L. (1975). Beneficial use of thermal effluent in lobster culture. *Proc. 6th Ann. Wkshop Wld Maricult. Soc.,* Seattle, Washington, USA, 27-31 January, pp. 509-19

Forster, J.R.M. (1970). Further studies on the culture of the prawn *Palaemon serratus* Pennant, with emphasis on the post-larval stages. *Fishery Invest., Lond., Ser. 2, 26, (6):* 40 pp.

—— (1974). Studies on nitrification in marine biological filters. *Aquaculture, 4:* 387-97

Forster, J.R.M. and Beard, T.W. (1974). Experiments to assess the suitability of nine species of prawns for intensive cultivation. *Aquaculture, 3:* 355-68

Forster, J.R.M. and Wickins, J.F. (1967). Experiments in the culture of the prawn *Palaemon serratus* (Pennant). ICES Fisheries Improvement Committee CM 1967, Doc E 13, 9 pp. (mimeo)

—— (1972). Prawn culture in the United Kingdom: its status and potential. Lab. Leafl. Fish. Lab. Lowestoft (New Series) No. 27, 32 pp.

Frost, J.V. (1975). Australia crayfish. In: Avault, J.W. Jr. (ed.), *Freshwater Crayfish, 2.* Louisiana State University Press, pp. 87-96

Gary, D.L. (1974). The commercial crawfish industry of South Louisiana. Baton Rouge: Louisiana State University Center for Wetland Resources, Publication No LSU-SG-74-01, 59 pp.

Gerhardt, H.V. (1978). The culture of *Penaeus indicus* Milne-Edwards in experimental closed systems with special reference to water quality. MSc thesis, Dept of Zoology and Entomology, Rhodes University, Grahamstown, South Africa, 86 pp.

Goddard, S. and Holdich, D. (1979). Explore the native potential first. *Fish Farmer, 2, (2):* 47

Godfriaux, B.L., Guerra, C.R. and Resh, R.E. (1977). Venture analysis for intensive waste heat aquaculture. *Proc. 8th Ann. Wkshop Wld Maricult. Soc.,* San José, Costa Rica, 9-13 January, pp. 707-22

Goldizen, V.C. (1970). Management of closed system marine aquariums. *Helgoländer Wiss. Meeresunters., 20:* 637-41

Goldman, C.R. (1973). Ecology and physiology of the California crayfish *Pacifastacus leniusculus* (Dana) in relation to its suitability for introduction into European waters. In: Abrahamsson, S. (ed.), *Freshwater Crayfish, 1.* Studentlitteratur, Lund, pp. 105-20

Guerra, C.R., Godfriaux, B.L., Eble, A.F. and Stolpe, N.F. (1975). Aquaculture in thermal effluents from power plants. In: Persoone, G. and Jaspers, E. (eds), *Tenth European Symposium on Marine Biology. I. Mariculture,* Ostend, 17-23 September, pp. 189-205

Hall, D.N.F. (1962). *Observations on the Taxonomy and Biology of Some Indo-West Pacific Penaeidae.* Fishery Publs. Colon. Off., No. 17, 229 pp.

Hanamura, N. (1976). Advances and problems in culture-based fisheries in Japan. FAO Tech. Conf. on Aquaculture, Kyoto, Japan, 26 May-2 June 1976. FIR:AQ/Conf/76 Doc R 18, 12 pp. (mimeo)

Hanson, J.A. and Goodwin, H.L. (1977). *Shrimp and Prawn Farming in the Western Hemisphere.* Dowden, Hutchinson and Ross Inc., Stroudsburg, Pennsylvania, 439 pp.

Harvey, G. (1979). Crayfish study gets under way in Hampshire. *Fish Farmer, 2 (5):* 37

Hedgecock, D., Nelson, K., Simons, J. and Shleser, R. (1977). Genic similarity of American and European species of the lobster *Homarus. Biol. Bull., 152:* 41-50

Hedgecock, D., Shleser, R.A. and Nelson, K. (1976). Applications of biochemical genetics to aquaculture. *J. Fish. Res. Bd Can., 33:* 1108-19

Heinen, J.M. (1976). An introduction to culture methods for larval and post-larval penaeid shrimp. *Proc. 7th Ann. Wkshop Wld Maricult. Soc.,* San Diego, California, 25-29 January, pp. 333-44

Hirasawa, Y. and Walford, J. (1976). The economics of Kuruma-Ebi *(Penaeus japonicus)* shrimp farming. FAO Technical Conference on Aquaculture. FIR:AQ/Conf/76/R.27, 21 pp. (mimeo)

Hirata, H., Mori, Y. and Watanabe, M. (1975). Rearing of prawn larvae, *Penaeus japonicus* fed Soy-cake particles and diatoms. *Mar. Biol., 29:* 9-13

Hirsch, G.C. and Buckmann, W. (1930). Der Arbeitsrhythmus der Mitteldarmdrüse von *Astacus leptodactylus.* III. Teil: Ein Oxydoreduktionssystem in Zusammenhang mit der Sekretion. *Z. Vergl. Physiol., 12:* 559-78

Hirsch, G.C. and Jacobs, W. (1930). Der Arbeitsrhythmus der Mitteldarmdrüse von *Astacus leptodactylus.* II. Teil: Wachstum als primärer Faktor des Rhythmus eines polyphasischen organigen Sekretionssystem. *Z. Vergl. Physiol., 12:* 524-58

HMSO (1978). Fisheries Research and Development Board. Third Report 1976/7. Ministry of Agriculture, Fisheries and Food, Department of Agriculture and Fisheries for Scotland. HMSO, London, 43 pp.

Hobbs, H.H. Jr. and Hall, E.T. Jr. (1974). Crayfishes (Decapoda: Astacidae). In: Hart, C.W. and Fuller, S.L.H. (eds), *Pollution Ecology of Freshwater Invertebrates.* Academic Press, London, pp. 195-214

Hofmann, J. (1980). *Die Flusskrebse,* 2nd edn. Paul Parey, Berlin and Hamburg, 108 pp.

Holdich, D.M., Jay, D. and Goddard, J.S. (1978). Crayfish in the British Isles. *Aquaculture, 15:* 91-7

Honn, K.V. and Chavin, W. (1975). Prototype design for a closed marine system employing quaternary water processing. *Mar. Biol., 31:* 293-8

—— (1976). Utility of ozone treatment in the maintenance of water quality in a closed marine system. *Mar. Biol., 34:* 201-9

Howard, A.E. and Bennett, D.B. (1976). The substrate preference and burrowing behaviour of juvenile lobsters *(Homarus gammarus* (L.)). ICES Shellfish and Benthos Committee, C.M. 1976, Doc. K 10, 8 pp. (mimeo)

Hudinaga, M. (1942). Reproduction, development and rearing of *Penaeus japonicus* Bate. *Jap. J. Zool., 10 (2):* 305-93

Hughes, J.T., Shleser, R.A. and Tchobanoglous, G. (1974). A rearing tank for lobster larvae and other aquatic species. *Progr. Fish. Cult., 36 (3):* 129-32

Hughes, J.T., Sullivan, J.J. and Shleser, R. (1972). Enhancement of lobster growth. *Science, 177:* 1110-11

Huner, J.V., Meyers, S.P. and Avault, J.W. Jr. (1975). Response and growth of freshwater crawfish to an extruded, water-stable diet. In: Avault, J.W. Jr. (ed.), *Freshwater Crayfish, 2.* Louisiana State University Press, pp. 149-57

Huner, J.V. (1977). Introductions of the Louisiana red swamp crayfish, *Procambarus clarkii*

(Girard): an update. In: Linqvist, O.V. (ed.), *Freshwater Crayfish, 3*. University of Kuopio, Finland, pp. 193-202

ICES (1978). Crustacean Working Groups' Reports 1977. ICES Cooperative Research Report, No. 83, 107 pp.

Idyll, C.P. (1971). Induced maturation of ovaries and ova in pink shrimp. *Comm. Fish. Rev., 33 (4):* 20

IPFC (1956). Symposium papers presented at the Sixth Session of the Indo-Pacific Fisheries Council, Tokyo, Japan, 30 September-14 October 1955, Section III, pp. 323-450

Jamieson, G.S., Apold, W.O and Heigh, I.C. (1976). A lobster culture development programme for New Brunswick, Report to New Brunswick Department of Fisheries, Fredericton, New Brunswick, Canada, December 1975, 53 pp.

Jay, D.J. and Holdich, D.M. (1977). The pH tolerance of the crayfish *Austropotamobius pallipes* (Lereboullet). In: Linqvist, D.V. (ed.), *Freshwater Crayfish, 3*. University of Kuopio, Finland, pp. 363-70

——— (1981). The distribution of the crayfish *Austropotamobius pallipes* (Lereboullet) in British waters. *Freshwater Biology, 11,* 121-9

Johnston, W.E. and Collinsworth, D.W. (1973). An annotated bibliography for economic evaluation of the aquaculture of selected crustaceans and mollusks. University of California, La Jolla, California, Sea Grant Publication No. 2, 25 pp.

Jones, D.A., Möller, T.H., Campbell, R.J., Munford, J.G. and Gabbott, P.A. (1975). Studies on the design and acceptability of micro-encapsulated diets for marine particle feeders. In: Persoone, G. and Jaspers, E. (eds), *Tenth European Symposium on Marine Biology*, Ostend, Belgium, 17-23 September, pp. 229-39

Jones, D.A., Kanazawa, A. and Rahman, S.A. (1979). Studies on the presentation of artificial diets for rearing of the larvae of *Penaeus japonicus* Bate. *Aquaculture, 17:* 33-43

Karlsson, A.S. (1977). The freshwater crayfish. *Fish Farming International, 4:* 2-6

——— (1978). Letters to the editor. *Fish Farmer, 1 (6):* 44

Kelly, R.O., Haseltine, A.W. and Ebert, E.E. (1977). Mariculture potential of the Spot prawn, *Pandalus platyceros* Brandt. *Aquaculture, 10:* 1-16

Kensler, C.B. (1970). The potential of lobster culture. *Am. Fish Fmr and Wld Aquacult. News, 1, (11):* 8-12 and 27

Kinne, O. (1976). Cultivation of marine organisms: water quality management and technology. In: Kinne, O. (ed.), *Marine Ecology, 3*. Wiley-Interscience, New York, pp. 19-300

Kossakowski, J. (1973). The freshwater crayfish in Poland. In: Abrahamsson, S. (ed.), *Freshwater Crayfish, 1*. Studentlitteratur, Lund, pp. 17-26

Kossakowski, J. and Orzechowski, B. (1975). Crayfish *Orconectes limosus* in Poland. In: Avault, J.W. Jr. (ed.), *Freshwater Crayfish, 2*. Louisiana State University Press, pp. 31-47

Krekorian, C.N., Sommerville, D.C. and Ford, R.F. (1974). Laboratory study of behavioural interactions between the American lobster, *Homarus americanus*, and the California Spiny lobster, *Panulirus interruptus*, with comparative observations on the rock crab, *Cancer antennarius*. *Fishery Bull, 72 (4):* 1146-59

Kurata, H. and Shigueno, K. (1976). Recent progress in the farming of penaeid shrimp. FAO Tech. Conf. on Aquaculture, Kyoto, Japan, 26 May-2 June 1976. FIR: AQ/Conf/76/Doc R. 17, 24 pp. (mimeo)

Laubier-Bonichon, A. (1978). Écophysiologie de la reproduction chez la crevette *Penaeus japonicus*. Trois années d'expérience en milieu contrôle. *Oceanol. Acta, 1 (2):* 135-50

Laurent, P. (1973). *Astacus* and *Cambarus* in France. In: Abrahamsson, S. (ed.), *Freshwater Crayfish, 1*. Studentlitteratur, Lund, pp. 69-78

Lee, C.S., Liang, S.R. and Liao, I.C. (1974). The effect of periodic starvation on prawns. II. Periodic starvation related to feeding amount and growth of Grass prawn, *Penaeus monodon* Fabricius. *J. Fish. Soc. Taiwan, 3 (2):* 93-110

——— (1976). The effect of periodic starvation on prawns. III. Periodic starvation related to feeding amount and growth of Kuruma prawn, *Penaeus japonicus* Bate. *J. Fish. Soc. Taiwan, 4 (2):* 11-20

Lee, D.L. (1971). Studies on the protein utilization related to growth of *Penaeus monodon* Fabricius. Reprinted from *Aquiculture, 1 (4):* 1-13, In: Tungkang Marine Laboratory, Collected Reprints, 1969-71, *1:* 191-203

Liao, I.C. (1969). Study on the feeding of 'Kuruma' prawn, *Penaeus japonicus* Bate. Reprinted from *China Fisheries Monthly, 197:* 17-18, In: Tungkang Marine Laboratory,

Collected Reprints, 1969-71, *1:* 17-24

Liao, I.C. and Huang, T.L. (1972). Experiments on the propagation and culture of prawns in Taiwan. In: Pillay, T.V.R. (ed.), *Coastal Aquaculture in the Indo-Pacific Region.* Fishing News (Books) Ltd, Farnham, pp. 328-54

Liao, I.C. and Lee, C.S. (1972). The effect of periodic starvation on prawns. I. Periodic starvation related to feeding amount and growth of red prawn, *Penaeus* sp. *J. Fish. Soc. Taiwan, 1 (1):* 99-120

Lightner, D.V. (1973). Normal post-mortem changes in the brown shrimp, *Penaeus aztecus. Fishery Bulletin, 72 (1):* 223-36

Ling, S.W. (1969). Methods of rearing and culturing *Macrobrachium rosenbergii* (de Man). *Fish. Rep. FAO, 57:* 607-19

Linqvist, O.V. and Louekari, K. (1975). Muscle and hepatopancreas weight in *Astacus astacus* L. (Crustacea: Astacidae) in the trapping season in Finland. *Ann. Zool. Fennici, 12:* 237-43

Lloyd, R. (1979). Toxicity tests with aquatic organisms. In: Sixth FAO/SIDA Workshop on Aquatic Pollution in Relation to Protection of Living Resources. Rome, FAO, TF-RAF 112 (SWE) – Suppl. 1: 165-78

Logan, D.T. and Epifanio, C.E. (1978). A laboratory energy balance for the larvae and juveniles of the American lobster, *Homarus americanus. Mar. Biol., 47:* 381-9

Lowery, R.S. and Mendes, A.J. (1977). The biology of *Procambarus clarkii* in Lake Naivasha, Kenya; with a note on its distribution. In: Lindqvist, O.V. (ed.), *Freshwater Crayfish, 3.* University of Kuopio, Finland, pp. 203-10

Lumare, F., Blundo, C.M., Gozzo, S. and Villani, P. (1973). Nuove prospettive per l'acqua cultura Italiana: l'allevamento del Crostaceo Decapode *Penaeus kerathurus* ('Mazzancolla'). *Riv. It. Piscic. Ittiop., 8 (1):* 9-15

Mason, J.C. (1975). Crayfish production in a small woodland stream. In: Avault, J.W. Jr. (ed.), *Freshwater Crayfish, 2.* Louisiana State University Press, pp. 449-79

―――― (1977). Artificial incubation of crayfish eggs *Pacifastacus leniusculus* (Dana). In: Lindqvist, O.V. (ed.), *Freshwater Crayfish, 3.* University of Kuopio, Finland, pp. 119-32

McAnuff, J.W. (1979). Towards a strategy for fish farming in the United Kingdom. *Food Policy, 4 (3):* 178-93

McCarty, P.L. and Haug, R.T. (1971). Nitrogen removal from waste waters by biological nitrification and denitrification. In: Sykes, G. and Skinner, F.A. (eds), *Microbial Aspects of Pollution.* Academic Press, London, pp. 215-32

McLeese, D.W. (1964). Oxygen consumption of the lobster, *Homarus americanus* Milne-Edwards. *Helgol. Wiss. Meeres, 10:* 7-18

Meade, T.L. (1974). The technology of closed system culture of salmonids. NOAA Sea Grant, Univ. of Rhode Island Mar. Tech. Rept, No. 30, 30 pp.

Mistakidis, M.N. (ed.) (1968). Proceedings of the world scientific conference on the biology and culture of shrimps and prawns. *FAO Fish. Rep., 57 (2):* 77-587

―――― (1969). Proceedings of the world scientific conference on the biology and culture of shrimps and prawns. *FAO Fish. Rep., 57 (3):* 589-1165

―――― (1970). Proceedings of the world scientific conference on the biology and culture of shrimps and prawns. *FAO Fish. Rep., 57 (4):* 1167-627

Mock, C.R. and Neal, R.A. (1974). Penaeid shrimp hatchery systems. FAO/CARPAS Symposium on Aquaculture in Latin America CARPAS/6/74, SE 29, 9 pp. (mimeo)

Möller, T.H. and Jones, D.A. (1975). Locomotory rhythms and burrowing habits of *Penaeus semisulcatus* (de Haan) and *P. monodon* (Fabricius) (Crustacea: Penaeidae). *J. Exp. Mar. Biol. Ecol., 18:* 61-77

Möller, T.H., Jones, D.A. and Gabbott, P.A. (1979). Further developments in the micro-encapsulation of diets for marine animals used in aquaculture. 3rd Int. Symp. Microencapsulation, Tokyo, 1977

Morrissy, L. (1978). *Western Australian Crayfish Cookery.* Claire Dane Publishers, Scarborough, Western Australia, 90 pp.

Morrissy, N.M. (1976). Aquaculture of Marron, *Cherax tenuimanus* (Smith). I. Site selection and the potential of Marron for aquaculture. *Fish. Res. Bull. West. Aust., 17 (1):* 1-27

―――― (1979). Experimental pond production of Marron, *Cherax tenuimanus* (Smith) (Decapoda: Parastacidae). *Aquaculture, 16,* 319-44

―――― (1980). Aquaculture. In: Williams, W.D. (ed.), *An Ecological Basis for Water Resourcement Management.* Australian National University Press, Canberra, pp. 215-26

Moshiri, G.A. and Goldman, C.R. (1969). Estimation of assimilation efficiency in the crayfish *Pacifastacus leniusculus* (Dana) (Crustacea: Decapoda). *Arch. Hydrobiol., 66 (3):* 298-306

Muench, K.A. (1976). The role of the electric utilities industry in developing the use of thermal effluent in aquaculture. *Proc. 7th Ann. Wkshop Wld Maricult. Soc.*, San Diego, California, 25-29 January, pp. 535-41

Muir, J.F. (1978). Studies in water treatment and re-use in aquaculture. PhD thesis, University of Strathclyde, 459 pp.

Nelson, S.G., Armstrong, D.A., Knight, A.W. and Li, H.W. (1977a). The effects of temperature and salinity on the metabolic rate of juvenile *Macrobrachium rosenbergii* (Crustacea: Palaemonidae). *Comp. Biochem. Physiol., 56A:* 533-7

Nelson, S.G., Knight, A.W. and Li, H.W. (1977b). The metabolic cost of food utilization and ammonia production by juvenile *Macrobrachium rosenbergii* (Crustacea: Palamonidae). *Comp. Biochem. Physiol., 57A:* 67-72

New, M.B. (1976). A review of dietary studies with shrimp and prawns. *Aquaculture, 9,* 101-44

Nichols, J.H. and Lawton, P. (1978). The occurrence of the larval stages of the lobster *Homarus gammarus* (Linnaeus, 1758) off the north east coast of England in 1976. *J. Cons. Int. Explor. Mar., 38:* 234-43

Niemi, A. (1977). Population studies on the crayfish *Astacus astacus* L. in the river Pyhäjoki, Finland. In: Linqvist, O.V. (ed.), *Freshwater Crayfish, 3.* University of Kuopio, Finland, pp. 81-94

Norfolk, J.R.W., Manns, T. and Wickins, J.F. (in prep). Growth and mortality in the prawn *Penaeus monodon* Fabricius.

Nose, T. (1964). Protein digestability of several test diets in cray and prawn fish. *Bull. Freshwater Fish. Res. Lab., Tokyo, 14:* 23-8

Otte, G. and Rosenthal, H. (1978). Water quality during a one year operation of a closed intensive fish culture system. ICES Mariculture Committee CM 1978, Doc. F.7, 18 pp. (mimeo)

Painter, H.A. (1970). A review of literature on inorganic nitrogen metabolism in micro-organisms. *Water Res., 4:* 393-450

Pantastico, J.B. (1979). 'Sugpo' cage farming in freshwater. Research Paper presented at the Technical Consultation on Available Aquaculture Technology in the Philippines, Southeast Asian Fisheries Development Centre, Aquaculture Department, Tigbauan, Iloilo, 8-10 February 1979, 5 pp. (mimeo)

Patwardhan, S.S. (1935a). On the structure and mechanism of the gastric mill in Decapoda. I. The structure of the gastric mill in *Paratelphusa guerini* (M. Edw.). *Proc. Indian Acad. Sci., 1B:* 183-96

—— (1935b). On the structure and mechanism of the gastric mill in Decapoda. II. A comparative account of the gastric mill in Brachyura. *Proc. Indian Acad. Sci., 1B:* 359-75

—— (1935c). On the structure and mechanism of the gastric mill in Decapoda. III. Structure of the gastric mill in Anomura. *Proc. Indian Acad. Sci., 1B:* 405-13

—— (1935d). On the structure and mechanism of the gastric mill in Decapoda. IV. The structure of the gastric mill in reptantous Macrura. *Proc. Indian Acad. Sci., 1B:* 414-22

—— (1935e). On the structure and mechanism of the gastric mill in Decapoda. V. The structure of the gastric mill in natantous Macrura – Caridea. *Proc. Indian Acad. Sci., 1B:* 693-704

Perkins, E.J. (1976). The evaluation of biological response by toxicity and water quality assessments. In: Johnston, R. (ed.), *Marine Pollution.* Academic Press, London, pp. 505-85

Perkins, H.C. (1972). Developmental rates at various temperatures of embryos of the Northern lobster *Homarus americanus* (Milne-Edwards). *Fishery Bull., 70 (1):* 95-9

Phillips, G. (1971). Incubation of the English prawn *Palaemon serratus. J. Mar. Biol. Ass. UK, 51:* 43-8

Pillay, T.V.R. (ed.) (1972). *Coastal Aquaculture in the Indo-Pacific Region.* Fishing News (Books) Ltd, Farnham, 497 pp.

Pillay, T.V.R. and Dill, W.A. (eds) (1979). *Advances in Aquaculture.* Papers presented at the FAO Technical Conference on Aquaculture, Kyoto, Japan, 26 May-2 June 1976. Fishing News (Books) Ltd, Farnham, 653 pp.

Platon, R.R. (1978). Design, operation and economics of a small-scale hatchery for the larval rearing of Sugpo, *Penaeus monodon* Fab. SEAFDEC Aquaculture Extension Manual No. 1, revised edn, 30 pp.

Price, V.A. and Chew, K.K. (1969). Shrimp larval culture. *Contr. Univ. Wash. College (Sch) Fish., 300:* 46

Puff, J.M. (1971). La compagnie Générale Transatlantique et l'Aquiculture. Exposé du 26 Sept 1969 Colloque Agquicultur de l'ASTEO, 22 pp. (mimeo)

Pullin, R.S.V. (1977). Coastal aquaculture in Europe. *Mairne Policy, 1 (3):* 205-14

Reddy, A.R. (1935). The structure, mechanism and development of the gastric armature in Stomatopoda with a discussion as to its evolution in Decapoda. *Proc. Indian Acad. Sci., B:* 650-75

———— (1938). The physiology of digestion and absorption in the crab *Paratelphusa (oziotelphusa) hydrodromus* (Herbst). *Proc. Indian Acad. Sci., 6B:* 170-93

Reeve, M.R. (1969). The laboratory culture of the prawn *Palaemon serratus. Fishery Invest., Lond., Ser. 2, 26 (1):* 38 pp.

Rensel, J.E. and Prentice, E.F. (1977). First record of a second mating and spawning of the Spot prawn *Pandalus platyceros* in captivity. *Fishery Bull. 75 (3):* 648-9

———— (1979). Growth of juvenile Spot prawn, *Pandalus platyceros,* in the laboratory and in net pens using different diets. *Fishery Bull., 76 (4):* 886-90

Rhodes, C.P. and Holdich, D.M. (1979). On size and sexual dimorphism in *Austropotamobius pallipes* (Lereboullet). *Aquaculture, 17:* 345-58

Richards, K. and Fuke, P. (1977). Freshwater crayfish: the first centre in Britain. *Fish Farming International, 4:* 6-8

Richards, P.R. (1981). Some aspects of growth and behaviour in the juvenile European lobster, *Homarus gammarus.* Doctorial thesis, University of Wales, Bangor, 209 pp.

Richards, P.R. and Wickins, J.F. (1979). Ministry of Agriculture, Fisheries and Food. Lobster Culture Research. Lab. Leafl., MAFF Direct. Fish. Res. Lowestoft., No. 47, 33 pp.

Richards, P.R., Beard, T.W. and Wickins, J.F. (in prep.). A laboratory-scale pilot plant for the continuous production of lobsters.

Rosenthal, H. (1974). Selected bibliography on ozone, its biological effects and technical applications. *Fish. Res. Bd Can., Tech. Rep., 456*

Rosenthal, H. and Otte, G. (1978). Reactions of a biological filter to salinity change. ICES Mariculture Committee CM 1978, Doc. F.8, 7 pp. (mimeo)

Rosenthal, H., Krüner, G. and Otte, G. (1978a). Effects of ozone treatment on recirculating water in a closed fish culture system. ICES Mariculture Committee C.M. 1978, Doc. F.9, 16 pp. (mimeo)

Rosenthal, H., Westernhagen, H.V. and Otte, G. (1978b). Maintaining water quality in laboratory scale sea water recycling systems. ICES Mariculture Committee C.M. 1978, Doc. F.10, 10 pp. (mimeo)

Salser, B., Mahler, L., Lightner, D., Ure, J., Danald, D., Brand, C., Stamp, N., Moore, D. and Colvin, B. (1978). Controlled environment aquaculture of penaeids. In: Kaul, P.N. and Sindermann, C.J. (eds), *Drugs and Food from the Sea – Myth or Reality?* University of Oklahoma, Norman, Oklahoma, pp. 345-55

Sander, E. and Rosenthal, H. (1975). Application of ozone in water treatment for home aquaria, public aquaria and for aquaculture purposes. In: Rice, R.G. and Browning, M.E. (eds), *Proc. First Symp. on Ozone for Water and Waste Water Treatment.* International Ozone Institute, Syracuse, New York, pp. 103-11

Sandifer, P.A. and Smith, T.I.J. (1977). Intensive rearing of post-larval Malaysian prawns (*Macrobrachium rosenbergii*) in a closed cycle nursery system. *Proc. 8th Ann. Wkshop Wld Maricult. Soc.,* San José, Costa Rica, 9-13 January, pp. 225-35

San Feliu, J.M. (1973). Present state of aquaculture in the Mediterranean and South Atlantic coasts of Spain. *Stud. Rev. GFCM, 52:* 1-24

San Feliu, J.M., Moñoz, F. and Alcaraz, M. (1973). Techniques of artificial rearing of crustaceans. *Stud. Rev. GFCM, 52:* 105-21

Santiago, A.C., Jr. (1977). Successful spawning of cultured *Penaeus monodon* Fabricius after eyestalk ablation. *Aquaculture, 11:* 185-96

Sastry, A.N. and Barton, S.E. (1975). The excretion of ammonia by larval and juvenile stages of American lobster, *Homarus americanus.* Abstract from: Sea Grant Lobster

Aquaculture Workshop Technical Session, Recent Advances in Lobster Aquaculture. W. Alton Jones Conference Center, Kingston, Rhode Island, USA, 21 April 1975, 1 p. (mimeo)

Sastry, A.N. and Zeitlin-Hale, L. (1977). Survival of communally reared larval and juvenile lobsters, *Homarus americanus. Mar. Biol., 39:* 297-303

Scarratt, D.J. (1973). Lobster populations on a man made rocky reef. ICES Shellfish and Benthos Committee C.M. 1973, Doc. K.47, 3 pp. (mimeo)

Schuur, A.M., Allen, P.G. and Botsford, L.W. (1974). An analysis of three facilities for the commercial production of *Homarus americanus*. American Society of Agricultural Engineers, 1974, Annual Meeting, Paper No. 74-5517, 19 pp.

Schuur, A., Fisher, W.S., Van Olst, J.C., Carlberg, J., Hughes, J.T., Shleser, R.A. and Ford, R.F. (1976). Hatchery methods for the production of juvenile lobsters. University of California, La Jolla, California, Sea Grant publication No. 48, 21 pp.

Sedgwick, R.W. (1979a). Effect of ration size and feeding frequency on the growth and food conversion of juvenile *Penaeus merguiensis* de Man. *Aquaculture, 16:* 279-98

—— (1979b). Influence of dietary protein and energy on growth, food consumption and food conversion efficiency in *Penaeus merguiensis* de Man. *Aquaculture, 16:* 7-30

Serfling, S.A., Van Olst, J.C. and Ford, R.F. (1974). A recirculating culture system for larvae of the American lobster, *Homarus americanus. Aquaculture, 3:* 303-9

Sharma, B. and Ahlert, R.C. (1977). Nitrification and nitrogen removal. *Wat. Res., 11:* 897-925

Sharma, M.L. (1966). Studies on the changes in the pattern of nitrogenous excretion of *Orconectes rusticus* under osmotic stress. *Comp. Biochem. Physiol., 19:* 681-90

Sheehy, D.J. (1976). Utilization of artificial shelters by the American lobster (*Homarus americanus*). *J. Fish. Res. Bd Can., 33:* 1615-22

Shell, E.W. (1978). Constraints to the development of mariculture. ICES Mariculture Committee C.M. 1978, Doc. F. 26, 10 pp. (mimeo)

Sherry, P.J. (1977). An investigation to assess the aquacultural and market potential of the freshwater crayfish *Pacifastacus leniusculus* in the United Kingdom. Thesis submitted for the degree of MSc in Applied Fish Biology, School of Environmental Sciences, Plymouth Polytechnic, Plymouth, UK, 127 pp.

Shigueno, K. (1975). *Shrimp Culture in Japan*. Ass. for Int. Tech. Promotion, Tokyo, Japan, 153 pp.

Shireman, J.V. (1973). Experimental introduction of the Australian crayfish (*Cherax tenuimanus*) into Louisiana. *Progr. Fish. Cult., 35 (2):* 107-9

Shleser, R.A. (1974). The effects of feeding frequency and space on the growth of the American lobster, *Homarus americanus. Proc. 5th Ann. Wkshop Wld Maricult. Soc.,* Charleston, South Carolina, 21-25 January, pp. 149-55

Shleser, R. and Tchobanoglous, G. (1974). The American lobster as a model for the continuous production of quality seafood through aquaculture. *Marine Technology Society Journal, 8 (1):* 4-8

Siddall, S.E. (1974). Studies of closed marine culture systems. *Progr. Fish Cult., 36 (1):* 8-15

Sindermann, C.J. (1971). Internal defenses of crustacea: a review. *Fishery Bull. Fish. Wildl. Serv. US, 69 (3):* 455-89

—— (ed.) (1977). *Disease Diagnosis and Control in North American Marine Aquaculture*. Developments in Aquaculture and Fisheries Science, No. 6. Elsevier, Amsterdam, 329 pp.

Snow, N.B. (1969). Studies on the nutrition, metabolism and energetics of the prawn *Palaemonetes varians* (Leach). PhD thesis, University of Southampton, 220 pp.

Söderhäll, K., Svensson, E. and Unestam, T. (1977). An inexpensive and effective method for elimination of the crayfish plague: barriers and biological control. In: Lindqvist, O.V. (ed.), *Freshwater Crayfish, 3*. University of Kuopio, Finland, pp. 333-42

Sorgeloos, P. (1979). List of commercial harvesters — distributors of *Artemia* cysts of different geographical origin. *Aquaculture, 16:* 87-8

Sorgeloos, P., Bossuyt, E., Laviña, E., Baeza-Meza, M. and Persoone, G. (1977). Decapsulation of *Artemia* cysts: a simple technique for the improvement of the use of brine shrimp in aquaculture. *Aquaculture, 12:* 311-15

Sousa, R.J., Meade, T.L. and Wolke, R.E. (1974). Reduction of ammonia toxicity by salinity and pH manipulation. *Proc. 5th Ann. Wkshop Wld Maricult. Soc.,* Charleston,

South Carolina, 21-25 January, pp. 343-54

Spoek, G.L. (1974). The relationship between blood haemocyanin level, oxygen uptake, and the heart beat and scaphognathite-beat frequencies in the lobster, *Homarus gammarus*. *Netherlands J. Sea Res., 8 (1):* 1-26

Spotte, S.H. (1970). *Fish and Invertebrate Culture*. Water Management in Closed Systems. Wiley, New York and London, 145 pp.

Sprague, J.B. (1969). Measurement of pollutant toxicity to fish. 1. Bioassay methods for acute toxicity. *Wat. Res., 3:* 793-821

———— (1970). Measurement of pollutant toxicity to fish. II. Utilizing and applying bioassay results. *Wat. Res., 4:* 3-32

———— (1971). Measurement of pollutant toxicity to fish. III. Sublethal effects and 'safe' concentrations. *Wat. Res., 5:* 245-66

Steenbergen, J.F. and Shapiro, H.C. (1974). Gaffkemia in California spiny lobsters. *Proc. 5th Ann. Wkshop Wld Maricult. Soc.*, Charleston, South Carolina, 21-25 January, pp. 139-43

Strempel, K.M. (1973). Edelkrebserbrütung in Zuger-gläsern und anfütterung der Krebsbrut. In: Abrahamsson, S. (ed.), *Freshwater Crayfish, 1*. Studentlitteratur, Lund, pp. 233-7

———— (1975). Künstliche erbrütung von Edelkrebsen in Zugergläsern und vergleichende beobachtungen im verhalten und abwachs von Edel-und Signalkrebsen. In: Avault, J.W. Jr. (ed.), *Freshwater Crayfish, 2*. Louisiana State University Press, pp. 393-403

———— (1976). Paarung, Erbrütung und Brutpflege des europäischen und amerikanischen Edelkrebsen (*Astacus astacus* und *Pacifastacus leniusculus*) unter dem gesichtspunkt der Künstlichen Aufzucht. *Arbeiten des Deutschen Fischerei-Verbandes, 19:* 125-9

Tabb, D.C., Yang, W.T., Hirono, Y. and Heinen, J. (1972). A manual for culture of pink shrimp, *Penaeus duorarum*, from eggs to post-larvae suitable for stocking. University of Miami, Florida, Sea Grant Special Bulletin No. 7, 59 pp.

Tcherkashina, N.Ya. (1977). Survival, growth and feeding dynamics of juvenile crayfish (*Astacus leptodactylus cubanicus*) in ponds and the river Don. In: Lindqvist, O.V. (ed.), *Freshwater Crayfish, 3*. University of Kuopio, Finland, pp. 95-100

Thomas, H.J. (1954). The oxygen uptake of the lobster, *Homarus vulgaris*, Edw. *J. Exp. Biol., 31:* 228-51

Tiews, K. (1978). Mariculture committee. ICES Administrative Report, C.M. 1978/Doc. F.1, 32 pp. (mimeo)

Ting, K.Y. (1973). Culture potential of spiny lobster. *Proc. 4th Ann. Wkshop Wld Maricult. Soc.*, Monterrey, Mexico, 23-26 January, pp. 165-70

TVA (1978). Waste heat utilization for agriculture and aquaculture. Tennessee Valley Authority and Electric Power Research Institute, Tech. Rep. B-12, 405 pp.

Unestam, T. (1973). Significance of diseases on freshwater crayfish. In: Abrahamsson, S. (ed.), *Freshwater Crayfish, 1*. Studentlitteratur, Lund, pp. 135-50

———— (1975a). The dangers of introducing new crayfish species. In: Avault, J.W. Jr. (ed.), *Freshwater Crayfish, 2*. Louisiana State University Press, pp. 557-61

———— (1975b). Defence reactions in and susceptibility of Australian and New Guinea freshwater crayfish to European-crayfish-plague fungus. *Austral. J. Exp. Biol. Med. Sci., 53:* 349-59

Unestam, T. and Weiss, D.W. (1970). The host-parasite relationship between freshwater crayfish and the crayfish disease fungus *Aphanomyces astaci:* responses to infection by a susceptible and a resistant species. *J. Gen. Microbiol., 60:* 77-90

Unestam, T., Söderhäll, L.M., Svensson, E. and Ajaxon, R. (1977). Specialization in crayfish defence and fungal aggressiveness upon crayfish plague infection. In: Lindqvist, O.V. (ed.), *Freshwater Crayfish, 3*. University of Kuopio, Finland, pp. 321-32

Van Olst, J.C. and Carlberg, J.M. (1978). The effects of container size and transparency on growth and survival of lobsters cultured individually. *Proc. 9th Ann. Wkshop Wld Maricult. Soc.*, Atlanta, Georgia, 3-6 January, pp. 469-79

Van Olst, J.C., Carlberg, J.M. and Ford, R.F. (1975). Effects of substrate type and other factors on the growth, survival and cannibalism of juvenile *Homarus americanus* in mass rearing systems. *Proc. 6th Ann. Wkshop Wld Maricult. Soc.*, Seattle, Washington, 27-31 January, pp. 261-74

———— (1977). A description of intensive culture systems for the American lobster (*Homarus americanus*) and other cannibalistic crustaceans. *Proc. 8th Ann. Wkshop Wld Maricult. Soc.*, San José, Costa Rica, 9-13 January, pp. 271-92

Van Olst, J.C., Ford, R.F., Carlberg, J.M. and Dorband, W.R. (1976). Use of thermal effluent in culturing the American lobster. Power Plant Waste Heat Utilization in Aquaculture, Workshop 1, Trenton, New Jersey, 6-7 November 1975. PSE and G Co., Newark, New Jersey, pp. 71-97

Van Weel, P.B. (1970). Digestion in Crustacea. In: Florkin, M. and Scheer, B.T. (eds), *Chemical Zoology V. Arthropoda A.* Academic Press, New York and London, pp. 97-115

Van Wormhoudt, A. (1973). Variation des protéases, des amylases et des protéines soluble au cours du développement larvaire chez *Palaemon serratus. Mar. Biol., 19:* 245-8

Van Wormhoudt, A. and Ceccaldi, H.J. (1975). Influence de la qualité de la lumière en élevage intensif de *Palaemon serratus* Pennant. In: Persoone, G. and Jaspers, E. (eds), *Tenth European Symposium on Marine Biology. I. Mariculture*, Ostend, Belgium, 17-23 September, pp. 505-21

Van Wormhoudt, A. and Malcoste, R. (1976). Influence d'éclairements brefs, à différentes longueurs d'onde, sur les variations circadiennes des activités enzymatiques digestives chez *Palaemon serratus* (Crustacea, Natantia). *J. Interdiscipl. Cycle Res., 7 (2):* 101-12

Vey, A. and Vago, C. (1973). Protozoan and fungal diseases of *Austropotamobius pallipes* Lereboullet in France. In: Abrahamsson, S. (ed.), *Freshwater Crayfish, 1.* Studentlitteratur, Lund, pp. 165-79

Vondruska, J. (1976). Aquacultural economics bibliography. US Dept of Commerce, NOAA Tech. Rep. No. NMFS-SSRF-703, 123 pp.

Wang, J.K. and Kuwabara, J.S. (1975). Engineering analysis of prawn larval culture. American Society of Agricultural Engineers, 1975 Annual Meeting, Paper No. 75-5015, 8 pp.

Wang, J.K. and Williamson, M.R. (1976). The design of a continuous juvenile production system for *Macrobrachium rosenbergii.* American Society of Agricultural Engineers, 1976 Annual Meeting, Paper No. 76-5033, 11 pp.

Wear, R.G. (1974). Incubation in British decapod Crustacea and the effects of temperature on the rate and success of embryonic development. *J. Mar. Biol. Ass. UK, 54:* 745-62

Westers, H. and Pratt, K.M. (1977). Rational design of hatcheries for intensive salmonid culture, based on metabolic characters. *Progr. Fish. Cult., 39:* 157-64

Westman, K. (1973). Cultivation of the American crayfish *Pacifastacus leniusculus.* In: Abrahamsson, S. (ed.), *Freshwater Crayfish, 1.* Studentlitteratur, Lund, pp. 211-20

Wickins, J.F. (1972a). Developments in the laboratory culture of the Common Prawn *Palaemon serratus* Pennant. *Fishery Invest., London, Ser. 2, 27 (4):* 23 pp.

——— (1972b). Experiments on the culture of the Spot prawn *Pandalus platyceros* Brandt and the giant freshwater prawn *Macrobrachium rosenbergii* (de Man). *Fishery Invest., London, Ser. 2, 27 (5):* 23 pp.

——— (1972c). The food value of brine shrimp, *Artemia salina* L. to larvae of the prawn *Palaemon serratus* Pennant. *J. Exp. Mar. Biol. Ecol., 10:* 151-70

——— (1976a). Prawn biology and culture. In: Barnes, H. (ed.), *Oceanography and Marine Biology: An Annual Review, 14.* Aberdeen University Press, pp. 435-507

——— (1976b). The tolerance of warm-water prawns to recirculated water. *Aquaculture, 9:* 19-37

——— (1981). Water quality requirements for intensive aquaculture: a review. In: Tiews, K. (ed.), *Aquaculture in Heated Effluents and Recirculation Systems.* Heenemann Verlagsgesellschaft, Berlin, vol. 1, pp. 17-37

Wickins, J.F. and Beard, T.W. (1974). Observations on the breeding and growth of the giant freshwater prawn *Macrobrachium rosenbergii* (de Man) in the laboratory. *Aquaculture, 3:* 159-74

——— (1978). Ministry of Agriculture, Fisheries and Food. Prawn culture research. Lab. Leafl., MAFF Direct. Fish. Res., Lowestoft, No. 42, 41 pp.

Wickins, J.F. and Helm, M.M. (1981). Sea water treatment. In: Hawkins, A.D. (ed.), *Aquarium Systems.* Academic Press, London, pp. 63-128

Williamson, D.I. (1968). The type of development of prawns as a factor determining suitability for farming. *Fish. Rep. FAO, 57 (2):* 77-84

Wood, P.C. and Ayres, P.A. (1977). Artificial sea water for shellfish tanks. Lab. Leafl., MAFF Direct. Fish. Res., Lowestoft, No. 39, 11 pp.

Yang, W.T. (1975). A manual for large-tank culture of penaeid shrimp to the post-larval stages. Sea Grant Technical Bull. No. 31, University of Miami, Coral Gables, Florida, 94 pp.

Zeitlin-Hale, L. and Sastry, A.N. (1978). Effects of environmental manipulation on the locomotor activity and agonistic behaviour of cultured juvenile American lobsters, *Homarus americanus. Mar. Biol., 47:* 369-79

3 SNAKEHEADS — THEIR BIOLOGY AND CULTURE

Kok Leong Wee

1. Introduction

Snakeheads are fish belonging to the family Channidae; they are also known as murrels and serpent-headed fish. They have long been regarded as valuable food fish in the Far East (Willey, 1910; Aldaba, 1931; Yapchiongo and Demonteverde, 1959) and their flesh is claimed to be rejuvenating, particularly during recuperation from serious illness and as a post-natal diet. In Thailand, they are one of the most popular food fish, and *Ophiocephalus striatus* is the commonest of the staple food fishes in that country (Smith, 1965).

They are very hardy and if kept moist, can remain alive for a long time out of water. This is very significant in the marketing of snakeheads, as live snakeheads fetch a considerably higher price than dead ones which are reputed to have a poorer taste. Snakeheads can remain out of water for a considerable time because of their possession of a pair of suprabranchial cavities. These cavities, communicating with the pharynx, are not as complicated as, for instance, those in members of the Anabantidae. There is no labyrinthine organ, but the cavity is lined with a thin epithelium. The obvious function of the cavity is to permit aerial respiration; this will be discussed in greater depth later in the chapter. They have been known to survive for months without water when buried in moist soil, and to survive droughts, living in a semi-fluid mud, lying torpid below the hard-baked crust at the bottom of ponds or wriggling overland in search of water.

Their peculiar habit of aestivating under hardened mud has led to a fascinating way of fishing for snakeheads in Thailand. The fishermen wade into the stiff mud up to their waists or above, and by slicing away the mud in layers using a long knife, reach the fish which are found singly or in clusters in cavities in the mud (Smith, 1965). Mittal and Banerji (1975) investigated the functional organisation of the skin in relation to the snakehead's peculiar mode of life by studying the structure and cytochemistry of the epidermal, dermal and subcuticular components of *Channa striata*. In the epidermis, they found a dense population of mitochondria within the malpighian cells, and strong phosphatase and succinnic dehydrogenase activity, indicating a high metabolic rate within the skin. Mucous cells were numerous, producing a thick coat of slime containing mucopolysaccharides, lipids and basic proteins which are probably important in keeping the skin moist for cutaneous respiration, retarding the rate of water loss by evaporation, facilitating burrowing in the mud and swimming movements in water and protecting the skin from bacterial and fungal attacks. The dermis consists of an outer layer of *stratum spongiosum* and an inner layer of *stratum compactum*. The *stratum spongiosum* is mainly composed of well-developed scales lodged in connective tissue pockets, which are characterised by the presence of huge deposits of lipids. These lipids may play important roles in supplying energy during the periods of fasting, acting as a barrier for water diffusion through the skin and serving as shock absorbers protecting the body

181

from mechanical injury during burrowing. The presence of acid-mucopoly-saccharides, in the *substantia amorpha* in the *stratum spongiosum*, has been described as an adaptation to prevent dessication.

Until recently, demands for snakeheads were met by catches from the wild in rivers, ponds, flooded rice fields and swamps. In the past, their voracious and piscivorous behaviour, except when young, made them undesirable in many forms of fish culture and they were sometimes considered as pests and became the subject of eradication measures (Chen, 1976). However, in recent years, the high market price for its firm, white, practically boneless and most agreeably flavoured flesh and hardiness when handled have made the culture of snakeheads economically viable. What was once farmed as a 'police fish' in polyculture with tilapia or with carp to supplement the income of carp farmers, has now developed into the object of a major farming industry in its own right.

2. Systematics

Under the strict rules of zoological nomenclature, the proper generic name for the snakeheads is *Channa*; however, the time-honoured name of *Ophicephalus* is still very commonly used (Smith, 1965). (In this review, the generic name of *Ophicephalus* and *Ophiocephalus* will be used as they appeared in the original reports.)

The snakeheads are native to the freshwater systems of tropical Asia and Africa, although the temperate species can also be found in northern parts of China, and *Ophicephalus striata* has been introduced into the Hawaiian islands (Figure 3.1). They can be found in almost any habitat where there is water, in lowland streams and canals, in upland and mountain streams and in lakes, rivers, ponds and swamps.

The body is elongate, more or less cylindrical, and compressed posteriorly. The mouth is large and protractile. The fins are without spines, dorsal and anal fins being long and running along the length of the posterior part of the fish. Ventral fins may be present or absent. The scales are of medium size, cycloid and striated, except on the upper surface of the head where they are larger and shield-like. The lateral line has a more or less developed curve in its anterior half or it may sometimes be interrupted. A swim-bladder is present, which continues into a prolongation of the abdominal cavity in the tail (Figure 3.2).

There are 30 named species of snakehead described in the literature. Smith (1965), in his book *The Freshwater Fishes of Siam or Thailand*, was able to differentiate the eight species of snakehead found in Thailand. According to Smith, the eight species of *Ophicephalus* from Thailand may be differentiated as follows:

1a. Vomer and palatines with a more or less continuous pluriserial band of small teeth, none of them canine.
2a. A conspicuous black light-edged ocellus at upper base of caudal fin.
3a. A posterior row of about 12 large conical teeth on each ramus of lower jaw; lateral line scales 60 to 70, dropping two rows at 16th to 18th perforated scale; scales in transverse series 4.5-1-11 to 13; rows of scales between eye and angle of preopercle 10; dorsal rays 45 to 55; anal rays 28 to 36; 4 or 5

Figure 3.1: World Distribution of the *Channa* spp.

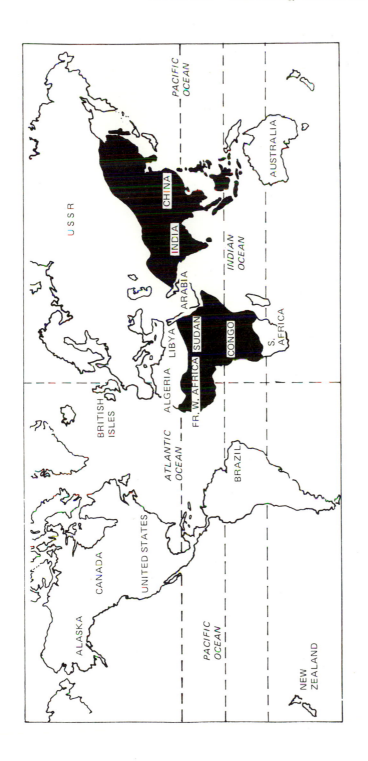

Source: After Sterba (1967).

Figure 3.2: The Red Snakehead, *Channa micropeltes*, exhibiting the typical features of the genus.

dark blotches alongside below lateral line — *marulius*.

3b. Posterior teeth in lower jaw uniserial, small, interspersed with widely separated blunt caniniform teeth; lateral line scales 55 to 58, dropping abruptly two rows at 17th to 20th perforated scale; scales in transverse series 35-1-10; rows of scales between eye and angle of preopercle 5; dorsal rays 45 to 47; anal rays 30 or 31; no dark blotches on side — *marulioides*.

2b. No ocellus at base of caudal fin; a posterior row of about five canine teeth on each ramus of lower jaw; lateral line scales 52 to 57, dropping abruptly two rows at 17th to 20th perforated scale; scales in transverse line 4 to 5.5-1-8 to 10; dorsal rays 37 to 45; anal rays 21 to 27.; back dark green, dark brown, or black; side usually with dark stripes running obliquely upward above, obliquely downward below; underparts white, irregularly blotched with black or brown — *striatus*.

1b. Vomer and palatines with one or two series of teeth which are mostly canine or caniniform.

4a. Lateral line scales 41 to 45, dropping one row at 10th to 13th perforated scale; scales in transverse series 3 or 4.5-1-7; rows of scales between eye and angle of preopercle 4 or 5; rows of scales on opercle 3; dorsal, caudal and anal fins with a narrow bright red margin (turning white in alcohol) — *gaucha*.

4b. Lateral line scales 50 to 65.

5a. Lateral line scales 50 to 55, dropping three rows at 14th to 16th perforated scale; scales in transverse series 4 or 4.5-1-8 or 9; rows of scales between eye and angle of preopercle 6 or 7; dark green or dark blue above, yellowish brown or reddish brown below; a red band from snout to caudal fin in young — *melasomus*.

5b. Lateral line scales 58 to 65; scales in transverse series 5 or 5.5-1-10 or 11; rows of scales between eye and angle of preopercle 10 to 13.

6a. Lateral line dropping two rows at 18th to 20th perforated scale; interorbital space less than length of snout; brown above, yellow below, a double row of dark spots on side with a zigzag light stripe between the spots — *lucius*.

6b. Lateral line dropping one row at 15th to 20th perforated scale; interorbital space greater than length of snout; olive green, with a light stripe from eye to caudal fin, and a series of dark blotches above light stripe and another below; three oblique brown bands on side of head — *siamensis*.

4c. Lateral line scales 82 and 95 without an abrupt drop; scales in transverse series 5.5 or 6.5-1-15 or 16; rows of scales on opercle 8; dark brown or dark blue above, white below; two narrow parallel black stripes extending from eye and angle of mouth to tip of caudal fin, interspace red (the stripes breaking up into irregular spots and blotches in older examples) — *micropeltes*.

The various species exhibit a wide range in size from about 10 cm maximum size in *Ophicephalus gachua* to over one metre in *Channa micropeltes* and *Channa marulius*. In Thailand, all the eight species are eaten while in the Indian Subcontinent only the larger species such as *O. marulius, O. striatus* and *O. punctatus* are extensively eaten. Table 3.1 shows the species of snakeheads commonly eaten and their size range and Table 3.2 their geographical distribution.

Table 3.1: The Size Range of the Commonly Eaten Species of Snakehead

Species	Size Range	Location
O. striatus	up to 1 m	Tropical Asia
O. marulius	up to 1.2 m	Tropical Asia
O. punctatus	up to 30 cm	Indian Subcontinent
O. maculatus	up to 22.5 cm	Taiwan, Hong Kong, China
C. micropeltes	up to 1 m	Tropical Asia

3. The Qualities of Snakeheads

In this section the qualities and the uses of snakeheads are considered. It will help to illustrate the popularity of snakeheads as food, sport and aquarium fish.

3.1 Nutritional Qualities

It has already been mentioned in Section 1 that the flesh is believed to have rejuvenating qualities. Sharma and Simlot (1971) determined the approximate chemical composition of two species of snake-heads *C. marulius* and *C. striata* (C. striata ≡ C. striatus). Their results and those for carp and tilapia are presented in Table 3.3 (after Tan, 1971).

Table 3.2: Distribution of Some Species of Snakehead

Species	Distribution	Source
O. marulius	India to south China	Smith (1965)
O. marulioides	Sumatra, Borneo, Isles of East Indies, Thailand	Smith (1965) Weber & Beaufort (1922)
O. striatus	From China to India, Ceylon, East Indies, Philippines; introduced into Hawaii	Smith (1965) Weber & Beaufort (1922)
O. melasomus	Indo-China through Thailand to Sumatra, Borneo and Palawan	Smith (1965) Weber & Beaufort (1922)
O. lucius	Java, Sumatra, Borneo, Islands of the Indo-Australian Archipelago, Malaysia, Indo-China, China	Smith (1965)
O. gachua	Java, Borneo, Sumatra and other East Indian Islands to Indo-China, Malaysia, Thailand, Andaman Isle, Ceylon, Burma, India, Baluchistan, Afghanistan	Smith (1965) Weber & Beaufort (1922)
O. micropeltes	Indo-China, Malaysia, East Indian Islands, Burma, India	Smith (1965)
O. punctatus	Indian Subcontinent, Burma, Ceylon, Malaysia, China, Tahiti, Polynesia	Misra (1959)
O. maculatus	China, Taiwan	Chen (1976)
O. argus warpachowskii	N. China, USSR especially River Amur watershed	Frank (1970) Nichols (1943)
O. obscurus	Africa from White Nile to West Africa	Sterba (1967)
O. africanus	West Africa, Lagos to the Cameroons	Sterba (1967)

It can be seen from Table 3.3 that *C. marulius* and *C. striata* both have a considerably higher protein and lower fat content than carp or tilapias. Snakeheads can therefore be considered to be more nutritive fish than carp or tilapia, in terms of a higher protein-to-fat ratio. Information on the dress-out weight of snakehead species is not available, but in the author's personal experience snakeheads have very little in the way of bony tissue and the dressing-out weight should be comparable to that of carp or trout. In any case, the eating habits of the people in the Far East regions who consume large amounts of snakeheads are peculiar to the regions, the fish being sold whole and eaten thus; in fact, the head of the fish is a particular fancy.

3.2 Economics of Snakehead Culture

In terms of desirable biological attributes, snakeheads would not appear to be a desirable species to culture. They are predators high in the food chain, extremely

Table 3.3: Proximate Chemical Composition of Carp, Tilapia, *C. marulius* and *C. striata*

Fish Species	Carp		Tilapia		*C. marulius*		*C. striata*	
	as % of whole dried pulverised fish	as % of whole fish	as % of whole dried pulverised fish	as % of whole fish	as % of whole dried pulverised fish	as % of whole fish	as % of whole dried pulverised fish	as % of whole fish
% crude protein	74.2	—	59.3	12.0	90.01	21.10	83.67	17.07
% crude fat	4.5	—	27.9	5.6	2.85	0.67	2.84	0.58
% Ash	21.3	—	12.8	2.6	5.03	1.18	5.58	1.14
% Moisture	—	—	—	80	—	76.56	—	79.60

Source: Sharma and Simlot (1971); Tan (1971).

carnivorous and in the adult stage subsist mainly on other fish. It follows therefore that within intensive culture systems they require a high protein diet. As such, the culture of snakehead is very capital intensive and mainly restricted to monoculture. However, culture can be very profitable because snakeheads can fetch such high prices. In Taiwan, they can fetch US $2.50/kg (1977 prices, Ling, 1977), in Hong Kong (US $1.00 = HK $5.50) HK $10-15/kg (1976 prices, S.W. Sin and Cheng 1976). These prices are comparable to the price of rainbow trout, still considered as a high-priced fish in the UK. In the past, in Hong Kong and Taiwan, in areas where fish are scarce and prices high, consumers have found it worthwhile to import snakeheads from Thailand and Kampuchea by air. The production economics of snakeheads and other species with different culture techniques are presented in Table 3.4.

3.3 Marketing of Snakeheads

Snakeheads are sold fresh and live fish fetch a considerably higher price than dead or frozen ones. In Thailand, the *C. striatus* price is reduced by 30-40 per cent when dead. Processed fish (i.e. salted or dried) command only 40-50 per cent of the price of live fish (Wee, 1981). Fish too stale for processing into salted or dried form are processed by fermentation and sold for around 20 per cent of the price of live fish. In the more developed Asian countries the trend is towards demands for portion-size fish. In Hong Kong, it is marketed at about 800 g, and in Taiwan it is marketed at a much bigger size of 600-1,000 g. In Thailand, the demand is for fish of less than one kilogram, but in the less-developed regions the fish are sold at any size, and prices in these regions are governed by supply and demand rather than by size.

3.4 Snakeheads as Food Fish

Snakeheads are both a 'luxury food fish' served up in restaurants and also a cheap source of protein for the poorer people. The fish is eaten whole, including the head, scales and appendages. It can be fried whole, can be filleted or can be cut up into steaks and steamed. Usually the fish are eaten fresh. In Thailand, however, whenever there is a surplus, the fish are sun-dried and can then be stored. They are decapitated, split, deeply gashed in regular lines and cured in a flat shape to facilitate packing and transportation.

3.5 Other Uses of Snakeheads

Snakeheads are often used as 'police fish' with tilapias to keep the population of the tilapias to manageable proportions, or with carp to keep out other extraneous pest fish in the pond system.

Snakeheads have also been popular as sport fish owing to their voracity and strength at the end of the line. They are also popular as an aquarium fish because of their brilliant colouration especially when young and the different colours that they possess as adults, and also for their ease in rearing, needing only minimal care, being air breathers and hardy. However, in an aquarium they have a restriction in that they cannot be kept with other species because of their extremely voracious behaviour. Popular species are the green snakehead, *C. striata,* and the red snakehead, *C. micropeltes.*

Table 3.4: Comparisons of the Economics of Fish Production with Different Culture Methods

Fish Species	Country of Culture	Method of Culture	Production Cost (US $/kg)	Net Income/Total Operation Cost %	References
Ictalurus punctatus	USA	freshwater floating net cages, intensive	0.71	76	Collins & Delmendo (1976)
Leptobarbus hoeveni, Barbus actus, Pangasius sp., *Channa micropeltes* and *C. striatus*	Lower Mekong River Basin	freshwater floating net cages, intensive	0.74-0.87	21.1-44.4	Pantulu (1976)
C. maculatus	Taiwan	freshwater ponds, intensive	1.2	67	Ling (1977)
C. striatus	Thailand	freshwater ponds, intensive	1.14	25	Wee (1981)
Trachinotus carolinus	USA	marine floating cages, intensive	–	10.3-57.9	Smith (1973)
Seriola quinqueradiata	Japan	marine floating cages, intensive	2.21	45.9	Fujiya (1976)
Epinephelus salmoides	Malaysia	marine floating cages, intensive	1.28-2.06	42.3-109.0	Chua & Teng (1980)

Source: Adapted from Chua and Teng (1980).

4. Ecological Requirements of Snakeheads

The success of any culture of endemic species depends on efficient management of the farming techniques. In order to do that successfully, the basic data on the biology and ecological requirements of the organism to be cultured should be known. This is all the more significant if a new species is being considered for culture. Although snakeheads have long been regarded as food fish, their culture is recent and the pioneer snakehead farmers have been proceeding on the basis of trial and error as the biology and the ecological requirements of snakeheads have not been looked at closely. In this section the available data on the biology and ecological requirements of snakeheads are presented.

4.1 Extremes of Physiological Tolerance

Temperature. Snakeheads are mainly tropical species and as such their preferred temperatures will be in the range 20-35°C. Snakehead fry of *O. striata* have been reported to tolerate a temperature of 14°C but a temperature of 11°C is fatal to them (Mookerjee *et al.*, 1948). Vivekanandan and Pandian (1977) reported that *O. striata* fry had a lower lethal temperature of 15°C. The higher lethal temperature for *C. striata* is about 40°C (Das, 1927). Vivekanandan and Pandian (1977) reported the upper lethal temperature of *O. striata* fry as 39°C.

pH. Studies on the pH tolerance of snakeheads have shown that the ranges of maximum survival (tolerance limits) for *C. punctatus* and *C. striata* are 4.25-9.40 and for *C. gachua* is 3.1-9.6. This suggests that they are generally ubiquitous and can survive in acidic and alkaline waters. They are however sensitive to sudden changes in pH of the medium, dying very quickly beyond their tolerance limits (Varma, 1979).

Salinity. *O. gachua* have been known to enter brackish waters (Weber and Beaufort, 1922). Javaid and Khan (1972) conducted an investigation on the tolerance of *C. punctatus* to artificial sea water with a view to culturing snakeheads in brackish swampy ponds. They found that *C. punctatus* was completely tolerant of 50 per cent seawater for the whole of the experimental period of 96 hours, and suffered 40, 55 and 100 per cent mortalities at 75, 90 and 100 per cent sea water respectively after 96 hours. This is a short-term experiment and the fish were not acclimatised to the new medium. Recent advances have shown that trout and tilapias, when fed a high-salt diet prior to transfer to sea water, have much greater resistance to the salt water. Therefore with more research on acclimatisation and the adaptability of snakeheads to changes in salinity, snakeheads may yet be able to be cultured in brackish water, as preliminary results have shown that they can completely tolerate 50 per cent sea water.

4.2 Toxic Substances

Susceptibility of Snakeheads to Pesticides. Fish farms utilising ground freshwater as the water source inevitably share the same water source with a range of agricultural farmers who, through necessity, usually apply pesticides to the crops and fertilisers to their pastures. The effects of such chemicals on snakeheads have been studied to at least a minimal extent. Konar (1969) investigated the lethal effects of the insecticide DDVP (O, O-dimethyl-2, 2-dichlorovinyl phosphate) on *Channa punctatus*. It was found that DDVP inhibits the development of fertilised eggs. Hatchlings are many times more susceptible to DDVP than eggs. Javaid and Waiz (1972) studied the acute toxicity of 5-chlorinate hydrocarbon insecticides to the fish *C. punctatus*. They found that the toxicity of these insecticides to the fish, in decreasing order, were Endrin, Dieldrin, DDT, Aldrin and BHC respectively. David *et al.* (1972), in their study on the eradication of unwanted air-breathing predatory fish and aquatic insects from nursery ponds by selective poisoning, found that benzene and turpentine were effective in killing *C. punctatus* without harming other species. In a study on the acute toxicity of organophosphorus insecticides to *C. punctatus* it was found that the insecticides tested, in order of their toxicity, were diazinon, malathion, methyl parathion and dimethoate respectively and their 96-hour median tolerance limit was in the following order: diazinon (0.455 ppm), malathion (0.920 ppm), methyl parathion (2.150 ppm) and dimethoate (20.0500 ppm) (Anees, 1975).

5. Air-Breathing Characteristics

Das (1927) described the bionomics, structure and development of the air-breathing organ of several species of *Ophiocephalus* and he showed that the air-breathing organs are developments of pouches of the pharynx. The first indication of their development is a thickening of the pharyngeal epithelium, on each side of the mid-dorsal line above the first gill arch, the thickening subsequently becoming hollowed to form a pocket. Hughes and Munshi (1973), using electron microscopy to describe the ultra-structure of the air-breathing organs of *C. punctatus* and *C. striata* and other air-breathing species, were able to show that the air-breathing organs of *Clarias* and *Saccobranchus* were modifications of gill structures. Evidence for a branchial origin of the accessory organs had been confirmed by the observation of pillar cells so characteristic of the gill secondary lamelae. Such pillar cells were not found in the accessory organs of the snakeheads and this suggested that the accessory organs in this genus were not modified gills and were thus very different from the accessory organs of other species.

Das (1927), in a series of simple experiments, showed the inefficiency of the oxygen uptake of the gills of snakeheads and their obligate free air access requirement. He placed his experimental fish in a series of water systems, all without surface access, and recorded the length of time they took to succumb. When placed in an artificial mixture of mud and water, *O. punctatus* asphyxiated in 15 minutes. In boiled water where all the oxygen had been expelled, *O. punctatus* survived for 8 hours. In fully oxygenated water, it survived for 3-8 hours only. In waters saturated with carbon dioxide gas, the fish drowned after 45 minutes. These results are somewhat ambiguous but nevertheless clearly demonstrate the need that snakeheads have for the opportunity to use their air-breathing organ.

Ojha *et al.* (1979) studied the oxygen uptake in juvenile and adult *C. marulius*. They found that the fish had a bimodal gas exchange mechanism, extracting oxygen from water through the gills, and used the suprabranchial chambers in exchanging respiratory gases with atmospheric air. Juveniles under 'surfacing prevented' conditions did not asphyxiate, whereas adults did succumb. It is presumed that the gills and skin of the juveniles are efficient enough to meet the minimum oxygen requirement for total metabolism of the fish so that it could survive even in submerged conditions with a continuous flow of normoxic water. In adults, on the other hand, it was presumed that the oxygen uptake efficiency of gills and skin decreased with increasing size and that therefore they could not cope with the increased oxygen demand. Thus fish of higher weight group succumbed in submerged conditions even in well-oxygenated water, confirming Das's earlier observations. From these findings, it could be concluded that the juveniles are facultative air-breathers, but adults are obligate breathers.

In tropical countries where availability of fresh water may be limited and dissolved oxygen tensions low, air-breathing fishes may have significant advantages for aquaculture as they survive and grow in shallow waters deficient of oxygen. However, the advantageous air-breathing habits of these fishes and the consequent need to surface more or less at regular intervals impose a considerable drain of energy which otherwise could have been channelled into flesh production. Pandian and Vivekanandan (1976) in a series of studies on this problem investigated energy loss associated with air-breathing. A specially designed experiment, where *O. striatus* were reared in tubular aquaria containing different depths of water, was set up. Since they are obligatory air-breathers, fish in different depths had to swim varying distances to enable them to exchange atmospheric air. The fish displayed the phenomenon of 'hanging' at the surface, i.e. they stayed on or near the surface for a period of time before returning to the bottom. Results from these experiments with starved and fed fish showed that starved fish surfaced less frequently than fed fish over the different depths of water they were reared in; however, the duration of 'hanging' was longer for starved fish than fed fish. The duration of 'hanging' increased with increasing intensity of starvation, as the starved fish became exhausted more quickly and more frequently. Feeding rate increased with increasing depth, but food absorption efficiencies did not vary appreciably between different depth groups. Both conversion rate and conversion efficiencies decreased with increasing depths. Consequently, cultivation of obligatory air-breathing fish in deep waters would result in slow growth rate and poor conversion efficiencies despite increased food consumption.

Therefore, although beneficial for survival in oxygen-deficient water, the advantageous air-breathing habits impose a very significant drain of energy. 'Hanging', therefore, appears to be unique adaptative behaviour of *O. striatus* and, most probably, the other ophiocephalids, permitting the fish to surface without involving vertical movements and consequent energy expense. 'Hanging' may be regarded as a condition in which the accumulation of oxygen debt and the resulting fatigue reach a maximum threshold, the exhausted fish 'hanging' on to the surface, repaying its oxygen debt and exchanging respiratory gases without swimming actively and perhaps without any active metabolic requirement for that function *per se*.

From the above considerations, air-breathing fishes are more suitable for culture in shallow waters. But shallow waters in the tropics undergo considerable diurnal and seasonal changes in temperature. Vivekanandan and Pandian (1977) found that *O. striatus*, enforced to swim longer distances in deeper aquaria at any tested temperature, clearly opted to increase food intake and to release more energy from the consumed food for such extra surfacing and swimming activity rather than to maintain a constant feeding rate and swimming activity by adjusting the surfacing frequency and/or 'hanging' duration. When input of food became a limiting factor, *O. striatus* reduced the energy cost of swimming by depressing activity via decreased surfacing frequency and prolonging 'hanging' duration. These are two gearing mechanisms available to *O. striatus* in which swimming activity is likely to be limited by the reduced input of food energy.

6. Biological Characteristics

6.1 Life History

Snakeheads can be found in almost every kind of tropical aquatic habitat, in ponds, swamps, rice fields, ditches, rivers, lakes, canals and occasionally even within tidal influence. Each species appears to have a preference for a particular habitat. Smith (1965) and Weber and Beaufort (1922) reported that *O. striatus* occurs in ponds, lakes, rivers, swamps and marshes. According to Raj (1916) *O. striatus* prefers margins of water overgrown with weeds and *O. punctatus* prefers stagnant and muddy to running water. Alikunhi (1953) considered, however, that *O. striatus* did not choose between muddy and clean water, but rather its presence depended on the availability of food and hiding places.

The nesting and breeding habits, together with the development of the egg and young of snakeheads, is best described by Willey (1910) for *Ophicephalus striatus* in Ceylon. The nests are crudely constructed by means of cut portions of aquatic vegetation placed over a slight depression created or found in shallow waters near the edge of a canal or lake. The nests are characteristic of each species and can be differentiated by the size of the diameter. Raj (1916) found nests of *O. striatus* of 10-11 inches in diameter and of *O. punctatus* 8-9 inches in diameter. The nest is merely a receptacle to receive the eggs and a place for courtship.

All species of *Ophicephalus* are monogamous and build nests for the deposition of eggs, although *O. striatus* (Alikunhi, 1953; Parameswaran and Murugesan, 1976b), *O. marulius, O. gachua* and *O. punctatus* (Parameswaran and Murugesan, 1976b) can still breed in ponds devoid of macro-vegetation. The process of spawning consists of the female lying belly-up in the nest and liberating eggs at regular intervals, the male shedding his milt over her at the same time. The act of spawning normally lasts for 15-45 minutes (Parameswaran and Murugesan, 1976b). The fertilised eggs, which are golden yellow or amber in colour, float and are spread like a film, flush with the surface, in a sub-circular area over the centre of the nest. Parameswaran and Murugesan (1976b) observed that the developing eggs of snakeheads could be identified by their size and general colouration. In their work on four species of murrels, it was noted that eggs of *O. marulius* have the maximum diameter size (1.84-2.14 mm), followed by *O. striatus* (1.15-1.46 mm), *O. punctatus* (1.03-1.31 mm) and *O. gachua* (0.82-1.19 mm). The buoyancy of the egg is provided by a single large oil globule in the yolk. There is some disagreement as to hatching times, although water temperature is probably the dominant factor. The periods of incubation of fertilised eggs of various species of snakeheads are given in Table 3.5.

Table 3.5: Periods of Incubation of Fertilised Eggs of Snakeheads

Species	Temperature of Incubation (°C)	Period of Incubation (hours)	Source
O. striatus	21.3-26.9	46-56.30	Parameswaran & Murugesan (1976b)
	16-26	54	Willey (1910)
	28-33	30	Willey (1910)
	—	24-40	Alikunhi (1953)
O. argus warpachowskii	23-25	36	Frank (1970)
O. gachua	26.7-30	19.40-27.15	Parameswaran & Murugesan (1976b)
O. punctatus	26.7-30	20.10-27.45	Parameswaran & Murugesan (1976b)
O. punctatus	26.5-28	24	Banerji (1974)

On the first day after hatching the larva is 3.5 mm long, while within four days after hatching it is 6.75 mm. At this time the larva leaves the surface and starts swimming freely at all levels. By the fourth week the larvae swarm and change direction in unison at the slightest disturbance. By the seventh week, larval development has finished and the larvae can live freely at the surface or at the bottom of the water column. By the end of the ninth week, when the fry have reached the size of about 17 mm, they migrate to the bottom and hide in the mud and assume the adult behaviour and habits. Raj (1916) observed the development of the eggs and young of *O. punctatus*, and reported that larval development resembles closely that of *O. striatus* but takes place more rapidly.

The colour of the juvenile and adult phases are different, and different species can be distinguished quite easily (Parameswaran and Murugesan, 1976a). Young of *O. striatus* retain the characteristic larval livery for nearly three months, after which the definitive markings begin to appear.

The snakeheads are sexually dimorphic (Dehadrai *et al.*, 1973; Parameswaran and Murugesan, 1976b). During the breeding season, the female and male of *O. punctatus* develop a deep yellow colour on the ventral side of the body up to the lateral line region. The males bear profusely distributed pinhead black spots, and the females are marked with diffused black blotches on the ventral, yellow surface. The females also show a distended abdomen. In the female, the urogenital opening is circular in shape and it becomes suffused with blood during the breeding season. In males it is, in contrast, elongated. These characters are however difficult to distinguish even for those familiar with the fish.

All species of *Ophicephalus* exhibit parental care. Willey (1910) and Raj (1916) reported that both sexes care for the newly hatched young in *O. striatus*. Later observations by Alikunhi (1953) confirmed the parental behaviour of both sexes, the larger female being more assiduous in brood-care. Similar behaviour during brood-care is reported to be exhibited by *O. marulius* and *O. gachua* (Khan, 1924). *Channa micropeltes*, a popular choice of food in Thailand, shows extremely savage behaviour during brood-care and will attack even human beings (Smith, 1965). Qayyum and Qasim (1962) gave a detailed account of the behaviour of *O. punctatus* during brood-care. They observed the parents to guard the young for about 15-20 days (when they were about 3.5 cm), after which the young adopted the demersal habits of the adults and began to occur singly and independently. This is probably also true for the other species. There is no evidence of parents cannibalising their own young, although cannibalism does occur in snakehead populations. However, Banerji (1974) observed to the contrary and reported no parental care in *O. punctatus* from brood stock after hypophysation. This observation was later confirmed by Parameswaran and Murugesan (1976b) for *O. punctatus, O. marulius, O. striatus* and *O. gachua*.

6.2 Fecundity

The fecundity of snakeheads when compared to other fishes, such as carp, is rather low. Raj (1916) reported that in *O. striatus* the number of eggs released ranged from a few hundred to a few thousand depending on the size of the female. In one study, fecundity in *O. marulius* ranged from 2,214 eggs in a fish 500 mm in length to 18,475 in another of 820 mm length. Fecundity in this species undergoes a rapid decline in older fishes (of four years old or more) (Devaraj, 1973b). Very small numbers of fry have been observed in brood batches. Rahimullah (1946) reported young fry in each batch numbering only 2,000-2,500. Chen (1976) reported fecundity of about 30,000 eggs for a 3 kg female *C. maculatus. O. punctatus* was reported to have a fecundity of 2,300-29,600 eggs by Qayyum and Qasim (1962). This was later confirmed by observations from Reddy (1979) showing a fecundity of from 2,200 in females of 12.1 cm total length to 33,873 in females of 22.2 cm total length for *C. punctata*. Amur snakeheads had been reported to have a fecundity of about 50,000 eggs (Frank, 1970), the highest fecundity for snakeheads ever reported anywhere.

6.3 Spawning – Natural

O. striatus breeds intensively during the rainy months (Willey, 1910; Khan, 1924; Rahimullah, 1946; Mookerjee *et al.*, 1948), and this breeding depends on the prevailing climatic conditions, a double peak in Peninsular and Southern India corresponding to the two monsoon seasons June-July and November and a single peak in North India, including the Punjab, extending from April to September. Breeding also occurs on a limited scale in the other months. Therefore, from these observations, it would appear that *O. striatus* breeds all year round in India. In Taiwan and Hong Kong, *Channa maculatus* breeds from April to September (Chen, 1976). *O. argus warpachowskii* spawns in the months of June and July (Frank, 1970).

Ophicephalus striatus attains maturity in about eleven months under natural conditions when it is about 10-12 inches long (Alikunhi, 1953), but Raj (1916) and Mookerjee *et al.* (1948) observed that *O. striatus* attains sexual maturity when about two years old; however, they appear to refer to specimens grown in artificial or adverse conditions. Devaraj (1973b), in his study on *O. marulius*, observed that maturity in both sexes begins at about 360 mm. *Channa maculatus* in Hong Kong and Taiwan matures in about two years, spawning when the temperatures range from 20 to 30° C (Chen, 1976). *O. argus warpachowskii* matures at about 2.5 years, its size being 30-35 cm (Frank, 1970).

6.4 Spawning – Artificial

At present, the supply of seed from natural sources appears adequate to meet the demand, but eventually controlled breeding of the fish would seem essential to meet seed requirements and to allow genetic developments to improve growth performances. Belsare (1966) obtained larvae of *Channa punctatus* by hypophysation for his study on the gonadal development, but he omitted to provide the details of his technique. Banerji (1974) successfully induced *Channa punctatus* to breed in the laboratory. He used autologous pituitary glands for hypophysation. These were homogenised in distilled water with dilution so adjusted that 0.2 ml of the solution contained 1 mg of the gland. The injection was administered intramuscularly, a little below the dorsal fin on the posterior regions, at a rate of 29-53 mg per kg body weight of males and 15-26 mg per kg body weight of females. Sustained temperatures below 29° C brought about by cool, rainy weather were found to be an important deciding factor in the spawning as well as the embryonic and larval development of the fish. The unusually higher dose to the male proved helpful in the breeding trials. The time lapse between the hormone injection and spawning was rather erratic, within a wide range of 6-25 hours compared to 6-9 hours in the case of Indian major carps.

Parameswaran and Murugesan (1976b) successfully bred four species of murrels through hypophysation using pituitary glands from carp and catfish. A dose range of 40-80 mg gland/kg female is adequate for the females of *O. marulius, O. striatus, O. punctatus* and *O. gachua* to spawn. The spawning took place 11-12 hours after the second injection. They observed that spawning by hypophysation yields more eggs than in natural spawning. The percentage hatching and survival rate of larvae were enhance if the fertilised eggs were transferred to plastic basins and provided with adequate clean water and aeration by dripping of water into the containers.

Chen (1976) reported the successful technique of artificially propagating *Channa maculatus* in Taiwan where mature fish of about two years old are used. The selected fish are kept in brood ponds for 2-3 months before the breeding season begins and they are given live food such as small fish and tadpoles. Generally, for a spawner of about 1 kg in body weight, the pituitaries of one or several common carp of a total weight of 2-3 kg require to be injected together with 20 rabbit units of Synahorin. The dose is divided into two equal portions, which are given at an interval of twelve hours. The male fish receives half the dose of the female in one injection. No second injection is necessary for the male. After hypophyseal treatment, the injected fish is placed into an ordinary fish pond for spawning and fertilisation to take place. One male and one female are placed in a compartment of 3-4 m³ formed by nylon netting. Sometimes 5-6 pairs are placed together in a small pond of 7-10 m³ without segregation. Depth of water is kept between 60 and 100 cm. Spawning and fertilisation usually take place during the next day or so.

6.5 Feeding Habits

Immediately after the mouth is formed, the larvae of *O. striatus* feed on protozoa and algae, while early fry apparently subsist on planktonic crustacea (Mookerjee *et al.*, 1948; Alikunhi, 1953). With further growth, the fry restrict their diet to purely animal foods such as shrimps, prawns, aquatic insects, young fish and tadpoles. Adults are extremely voracious, carnivorous predators feeding on larger aquatic animals such as frogs, other fish and even small aquatic snakes. *Channa punctatus* has been reported to feed by smell; hence the well-developed olfactory organ, nasal accessory sac and taste buds whose distribution extends into the oesophagus (Panday and Dwivedi, 1974). *Channa marulius* is reputed to feed by sight with a well-developed tongue acting as a tactile organ (Singh, 1967). Tandon and Goswani (1968) expressed the view that the tongue of murrels, apart from supplementing the function of the teeth in the retention of prey, may also compensate for the absence of barbels and other integumentary sense organs. Studies on the morphology and histology of oesophagus and stomach by Singh (1967) on *Channa marulius* and by Mehrota and Khanna (1969) on *O. striatus* show that the oseophagus is a short, well-developed tubular structure with prominent internal folds running parallel to each other. Taste buds comprise mainly gustatory cells and are present in the intestinal mucosa. The stomach is a large thick-walled sac, wider anteriorly, and can be differentiated into cardiac and pyloric regions. *Muscularis mucosa* is present to provide additional strength to the muscular stomach wall. The intestine is a long narrow tube with a well-developed pyloric valve at the junction of the stomach and intestine. A pair of long and tubular pyloric caeca are present at this junction. The rectum opens to the exterior through a slit-like anus. Little information is available on feeding regimes but Javaid (1970) found that *Channa punctatus* fed mainly in daylight around midday.

6.6 Food Utilisation

Food intake, digestion, absorption and conversion are the four successive steps involved in the transformation of food into animal tissues. Therefore, factors affecting or controlling these steps will eventually affect the efficiency of the conversion and it is of utmost importance that these factors be thoroughly investigated so that the culture of organisms will be at its most efficient.

Effects of Temperature. The effects of temperature on food utilisation have been investigated by Gerald (1973, 1976a, b) in *C. punctatus* and in *C. striata* by Vivekanandan and Pandian (1977). Gerald (1973) in her study on the rate of digestion in *C. punctatus* showed that the time taken for 100 per cent digestion is inversely related to temperature; the peaks of digestion, however, occur eight hours after feeding irrespective of temperature. In another experiment Gerald (1976b) chose three temperatures 20°C, 28°C, 33°C and three size groups: A (11 g); B (17 g); C (25-31 g). She found that the overall feeding rate is maximal at 28°C and decreased at 20°C and 30°C for all sizes. The feeding rates at 28°C were 12.08, 6.5 and 3.71 per cent body weight/day for the groups A, B, C, respectively. At 28°C, the normal preferred temperature, it would be expected that the fish, not under any thermal stress, would be able to take more food than at other temperatures irrespective of size. The smaller fish (A group) are more cold-sensitive in that the reduction in food intake far exceeded that of the warmer temperature and the larger fish (C group) more heat-sensitive. At 28°C, there was a progressive decrease in food intake in all sizes from the first day to the last day (over the 50-day period of the study). The absorption efficiencies were always high (90-98 per cent) and were unaffected by size or temperature. There is also a direct relation between the rate of feeding and the conversion efficiency at 28°C and 33°C. The rate of absorption and the rate of conversion is governed by the rate of feeding.

Vivekanandan and Pandian (1977) used *O. striata* and tested it over five temperatures of 17, 22, 27, 32 and 37°C, using fed and starved fish kept at different depths. At all temperatures, the feeding rate of the fish steadily increased with increasing depth, statistical tests showing that the different feeding rates at different depths were significant, indicating that feeding rate is a depth-dependent activity. The feeding rate increased steeply with increasing temperature up to 27°C and then levelled off at 32°C and 37°C; this is comparable to the maximal feeding rate at 28°C for *C. punctatus* obtained by Gerald (1976b). The absorption efficiencies were not significantly different at the different temperatures or at the different depths, indicating that efficiency is not affected by temperature or depth. Conversion efficiency was highest in those fish placed in the minimum depth at all temperatures except 37°C. This group also consumed the lowest amount of food and still exhibited the best conversion efficiency because it minimised its energy costs by less swimming and surfacing activities in the shallowest aquarium. The mean conversion efficiency was the least at the lowest temperature of 17°C and maximum at 32°C, a temperature level where fish at all depths had the highest feeding rate, i.e. *O. striata* converted the highest rations more efficiently at higher temperatures and the lowest rations with the highest efficiency at 32°C. It would appear therefore that the lower rations are converted more efficiently by the fish at temperatures prevailing in their respective habitats. The rate of conversion was least at 17°C reaching a maximal level at 32°C and decreased at 37°C; therefore at the optimum temperature of 32°C, not only was the maximum food consumed and the highest efficiencies of conversion exhibited, but also the highest growth rate was displayed.

Effects of Size. The daily quantity of food consumed by *O. striata* of different size groups varies considerably, although certain trends have been observed. Frequently a day of intensive feeding can be followed by a day of little feeding (Pandian, 1967a). Similar trends have also been observed in voluntary feeding experiments with *C. micropeltes* (Wee, 1980, unpublished). Pandian (1967b) found that feeding rate at 28°C was fastest in the group of fish of smallest size (weighing less than 2 g) at 9.2 per cent body weight (b.w.)/day, falling to 3.1 per cent b.w./day and 1.8 per cent b.w./day in fish weighing 13 g and 123.8 g body weight respectively. Using data from this experiment, the inverse relationship was found to be linear for individuals of 10-150 g body weight. The regression line was y = 2.897 – 0.0098x, i.e., for 1 g increase in body weight, the decrease in feeding rate was 0.0098 per cent b.w./day. Gerald (1976a) observed such a similar trend; however, her feeding rate for *O. punctatus* was 12.05 per cent b.w./day for 5.57 g fish and 6.5 and 3.71 per cent b.w./day for 15.41 g and 30.39 g body weight respectively.

Size of body does not appear to influence the absorption efficiency of food. On average the absorption efficiency for *O. punctatus* is 95.5 per cent (Gerald, 1976a) and for *O. striata* the lower value of 90.6 per cent (Pandian, 1967b).

In both species, *O. striata* and *O. punctatus*, the rate of absorption, conversion efficiency and conversion rate are inversely related to size. The decrease in absorption rate and conversion rate with increase in size appears to be due to the lower feeding rate of larger fish (Gerald, 1976a) and according to Pandian (1976b) the inverse relationship is due to old age (senility), when intermediary metabolism becomes less efficient because of changes in digestive enzymes, either qualitatively or quantitatively.

Vivekanandan (1977b) investigated the surfacing activity and food utilisation of five weight classes (0.1, 0.75, 10, 20 and 41 g) of *O. striatus* exposed to different aquarium depths (2.5-70 cm). He found that surfacing frequency was depth dependent for fish of less than 20 g body weight, the frequency being twice as high in fish fed *ad lib* than in starved ones. Feeding rate was observed to be depth (up to 30 cm depth) and size dependent, decreasing as depth and body weight increases. Rate and efficiency of conversion also decreased with increasing body weight. An interesting finding in this study is the shift observed in the surfacing activity and food utilisation to volume dependency from depth dependency in fish of the 0.1 g weight class and when fish grow over 20 g. In fish of the 0.1 g class, the rate and efficiency of conversion were significantly higher in deeper aquarias and hence the growth is volume dependent as deeper aquarias contain larger volumes of water. This is because fish at this size have not yet fully completed the ontogenetic development of the air breathing organ (Das, 1927) and the regular surfacing behaviour (Vivekanandan, 1977c). From these data, Vivekanandan recommended that fish of less than 0.75 g be cultured in nurseries containing a larger volume of water, those weight classes between 0.75-20 g in shallow waters and the larger fish in big aquatic systems.

Effects of Feeding. One of the effects of increasing feeding rates is that *O. striata* is required to visit the surface more often because of increased metabolic oxygen demand (Vivekanandan, 1976). Vivekanandan also found that a feeding level of 43.3 per cent b.w./day may be the maximum which a 0.9 g fish can consume in an aquarium 15.5 cm deep at $28°C$. Food absorption efficiencies averaged 85.6 per cent and did not vary appreciably between groups receiving different rations. Food conversion efficiency, however, increased with increase in ration, reaching a maximum in a group receiving 25.5 per cent b.w./day ration. Conversion rate increased with increase in ration. Geometric derivation of ration-conversion rate suggested that 10.5, 28 and 43.3 per cent b.w./day represented the maintenance, optimum and maximum rates respectively.

Effects of pO_2. A reduction of pO_2 from 130 mm Hg to 74 mm Hg did not cause any significant change in the surfacing activity of *O. striata*. However, when pO_2 dropped from 72 to 42 mm Hg, the surfacing frequency increased from 905 times/day to 1,183 times/day. From these observations, Vivekanandan (1977a) concluded that a pO_2 below the range of 80-60 mm Hg may be the critical level at which air-breathing fishes switch to a more frequent surfacing to obtain proportionately more oxygen from the atmosphere. Continuous aeration did not influence food intake significantly; the rate and efficiency of absorption also did not vary appreciably between those fish in aerated and those in non-aerated waters. The conversion rate and conversion efficiency between the two series were different, *O. striata* being slightly the more efficient convertor in non-aerated water. This was also observed by Herrmann *et al.* (1962) in the juvenile of the coho salmon *Oncorhynchus kisutch*. They concluded that at low pO_2 levels, the fish converted more efficiently as they required a lower level of maintenance requirements.

Effects of Stocking Density. There is a decrease in food consumption and a reduction in growth rate of fish at higher densities. Brown (1946) attributed this to competition for a limited supply of food, mutual mechanical disturbance leading to increased activity, imposing a greater demand for food, restriction of space per individual and accumulation of excreta and other metabolic products in the water, and all subsequent intensive fish culture research has tended to support this view.

Sampath and Pandian (1980) investigated these problems with *O. striata* fry weighing 672 mg in 7-litre aquaria. They used five groups of fish (1, 2, 4, 8 and 16 fish per aquarium). They found that an increase in density increased the food consumption, with *ad lib* feeding until a maximum feeding rate was achieved in the group with four fish per aquarium. Thereafter the rate decreased. Absorption efficiency, conversion rate and conversion efficiency were all maximal in the four fish per aquarium group. Another important effect of density stress is the establishment of a hierarchy that will lead to a size dispersion between the smallest and the largest. Size dispersion in turn leads to inhibition of growth in the smaller individuals in that the smaller fish are constantly under social stress in the presence of larger individuals and therefore do not feed. Thus it is very important to grade and separate different sizes of fish to prevent the establishment of a hierarchy. This is particularly important in the case of snakeheads as they are highly carnivorous and cannibalistic. At the density of four fish per aquarium, the difference in conversion rates between the largest and the smallest fish in Sampath and Pandian's study was not significant. The surfacing frequency was the lowest in the four fish/aquarium group and this probably explains why the fish in this group had the highest efficiency since the role of minimal surfacing frequency in conservation of energy is particularly important, as has already been discussed. Production was highest in the highest density group but the mean gain in body weight was only 50 per cent of that gained by the fish in the other density groups. As the ideal practice of fish farmers is to culture fish for maximum production without sacrificing mean body weight, a high stocking density may not necessarily give the most desired results.

Effects of Starvation. Nikolsky (1963) reported that as a general rule partial or complete starvation, in tropical fish species, is followed by a voracious feeding period and subsequent restoration of growth during the ensuing more favourable season. As snakeheads do encounter adverse conditions such as drought, when they may be required to aestivate underneath the mud, they will undergo periods of starvation. Several workers have looked at the effects of such starvation regimes on food utilisation in snakeheads. Pandian (1967a) starved *O. striata* for 0, 10, 20, 30 and 40 days following which he fed the fish for as many days as they were starved. For the control group, the food intake was 2.7 per cent b.w./day. The effect of the starvation on the feeding rate depended on the duration or the length of fasting. On resumption of feeding, the rate increased, the magnitude of this increase depending on the lenth of fasting.

A period of fasting of less than 20 days increased the feeding rate 2-3 times and a period of more than 20 days' fasting increased the rate 0.5-0.8 times over the normal feeding rate. However, these fasting effects only lasted for a few days before they ceased, so that a 40-day fast was not followed by 40 days of increased feeding activity. The absorption efficiencies were very similar to the control group, but the conversion efficiencies were however higher in the groups starved for less than 20 days than those starved for more than 20 days. Thus the conversion efficiency is a function of feeding rate, as those fish starved for less than 20 days were feeding at a higher rate and yet were able to convert as efficiently as those fed regularly. Therefore the main conclusion from the experiment was that upon resumption of feeding, food intake as well as metabolic rate are restored immediately to the pre-starvation rate or even higher levels and are accompanied by a corresponding increase in conversion efficiency for a limited period.

The above studies on the effects of environmental factors on food utilisation provide a baseline of information on the snakehead's response to varying environmental factors of importance to any consideration of their intensive culture. Thus it has been shown that *C. punctatus* thrives in fresh water at an optimum temperature of 28°C irrespective of size, while for *O. striata* the optimum temperature for culture is 32°C, at which feeding rate is maximum and the highest efficiency and best growth rate are obtained. For *O. striata* a feeding rate of 25.5 per cent b.w./day is the optimum feeding rate, giving the best conversion efficiency for 0.9 g fish. Any feeding rate higher than 25.5 per cent b.w./day will be wasted because food absorption rate is influenced by feeding rate up to a maximum and then it decreases. It has also been shown that continuous aeration of water does not appear to help to increase the efficiency of food utilisation and provided the pO_2 does not drop below 60 mm Hg, continuous aeration is not necessary. It may be suggested from the information available that the best stocking density for *O. striata* fry is 1 g/1 litre of water, but this, like the other observations recorded, is based on single or at best duplicated studies and there is a need for a major replicated study of the environmental factors effecting the optimal conditions for snakehead culture in intensive circumstances.

6.7 Growth Rate Comparisons

From Table 3.6, which summarises the recorded information on the sizes attained by the various species of snakeheads in India (except *C. micropeltes*), it can be seen that fish of different species appear to attain different sizes in the same period in different regions of the country, this is true even of fish of the same species. Such differences in growth rates are probably attributable to differences in water quality, soil, food organisms in the ponds, size of ponds, stocking densities and other variables. As such, the growth rates from these data are not strictly comparable. However, a definite pattern can be detected in that *C. marulius* and *C. micropeltes* grow faster and attain a bigger size. From his experiment on the growth rates of *C. marulius* and *C. striata*, Murugesan (1978) concluded that *C. marulius* and *C. striata* are the two species with considerable growth potential and were the optimal species to be utilised for cultures. Devaraj (1973b), in his study on the biology of the giant snakehead *C. marulius*, compiled a table using his and other workers' results to compare the total length attained at different ages by three species of *Ophicephalus* (Table 3.7). The table shows that growth in *O. punctatus* is the least rapid. *O. striata* is moderately better but *C. marulius* is the best. These results are compatible with those of other workers as the locations from which speciments were collected are in the same geographical regions as those of other workers. For comparison's sake, data on the growth of the Amur snakehead *O. argus warpachowskii* from Czechoslovakia were included. Its growth rate was slightly better than that of *O. punctatus* but worse than that of *O. striata*, yet it should be borne in mind that the Amur snakehead was grown in very much lower temperatures than those available to the other species and so proper comparison is not possible.

Table 3.6: Sizes (mm) Attained by *Ophicephalus* spp. after Periods of Growth

Species	Months					
	7	8	9.5	12	13.5	24
C. marulius	528[a]			386-480[b] 500[b]		
C. striatus		401[a]	233-317[b]	152.4[d] 252[e]	250-270[b]	320[c] 304.8[d] 320[e]
C. micropeltes				300.8[f]		

Sources: a. Murugesan (1978); b. Chacko and Kuriyan (1947); c. Mookerjee *et al.* (1948); d. Bhatt (1970); e. Raj (1916); f. Wee (1980, unpublished).

7. Culture of Snakeheads

Although they have long been regarded as valuable food fishes, snakeheads were not cultured on any scientific basis until very recently. Demand for the food market was met entirely by capture fisheries from the wild, using rod and line and ingenious traps devised by local fishermen fishing in rivers, canals, lakes and flooded rice fields. As the popularity of the fish has increased, the catches from the wild could not meet the demand, and the resulting high market price has created much interest in the culture of snakeheads on their own.

The Indo-Chinese were among the first people to culture snakeheads, especially in the Mekong Basin and Tonle Sap area, i.e. mainly in Kampuchea and Vietnam. Culture of snakeheads in these areas is mainly in cages (Pantulu, 1976). It is considered a lucrative enterprise and has thus progressed very rapidly. Traditional cage culture of snakeheads developed from the well-known practice of holding valuable species of live fish in good condition in bamboo cages before transporting them to market to be sold live. The Vietnamese further improved the design and construction by learning from immigrant Kampucheans who floated their cages down the Mekong into Vietnam.

In Kampuchea, the cages are normally trailed behind boats. Those moored near shores are box-shaped, while cages that form part of a fisherman's boat are

Table 3.7: Comparison of Observed Lengths of *Ophicephalus* sp.

Species (Source)	Method of Growth Determination	Age (Years)													Locations
		0	0+	1	1+	2	2+	3	3+	4	4+	5	5+	6	
O. punctatus (Qayyum & Qasim, 1964; Qasim & Bhatt, 1966)	Length frequency (cm)		10.27		17.32		21.65		24.95						Aligarh (ponds)
O. striatus (Bhatt, 1969, 1970)	Length frequency (cm)		20.90		35.10		41.10		45.10		48.10		53.10		Aligarh (rivers, channels, ponds)
O. marulius (Devaraj, 1973b)	Length frequency (cm)	10.6		38.60		53.30		65.30		76.60					Poongar swamp
O. argus warpachowskii (Frank, 1970)	Measured (cm)				21.1		31.1								Czechoslovakia

Source: Devaraj (1973b).

streamlined to fit the shape of the boat. Cages vary in size from 40 to 625 m^3. The fry for stocking are obtained from natural sources. They are fed on cooked pumpkin, banana and a combination of cooked, broken and glutinous rice and rice bran. The bigger fish are fed on pieces of raw fish, small live fish and kitchen refuse. Generally, monoculture is practised, but occasionally cyprinids such as carp and minnows are stocked with the snakeheads. The stocking rate for a large cage measuring 5 x 50 x 2.5 m^3 is between 6,000 and 10,000 fry. The stock is harvested nine months later when the fish are 1.5-2.5 kg in weight.

In Vietnam, cage culture is of relatively recent origin. Snakehead culture formed 17.9 per cent of all fish cultured in a survey of 173 cages carried out by Rainboth *et al.* (as cited by Pantulu, 1976). The cages used vary in size and average 125 m^3. The two species of snakeheads cultured are *Channa micropeltes* and *Channa striata*. Generally, monoculture is practised, but if the fry of the desired species are not available in the required numbers, the shortage is made up by stock with more readily-available fry of other species. The cages are stocked with 38-63 mm fry obtained from natural sources. The fry are either sold directly to cage-owners or are held by wholesalers who raise the fry further to fingerling or juvenile size for resale. The stocking rate is about 80 fry/m^3. They are fed on forage fish, and as the supply of forage fish can be expensive and irregular, the culture of snakeheads is practised by rich entrepreneurs only. Supplemental feeds such as cooked or uncooked flesh of snails and mussels are fed, and the fish are generally harvested after nine months of culture.

In Thailand (Anon, Thailand), snakeheads, *O. striatus*, are recommended to be grown in square ponds of area ranging from 800 to 1,600 m^3 with good inlets and outlets. The depth of the water is generally kept at 1.5-2 m. The pond is prepared by digging up the bottom mud and the pond edge repaired. Lime is applied at 1 kg/m^2 and left to dry for 5-10 days. The recommended stocking rate is 40-60 fish/m^2 for fingerlings. The feed is forage fish, rice bran and broken rice in an 8:1:1 ratio. The feeding is done twice daily, once in the morning and once in the afternoon. Ideally water should be kept clean and changed three times daily, but in most Thai systems this is not possible. Fish can be expected to grow to marketable size within 7 or 8 months. Recent reports from Thailand indicate that snakeheads are being farmed in conjunction with pig or poultry in mixed culture farms with the waste from the pigs and poultry forming the food for the snakeheads. A particularly important requirement in such ponds is a small mesh fence round the ponds to keep the fish inside and predators out.

In Hong Kong (Anon, Hong Kong), monoculture of snakeheads, *C. maculatus*, is practised. They are cultured in ponds that are usually less than 0.5 ha and about 1 metre deep. These are normally converted agricultural plots. The water supply is usually rain-fed, but may be supplemented by water from wells and drains. Fine-meshed wire fencing is required to prevent the escape of fish. These fences are erected on top of bunds formed by rectangular mud blocks. The stocking rate is 150,000-300,000 fry/hectare of 2-4 cm size fish and stocking is usually undertaken in spring. They are fed *ad lib* daily with minced fish, and conversion ratios of between 6:1 and 10:1 are reported.

In Taiwan, *C. maculatus* is usually reared in polyculture with Chinese carp where it serves as a 'police fish' in these ponds, eliminating other extraneous or pest fish. It is recommended that not more than 500 snakeheads of 10 cm each in length be stocked in a 1-hectare pond, and that the carp should be at least 10 cm in length before stocking the snakeheads (Chen, 1976). No special feeding is provided. Snakeheads are also cultured in polyculture with tilapia, which serve as forage fish. The recommended stocking rate is 90,000 fingerlings of 10 cm per hectare, which should be graded 2-3 times per year and restocked at a lower density to avoid cannibalism. The final stocking density should be 15,000-24,000 fish per hectare.

Monoculture of snakeheads (*C. maculatus*) in Taiwan is not common because of high production costs. Monoculture of snakeheads requires supplementary feeding. This is provided by a mixture of 80 per cent minced trash fish and 20 per cent wheat flour or formulated eel feed. Provided adequate natural feed is present, the growth of the fish is rapid, 10 cm fingerlings achieving 600-1,000 g weight in 9-10 months and over 1 kg in one year. It is claimed that a 90 per cent survival rate can be achieved if proper care is taken.

Reports of snakeheads cultured in irrigation wells and canals in the Indian Subcontinent are available, but the techniques used have not been described in detail (Anon., 1957, 1958).

Preliminary experiments on the culture of the large snakehead, *Ophicephalus marulius*, have been undertaken by Chacko and Kuriyan (1947) and Devaraj (1973a). Chacko and Kuriyan collected fingerlings from wells and stocked 100 fingerlings to each irrigation well, and fed them with live minnows, frogs, dead birds and rats and kitchen refuse. A very poor survival rate of only 10 per cent was reported. This low figure was attributed to the intense competition for food and cannibalism. The survivors achieved a growth of 1.5 feet (0.5 m) in the first year.

Devaraj (1973a) observed that the growth rate of *O. marulius* fry and fingerlings was 32 mm per month in swamps, 15.6 mm in a nursery (cement tanks), 6.4 mm in backyard wells and up to 29 mm in irrigation canals under different feeding conditions. The experiment was extended to farm ponds where the growth rate of the one- and two-year-old fish was 12 mm and 8 mm per month respectively. Eutrophication of the pond during the second year increased the growth rate to 33 mm per month in fingerlings and 27 mm per month in the one-year-old fish. The survival rate was very poor. During the first year it was 1.3 per cent in nursery ponds, 13.6 per cent in backyard wells and 34.0 per cent in irrigation canals. Recovery from the farm ponds varied from 57 to 100 per cent, but the apparently higher survival rate was due to the breeding in the farm pond. Since the reason for the high mortality rate was principally cannibalism, for culture in ponds of sizes of 0.25-0.5 hectare a stocking rate of 1,000-4,000 fingerlings is suggested. The disparity of growth rate shown also greatly emphasised the need for periodic grading and sorting.

A summary of the culture of snakeheads in the countries described in

Table 3.8: Summary of Snakehead Culture

Country	Culture	Species	Size of Ponds/Cages	Stocking Rates	Type of Feed	Source
Thailand	monoculture	*Channa striatus*	800-1,600 m² 1.5-2 m depth	40-60/m²	Forage fish, rice bran, broken rice in 8:1:1 ratio	Anon (Thailand)
Taiwan	monoculture	*Channa maculatus*			Minced trash fish and eel feed or wheat-flour in 8:2 ratio	Chen (1976)
	polyculture	*C. maculatus* Tilapia		Initially 90,000 of 10 cm fish per hectare. Final stocking rate of 1,500-2,400/ha	Tilapia	Chen (1976)
	polyculture	*C. maculatus* Chinese carp		Not more than 500 fish over 10 cm long per hectare pond	Wild fish and young tilapia	Chen (1976)
Hong Kong	monoculture	*C. maculatus*	< 0.5 ha ponds of 1 m deep	150,000-300,000 per hectare	Minced fish	Anon (Hong Kong)
India	monoculture	*C. marulius*	0.25-0.5 ha ponds	Recommended 1,000-4,000 fingerlings ponds	Tilapia	Devaraj (1973a)
Kampuchea	monoculture (cage culture)	*C. micropeltes* *C. striatus*	40-625 m³	6,000-10,000 fry for cage 5 x 50 x 2.5 m	Vegetable and animal products larger fish, pieces of raw fish	Pantulu (1976)
Vietnam (Mekong Basin)	monoculture (cage culture)	*C. micropeltes* *C. striatus*	60-181 m³	80 fry/m³ of 4-6 cm long	Forage fish supplemented by cooked or raw flesh of mussels and snails	Pantulu (1976)

the text is presented in Table 3.8.

8. Diseases of Snakeheads

There are relatively few reports of outbreaks of disease affecting snakeheads. Pantulu (1976) reported that specimens cultured in cages in Kampuchea were sometimes afflicted by parasites and fungal disease, but did not name the infecting organisms. The treatment for these diseases consists of dumping cow dung and the bark of certain trees or a mixture of salt, mud and leaves of plants into the cage. The efficacy of these treatments has not been systematically evaluated.

There are, however, numerous reports of the incidence of organisms found infesting snakeheads. Most of them do not describe any pathological effects of these infestations, but are merely reports of incidence. Organisms found to infest snakeheads include: a newly described nematode parasite from the intestine of *Channa striatus, Metaquimperia madhuai* (Sood, 1973); three trematodes from the intestines of *C. punctatus* and *C. marulius,* i.e. *Eucreadium daccai, Crowcrocaecum channai* and *Neopecoelina saharanpuriensis* (Bashrimullah and Mustaque Elahi, 1972); the ectoparasitic fluke *Cercaria suparkari* between the scales and musculature of *C. punctata* (Pande and Shukla, 1972); metacecaria of the trematode *Proalariodes tripidonotis* (Karyakarte, 1970); the trematode *Genarchopsis goppo* from the stomach of *C. punctatus* (Madhavi, 1978); two new parasitic helminths, *Gorgorhynchus ophiocephalii* and *Senga phangenis* from *Channa micropeltes* (Furtado and Lau, 1971). A parasite *Isoparorchis hypeselobagri,* normally found in the swim-bladder of siluroids, was found in the swim-bladder and in all other organs (Mahajan *et al.,* 1978).

Snakeheads have proved to be good hosts for trypanosomes. Among the Ophicephalids, *Ophicephalus striata, O. punctatus, O. maculatus* and *O. obscurus* are known hosts of trypanosomes. *Trypanosoma striati, T. ophicephali, T. punctati* and *T. elongatus* and *T. muknudi* are parasites of Ophicephalids. The first two are recorded from *O. striata* and the latter three are recorded from *O. punctatus* (Quadri, 1955; Pearse, 1933; Hasan and Qasin, 1962; Raychaudhuri and Misra, 1973, all as cited by Misra *et al.,* 1973). *Ophicephalus gachua* has also been proved a sensitive host for *Trypanosoma gachuii.*

In studies of intensively cultured snakeheads Wee (1980, unpublished) found that the fry of *Channa micropeltes* and *C. striata* were very susceptible to ectoparasites, *Costia* spp., *Chilodonella* spp. and *Trichodina* sp. being recorded frequently from *C. micropeltes. Ichthyophthirius multifilis* was also recorded in *C. micropeltes* and *C. striata.* A treatment by long-term bathing in methylene blue proved to be therapeutic with little in the way of loss of fish (dosage: 0.3 cm^3 of a stock solution of methylene blue (10 g/litre) per litre of aquarium water, continuing the treatment for three weeks).

It would appear from the literature that parasites are the most prevalent organisms affecting snakeheads, but this is probably because parasites were generally the only organisms studied extensively. Misra *et al.* (1972) observed a blood-inhabiting protozoan *Mesnilium malariae* from *O. punctata* which produced certain pathogenic effects upon the host (extreme parasitism caused death). Srivastava and Srivastava (1976) recorded that adult individuals of

Channa marulius suffered from branchiomycosis. Infected fish gasped for air on the surface of the water for a certain period of time after which they usually died. Gills of infected fish revealed destruction and discolouration of gill lamellae. The pathogen was identified as a phycomycete – *Branchiomyces sanguinis*. This is the first report of such gill rot in *C. marulius*. Srivastava and Srivastava also reported fungal infection on the adults of *Channa punctatus*, the signs being white cotton-like patches scattered on the body. The fungi were isolated and identified as *Dictyuchus anomalous*. Infected fish usually die within 24 hours of the appearance of fungal tufts. Injury to fish greatly lowers the resistance to fungal infection. In studies of intensively-cultured snakeheads, Wee (1980, unpublished) has regularly associated haemolytic *Aeromonas hydrophila* with acute septicaemic disease in young *O. striata* and recently has recorded a central and peripheral neurological condition showing many features of neurotropic virus infection (Wee and Roberts, 1980, unpublished).

9. Research Areas for Improvements

From the available literature it would appear that there are several areas where significant research effort should be applied to allow improvement in snakehead culture. The main requirements can be categorised as follows:

(1) Formulation of a standardised feed to be fed in pellet form to allow intensification of culture and removal of dependency on trash fish.
(2) Disease control.
(3) Breeding and seed fish production.
(4) Genetic selection and hybridisation.
(5) Aqua-farm management technology systems to allow advantage to be taken of the snakehead's particular advantages.

The formulation of a standard diet available in pelleted rations is one of the most vital factors in the development and progress of any fish farming system. The success of the catfish industry in the USA is due to a variety of favourable factors, but it can be reasonably asserted that one of the most vital of these is the availability of pelleted rations formulated in accordance with the basic nutritional requirements of the fish (Cowey and Sargent, 1972). This could similarly be said for the state of trout and salmon culture in the West today.

Advantages in using pelleted foods include continuous availability and uniformity of food, ease of transport and of storage, ease of feeding with controlled rations by automated, timed or demand feeders, water stability of the pellet (implying minimal nutrient loss by leaching and maintenance of water quality), and reduction in the risks of transmitting disease by feeding trash wild fish to hatchery fish (Cowey and Sargent, 1972).

The present snakehead culture systems do not use pelleted feeds. Their food is usually trash fish, vegetable or animal products or a combination of these. These types of feed have significant disadvantages in that the food will not keep, and in the case of trash fish, the supply is not unlimited and often the availability is seasonal and unreliable. Therefore, feeding by trash fish is only practised by rich entrepreneurs who can afford huge freezers to store trash fish and can provide the capital for storage. Feeding of wet diets also seriously pollutes the water and poor water quality usually leads to poor growth and increased incidence of disease.

However, although the snakehead farmers utilising such wet diets will often insist that their feed is adequate, their knowledge of the feed requirements is derived from trial and error experiments. Although their feed may apparently seem to be adequate, producing apparently healthy and good-condition fish, there is almost certainly scope for improved production when the detailed nutritional requirements of snakeheads are known and can be met in pelleted feeds. Therefore the formulation of a standard diet to be fed in pelleted form is a top priority on the list.

Preliminary work on the protein requirements of *C. micropeltes* has commenced in the Institute of Aquaculture, Stirling (Wee, 1980, unpublished) with semi-purified diets and varying contents of proteins. Preliminary results have shown that *C. micropeltes* requires quite high levels of protein, and there has been no levelling off of the curve of specific growth rate against percentage protein in diets even at the highest level of 60 per cent. This could be due to the fact that the diets are not totally isocaloric and further analysis is being done to determine this, but it certainly appears that the snakehead can efficiently utilise protein at remarkably high levels.

9.1 Disease Control

As mentioned in the section on diseases of snakeheads, there has been little clinical work on their diseases in culture. As snakehead farming technology continues to progress and expand, the trend of culture techniques is inevitably towards one of further intensification and with this advance the problems associated with intensive systems due to overcrowding, accumulation of wastes and metabolites will become more acute. Another area of research should be on nutritional diseases, as the ultimate solution to cut down the feeding costs will, as indicated above, be the development of a completely artificial diet fed in pellet form, using low-cost locally available materials.

9.2 Breeding and Seed Production

Considerable breakthroughs in artificial propagation have been achieved with *C. maculatus* in Taiwan and Hong Kong and with *C. punctatus* in India. The need now is to refine the techniques so that the snakehead farmers themselves can produce the fry.

9.3 Genetic Selection and Hybridisation

With the achievement of artificial propation, the next step will be to develop pure lines of fast-growing stock to obtain the desirable attributes such as improved growth performance and disease resistance.

9.4 Aqua-farming Technology

From the previous section on the culture of snakeheads, it can be clearly seen that different countries utilise different techniques for culture. There is wide variation in, for example, the optimum stocking density. The frequency of grading also needs to be worked out. A successful method of harvesting also needs to be perfected as snakeheads can be very elusive. Also, with further intensification of culture techniques, higher numbers of fish being stocked with supplementary feeding will render the maintenance of optimal water quality critical. As yet there is no information as to what the tolerance limits of snakeheads to accumulated waste metabolites such as dissolved ammonia, nitrate or nitrite are.

Acknowledgements

The author would like to thank Dr J.F. Muir for his assistance with this review and Dr A.G. Tacon for his comments on the manuscript. The programme of tropical fish culture research of the Institute of Aquaculture, of which this forms a part, is funded by the Overseas Development Administration.

References

Aldaba, V.C. (1931). The Dalag Fishery of Laguna de Bay. *Phil. Jour. Sci., 45 (1):* 41-59
Alikunhi, K.H. (1953). Notes on the bionomics, breedings and growth of the murrel, *Ophicephalus striatus* (Bloch). *Proc. Indian Acad. Sci. (B), 38 (1):* 10-20
Anees, M.A. (1975). Acute toxicity of four organophosphorus insecticides to a freshwater teleost *Channa punctatus* (Bloch). *Pakistan J. Zool., 7 (2):* 135-41
Anon. (Thailand). Culture of *Channa striata.* Leaflet produced by the National Inland Fisheries Institute, Bangkok, Thailand
———— Hong Kong. Monoculture of snakehead in Hong Kong. Leaflet produced by Au Tau Fisheries Office, Hong Kong
———— (1957). Fish culture in backyard wells. *Adm. Rep. Fish. Dept, Madras., 1955-6:* 91
———— (1958). Fish culture in wells. *Adm. Rep. Fish. Dept, Madras., 1956-7:* 88
Banerji, S.R. (1974). Hypophysation and life history of *Channa punctatus* (Bloch). *J. Inland Fish. Soc. India, VI:* 62-73
Bashirullah, A.K.M. and Mustaque Elahi, K. (1972). Three trematodes (Allocreadiidae) from the freshwater fishes of Dacca, Bangladesh. *Noweg. J. Zool., 20 (3):* 205-8
Belsare, D.K. (1966). Development of the gonads in *Channa punctatus* (Bloch). (Osteichtyes: Channidae). *J. Morph., 119:* 467-76
Bhatt, V.S. (1969). Age determination of *Ophicephalus striatus* (Bloch). *Curr. Sci., 38 (2):* 41-3
———— (1970). Studies on the growth of *Ophiocephalus striatus* (Bloch). *Hydrobiologia, 36 (1):* 165-77
Brown, M.E. (1946). The growth of brown trout (*Salmo trutta*, Linn) I. Factors influencing growth of trout fry. *J. Exp. Biol., 22:* 118-29
Chacko, P.I. and Kuriyan, G.K. (1947). On the culture of *Ophicephalus marulius*, Hamilton, in the Coimbatore and Salem districts, Madras. *Proc. 34th Ind. Sc. Cong.*, Pt 3, Abstracts: 180
Chen, T.P. (1976). *Aquaculture Practices in Taiwan.* Fishing News (Books) Limited, Farnham, England
Chua, T.E. and Teng, S.K. (1980). Economic production of Estuary Grouper, *Epinephelus salmoides* Maxwell, reared in floating net cages. *Aquaculture, 20:* 187-228
Collins, R.A. and Delmendo, M.N. (1976). Comparative economics of aquaculture in cages, raceways and enclosures. In: *FAO Technical Conference on Aquaculture*, Kyoto, Japan, 26 May-2 June 1976, FIR: AQ/CONF/76/R.37
Cowey, C.B. and Sargent, J.R. (1972). Fish nutrition. *Adv. Mar. Biol., 10:* 383-492
Das, B.K. (1927). III. The bionomics of certain air-breathing fishes of India, together with an account of the development of their air-breathing organs. *Phil. Trans. Roy. Soc. Lond., 216:* 183-218

David, A., Panicker, G.C. and Chakraborty, D.P. (1972). Eradication of unwanted air-breathing predatory fish and aquatic insects from nursery ponds by selective poisoning. *J. Inland Fish. Soc. India, IV:* 189-93

Dehadrai, P.V., Banerji, S.R., Thakur, N.K. and Das, N.K. (1973). Sexual dimorphism in certain air-breathing teleosts. *J. Inland Fish. Soc. India, V:* 71-7

Devaraj, M. (1973a). Experiments on the culture of the large snakehead *Ophicephalus marulius* (Hamilton). *Indian J. Fish., 20 (1):* 138-47

——— (1973b). Biology of the large snakehead *Ophicephalus marulius* (Ham.) in Bhavanisagar waters. *Indian J. Fish., 20 (2):* 280-301

Frank, S. (1970). Acclimatization experiments with Amur Snakehead, *Ophiocephalus argus warparchowskii,* Berg, 1909 in Czechoslovakia. *Cesk. Spolecnosti Zool., 34:* 277-83

Fujiya, M. (1976). Coastal culture of yellowtail (*Seriola quinqueradiata*) and red sea bream (*Sparus major*) in Japan. In: *FAO Technical Conference on Aquaculture,* Kyoto, Japan, 26 May-2 June 1976, FIR: AQ/CONF/76/E.53

Furtado, J.I. and Lau, C.L. (1971). Two new helminth species from the fish *Channa micropeltes* Cuvier (Ophiocephalidae) of Malaysia. *Folia Parasitologica (Praha), 18:* 365-72

Gerald, V.M. (1973). Rate of digestion in *Ophiocephalus punctatus* (Bloch). *Comp. Biochem. Physiol., 46A:* 195-205

——— (1976a). The effect of size on the consumption, absorption and conversion of food in *Ophiocephalus punctatus* (Bloch). *Hydrobiologia, 49 (1):* 77-85

——— (1976b). The effect of temperature on the consumption, absorption and conversion of food in *Ophiocephalus punctatus* (Bloch). *Hydrobiologia, 49 (1):* 87-93

Herrmann, R.B., Warren, C.E. and Doudoroff, P. (1962). Influence of oxygen concentration on the growth of juvenile Coho Salmon. *Trans. Am. Fish. Soc., 91:* 155-67

Hughes, G.M. and Munshi, D.J.S. (1973). Nature of the air-breathing organs of the Indian fishes *Channa, Amphipnous, Clarias* and *Saccobranchus* as shown by electron microscopy. *J. Zool. Lond., 170:* 245-70

Javaid, M.Y. (1970). Diurnal periodicity in the feeding activity of some freshwater fishes of West Pakistan. I. Studies on *Channa punctatus* and *Mystus vittatus. Pakistan J. Zool., 2 (1):* 101-11

Javaid, M.Y. and Khan, Z.U. (1972). Salinity tolerance of two species of Teleosts of Pakistan. *Pak. J. Sci. Res., 24 (3-4):* 291-8

Javaid, M.Y. and Waiz, A. (1972). Acute toxicity of five chlorinated hydrocarbon insecticides to the fish, *Channa punctatus. Pakistan J. Sci. Ind. Res., 15 (4-5):* 291-3

Karyakarte, P.P. (1970). Metacercaria of *Proalariodes tropidonotis* Vidjarthe, 1937 (Trematoda: Proterodiplostomidae) from the fish *Ophiocephalus marulius* (Harn) in India. *Rivista di Parassitologia, 31 (1):* 69-70

Khan, M.H. (1924). Observations on the breeding habits of some freshwater fishes in the Punjab. *Journ. Bombay Nat. Hist. Soc., 29:* 958-62

Konar, S.K. (1969). Lethal effects of the insecticide DDVP on the eggs and hatchlings of the snakehead, *Channa punctatus* (Bl.) (Ophiocephliformes: Ophiocephalidae). *Japanese Journal of Ichthyology, 15 (3):* 130-3

Ling, S.W. (1977). *Aquaculture in South East Asia — A Historical Overview.* Washington Sea Grant Publc., University of Washington Press, Seattle and London

Madhavi, R. (1978). Life history of *Genarchopsis goppo* Ozaki, 1925 (Trematoda: Hemiuridae) from the freshwater fish *Channa punctata. Journal of Helminthology, 52:* 251-9

Mahajan, C.L., Agrawal, N.K., John, M.J. and Katta, V.P. (1978). Parasitization of *Isoparorchis hypselobagri* Billet in *Channa punctatus* (Bloch). *Curr. Sci., 47 (21):* 835-6

Mehrota, B.K. and Khanna, S.S. (1969). Histomorphology of the oesophagus and the stomach in some Indian teleosts with inference on their adaptational features. *Zoologische Beitrage, 15 (2-3):* 375-91

Misra, K.K., Chandra, A.K. and Choudhury, A. (1973). *Trypanosoma gachuii* n.sp. from a freshwater teleost fish *Ophicephalus gachua* Hami. *Arch. Protisterik., 115:* 18-21

Misra, K.K., Haldar, D.P. and Chakravarty, M.M. (1972). Observations on *Mesnillium malariae* gen. nov., spec. nov., (Haemosporidia, Sporozoa) from the freshwater teleost, *Ophiocephalus punctata,* (Bloch). *Arch. Protistenk. Bd, 114:* 444-52

Misra, K.S. (1959). An aid to commercial fishes. *Records of the Indian Museum, 57:* 218-20

Mittal, A.K. and Banerji, T.K. (1975). Histochemistry and the structure of the skin of a Murrel, *Channa striata* (Bloch) 1797. I. Epidermis. II. Dermis and Subcutical. *Can. J. Zool., 53 (6):* 833-52

Mookerjee, H.K., Ganguly, D.N. and Bhattacharya, R.N. (1948). On the bionomics, breeding habits and development of *Ophicephalus striatus* Bloch. *Proc. Zool. Soc. Bengal, 1 (1):* 58-64

Murugesan, V.K. (1978). The growth potential of the murrels, *Channa marulius* (Hamilton) and *Channa striatus* (Bloch). *J. Inland Fish. Soc. India, 10:* 169-70

Nichols, J.T. (1943). *The Fresh Water Fishes of China, Natural History of Central Asia.* The Amer. Mus. of Nat. Hist., New York, 9, 321 pp.

Nikolsky, G.V. (1963). The Ecology of Fishes. Transl. from Russian. Academic Press, New York, 352 pp.

Ojha, J., Mishra, N., Saha, M.P. and Munshi, J.S.D. (1979). Bimodal oxygen uptake in juveniles and adults amphibious fish *Channa* (= *Ophiocephalus*) *marulius*. *Hydrobiologia, 63 (2):* 153-9

Panday, J.P. and Dwivedi, A.S. (1974). Studies of morphology and physiology of olfactory organs in a murrel *Channa punctatus* (Bloch). *Indian J. Zootomy, 14 (1):* 59-66

Pande, B.P. and Shukla, R.P. (1972). On the juvenile and adult of an ectoparasitic fluke in some of our freshwater fishes. *Curr. Sci., 41 (18):* 682-3

Pandian, T.J. (1967a). Intake and digestion, absorption and conversion of food in the fishes *Megalops cyprinoides* and *Ophiocephalus striatus. Mar. Biol., 1:* 16-32

―――― (1967b). Food intake, absorption and conversion in the fish *Ophiocephalus striatus. Helgolander Wiss. Meeresunters., 15:* 637-47

Pandian, T.J. and Vivekanandan, E. (1976). Effects of feeding and starvation on growth and swimming activity in an obligatory air-breathing fish, *Ophiocephalus striatus. Hydrobiologia, 49 (1):* 33

Pantulu, V.R. (1976). Floating cage culture of fish in the lower Mekong basin. In: *FAO Technical Conference on Aquaculture*, Kyoto, Japan, 26 May-2 June 1976, FIR:AQ/CONF/76/E10

Parameswaran, S. and Murugesan, V.K. (1976a). Breeding season and seed resources of murrels in swamps of Karnataka State. *J. Inland Fish. Soc. India, 8:* 60-7

―――― (1976b). Observations on the hypophysation of murrels (Ophiocephalidae). *Hydrobiologia, 50 (1):* 81-2

Qasim, S.Z. and Bhatt, V.S. (1966). The growth of the freshwater murrel, *Ophiocephalus punctatus* Bloch. *Hydrobiologia, 27:* 289-316

Qayyum, A. and Qasim, S.Z. (1962). Behaviour of the Indian murrel, *Ophicephalus punctatus,* during brood-care. *Copeia, 2:* 465-7

―――― (1964). Studies on the biology of some freshwater fishes. Part I. *Ophicephalus punctatus* Bloch. *J. Bombay Nat. Hist. Soc., 61 (1):* 74-98

Rahimullah, M. (1946). Observations on the breeding habits of *Ophicephalus striatus* Bl. In: *Proc. 33rd Ind. Sci. Congr.,* Bangalore, 111, Abstracts: 129

Raj, B.S. (1916). Notes on the freshwater fish of Madras. *Rec. Ind. Mus., Calcutta, 12:* 249-94

Reddy, B.P. (1979). The fecundity of *Channa punctata* (Bloch, 1793) (Pisces, Teleostei, Channidae) from Guntur, India. *Proc. Indian Acad. Sci., 88B (1) 2:* 95-8

Sampath, K. and Pandian, T.J. (1980). Effects of density on food utilization and surfacing behaviour in the obligatory air breathing fish *Channa striatus. Hydrobiologia, 68 (2):* 113-17

Sharma, K.P. and Simlot, M.M. (1971). Chemical composition of some commercially important fishes of Jaisamand Lake, Udaipur. *J. Inland Fish. Soc. India, 111:* 121-2

Sin, A.W. and Cheng, K.W.J. (1976). Management systems of inland fish culture in Hong Kong. In: *Symposium on the Development and Utilization of Inland Fishery Resources,* Colombo, Sri Lanka, 27-29 October 1976, 10 pp.

Singh, R. (1967). Studies on the morphology and histology of alimentary canal of *Ophicephalus marulius. Agra Univ. J. Res., 16:* 27-38

Smith, H.M. (1965). Freshwater fishes of Siam or Thailand. *Bulletin, 188,* United States National Museum: 465-74

Smith, T.I.J. (1973). The commercial feasibility of rearing pompano, *Trachinotus carolinus* (Linnaeus) in cages. Ph.D. dissertation, University of Miami, Coral Gables, Florida, 62 pp.

Sood, M.L. (1973). A new piscine nematode parasite *Metaquimperia madhuai* n.sp. from *Channa striatus* from India. *Zool. Anz. Leipzig, 190 (5/6):* 347-9

Srivastava, G.C. and Srivastava, R.C. (1976). A new host record for *Branchiomyces sanguinis* Plehn. *Curr. Sci, 45 (24):* 874

——— (1977). *Dictyuchus anomalous* (Nagai) a new pathogen of freshwater teleost. *Curr. Sci., 46 (4):* 118

Sterba, G. (1967). *Freshwater Fishes of the World.* Studio Vista, E. Germany, 877 pp.

Tan, Y.T. (1971). Proximate composition of freshwater fish. Grass carp, *Puntius gonionotus* and *Tilapia. Hydrobiologia, 37:* 361-6

Tandon, K.K. and Goswami, S.C. (1968). A comparative study of the digestive system of Channa species. *Research Bulletin (NS) of the Punjab University, 19:* 13-31

Varma, B.R. (1979). Studies on the pH tolerance of certain freshwater teleosts. *Comp. Physiol. Ecol., 4 (2):* 116-17

Vivekanandan, E. (1976). Effects of feeding on the swimming activity and growth of *Ophiocephalus striatus. J. Fish. Biol., 8:* 321-30

——— (1977a). Effects of the pO_2 on swimming activity and food utilization in *Ophiocephalus striatus. Hydrobiologia, 52 (2-3):* 165-9

——— (1977b). Surfacing activity and food utilization in the obligatory air-breathing fish *Ophiocephalus striatus* as a function of body weight. *Hydrobiologia, 55 (2):* 99-112

——— (1977c). Ontogenetic development of surfacing behaviour in the obligatory air-breathing fish *Channa (Ophiocephalus) striatus. Physiol. Behav., 18:* 559-62

Vivekanandan, E. and Pandian, T.J. (1977). Surfacing activity and food utilization in a tropical air-breathing fish exposed to different temperatures. *Hydrobiologia, 54 (2):* 145-60

Weber, M. and Beaufort, L.F. (1922). *The Fishes of the Indo-Australian Archipelago IV,* pp. 312-30

Wee, K.L. (1981). A case study of snakehead (*Channa striatus*) farming in Thailand. Internal report of Institute of Aquaculture, University of Stirling

Willey, A. (1910). Observations on the nests, eggs and larvae of *Ophiocephalus striatus. Spol. Zeyl., 6:* 108-23

Yapchiongo, J.V. and Demonteverde, L.C. (1959). The biology of Dalag (*Ophicephalus striatus* Bloch). *The Philippine Journal of Fisheries, 7 (1-2):* 105-40

4 CARP *(CYPRINUS CARPIO L.)* NUTRITION — A REVIEW

Kim Jauncey

1. Introduction

The past fifty years have seen the understanding of the food requirements of man and his food animals advance to the stage where the study of comparative nutrition is a recognised activity. This has been greatly enhanced by increased understanding of the underlying physiological and biochemical processes involved. In so far as the terrestial animals are concerned, advances such as the discovery of vitamins, definition of the trace elements and elucidation of essential amino and fatty acids have allowed the virtual establishment of the nutrient requirements for many species. With the increasing food demands of an expanding population man has a growing need to understand the nutrition of those species upon which he feeds. As interest in fish culture, as a means of producing high quality protein for human consumption, has increased, this has stimulated great interest in the nutrition of fish.

The rearing of large numbers of animals in relatively confined conditions, whether terrestrial or aquatic, necessitates a detailed understanding of their nutrition in order to provide a diet that is adequate for their growth and well-being. In the past, large quantities of fish, enough to satisfy demand, were available from natural sources, but more recently the scientific intensive farming of fish has become economically viable due to a growing demand coupled with a depletion of natural resources. Practical, economic fish husbandry has become essential in those areas of the world with a low dietary protein intake. Incomplete or incompletely applied knowledge of fish nutrition has often meant that the potential of streams and impoundments suitable for fish culture in these areas has not been reached.

Adequate diets for fish husbandry are the foundation on which fish farming is built, with the success or failure of a fish culture project dependent upon the nutritional status of the fish. Whilst the degree of knowledge relating to the feeding of the majority of commercially exploited species of animal has reached a high level, the provision of reliable results on fish nutrition is still in its infancy.

Until recently the emphasis was on natural feedstuffs and, consequently, little or no data existed on the nutrient requirements of fish. Fish farming, on a global scale, still largely depends upon natural food with some supplementation with the by-products of other forms of agriculture or industry, such as slaughter house wastes, grain wastes, silk-worm pupae, night soil, etc. At low culture densities these diets are adequate as most of the nutrient requirements of the fish are satisfied from natural sources. However, at high densities fish are dependent on artificial feeds, only benefitting slightly, if at all, from natural food. Thus at higher densities inadequate supplementary feeding leads to poor growth, nutritional disease and, due to poor fish condition, increased susceptibility to parasitic and bacterial infestations.

Due to the generally more intensive culture practices used with these species, research in fish nutrition has tended to centre on salmonids. However, more recently, intensification of fish culture systems for other species has led to diversification in fish studied. In this review research into carp nutrition will be the main subject of discussion, but where there are omissions in this field or where analogies may be drawn from research upon other species these will also be discussed. In order to evaluate nutritionally balanced diets properly it is necessary to understand the biochemical roles and interactions of the various food components. The biochemistry of fish nutrition is a field too vast and detailed to be covered here and reference should be made to other works on the subject (Cowey and Sargent, 1972, 1979; Halver, 1972). However, a broad outline of the role of each group of nutrients and a review of the research into the requirements for each is given.

2. Proteins and Amino Acids

The protein component of fish feeds is given first consideration as it is generally accepted that this contributes most significantly towards the cost. Proteins are a class of complex chemical substances which may be regarded as being built up of a number of simpler subunits; these are the amino acids. Dietary protein is required for three basic functions:

(1) maintenance – the making good of tissue wear and tear;
(2) repletion of depleted tissues;
(3) growth or formation of new additional protein (Cowey and Sargent, 1972).

No single protein is ideal in that it satisfies all of the requirements of an animal. The amino acids, of which proteins are composed, are utilised for a variety of metabolic processes and thus, even for a single fish species, the quantitative or qualitative requirements may not be constant. Approximately 24 amino acids are common to all proteins, the biological value of a protein depending upon the relative proportions of its amino acid constituents. Utilisation of a protein is affected by its amino acid profile, amino acid availability, protein intake level, energy content of the diet and the physiological state of the animal. The amino acid profile of a protein describes the relative proportions of the essential amino acids within that protein. A protein with a 'good' profile is one whose composition, in terms of proportions of essential amino acids, approximates to the quantitative essential amino acid requirements of the fish.

Amino acids fall into two basic categories, those that are essential and cannot be synthesised by the fish and those that are non-essential and can be synthesised provided that an alternative source of nitrogen is available. There is also a third vague category of semi-essential amino acids which can be synthesised but not a a rate fast enough to satisfy metabolic demands; such amino acids are usually classed as essential.

2.1 Qualitative Amino Acid Requirements

Test diets used to establish the amino acid requirements of fish generally follow that of Halver (1957b) who formulated the first successful purified (or defined) diet and established the essential amino acid requirements of chinook salmon (*Oncorhynchus tschawytscha*). Other workers have attempted to adapt such a diet to other species with varying degrees of success.

Aoe *et al.* (1970), using Halver's test diet, failed to get carp (*Cyprinus carpio*) to grow unless the diet was supplemented with casein, even though 90 per cent of the nitrogen in the test diet was absorbed. Nose *et al.* (1974), in a subsequent and more successful experiment, attributed this to the low feeding rate employed by Aoe *et al.* (3 per cent per day) and the low feed efficiency that they experienced.

Nose *et al.* (1974) managed to obtain limited growth of carp on a test diet neutralised to a pH of 6.5-6.7 by feeding *ad libitum* six times per day. It was proposed that as carp is a stomachless fish, digestion normally proceeds in neutral or slightly alkaline conditions and that the strong acidic buffering of free amino acids and amino acid hydrochlorides disturbed their assimilation. However, even neutralisation of the test diet resulted in growth rates of less than 60 per cent of that of comparable casein controls indicating some peculiarity in the nutritional value of free amino acids to carp. Despite the low growth rates obtained Nose *et al.* (1974) demonstrated that fish fed diets deficient in each of arginine, histidine, isoleucine, leucine, lysine, methionine, phenylalanine, threonine, tryptophan and valine failed to grow until the deleted amino acid was replaced. It would thus appear that carp require the same ten amino acids reported to be essential for other species (Table 4.1).

Table 4.1: Species Requiring the Same Ten Essential Amino Acids as Carp

Species		Reference
Carp	(*Cyprinus carpio*)	Nose *et al.* (1974)
Rainbow trout	(*Salmo gairdneri*)	Shanks *et al.* (1962)
Channel catfish	(*Ictalurus punctatus*)	Dupree & Halver (1970)
Eel	(*Anguilla japonica*)	Nose (1970)
Chinook salmon	(*Oncorhynchus tschawytscha*)	Halver (1957b)
Sockeye salmon	(*Oncorhynchus nerka*)	Halver & Shanks (1960)
Plaice	(*Pleuronectes platessa*)	Cowey *et al.* (1970)
Sole	(*Solea solea*)	Cowey *et al.* (1970)
Sea bass	(*Dicentrarchus labrax*)	Metailler *et al.* (1973)
Tilapia	(*Tilapia zillii*)	Mazid *et al.* (1978)

2.2 *Quantitative Essential Amino Acid Requirements*

The quantitave essential amino acid requirements of fish were first elucidated in the chinook salmon by Halver and his colleagues (Mertz, 1969) using a modification of the test diet first used by Halver in 1957. This diet contained a small portion of whole proteins (casein/gelatin) and a larger component of crystalline amino acids. Diets containing graded levels of the essential amino acids were fed and an Almquist dose/response curve plotted.

More recently the quantitative essential amino acid requirements for carp have been defined (Nose, 1978). The test diet employed contained 48 per cent L-amino acids, 18 per cent dextrin, 7 per cent corn oil, 2 per cent pollack liver oil, 9 per cent vitaminised cellulose powder, 4 per cent minerals, 2 per cent cellulose powder and 10 per cent carboxymethylcellulose (binder). The level of the amino acid under investigation was varied by isonitrogenous replacement of L-alanine maintaining a crude dietary protein (N x 6.25) level of 38.5 per cent. This protein level was chosen as it had been shown in previous experiments (Ogino and Saito, 1970) to be the optimum for growth of carp and essential amino acid requirement increases with increasing dietary protein level up to the optimum. The requirements of carp for essential amino acids were found to be as presented in Table 4.2 in comparison to other species.

Ogino (1980) has employed a novel approach to the elucidation of the quantitative essential amino acid requirements of carp and rainbow trout. The method depends on the assumption that the daily increase in the essential amino acid content of 100 g of fish, is supplied from the dietary protein. Hence, analysis of the fish tissues will enable calculation of the essential amino acid requirements. The results obtained when feeding these species a diet containing 40 per cent protein, that is 80 per cent digestible, at 3 per cent of the body weight per day are also presented in Table 4.2. If the feeding rate, dietary protein level and protein digestibility are known, these data permit calculation of the required dietary amino acid profile assuming that all of the amino acids are equally digestible.

Many of the comparisons between the essential amino acid requirements listed in Table 4.2 are self-evident; however some require further comment. The arginine requirement of fish is considerably higher than that of the rat and similar to that of the chick possibly, as Mertz (1969) pointed out, because both fish and birds have a poorly developed urea cycle which in mammals supplies 75 per cent of the requirement for arginine.

The deficiency symptoms, of scoliosis and lordosis, recorded in sockeye salmon and rainbow trout fed diets deficient in tryptophan (Halver and Shanks, 1960; Shanks *et al.*, 1962) were not recorded in carp. Tyrosine was reported to be dispensable for growth in carp (Nose *et al.*, 1974). However, diets low in both phenylalanine and tyrosine caused growth depression; thus the tyrosine requirement for a low (1 per cent) phenylalanine diet was determined and was found to be 0.8 per cent of the diet and 2.1 per cent of the dietary protein.

The variation in essential amino acid requirements for the fish species so far studied is sufficiently great as to result in a need to formulate the essential amino acid levels of diets on a species basis. For those species whose quantitative requirements are, as yet, undetermined it would be advisable to formulate diets to contain the highest levels of essential amino acids required by any species so far evaluated.

There are some incomplete data to suggest that the essential amino acid requirements of carp in extensive culture may, in part, be provided from intestinal microflora. Syvokiene *et al.* (1974, 1975) and Lesauskiene *et al.* (1974, 1975) (cited by Dabrowski, 1979) found that in carp fry as well as two and three-year-old fish that some essential and non-essential amino acids were being derived from gut bacteria. The essential amino acids valine, lysine, threonine, phenylalanine, leucine, methionine and histidine were synthesised by the gut bacteria at varying rates and were concluded to be contributing to the requirement of carp for these amino acids.

Amino Acid Availability. It is necessary not only to know the amino acid profile of feed proteins but also to determine the availabilities of the amino acids

Table 4.2: The Essential Amino Acid Requirements of Four Species of Fish, the Rat and the Chick[e]

Amino Acid	Chinook Salmon[a]	Japanese Eel[b]	Carp[b]	Carp[c]	Rainbow Trout[c]	Rat[a]	Chick[d]
Arginine	2.4 (6.0/40)	1.7 (4.0/42)	1.6 (4.3/38.5)	1.52 (3.8/40)	1.40 (3.5/40)	0.2 (1.5/13.19)	1.1 (6.1/18)
Histidine	0.7 (1.8/40)	0.8 (1.9/42)	0.8 (2.1/38.5)	0.56 (1.4/40)	0.64 (1.6/40)	0.4 (3.0/13.19)	0.3 (1.7/18)
Isoleucine	0.9 (2.2/41)	1.5 (3.6/42)	0.9 (2.5/38.5)	0.92 (2.3/40)	0.96 (2.4/40)	0.5 (3.8/13.19)	0.8 (4.4/18)
Leucine	1.6 (3.9/41)	2.0 (4.8/42)	1.3 (3.3/38.5)	1.64 (4.1/40)	1.76 (4.4/40)	0.9 (6.8/13.19)	1.2 (6.7/18)
Lysine	2.0 (5.0/40)	2.0 (4.8/42)	2.2 (5.7/38.5)	2.12 (5.3/40)	2.12 (5.3/40)	1.0 (7.6/13.19)	1.1 (6.1/18)
Methionine	0.6 (1.5/40)	1.2 (2.9/42)	1.2 (3.1/38.5)	0.64 (1.6/40)	0.72 (1.8/40)	0.6 (4.6/13.19)	0.8 (4.4/18)
	Cys = 1%	Cys = 0%	Cys = 0%	Cys = +	Cys = +	Cys = 0.6	
		0.9 (2.1/42)	0.8 (2.1/38.5)				
		Cys = 1%	Cys = 2%				
Phenylalanine	1.7 (4.1/41)	2.2 (5.2/42)	2.5 (6.5/38.5)	1.16 (2.9/40)	1.24 (3.1/40)	0.9 (6.8/13.19)	7.3 (7.2/18)
	Tyr = 0.4%	Tyr = 0%	Tyr = 0%	Tyr = +	Tyr = +	Tyr = 0%	
		1.2 (2.9/42)	1.3 (3.4/38.5)				
		Tyr = 2%	Tyr = 1%				
Threonine	0.9 (2.2/40)	1.5 (3.6/42)	1.5 (3.9/38.5)	1.32 (3.3/40)	1.36 (3.4/40)	0.5 (3.8/13.19)	0.6 (3.3/18)
Tryptophan	0.2 (0.5/40)	0.4 (1.0/42)	0.3 (0.8/38.5)	0.24 (0.6/40)	0.20 (0.5/40)	0.2 (1.5/13.19)	0.2 (1.1/18)
Valine	1.3 (3.29/40)	1.5 (3.6/42)	1.4 (3.6/38.5)	1.16 (2.9/40)	1.24 (3.1/40)	0.4 (3.0/13.19)	0.8 (4.4/18)
Total	12.3 (30.49)	14.8 (35.4)	13.7 (35.8)	11.28 (28.2)	11.64 (29.1)	5.6 (42.4)	7.4 (45.5)

Notes: a. Data for chinook salmon and rat from Mertz (1969).
b. Data for Japanese eel and carp from Nose (1978).
c. Data for rainbow trout and carp from Ogino (1980).
d. Data for chick from NAC (1977).
e. Values are expressed as grams per 100 g of dry diet. In parentheses the numerators are grams per 100 g of protein and the denominators are the per cent total dietary protein.

they contain. In particular two of the essential amino acids, lysine and methionine, readily undergo changes during the processing of feedstuffs that may render them unavailable to the fish.

Lysine is a basic amino acid, in addition to the α-amino group normally found in a peptide linkage; it also contains a second, ϵ-amino group which must be free and reactive or the lysine, although chemically measurable, will not be biologically available (Cowey and Sargent, 1972; Cowey, 1978). Methods for measuring 'available' lysine have been published (Carpenter and Ellinger, 1955) and determinations have been found to correlate well with experimentally determined biological values in birds and mammals (Cowey, 1978).

Methionine is difficult to measure in feedstuffs as it readily undergoes oxidation during processing to form a sulphoxide or sulphone. Methionine is usually determined by performate oxidation to methionine sulphone. However, this does not reveal what proportion of the sulphone was originally present. In addition methionine sulphoxide may have some biological value to fish if they are able to reduce it to methionine. Njaa (1977) has proposed a method that measures both methionine and methionine sulphoxide in the feed and Ellinger and Duncan (1976) have described a specific method for methionine — these methods remain to be evaluated in a practical context.

As remarked by Cowey (1978) 'measurement of sulphur amino acids in protein feeds is identified as a need in assessment of nutrients for fish. The biological value of methionine sulphoxide (if any) also ought to be evaluated.'

2.3 Dietary Protein Requirements

Protein is the basic building nutrient of any growing animal and with muscle being, anatomically, by far the major component of the fish body, protein usually accounts for 65-85 per cent of the dry matter. The level of dietary protein producing maximum growth of fish depends upon the proteins' amino acid profile and amino acid availability (as already discussed), the caloric content of the diet and the physiological state of the animal.

The efficiency with which fish are able to utilise dietary protein is usually determined by measurement of one of the following parameters. Protein Efficiency Ratio (PER) is defined as the gain in wet weight of the animal per gram of crude protein consumed (Osborne *et al.*, 1919) thus:

$$PER = \frac{\text{weight gain, g wet fish}}{\text{g crude protein fed}}$$

Although PER values give a somewhat better indication of the nutritional status of the fish, with respect to dietary protein, than Food Conversion Ratios (FCRs, grams of diet fed per gram of weight gained), they do not take into account the proportion of ingested protein used for maintenance and are based on the assumption that the growth of the fish consists of tissues with identical composition in all groups. An improved assessment of the nutritional status of the fish with respect to dietary protein utilisation is the apparent efficiency of deposition of dietary protein as body tissue, the Net Protein Utilisation (NPU). NPU in fish is generally determined by the carcass analysis method of Bender and Miller (1953) and Miller and Bender (1955). If no corrections are made for endogenous nitrogen losses the results are expressed as Apparent NPU:

$$\text{Apparent NPU} (\%) = \frac{Nb - Na}{Ni} \times 100$$

where Nb is the body nitrogen at the end of the test, Na the body nitrogen at the beginning of the test and Ni the amount of nitrogen ingested.

Estimation of endogenous nitrogen losses will permit the calculation of True NPU. These can be measured by finding the body nitrogen loss on a zero protein diet. However, this poses acceptability (palatability) problems that are usually overcome by the feeding of a low protein diet as proposed by Cowey *et al.* (1974). Hence:

$$\text{True NPU} (\%) = \frac{B - (Bk - Ik)}{I} \times 100$$

where B is the total body nitrogen of fish fed the test diet and Bk the total body nitrogen of fish fed the low protein diet, with nitrogen intakes of I and Ik respectively.

Estimates of endogenous nitrogen excretion (ENE, nitrogen excreted by fish fed a zero protein diet) have been made for carp. Ogino *et al.* (1973) reported the ENE of carp to be 7.2 mgN/100 g of fish/day at 20°C and 8.6 mgN at 27°C. A more recent publication (Ogino *et al.*, 1980) records the ENE of carp as 14 mg/100 g of body weight/day at 22°C. The earlier study employed fish of 78-370 g at 20°C and 133-215 g at 27°C whilst the latter study used fish of 1.5-11.9 g. It may be postulated that the differences in ENE values may be due to the higher metabolic rate (with concomitantly higher ENE) of the smaller fish.

Obtaining a value for True NPU in this way, together with the determination of true protein digestibility, permits calculation of the Biological Value (BV) of the protein which is a measure of the percentage of absorbed nitrogen retained as body tissue.

$$\text{BV} (\%) = \frac{\text{True NPU}}{\text{True Digestibilty}} \times 100 \text{ (Bender and Miller, 1953)}$$

As the level of dietary protein markedly affects its utilisation and, consequently, the previously described parameters, it is necessary to determine the 'optimum' dietary protein level for a particular species and to perform studies comparing different dietary protein sources at this level.

Definition of the minimum dietary protein level giving optimum weight gain of a species of fish was first investigated in chinook salmon in 1958 by DeLong *et al.* and this type of experiment has since been repeated on many fish species. Ogino and Saito (1970) were the first to attempt definition of the optimum dietary protein level for carp using diets containing varying proportions of casein to give crude protein levels of 0.4-55 per cent. Over the whole range they found that weight gain was directly proportional to dietary protein level. However, the accumulation of body protein was found to reach a maximum at a dietary protein level of 38 per cent. This study was conducted at 23°C with diets containing a metabolisable energy content of approximately 3.7 kcal/g. These authors also found that both PER and NPU decreased linearly with increasing dietary protein level. That this should be so is contrary to the results of Cowey *et al.* (1972) with plaice, Zeitoun *et al.* (1974) with rainbow trout and Dabrowski (1977) with grass carp. However, if absolute cases are considered at the maintenance level of dietary protein (where nitrogen ingested = nitrogen excreted) both PER and NPU must, by definition, be zero. At still lower levels of dietary protein the fish would lose weight. Fish fed a zero protein diet would, if anything, exhibit a negative PER and NPU approximating to the endogenous nitrogen excretion.

In a subsequent study Ogino *et al.* (1976) reported an optimum dietary protein (casein) content of 35 per cent at $20°C$ with a metabolisable energy content of 3.4 kcal/g. Sin (1973a, b) also determined the optimum dietary protein level for carp using a mixture of dietary proteins (corn, fishmeal, wheat gluten and corn gluten). Fish of 4-7 g at $25°C$ exhibited an optimum dietary protein requirement of 38.4 per cent on diets having a metabolisable energy (ME) content of 2.7 kcal/g and 33 per cent with an ME of 3.06 kcal/g.

In experiments conducted by the author (Jauncey, 1979) the dietary protein level (using fishmeal) producing maximum growth of 60-70 g carp maintained in a thermal effluent at $28°C$ was 35 per cent with a diet containing 3.82 kcal of metabolisable energy per gram. Dietary protein levels have been recommended for carp of various sizes (NAC, 1977). A level of 43-47 per cent was recommended for fry and fingerlings, 37-42 per cent for fingerlings and sub-adults and 28-32 per cent for adults and brood fish.

2.4 Sparing of Dietary Protein for Growth by Elevated Levels of Dietary Lipid and Carbohydrate

Fish are efficient converters of food to flesh, with food conversion ratios frequently in the range 0.6-1.5 as compared to values of 3 or more obtained for conventional livestock such as pigs and poultry. This improvement in food conversion is outweighed, in economic terms, by the fact that fish feeds contain three times as much protein as conventional livestock feeds. A large proportion of the protein in fish diets must, therefore, be catabolised rather than being incorporated within the tissues as measurable growth.

In the nutrition of fish it is apparent that proteins are not only important as amino acid sources providing the enzymatic and structural components of cells but also as a source of energy. Under conditions where energy intake is inadequate dietary protein will be used as an energy source, and in fact protein synthesis within the animal will only reflect the quantity and quality of the dietary protein when a sufficient energy intake occurs (Cowey, 1978). At high levels of dietary protein a proportion is deaminated and the carbon residues burned as energy. Conversely an excessive energy intake at moderate levels of dietary protein will lead to deposition of fat in the fish leading to undesirable changes in carcass composition. Thus the design of practical diets is a compromise between a protein level that will permit good growth with little conversion to energy and an energy level concomitant with high rates of protein synthesis but not such as to lead to greater deposition of carcass lipid than within wild individuals of the same species.

In order to satisfy the energy requirements of fish it is wasteful to use dietary protein since, per kilocalorie, protein is an expensive energy source. In addition the excretory end-products of protein catabolism, principally ammonia, are toxic to fish and cause growth depression when they accumulate in the culture water. As carp is a warm-water species with a temperature optimum of 28° C (Section 8) its metabolic energy requirements are far greater than those of the salmonids reared at 12-15°C and thus the sparing of dietary protein for growth becomes even more significant.

Protein Sparing by Dietary Lipid. The total energy contents of protein, lipid and carbohydrate have been estimated as 5.5, 9.1 and 4.1 kcal/g respectively (Brody, 1945). Thus in terms of energy supplied per gram, dietary lipid should have the greatest protein-sparing effect. Numerous studies have been made of the effects of increasing the dietary energy intake by increasing levels of lipid, on food conversion, protein utilisation and growth of various species (Tiemeier *et al.*, 1965; Stickney and Andrews, 1972; Lee and Putnam, 1973; Sin, 1973a, b; Adron *et al.*, 1976; Higuera *et al.*, 1977; Viola and Rappaport, 1978; Reinitz, 1978; Takeuchi *et al.*, 1978a, b, c, 1979a).

Studies on warm-water species should prove especially valuable as several authors have reported that lipid metabolism (digestion, absorption and utilisation) is improved with increasing environmental temperatures (Kayama and Tsuchiya, 1959; Atherton and Aitken, 1970; Shcherbina and Kazlauskene, 1971; Stickney and Andrews, 1972; Andrews *et al.*, 1978).

Experiments were conducted by the author (Jauncey, 1979) using diets containing 21, 29, 37 and 45 per cent protein with levels of dietary lipid of 6, 12 and 18 per cent at each protein level. The protein sparing of dietary lipid for mirror carp at 20° C was evident from the increased values for Specific Growth Rate (SGR), Protein Efficiency Ratio (PER) and Apparent Net Protein Utilisation (NPU) at each protein level with increasing levels of dietary lipid. It proved possible to reduce the protein content of diets containing 18 per cent lipid from 45 to 29 per cent with no diminution of weight gain and with improved utilisation of dietary protein. This finding is in general agreement with those of others. Sin (1973a, b) showed that the 'optimum' protein content of diets for mirror carp was reduced from 38 to 33 per cent by increasing the metabolisable energy from 2.8 to 3.1 kcal/g by the addition of 5 per cent soybean oil to the diets.

An experiment demonstrating an 'extra-caloric' effect of an oil supplement to carp diets has been reported (Viola and Rappaport, 1979). The energy content of a basal diet was increased by 12 per cent by replacement of 6 per cent bentonite with oil (acid soybean oil) and by replacing 13 per cent bentonite with grains (sorghum and wheat). Weight gains on the oil supplemented pellets exceeded gains on the grain supplemented pellets by 50 per cent. Addition of 60 calories to the pellets, in the form of oil, resulted in the retention of 105 additional calories in the body, demonstrating the 'extra-caloric' or protein-sparing effect of the oil.

Contrary to all of the above results Takeuchi *et al.* (1979b) found that increasing the lipid content from 5 to 15 per cent in diets containing 22, 32 and 41 per cent protein (from casein) had little or no effect on the growth rate, food conversion, energy retention or protein utilisation of carp.

Fish oil, used as the sole lipid source, was found to spare dietary protein for growth in rainbow trout (Phillips, 1969). More recently Takeuchi (1978b) found that the protein level, in trout diets, could be reduced from 48 to 35 per cent with no loss of weight gain given dietary lipid levels of 15-20 per cent. Results obtained with turbot (*Scophthalmus maximus*) present a similar picture

(Adron *et al.*, 1976). Protein levels were reduced from 50 to 35 per cent with no loss of weight gain and with improved protein utilisation provided that the diets had a gross energy content of 3 kcal/g or more.

In the experiment conducted by the author (Jauncey, 1979) the maximum amount of dietary lipid fed (18 per cent) did not depress growth. Results for channel catfish, however, showed depression of growth when dietary lipid levels were raised from 12 to 16 per cent of the dry diet (Dupree, 1969). Other authors report no growth-depressing or pathological effects of very high levels of dietary lipid. Higashi *et al.* (1964) fed diets containing 25 per cent marine oil to rainbow trout with no ill-effects and Kitamikado *et al.* (1964) fed diets containing up to 30 per cent lipid. More recently Takeuchi (1978a) fed diets containing 54 per cent protein and 20 per cent lipid to rainbow trout with no observed pathological disturbances. Higuera *et al.* (1977) fed a diet containing 18 per cent lipid to rainbow trout for six months with no pathological disturbances and, in fact found functional adaptation to the 'high fat' diet as the excretion of bile salts increased by approximately 27 per cent.

Varying the dietary lipid content also affects the carcass composition of the fish. Increasing the dietary lipid content increases the carcass lipid content with a concomitant decrease in the carcass moisture content (Brett *et al.*, 1969; Andrews and Stickney, 1972; Papoutsoglou and Papoutsoglou, 1978; Takeuchi, 1978a; Jauncey, 1979). Increased carcass lipid contents with increasing levels of dietary lipid have been reported for channel catfish (Page and Andrews, 1973; Murray *et al.*, 1977; Garling and Wilson, 1976), chinook salmon (Buhler and Halver, 1961), turbot (Adron *et al.*, 1976), rainbow trout (Lee and Putnam, 1973; Austreng, 1976; Takeuchi, 1978d; Reinitz *et al.*, 1978) and mirror carp (Sin, 1973a, b; Jauncey, 1979). Even when fed a diet containing 45 per cent protein and 18 per cent lipid the maximum level of carcass lipid reported in carp (Jauncey, 1979) was 9.5 per cent of the wet weight. This is considered to be perfectly acceptable to the consumer as carp is generally reported to be a fatty fish with carcass lipid levels of 20 per cent reported for pond-reared specimens (Meske and Pfeffer, 1978). In intensive culture systems using high culture temperatures carcass lipid levels of 10 per cent are more generally obtained (Sin, 1973a, b; Takeuchi, 1978c).

As the maximum level of dietary lipid tolerated by carp would appear to be in excess of 18 per cent it might appear possible to incorporate levels above this. However, reducing the ratio of dietary protein to energy even further by increasing levels of dietary lipid might result in a reduction in food intake if carp feed to a set dietary energy intake. Lee and Putnam (1973) reported that rainbow trout adjust their total food intake to a set energy level and thus diets with very low protein to energy ratios resulted in reduced growth due to reduced protein intake. Rozin and Mayer (1964) reported that increasing the caloric density of diets reduced the voluntary food intake of goldfish (*Carassius auratus*).

Further difficulties in the formulation of feeds containing high levels of dietary lipid are encountered with the mechanics of the pelletising process and the difficulty in storing such diets so as to prevent oxidative rancidity of the lipid. These obstacles might be reduced by the use of saturated fats in carp feeds, a possibility which is discussed with respect to essential fatty acid requirements in Section 3.

Some general conclusions may be drawn from the discussion presented here.

It has proved possible to reduce the protein content of diets for mirror carp from 45 to 29 per cent, at a dietary lipid level of 18 per cent, with no reduction in growth performance, with improved protein utilisation and with no unacceptable accumulation of carcass lipid (Jauncey, 1979). This demonstrates the beneficial effects, both economic and nutritional, of increasing the level of dietary lipid as compared to conventional 'low fat' fish feeds.

Protein Sparing by Dietary Carbohydrate. As previously stated, in the formulation of commercial fish rations an attempt should be made to maximise the use of dietary protein for growth by including levels of dietary lipid and carbohydrate commensurate with this aim. The use of carbohydrate as a protein-sparing energy source has received much less attention than the use of dietary lipid for the same purpose and there is still some dissension as to the levels of dietary carbohydrate that should be included in commercial fish rations.

There is also disagreement as to the possibly deleterious effects that high levels of dietary carbohydrate may have. More than 12 per cent carbohydrate was reported to cause poor growth, liver abnormalities and mortalities in rainbow trout (Phillips *et al.*, 1948). Dupree and Sneed (1966) found, with channel catfish, that increasing the level of dextrin in the diet from 2.5 to 10 per cent increased weight gains but further increase to levels of 15-20 per cent depressed growth. Edwards *et al.* (1977) reported that the growth of rainbow trout was depressed by increasing the level of dietary carbohydrate from 17 to 35 per cent. Contradictory results have been presented for channel catfish (Garling and Wilson, 1976, 1977) and salmonids (Buhler and Halver, 1961; DeLong *et al.*, 1958; Lee and Wales, 1973; McLaren *et al.*, 1947a) which tolerated dietary carbohydrate levels up to 61.5 per cent of the dry diet.

As carp is an omnivorous fish species whose natural diet can contain large amounts of complex carbohydrates, efficient utilisation of such an energy source might be expected. Although the ratio of the metabolisable energy contents of lipid and carbohydrate is approximately 2.25:1 (Garling and Wilson, 1977), cheap dietary ingredients, such as cereals, contain a large proportion of carbohydrates and thus their use is desirable. Adron *et al.* (1976) report definite protein sparing by carbohydrate, as reflected by increased PER and NPU, in turbot when dietary carbohydrate was increased from 9 to 18 per cent by addition of equal amounts of dextrin and glucose. Protein sparing by the addition of glucose to diets of plaice has also been reported (Cowey *et al.*, 1975). Pieper and Pfeffer (1978) reported that replacement of a basic diet with 30 per cent glucose, gelatinised starch or sucrose spared dietary protein in rainbow trout. They found that the efficiency of energy retention was greater in diets containing gelatinised starch or sucrose than in diets containing glucose and postulated that the slower influx of monosaccharides from the digestion of sucrose and starch might be beneficial. These authors also report a trebling of the hepatosomatic index (per cent liver weight of the final fish wet weight) when gelatinised starch was added to the basal ration and conclude that most of this was due to glycogen accumulation. Edwards *et al.* (1977), also with rainbow trout, fed three diets that were isonitrogenous and isoenergetic but with differing proportions of the energy supplied by carbohydrate. They found, conversely to the above, best growth on the lowest carbohydrate diet (17 per cent) and worst growth on the highest carbohydrate diet (38 per cent).

Results for chinook salmon (Buhler and Halver, 1961) showed that increasing the level of dextrin, in diets containing 38 per cent protein, from 0 to 48 per cent increased PER from 1.65 to 2.37, thus demonstrating the protein-sparing action of dietary carbohydrate in this species. As the dietary level of dextrin was raised the liver size and glycogen content increased without apparent pathological effects. Lin *et al.* (1977) reported that coho salmon grew as well on a diet where 46 per cent of the total metabolisable energy was supplied by carbohydrate as they did on a diet where the same quantity of energy was supplied by dietary lipid.

Results for channel catfish (Page and Andrews, 1973) indicate that lipid and corn starch are equally effective as dietary energy sources. These authors found that 25 per cent corn starch was equivalent, in metabolisable energy, to 12 per cent lipid and thus assigned them digestible energy values of 2.6 and 6.8 kcal/g respectively. Garling and Wilson (1977) continued these studies by replacing 15 per cent lipid stepwise by substitution with dextrin in a ratio of 2.25 dextrin:1 lipid. Replacement of the dietary lipid with carbohydrate to a level of 22.5 per cent dextrin, 5 per cent lipid did not affect weight gain, food conversion ratio, protein utilisation or energy retention. However, increasing the proportion of dietary carbohydrate resulted in increased liver size and liver glycogen contents although no pathological effects were observed.

Cowey and Sargent (1979) summarised experimentation on channel catfish, rainbow trout and plaice by concluding that levels of dietary carbohydrate up to about 25 per cent are as effective as an isocalorific amount of fat as an energy source for these species. This they relate to the work of Singh and Nose (1967) who found that the digestibility of starch in rainbow trout decreased from 68 per cent in diets containing 20 per cent starch to 26 per cent in diets containing 60 per cent starch.

Chiou and Ogino (1975) found that, in contrast to rainbow trout, carp were able on average to digest 85 per cent of the ingested starch at dietary levels from 19 to 48 per cent. This would seem to suggest better utilisation by carp of higher levels of dietary carbohydrate, although from the following discussion it will be seen that this is not necessarily the case.

Growth trials investigating the effects of varying carbohydrate levels on carp are inconclusive. Ogino *et al.* (1976) fed carp diets containing casein (60-0 per cent) the level of which was varied by replacement with dextrin (0-60 per cent), thus the effects of varying carbohydrate and protein levels were confounded and the conclusion of these authors, that carbohydrate was effectively used by carp, is not supported. A similarly inconclusive experiment was conducted by Sen *et al.* (1978) who replaced dextrin (71-26 per cent) with casein (0-45 per cent). They suggested that optimum growth of carp occurred with a diet containing 45 per cent protein and 26 per cent carbohydrate. This result is not surprising as 45 per cent was the highest protein level and 26 per cent the lowest carbohydrate level that were fed, and thus the range of treatment was insufficient.

Takeuchi *et al.* (1979b) increased the digestible carbohydrate levels of carp diets, at varying dietary protein levels, by replacing α-cellulose with starch and dextrin. They found no increase in growth, food conversion or protein utilisation when the digestible carbohydrate level was raised at protein levels of 22, 32, 37 and 45 per cent.

Erman (1969) reported on the 'nitrogen saving' effect of carbohydrate in carp diets as determined by measuring nitrogen excretion. This author found that the quantity of nitrogen excreted by fish fed a commercial ration was halved by the addition of 14 per cent boiled starch to the diet. However, the addition of starch to the diet reduced the protein content from 49 to 33 per cent. This reduction in protein content would have led to a higher NPU (Ogino and Saito, 1970) which may account for the observed reduction in nitrogen excretion.

Further, more biochemical, evidence of anomalies in the utilisation of dietary carbohydrate by carp is given in Section 4. It suffices to say at this point that no definitive study has, to date, adequately demonstrated the efficacy of dietary carbohydrate as an energy source in carp.

2.5 Dietary Protein Sources

Up until the present time fishmeal has been the principal source of protein in complete commercial fish rations. However, in recent years the supply of fishmeal has become increasingly uncertain and the price has risen rapidly. The most prominent event in this respect was the fishmeal crisis in 1972/3 when the fishery for the Peruvian anchovy (*Engraulis ringens*) failed; at this time it was estimated to supply more than 80 per cent of the worldwide production of saleable fishmeal (Anon, 1973). For this reason it has become of great importance to either replace fishmeal in commercial fish rations or, at least, reduce its use to a minimum.

Much of the research into the possibility of using unconventional protein sources as replacements for fishmeal in compounded fish feeds has centred upon salmonids, as this family includes the species most commonly reared on completely artificial diets. Many of the results obtained so far have proved encouraging. Replacement of fishmeal in the diets of warm-water fish species has received less attention since complete compounded rations have only recently been employed in the commercial culture of such species. The use of plant proteins would seem to be the most appropriate and significant, as such proteins are likely to be more constantly available, and cheaper to produce, than fishmeal. Soybean meal is an obvious choice as it is the dominant oilseed protein on a worldwide basis and is readily available at a cost per tonne half that of Peruvian or menhaden fishmeal (Anon, 1978). Assuming a protein content for fishmeal of 70 per cent and 50 per cent for defatted soybean meal, the relative cost per unit of protein from soybean meal is 70 per cent of that of fishmeal.

Cowey *et al.* (1971) replaced approximately half of the protein, in a 40 per cent codmeal protein diet, with soybean meal and found that this depressed both growth and protein utilisation of plaice. Similar results have been reported for channel catfish (Andrew and Page, 1974) where the isonitrogenous replacement of dietary menhaden meal with soybean meal depressed growth and food utilisation even when the soybean meal was supplemented with methionine, cystine and lysine to the levels found in the fishmeal control. These three amino acids are generally considered to be the most limiting in soybean (NAC, 1973). However, Krishnandhi and Shell (1967) found that channel catfish grew as well on a 30 per cent protein diet containing a 50:50 mixture of soybean and casein proteins as they did on a diet containing casein alone.

Koops *et al.* (1976) evaluated isonitrogenous replacement of fishmeal, at two protein levels (39 and 47 per cent), with either soybean meal or soybean protein in diets of rainbow trout. They found that 25 per cent of the dietary fishmeal could be replaced by soybean protein, but that higher levels of replacement resulted in depressed growth. Rumsey and Ketola (1975) found that the growth of rainbow trout fingerlings, fed diets containing 40 per cent protein with soybean as the sole protein source, was significantly improved by supplementation with amino acids in certain combinations. A comprehensive, recent study of the replacement of fishmeal with soybean meal in tilapia (*Tilapia aurea*) diets at various protein levels found that at a level of 36 per cent protein, tilapia grew as well on all soybean protein diet as they did on an all fishmeal protein diet (Davis and Stickney, 1978). All of the diets in this study were supplemented with methionine to bring the total in the diet to 1.1 per cent.

Information on the effects of substitution of fishmeal with soybean meal in carp diets is scarce. Kaneko (1969) reported unpublished Japanese data that one-third of the white fishmeal, in a carp diet, could be replaced by soybean oil meal with no depression of growth. Hepher *et al.* (1971) reported that, in pond carp feeds, those diets containing fishmeal produced better growth than those containing soybean meal. Viola (1975) isonitrogenously reduced the fishmeal content of a 25 per cent protein carp feed from 15 to 5 per cent by replacement with soybean meal supplemented with amino acids, vitamins and minerals. The group fed the soybean diet did not perform as well as the controls despite these supplementations. Atack *et al.* (1979) reported that a soybean protein concentrate was poorly utilised as the sole protein source in 30 per cent protein rations for carp. Experiments were conducted by the author (Jauncey, 1979) to determine the effects of isonitrogenous substitution of fishmeal, in a 30 per cent protein carp diet, with soybean protein concentrate. Substitution of only one-third of the dietary protein with soybean protein caused a significant decrease in growth rate and food utilisation. This corresponded to a decrease in the calculated methionine contents of the diets from 2.9 to 2.4 per cent where the diets contained 1.5 per cent cystine. These figures correspond to the essential amino acid requirements reported by Nose (1978) who found the requirement for methionine to be 3.1 per cent in the absence of cystine and 2.1 per cent in the presence of 2 per cent cystine.

Amino acid supplementation should, theoretically, improve the utilisation of essential amino acid deficient proteins such as soybean. However supplementation of diets for warm-water fish has generally proved unable to elicit a growth response (Hepher *et al.*, 1971; Andrews and Page 1974; Viola, 1975; Hepher, 1978) possibly due to an inability of these fish to utilise free amino acids or peptides (Aoe *et al.*, 1970, 1974; Page 1974).

In addition to soybean meal several other potential dietary protein sources have been examined in carp feeds. Atack *et al.* (1979) fed a variety of novel proteins to carp, in addition to soybean protein, as the sole protein sources in 30 per cent protein feeds. The results of this experiment are presented in Table 4.3.

Slight variations in the levels of the dietary proteins in this experiment make absolute comparisons impossible. However, the results indicate that carp, like salmonids, require diets with high protein quality for optimal growth in tanks. A major portion of this protein could be provided by bacterial and yeast proteins, but the plant proteins tested were poorly utilised and it is unlikely that they could be used to contribute significant levels of dietary protein in intensive culture systems.

Table 4.3: Utilisation of Some Novel Proteins in Carp Diets

Protein Source	Food Conversion	PER	NPU (%)	Digestibility (%)	BV (%)
Herring meal	1.42[c]	2.82[f]	64[j]	80.3	79[n]
Methanophilic bacterium (ICI 'Pruteen')	1.14[d]	2.54[g]	49[k]	95.5	52[p]
Casein	1.39[c]	2.48[g]	49[k]	93.0	52[p]
Petroleum yeast	1.55[c]	2.08[h]	47[k]	96.6	49[q]
Soybean protein	2.86[a]	1.35[i]	42[l]	83.7	51[q]

Note: Figures in each column with common superscripts are insignificantly different ($P < 0.05$).

Experiments were conducted by Ogino and Chen (1973a) that determined the biological values of several different protein sources to carp. The biological values of proteins from animal sources such as egg yolk (89), casein (80) and white fishmeal (76) were found to be higher than those from vegetable sources such as soybean meal (74) and corn gluten meal (55). The same authors (Ogino and Chen, 1973b) determined the true digestibilities of a range of dietary proteins in carp. Casein, white fishmeal, dried egg yolk, gelatin, corn gluten meal, soybean meal and wheat germ meal proteins all had digestibilities in excess of 92 per cent.

Unorthodox sources of dietary protein for fish worthy of investigation include algal meals, the commonest of which are *Chlorella, Scenedesmus* and *Spirulina* (to which reference is made above). Hepher *et al.* (1978) have discussed this possibility in some detail pointing out that as methods of algal culture and harvesting are improved the economic provision of algal meals as dietary protein sources becomes more probable. These authors reported the digestibility (per cent) of several algae as follows:

Spirulina	90	*Scenedesmus*	81
Oocystis	85	*Euglena*	80

These authors also reported higher digestibilities for drum-dried versions of the same product. Growth trials were performed using algal meal (an unspecified mixture of *Chlorella* and *Euglena*) isonitrogenously to replace soybean meal in a soybean/fishmeal diet. It was found that diets containing the algal meal produced a better growth response than soybean meal.

Meske and Pfeffer (1977) report on trials to produce a fishmeal-free dry food for carp. The diet developed consisted of a mixture of soybean meal, acid whey powder and fat in a ratio of 4:5:1 plus vitamins and trace elements. This feed (designated SM3) was found to be superior, in terms of growth and food utilisation, to a commercial fishmeal based trout diet. Meske and Pruss (1977) continued this research, investigating the incorporation of a single-celled alga (*Scenedesmus obliquus*) in the fishmeal-free ration. They found that a diet containing soybean meal (26.3 per cent), acid whey powder (34 per cent), fat (6.7 per cent) and *Scenedesmus* (32 per cent) produced growth equivalent to that of a commercial, fishmeal-based, trout feed and a PER of 1.3 as compared to 1.28 with the trout feed. Further experiments (Meske and Pfeffer, 1977) modifying this basic diet were carried out replacing the algae with casein and yeast. The growth rates of fish fed 68 per cent SM3 plus 32 per cent casein or algae were very similar to those of fish fed the commercial trout diet. Fish fed 68 per cent SM3 plus 32 per cent casein had a greatly depressed growth rate whilst the growth rate of fish fed SM3 alone were much the lowest.

Experiments were conducted by Ogino *et al.* (1978) to determine the nutritional quality of another novel plant protein, leaf protein concentrate (LPC) mechanically extracted from rye grass. Casein, as the reference protein source, was partially and wholly replaced by LPC in diets for carp and rainbow trout. The digestibility of the LPC was high (86 per cent) and the digestibility of diets containing mixtures of LPC and casein was generally greater than 90 per cent. Values of PER and NPU for LPC were also relatively high and improved by combination with casein. PER and NPU were higher in carp fed 43 per cent LPC (of total protein) and 57 per cent casein (of total protein) than when either protein source was fed singly. The LPC evaluated contained 57 per cent protein which makes its incorporation in relatively high protein fish diets feasible. In both LPC and casein, methionine might be assumed to be the first limiting amino acid. However, it appears that the overall balance of amino acids in a combination of proteins is superior to that of casein alone. Kim (1974) reported on the digestibility of a whole range of proteins by carp although giving no growth or food utilisation data (Table 4.4).

Table 4.4: The Digestibility of Various Protein Sources by Carp

Protein Source	% Protein in Test Diet	Apparent Digestibility	True Digestibility
White fishmeal	7.8	83.5	90.3
White fishmeal	24.6	86.9	89.2
White fishmeal	42.4	88.8	90.2
Silkworm pupae	20.9	63.9	66.3
Wheat germ	10.3	85.0	90.2
Brown fishmeal	13.9	73.9	77.6
Purified soybean protein	12.3	86.8	91.6

Source: Kim (1974).

Using a total faecal collection method Bondi *et al.* (1957) examined the digestible protein contents of a number of feeds for carp, the results of which are presented in Table 4.5. These authors also claim, although giving no supporting evidence, that the digestion coefficients of fibre in these feeds ranged from 0 to 89 per cent. Such high digestibilities for fibre are not substantiated by other authors. This subject is more fully discussed in Section 4. The figures in Table 4.5 are also very difficult to compare to others as the results are presented as percentage digestible protein in the feeds with no reference to the protein contents of the feeds.

Schlumberger and Labat (1978) fed carp fry on moist diets prepared from industrial wastes. Paper mill waste (56 per cent fibre, 44 per cent minerals) and two different powder wheys (70 per cent and 13 per cent protein respectively)

were mixed with fishmeal, cod liver oil, urea, starch and vitamins. The test diets were compared to a 51 per cent protein commercial dry pelleted feed. Both whey proteins were very poorly assimilated and the incorporation of urea and other additives in the feed did not improve this. The high mineral and fibre content of the paper mill waste also resulted in extremely poor growth and food conversion.

Table 4.5: The Digestible Protein Contents of a Number of Feeds for Carp

Food	% Digestible Protein	% Total Digestible Nutrients
Peanut oil meal	41	73.5
Soybean oil meal	34	74.4
Cottonseed oil meal (Grade A)	38.7	73.6
Cottonseed oil meal (Grade B)	32.3	58.7
Sweet lupin seed	30.4	75.2
Condemned peas	23	71.8
Rice meal	9.6	83.0
Wheat bran	13.7	61.4
Barley feed	10.2	43.6
Barley grain	6.2	60.9
Corn grain	7.4	74.0
Alfalfa meal	13.6	49.1

Source: Bondi *et al.* (1957).

Lukowicz (1978) investigated the replacement of fishmeal in carp diets with meal of Antarctic ocean krill (*Euphasia superba*). The krill compared very favourably with that of fishmeal and three diets were prepared containing 42 per cent crude protein viz:

A. No animal protein except krill.
B. Fifty per cent krill, 50 per cent miscellaneous animal proteins.
C. Fifty per cent fishmeal, 50 per cent miscellaneous animal proteins.

Diet A was found to produce superior growth rates, food conversion ratios and protein utilisation to diets B and C, demonstrating that in a compounded commercial feed all of the animal protein sources could be replaced by krill with concomitant improvements in the feed.

Studies on the replacement of casein in carp diets by krill meal were performed by Meske and Pfeffer (1978). In a preliminary experiment three dietary protein contents (20, 30 and 40 per cent) were fed with three ratios of casein:krill protein (10:0, 5:5, 0:10) at each dietary protein level. In a second experiment with a dietary protein level of 40 per cent casein:krill ratios of 10:0, 9:1, 8:2, 7:3, 6:4 and 5:5 were studied and compared to a commercial trout feed. Mixtures containing casein alone caused mortalities and cessation of growth after three weeks. Optimum growth occurred with diets containing 40 per cent protein and a casein to krill ratio of 5:5. Increasing the proportion of krill in the diet caused a linear improvement in growth up to this ratio.

Non-protein Nitrogen. Other feedstuffs that may be considered as possible dietary protein replacements for fish are non-protein nitrogen (NPN) compounds. It is well known that ruminants can utilise NPN, usually in the form of urea, in lieu of true dietary protein. Urea is reduced in the rumen to ammonia which is then used for protein synthesis by rumen bacteria which are, in turn, used as a protein source by the animal. Cowey and Sargent (1972) quoted Vallet (1970) who compared the growth of grey mullet receiving a diet containing 4 per cent nitrogen as protein (equivalent to a dietary protein level of 25 per cent) with that of fish receiving 2 per cent protein nitrogen plus 2 per cent urea nitrogen and with that of fish receiving 4 per cent protein nitrogen plus 4 per cent urea nitrogen (equivalent to a dietary protein level of 50 per cent). The weight increments were 8.88, 8.36 and 10.13 per cent respectively, demonstrating that, in this species at least, urea can replace a proportion of the protein intake of the fish.

Warm-water fish possess large numbers of intestinal bacteria; Teshima and Kashiwada (1967, 1969) found 209 species of bacteria in carp digestive tracts. Hepher *et al.* (1978) examined the possibility of substituting urea and urea phosphate isonitrogenously for protein in carp diets. Their results conclusively demonstrated that carp cannot utilise urea as a non-protein nitrogen source. Kerns and Roelofs (1977) isonitrogenously and isoenergetically substituted dried poultry wastes (of which 70 per cent of the nitrogen is NPN, mainly uric acid and ammonium salts) in carp diets. The poultry waste diets resulted in greatly depressed growth rates.

Dabrowski and Wojno (1978a, b, c, d), however, report contradictory results for carp fed ammonium citrate and urea. A control feed containing 42 per cent protein was compared with test diets containing 27-32 per cent and supplemented with non-protein nitrogen (NPN). Feed conversions ranged from 1.3 to 1.5 for both control and test diets and these authors conclude that NPN is being utilised by the fish. Net protein utilisation (NPU) of NPN-supplemented feeds ranged from 59 to 95 per cent and some NPN appeared to be used for growth improving nitrogen balance. The best results were obtained with diets supplemented with 2.5 per cent urea or 9.3 per cent ammonium citrate. Addition of NPN to the diets improved the ratio of true to crude protein in the carcasses and also resulted in lower levels of ammonia and urea in the blood.

3. Lipids

There are two principal requirements for dietary lipids, first as a source of metabolic energy (ATP) and secondly to maintain the structure and integrity of cellular membranes in the form of phospholipids. The requirements for dietary lipids as energy sources have been discussed under the heading of protein-sparing by dietary lipid and the arguments will not be reiterated here. Studies of lipids in fish have been largely stimulated by the fact that fish, especially marine species, contain large quantities of polyunsaturated fatty acids (PUFA) which are also rich in fat-soluble vitamins and are commercially exploitable. A number of reviews on fish nutrition have been published which contain information on the lipid requirements of fish (Castell, 1978; Cowey and Sargent, 1972, 1977, 1979; Halver, 1975; Lee and Sinnhuber, 1972; Sinnhuber, 1969). The main emphasis of studies on lipid requirements has been with salmonids.

Fish tissues contain predominantly fatty acids of the $\omega3$* series rather than those of the $\omega6$ series present in terrestrial animals (Cowey and Sargent, 1979). In fact $20:5\omega3$ are the major fatty acids present in fish as opposed to $18:2\omega6$ (linoleic) and $20:4\omega6$ (arachidonic) which are the principal fatty acids of terrestrial animals. Biochemical evidence (presented in detail by Castell, 1978 and Cowey and Sargent, 1979) suggests that the higher degree of unsaturation is due to the lower temperatures prevailing in the aquatic environment. This is explained by consideration of biomembrane fluidity where biomembranes are described as proteins (including enzymes) floating in a sea of lipid. The fluidity of the phospholipid phase of biomembranes is directly related to the degree of unsaturation of the lipids which, in turn, affects the melting point of the lipids (Cowey, 1979). Farkas and Csengari (1976) injected (^{14}C)-acetate into mirror carp and found that the degree of incorporation of the label into saturated, monounsaturated and polyunsaturated fatty acids of total lipids and phospholipids depended on the immediate experimental temperature.

Linolenic acid ($18:3\omega3$) has been demonstrated to have essential fatty acid activity in rainbow trout (Higashi *et al.*, 1964, 1966; Nicolaides and Woodall, 1962; Sinnhuber, 1969) as opposed to linoleic acid ($18:2\omega6$) which has essential fatty acid activity in most mammals. Later work by Castell *et al.* (1972a, b, c) showed linolenic acid to be superior to linoleic acid in promoting growth of rainbow trout fed a diet lacking in PUFA. The requirement for linolenic acid was found to be 1 per cent of the diet or 2.7 per cent of the total dietary calories. Essential fatty acid deficiency in rainbow trout caused cessation of growth, caudal fin erosion, a shock syndrome, occasional heart myopathy and fatty livers. Yu and Sinnhuber (1972) showed that $22:6\omega3$ was equally effective as $18:3\omega3$ (linolenic acid) in reversing the effects of essential fatty acid deficiency in rainbow trout. Watanabe *et al.* (1974a, b, c) found 0.88-1.66 per cent dietary linolenic acid to be the requirement for rainbow trout.

In an initial 22-week experiment on 2.5 g mirror carp Watanabe *et al.* (1975a) were unable to obtain any signs of essential fatty acid deficiency when fish were fed a lipid-free diet. However, in a subsequent experiment (Watanabe *et al.*, 1975b) on 0.65 g mirror carp that had been maintained on a fat-free diet for four months prior to the experiment, essential fatty acid deficiencies in fish fed a lipid-free diet occurred. Addition of methyl linoleate or methyl linolenate to fat-free diets improved growth.

It has been noted in small mammals and fish that diets deficient in PUFA result in chain elongation and desaturation of oleic acid ($18:1\omega9$) to $20:3\omega9$ (Cowey and Sargent, 1979) and, in line with studies on small mammals, Castell *et al.* (1972a, b, c) proposed that the ratio of $20:3\omega9/20:5\omega3$ in liver phospholipids be used as an index of $\omega3$ PUFA deficiency in rainbow trout, with a value of less than 0.4 for this ratio indicating a satisfactory diet.

Watanabe *et al.* (1975b) found elevated levels of $20:3\omega9$ in PUFA deficient fish, but that the level was lowered by additions of linoleate or linolenate to the

---

* The recognised shorthand notation for fatty acids will be used throughout. For example linolenic acid would be written $18:3\omega3$, the first number identifying the number of carbons, the second the number of double bonds and the last (prefixed by ω) the position of the first double bond numbering from the methyl end.

diet. Addition of linoleate resulted in elevated levels of $20:4\omega6$ and addition of linolenate in elevated levels of $22:6\omega3$. These authors, therefore, suggested that the ratios of $20:3\omega9/20:4\omega6$ and $20:3\omega9/22:6\omega3$ be used to assess essential fatty acid deficiency in carp.

Takeuchi and Watanabe (1977) continued this study in a 14-week experiment on 0.9 g mirror carp to determine the requirements for linoleic and linolenic acids. They also compared those fatty acids to the growth response obtained by using $22:6\omega3$ and $\omega3$-HUFA (highly unsaturated fatty acids). Fat-free and essential fatty acid deficient diets resulted in retarded growth. Addition of methyl linoleate and methyl linolenate to the diets improved growth with methyl linolenate being the most effective. However, the greatest improvement in growth was obtained by supplementation with 1 per cent linolenate and 1 per cent linoleate.

Addition of $\omega3$ fatty acids, both $22:6\omega3$ and $\omega3$-HUFA, resulted in vastly improved growth and food conversion with 0.5 per cent $\omega3$-HUFA giving a response slightly greater than that of 1 per cent $18:3\omega3$. Essential fatty acid deficient fish accumulated $20:3\omega9$ and the monoethylenic fatty acids $16:1$ and $18:1$. These levels were lowered by additions of both $\omega3$ and $\omega6$ fatty acids to the diets with $\omega3$-HUFA and $22:6\omega3$ appearing to be more effective than either of these. As previously proposed (Watanabe *et al.*, 1975b) the ratios of $20:3\omega9/20:4\omega6$ and $20:3\omega9/22:6\omega3$ were used as essential fatty acid indices. If the former is less than 0.4 and the latter less than 0.6 these authors propose that carp are receiving sufficient $\omega6$ and $\omega3$ fatty acids.

Slightly differing results were obtained by Farkas *et al.* (1977) who fed carp diets containing 0.05, 0.1, 0.34 and 1.1 per cent linolenic acid for four weeks and determined the fatty acid composition of total liver lipids as well as that of liver triglycerides and liver phospholipids. The fish fed the diet containing 1.1 per cent linolenate accumulated negligible amounts of $20:3\omega9$ and exhibited a very low ratio of $20:3\omega9/22:6\omega3$ of 0.07 in the total liver lipid. This ratio increased from 0.33 to 2.00 in the total liver lipids of fish ingesting 0.34, 0.1 and 0.05 per cent linolenic acid. These authors stated a requirement of 1 per cent dietary linolenic acid for carp although in a subsequent review of this data (Csengari *et al.*, 1978) the requirement was stated as 1.5 per cent. The level of oleic acid in liver triglycerides was found to be inversely related to the concentration of linolenic acid in the diet. Linoleic acid at concentrations of 1.3-1.6 per cent was unable to prevent accumulation of $20:3\omega9$ in the lipids of fish fed diets low in linolenic acid.

Castell (1978) has attempted to explain the discrepancy between these results and those of Takeuchi and Watanabe (1977). As stated by Takeuchi and Watanabe (1977), $20:2\omega6$ and $20:3\omega9$ are poorly resolved by gas-liquid chromatography. It is possible that the $20:3\omega9$ identified by Farkas *et al.* (1977) in fishes fed 1.3-1.6 per cent linoleic acid was, in fact, at least partially $20:2\omega6$ thus masking the reduction in $20:3\omega9$ obtained by feeding linoleic acid, one principal product of which is $20:2\omega6$.

Farkas *et al.* (1978) injected (^{14}C)-acetate into carp fed three diets in which the level of linolenic acid varied inversely with the level of carbohydrate whilst the level of linoleic acid remained virtually constant. The highest rates of fatty acid biosynthesis were noted in fish fed the high carbohydrate, low linolenic acid diet. Such an increased rate of lipogenesis has been recognised as the earliest sign of essential fatty acid deficiency in rats and mice (Farkas *et al.*, 1978). One per cent linolenic acid in the diet was found to be sufficient to depress lipogenesis to a low level preventing the formation of fatty livers and, presumably, fatty carcasses. Carp are frequently fed supplementary feeds high in carbohydrate and low in PUFA which could lead to unhealthy fish through liver lipid accumulation and also to undesirably fatty carcasses; thus attention should be paid to dietary levels of PUFA.

In continuation of the previously presented theory of the effects of temperature on the essential fatty acid requirement of fish there is the implication that fish raised at higher temperatures may have different require-ments for PUFA. Stickney and Andrews (1972) showed that channel catfish grew equally well on lipid sources of either beef tallow, olive oil or menaden oil, the first two being sources of $\omega 6$ fatty acids and the last a source of $\omega 3$. Takeuchi *et al.* (1978) looked at hydrogenated beef tallow and hydrogenated fish oil as energy sources in 10 per cent lipid diets for carp and trout. When hydrogenated oil was used as the sole lipid source it induced essential fatty acid deficiency symptoms in both species. However, if 4-6 per cent of the dietary oil is supplied by pollock oil or cuttlefish liver oil (rich in essential fatty acids) the best weight gains and food conversions were obtained. Therefore hydrogenated oil (totally saturated) proved sufficient as an energy source but some marine lipid was required to provide essential fatty acids. More research is required to determine whether less unsaturated fatty acids than the $\omega 3$ series have essential fatty activity in carp reared at temperatures of 25-30°C. Continuing these studies Takeuchi *et al.* (1979a) examined the digestibility of hydrogenated fish oils in carp and rainbow trout. The digestibility of hydrogenated fish oils was found to increase as the melting point decreased. Oils with a melting point of 53°C had a very low digestibility in both species. Beef tallow and hydrogenated fish oils with a melting point of 38°C had a digestibility of 70 per cent regardless of fish size and water temperature.

The requirement of fish for PUFA of the $\omega 3$ and $\omega 6$ series creates problems with respect to feed storage as these types of fatty acid are very susceptible to oxidation. The products of such oxidation may react with other nutrients reducing their availability or the oxidation products themselves may be toxic. Addition of vitamin E to diets has been found to alleviate the toxicity symptoms caused by feeding rancid dietary fats in rainbow trout (Sinnhuber *et al.*, 1968) and carp (Watanabe and Hashimoto, 1968). In diet production only fresh oils with low peroxide values should be used and ingredients such as fishmeals should be protected against oxidation. As the level of dietary PUFA is increased so should the level of vitamin E be raised. The levels of vitamin E to be included in carp rations are discussed in the section (Section 5) on vitamins.

4. Carbohydrates

The principal function of dietary carbohydrate is as an energy source. This aspect of carbohydrate utilisation in carp has been covered in Section 2.4.2. The present discussion will therefore restrict itself to those aspects not previously mentioned. The biochemistry of carbohydrate metabolism in fish has been excellently reviewed by Cowey and Sargent (1979) and it is evident from the review that this is an area in which a great deal more research is required in order to obtain a clear understanding of the biochemical processes involved.

The possible utilisation of dietary cellulose as an energy source by fish presents interesting possibilities in the utilisation of plant materials as sources of nutrients. Cellulase activity has been demonstrated (Stickney and Shumway, 1974) in the digestive tracts of 16 estuarine fish from the south-eastern coast of the USA. In addition cellulase activity was found in channel catfish reared in an intensive outdoor culture system and was directly attributable to cellulase producing gut microflora. Shcherbina and Kazlauskene (1971) reported a digestibility of 50 per cent for crude fibre in 200-300 g pond-raised mirror carp at 27° C. In addition Syvokiené and Jankevicius (1973, cited by Dabrowski, 1979) identified cellulase-producing bacteria in the intestine of carp which however had a much lower activity than that in grass carp. In contrast, unpublished experiments by the author were unable to show any digestion of cellulose by 12 g carp held in a laboratory recycling system. The high digestibility of cellulose by carp reported by Shcherbina and Kazlauskene (1971) is not supported by the research findings of other workers in the field.

In omnivorous mammals deprived of food the carbohydrates of the body are rapidly metabolised for energy with glycogen being swiftly converted to glucose. By contrast fish do not rapidly metabolise glycogen when starved. Nagai and Ikeda (1971a) fed carp diets containing up to 90 per cent starch by replacement of casein. They report that even the highest levels of dietary carbohydrate resulted in virtually no accumulation of carcass lipid whilst high dietary protein levels resulted in large accumulations of carcass lipid. In addition the blood glucose and liver glycogen levels of carp starved for 22 days were not significantly different from those of carp fed the various diets. Even after 100 days of starvation appreciable amounts of glycogen remained in the liver. These results suggest that the capacity of carp to oxidise glucose is limited and that the demands of glucose-fuelled tissues (brain and nervous tissue) are met by gluconeogenesis rather than glycogenolysis (Cowey and Sargent, 1979).

Nagai and Ikeda (1971b) concluded, from the injection of (^{14}C)-glucose, that glucose oxidation in carp fed diets containing 50 per cent protein was significantly lower than in carp fed 10 per cent protein high-starch diets. In further studies (Nagai and Ikeda, 1973) they replaced casein with levels of starch from 12 to 75 per cent and followed the metabolism of isotopically labelled ($U^{14}C$)-acetate and ($U^{14}C$)-L-alanine. From this study these authors concluded that the depressed utilisation of carbohydrate in carp resembled the diabetic system of higher animals and that amino acids are superior to glucose as an energy source.

5. Vitamins

Modern definitions of the term 'vitamin' usually include the following points (Cowey and Sargent, 1972):

(1) the organic nature of the substance concerned;
(2) the presence of these substances in natural food in extremely small amounts;

(3) these substances are distinct from the major components of the food, i.e. protein, carbohydrate and fat;

(4) animals have an absolute dietary requirement for them – they cannot themselves synthesise the compounds concerned;

(5) specific deficiency diseases occur when these compounds are totally absent from the diet.

The early work leading to the formulation of a suitable vitamin test diet for fish has been traced by Halver (1969). Halver and Coates (1957) improved and modified a test diet first developed by Wolfe (1951) and applied it, successfully, to chinook salmon. This diet contains sixteen vitamins and has formed the basis of most modern research into the vitamin requirements and deficiency diseases of fish.

Values for the dietary requirement of certain vitamins may depend on the method of assessment used (growth rate or tissue level for example), and where certain vitamins fulfil more than one metabolic role the requirements for each may differ. Because many vitamins function as coenzymes one might logically regard the vitamin requirement as the dietary level of a vitamin which permits optimal activity of all those enzymes for which the vitamin serves (possibly in a modified form) as a coenzyme. Thus correlation between vitamin intake and the activity of the related enzymes in controlled experiments would be the ideal way to establish quantitative vitamin requirements. However, the type of experiment performed to date usually relates to the growth and tissue level of the vitamin as these are somewhat easier to measure. A useful guide to vitamins as nutrients for fish has been compiled by Hashimoto and Okaichi (1969) and an English translation is available, published by Roche. An informative review of the vitamin requirements of finfish has been prepared by Halver (1978).

Aoe and his co-workers (Aoe *et al.*, 1967a, b, c, d, 1968, 1969, 1971; Aoe and Masuda, 1967) are principally responsible for a programme of research investigating the vitamin requirements of carp. Carp were kept in 100-litre plastic troughs supplied with 1 litre/minute of aerated well water at $22\text{-}25^\circ$ C and fed 10 per cent of their body weight per day in five daily feeds. The test diet devised by Halver and Coates (1957) was adopted and the results are summarized in the next sections.

5.1 Riboflavin (Vitamin B_2)

Aoe *et al.* (1967a) investigated the requirements of carp for riboflavin. Deficiency symptoms developed in groups of carp fed on diets free of, or poor in, riboflavin. In such groups anorexia was observed after three weeks and the fish floated on the surface, responding to the sound of feeding but unable to catch their food properly. The fish then became extremely thin and showed high mortality. Haemorrhages or damaged epidermis were also observed by the fourth week. Carp receiving 0.06 mg of vitamin B_2 per kg of body weight per day showed no obvious deficiency symptoms except retarded growth.

In histological investigation, intra-myocardial haemorrhage in the atrium and ventricle was pronounced, especially in the latter. The kidney showed atrophy and partial necrosis of the kidney tubules. In addition, atrophy and oedema were noticeable in the muscle, and focal necrosis and lymphoid cell infiltration were conspicuous in the stomach epithelium.

The quantitative requirement for vitamin B_2 was estimated from body weight gain, vitamin concentration in the hepatopancreas, and feed conversion at levels of 1, 2, 4, 8 and 20 mg of vitamin B_2 per kilogram of feed. The same figure was obtained in respect of both weight gain and feed conversion. This figure was 0.11 mg per kg of body weight per day, or 4 mg per kg of feed. According to the storage level in the hepatopancreas the level required was 1.7 mg per kg of body weight per day, or 6.2 mg per kg of diet.

Ogino (1967) found the riboflavin requirement of carp to be 7.10 mg per kg of diet, only slightly higher than the above. Takeuchi *et al.* (1980) investigated the relative potencies of riboflavin and riboflavin tetrabutyrate in carp diets and found both to be equally effective with a dietary level of 7 mg/kg providing maximum storage of the vitamin in hepatic tissues. The vitamin B_2 requirements of carp are thus similar to those reported for rainbow trout (Phillips and Brockway, 1957; McLaren *et al.*, 1947a) and for chinook salmon (Halver, 1971).

5.2 Para-amino-benzoic acid (PABA)

Aoe and Masuda (1967) investigated the requirement of carp for para-amino-benzoic acid. In a preliminary experiment with one PABA-deficient and one control group with 400 mg/kg PABA in the diet, the fish did not exhibit any deficiency symptoms and had good appetites and normal growth. After 16 weeks a group of carp on a PABA-deficient diet were changed onto a diet which was still deficient but also contained the PABA anti-vitamin sulphanilamide. The fish in all groups still remained healthy and normal indicating that neither PABA nor sulphanilamide have any observable effect on the growth of young carp. Haematological examination showed no difference in erythrocyte counts and thus PABA and sulphanilamide have no effect on erythropoiesis in carp.

5.3 Inositol

The requirement of carp for inositol was also determined by Aoe and Masuda (1967). In contrast to the results with PABA, diets poor in inositol produced clear deficiency symptoms and reduced weight gains after 4-7 weeks. Deficiency led to loss of appetite, and thus reduced weight gains, and the appearance of skin lesions. Skin lesions appeared as haemorrhages in the dorsal and lateral skin especially around the base of the dorsal fin. Decomposition of the mucosa was also observed. Affected areas developed gradually until, in severe cases, scales, fin and epidermis sloughed off and muscle and bones were exposed. The hepatopancreas and kidneys of these diseased fish appeared normal on examination.

The quantitative requirement for inositol was estimated from average body weight gains using levels of 0.02, 0.1, 0.26, 0.58, and 4 g/kg of inositol. From the results obtained the requirements for inositol were estimated to be approximately 44 mg per 100 g of diet or 7-10 mg per kg of body weight per day.

5.4 Niacin

The requirements of carp for dietary niacin were determined by Aoe *et al.* (1967b). Diets deficient in niacin led to loss of appetite, poor growth, haemorrhages in the skin and high mortality. Diseased fish generally appeared reddish as a result of skin haemorrhages.

To determine the quantitative requirements for niacin the effects of levels of 0, 15, 20, 25, 30, 35, 60 and 100 mg/kg of diet on growth and food conversion were investigated. Optimum weight gains were attained when the diet contained 25 mg/kg of added niacin. The requirement of young carp, therefore, was calculated to be about 28 mg/kg of diet (0.55 mg/kg body weight) as the basal diet was not completely free of niacin.

5.5 Thiamine (Vitamin B_1)

The requirement of young carp for thiamine was investigated (Aoe *et al.*, 1967c) using two diets, the test diet used previously and a short-necked clam diet made thiamine-free by autolysis. On the B_1-deficient test diet carp showed no deficiency symptoms after 16 weeks and on the thiamine-free clam diet no deficiency symptoms after 8 weeks. As it is known that the thiamine requirements of warm-blooded animals are closely linked to the dietary level of carbohydrate and the previous diets were carbohydrate poor, the experiment was therefore repeated using a carbohydrate-rich diet (Aoe *et al.*, 1969). The elimination of thiamine from this diet led to poor growth and deficiency symptoms after 6-8 weeks; these were anorexia, fading of the body colour, congestion of the fins and congestion of the skin.

The same authors also considered the addition of thiaminases to a thiamine-deficient, high-carbohydrate, diet. With added amprolium the deficiency symptoms were severe by the fifth week and skin congestion made the fish red in colour. The addition of pyrithiamine or oxythiamine produced rather different symptoms of pronounced nervousness with abnormal swimming behaviour.

A requirement for thiamine, dependent on the dietary carbohydrate level, and its deficiency symptoms were thus demonstrated in carp. However it is apparent that carp can withstand thiamine deficiency for long periods of time and for this reason no definite dietary level was recommended. Dabrowski (1979), however, suggests a level of 60 ppm in the diet.

5.6 Folic Acid

The elimination of folic acid from the diet of carp (Aoe *et al.*, 1967d) had no appreciable effect on growth, mortality, concentration of the vitamin in the hepato-pancreas, or erythrocyte numbers. The combined elimination of both folic acid and PABA also had no effect. No conclusive results can thus be drawn about the requirement of young carp for folic acid. However, Dabrowski (1978) suggests a level of 15 mg/kg in the diet and quotes the findings of Kashiwada *et al.* (1970, 1971) that the intestinal microflora of carp are capable of synthesising folic acid.

5.7 Pyridoxine (Vitamin B_6)

Ogino (1965) studied the requirements of carp for vitamin B_6 and pantothenic acid using the same experimental test diets and conditions as Aoe *et al.* above. Vitamin B_6 was present in the diets at levels of 0.40, 0.65, 0.90, 1.4, 5.4, 10.4 and 40.4 mg/kg of diet. The groups fed diets containing 1.4 mg/kg or less showed retarded growth after four weeks with high mortalities. The deficiency symptoms were characteristic of nervous disorders – loss of balance, epileptic fits and abnormal swimming. After developing such symptoms most fish died within a few days. Those that survived slightly longer also developed oedema and exophthalmia. The requirement of vitamin B_6 for growth of young carp under these conditions was estimated to be approximately 0.15 mg/kg of carp per day based on optimal growth, or to be in excess of 0.15 mg/kg/day based on the criterion of storage in the hepatopancreas.

5.8 Pantothenic acid

Ogino (1967) investigated the requirements of carp for pantothenic acid. Carp fed a ration lacking in pantothenic acid suffered reduced appetite after ten days and body weights began to decrease. Swimming movements of the fish became feeble and the fish remained close to the surface of the water. After about five weeks, exophthalmia and haemorrhages appeared in some individuals. Anaemic conditions with whitish scales were also observed.

Weight gains increased with pantothenic acid levels up to 50 mg/kg of diet and above this level no further differences were found in the pantothenic acid contents of the liver or pancreas. In a further experiment optimum growth occurred with a level of 30 mg of pantothenic acid per kg of diet. It was concluded that carp require 30-40 mg per kg of feed corresponding to approximately 1.0-1.4 mg per kg of body weight per day.

5.9 Cyanocobalamin (Vitamin B_{12})

In a study by Teshima and Kashiwada (1967) it was shown that vitamin B_{12} is synthesised by bacteria in the intestinal tract of carp. Some 198 strains of bacteria were isolated approximately half of which were capable of synthesising B_{12} and these authors theorised that this may be sufficient to meet the requirements of the fish.

5.10 Biotin

Ogino *et al.* (1970a) fed carp diets containing graded amounts of biotin. Biotin requirements were assessed in terms of growth, mortality, vitamin content in the hepatopancreas and the appearance of deficiency symptoms. Biotin deficiency resulted in reduced growth and food intake and decreased activity. It was concluded that carp required 10 mg of biotin per kg of feed corresponding to an intake of 0.02-0.03 mg per kg of body weight per day.

5.11 Choline

Ogino *et al.* (1970b) maintained carp on diets containing varying levels of choline. Choline was assessed in terms of accumulation of neutral fats in the hepatopancreas. Carp fed diets low in choline accumulated excessive amounts of such fat and it was concluded that they require 2,000-4,000 mg of choline per kg of feed corresponding to an intake of 60-120 mg per kg of body weight per day.

5.12 Ascorbic Acid (Vitamin C)

That carp are able to synthesise vitamin C was shown by the studies of Ikeda and Sato (1964) who followed the incorporation of (^{14}C)-glucose and (^{14}C)-glucuronic acid. The essential requirement of ascorbic acid for channel catfish has been conclusively demonstrated by growth studies (Lovell, 1973; Andrews and Murai, 1975). Halver *et al.* (1975) state that an absolute requirement for vitamin C has been 'previously shown' for carp, yellowtail, coho salmon and rainbow trout. One problem with vitamin C is its liability to oxidation, although a possible solution is the use of ascorbate-2-sulphate which has been demonstrated to have vitamin activity in rainbow trout (Halver *et al.*, 1975) and which is stable for weeks even in aqueous solution. Halver *et al.* (1975) used ascorbate-2-sulphate at a level of 160 mg per kg of diet which is fairly close to the 200 mg per kg of diet found to be the vitamin C requirement of rainbow trout by Kitamura *et al.* (1967).

5.13 Vitamin A

Following studies on the water-soluble vitamins, the role of vitamin A in the nutrition of carp was investigated by Aoe *et al.* (1968). The exclusion of vitamin A from the diet led to deficiency symptoms which became apparent after eight weeks. The fish suffered from anorexia with consequent weight loss, the body colour faded and haemorrhages developed in the fins and skin. The affected carp also showed warped gill operculae and exophthalmia.

Analysis of vitamin concentrations in the hepatopancreas indicated that storage of the vitamin only occurred with dietary levels higher than 2,000 i.u. per kg. When this vitamin A level was added to the diet of a vitamin A deficient group recovery was very slow. Warped gills, haemorrhages and exophthalmia did not completely regress even after 13 weeks.

From these results it is difficult to assess the exact vitamin A requirements of young carp but it may be concluded that the requirement is in the range 4,000-20,000 i.u. per kg of body weight per day. However, in fresh-water fish vitamin A_2, as well as vitamin A_1, is important. In the study performed above only A_1 was administered in the diet but the measurement of levels in the hepatopancreas included A_2. Further studies are necessary, especially on the relative importance of the vitamins A_1 and A_2.

5.14 α-Tocopherol (Vitamin E)

The requirement of carp for vitamin E has been studied by Watanabe *et al.* (1970a, b). Management and feeding methods were similar to those of Aoe *et al.* (above). The basic test diet of Halver and Coates (1957) was once again employed but with the total elimination of α-tocopherol.

For a feeding test a completely tocopherol-free ration was compared with the basal ration which was supplemented with 500 mg of D, L,-α-tocopherol per kg of dry diet. The fish receiving the diet deficient in tocopherol gained weight at a significantly lower rate than those fed a complete ration. Apparent muscular dystrophy, characterised by marked loss of muscle tissue in dorsal myotomes appeared after 90 days in the deficient group and affected 68 per cent of the fish by the end of the experiment. Histopathological investigation showed that tocopherol deficiency caused degenerative changes in the lateral muscle, kidney, epithelial structures of the skin and the cornea.

A second experiment was designed to determine the vitamin E requirement of carp and diets containing 0, 100, 300 and 500 mg per kg were used. Vitamin E deficient fish in this experiment exhibited exophthalmia and lordosis in addition to the symptoms previously described. A level of 100 mg of tocopherol per kg of dry diet appeared to satisfy the requirements of carp.

Watanabe *et al.* (1977a, b) conducted experiments to examine the relationship between the vitamin E requirement and levels of dietary linoleate in carp.

Elevated levels of dietary linoleate increased the requirement of carp for tocopherol judging by the appearance of apparent muscular dystrophy. Five per cent linoleate in the diet increased the requirement of carp for vitamin E to 300 mg per kg of diet.

A summary of the vitamin requirements of carp is presented in Table 4.6. That problems occur in formulation of a vitamin mixture for practical feeds is evident from the work of Aoe *et al.* (1971). They fed carp a diet containing the estimated minimum levels of all the vitamins whose requirements had been established at that time. This diet failed to support the growth of carp unless supplemented with thiamine, riboflavin and nicotinic acid. These authors suggest possible relationships between the dietary levels of some vitamins as the cause of this anomaly.

6. Minerals

The mineral requirements of fish are complicated by the fact that minerals are not only available to the fish from dietary sources but also from those dissolved in the surrounding water be it fresh or saline. This ability to obtain inorganic ions from sources other than the diet makes it difficult to elucidate the functions of dietary minerals. Ions absorbed from the external medium have both nutritional and osmoregulatory functions, and the interdependence of both functions complicates any study on the uptake and excretion of inorganic elements in fish. Previous studies on the metabolism of minerals by fish have been largely concerned with osmoregulation and only relatively recently have the nutritional aspects come under investigation. As there is a net outward flux of ions from freshwater fish, the animal is very dependent on an adequate mineral supply from its food (Cowey and Sargent, 1979).

The nutritional requirements of fish for dietary minerals have been reviewed by Nose and Arai (1976), Cowey and Sargent (1972, 1979) and Lall (1978). Ogino and Kamizono (1975) investigated supplementation of diets for rainbow trout and carp with McCollum salt mixture No. 185 (formulated for rats) supplemented with trace elements (Halver and Coates, 1957). Carp were fed casein/gelatin diets supplemented with from 0 to 8 per cent minerals and rainbow trout diets with from 0 to 4 per cent minerals. Rainbow trout fed a diet devoid of minerals lost their appetite and became sluggish within two weeks. After 50 days more than 10 per cent of the fish had died following convulsions. Mineral deficiency in rainbow trout manifested itself as lordosis, scoliosis, cranial deformity, depressed erythrocyte counts and reduced growth. For maximum growth a level of 4 per cent mineral mixture was found to be adequate.

Carp were found to be not as sensitive to a deficiency of dietary minerals. No symptoms of deficiency were apparent in carp fed diets devoid of minerals for 50 days and haemoglobin contents were normal. However, additions of mineral mixture up to 4 per cent caused a slight improvement in growth but levels above this caused depressed growth. Pfeffer and Meske (1979) fed carp diets containing between 0 and 8 per cent ash from fishmeal as the only dietary sources of minerals. Feed conversion, growth and mineral retention showed no differences due to differences in dietary mineral content. However, recent studies have shown that carp maintained on a mineral-free synthetic diet

Table 4.6: A Summary of the Vitamin Requirements of Carp

Vitamin	Avitaminosis Signs	Suggested Dietary Level	Recommended Intake (mg/kg/day)	Reference
Thiamine (B$_1$)	poor growth, anorexia, loss of colour, hyperaemia of fins and skin	60 ppm	–	Aoe *et al.* (1967c, 1969)
Riboflavin (B$_2$)	anorexia, disorientation, mortalities, skin and heart haemorrhages, necrotic kidney	40-62 ppm 70-100 ppm	0.11-0.17 0.23-0.33	Aoe *et al.* (1967a) Ogino (1967)
Pyridoxine (B$_6$)	poor growth, loss of balance, epilepsy, abnormal swimming, oedema, exophthalmia	20 ppm	0.15-0.20	Ogino (1965)
Pantothenic Acid	anorexia, weight loss, inactivity, exophthalmia, haemorrhages, anaemia	30-40 ppm	1.0-1.4	Ogino (1967)
Inositol	anorexia, poor growth, skin lesions, haemorrhages, skin erosion	440 ppm	7-10	Aoe and Masuda (1967)
Biotin	poor growth, changes in haemopoietic tissues	10 ppm	0.02-0.03	Ogino *et al.* (1970)
Folic Acid	no signs	15 ppm	–	Dabrowski (1979)
Para-amino-benzoic Acid	no signs	200 ppm	–	Dabrowski (1979)
Choline	poor growth, accumulated hepatopancreatic fat	2,000-4,000 ppm	60-120	Ogino *et al.* (1970)
Niacin	anorexia, poor growth, skin haemorrhages, high mortality	28 ppm	0.55	Aoe *et al.* (1967b)
Cyanocobalamin (B$_{12}$)	no signs	0.09 ppm	–	Dabrowski (1979)
Vitamin A	anorexia, weight loss, loss of colour, fin and skin haemorrhages, warped operculae, exophthalmia	2,000 i.u./kg	4,000-20,000 i.u./kg/day	Aoe *et al.* (1968)
α-Tocopherol (E)	low weight gain, muscular dystrophy, exophthalmia, lordosis	100 ppm 300 ppm	–	Watanabe *et al.* (1970a, b)
Ascorbic Acid (C)	no data	2,000 ppm	–	Dabrowski (1979)
Menadione (K)	no data	40 ppm	–	Dabrowski (1979)

displayed deficiency symptoms after eight weeks, including lordosis, loss of muscle tone, reduced growth and food conversion efficiency, and reduced haemoglobin, haematocrit and mean corpuscle haemoglobin concentrations (Tacon, 1980, pers. comm.).

Minerals which have demonstrable biological functions either in elemental form or incorporated into specific compounds include calcium, phosphorus, magnesium, sodium, potassium, sulphur, chloride, iron, copper, cobalt, iodine, manganese, zinc, molybdenum, selenium and fluorine (Lall, 1978).

6.1 Calcium and Phosphorus

Calcium and phosphorus are considered the major mineral nutrients as they are the most abundant elements in the animal body being principally located in the skeletal tissues and responsible for acid-base equilibrium. Calcium ions also take part in other physiological and biochemical processes including muscle contraction, blood-clot formation, nerve transmission and maintenance of membrane integrity. Phosphorus also has functions in lipid and carbohydrate metabolism. Nose and Arai (1976) report that in Japanese carp hatcheries some feeds gave rise to cranial and skeletal deformities that could be alleviated or healed by the addition of 5 per cent calcium monohydrogen phosphate or McCollum's salt mixture to the diet. These authors also cite the unpublished data of Ogino and Takeda (1974) that both rainbow trout and carp grow well on a purified test diet containing only 300 ppm calcium provided that adequate dietary phosphorus was present. Thus both species appear to have been able to absorb sufficient calcium from culture water containing 16-20 ppm calcium.

When carp were fed diets containing varying levels of calcium and phosphorus (Ogino and Takeda, 1976) growth rate was found to be positively correlated with dietary phosphorus concentration but not with dietary calcium concentration. The concentrations of phosphorus and calcium in the culture water were 0.002 and 20 ppm respectively and these authors suggest that carp were able to compensate for low dietary calcium levels by uptake from the water. Maximum growth of carp occurred on diets containing 0.6-0.7 per cent phosphorus. Diets deficient in phosphorus also caused deformity of the head, poor feed efficiency and low ash, calcium and phosphorus contents of the vertebrae. Phosphorus must, principally, be obtained from the diet since it is not a major component of either fresh or sea water.

The ratio of calcium to phosphorus in the diet has been shown to influence the growth of Red Sea bream (Sakamoto and Yone, 1973) where a Ca:P ratio of 1:2 (0.34 per cent Ca:0.68 per cent P) was found to be optimal. Another consideration is the availability of the minerals present in the diet. Nose and Arai (1976) report unpublished data of Takeda and Ogino (1975) on the availability of dietary phosphorus to trout and carp (Table 4.7).

The differences in availability of dietary phosphorus between the two species may, in part, be due to the low pH of the trout stomach rendering certain forms of phosphorus more available than in the stomachless carp. Most commercial diets contain fishmeal which contains high levels of calcium and phosphorus. Substitution of fishmeal in commercial rations will almost certainly result in a need to vary the mineral mixture to compensate for reduced levels of these two elements.

Ogino *et al.* (1979) have reported on the availability of dietary phosphorus to carp and rainbow trout (Table 4.8) and observed that carp fed diets low in available phosphorus had increased visceral lipid contents.

Table 4.7: The Availability of Dietary Phosphorous to Rainbow Trout and Carp

	Phosphorus Content of Diet (%)	Availability	
		Carp (%)	Trout (%)
Tribasic calcium phosphate	0.65	3	51
Monobasic calcium phosphate	0.79	80	65
Phytin	1.65	8	19
Casein	0.47	106	90
Fishmeal	0.99	26	60
Hydrocarbon yeast	0.46	99	91
Wheat germ	0.58	57	58
Activated sludge	0.84	12	49
Rice bran	0.79	25	19

Table 4.8: Available Phosphorus in Feed Ingredients for Carp and Trout

Source	Carp (%)	Rainbow Trout (%)
Casein	97	90
White fishmeal	26	60
White fishmeal	18	—
White fishmeal	10	72
Brown fishmeal	13	72
Brown fishmeal	25	81
Brown fishmeal	33	70
Yeast	93	91
Rice bran	25	19
Wheat germ	57	58
Phytin	8	19
Phytin	38	—

A large part (60-80 per cent) of the total phosphorus of cereal grains and oils exists organically bound as phytic acid (Lall, 1978). Such phytin-bound phosphorus is unavailable to fish as they do not possess the gastro-intestinal enzyme phytase.

6.2 Magnesium

It has been demonstrated in trout that a deficiency of dietary magnesium will cause renal calcinosis with dietary calcium levels of 2.6 per cent or more (Cowey *et al.*, 1977). Ogino and Chiou (1976) examined the magnesium requirement of carp and found it to be 0.04-0.05 per cent of the dry diet. Carp deficient in dietary magnesium showed loss of appetite, poor growth, high mortality, sluggishness and convulsions. Evidence indicates that the severity of magnesium deficiency in terrestrial animals is influenced by the calcium and phosphorus levels of the diet. A large dietary intake of calcium aggravates the severity of magnesium deficiency, but when dietary calcium and phosphorus are reduced, along with magnesium intake, magnesium deficiency does not occur. Conversely, consumption of excess dietary calcium and phosphorus increases the magnesium requirement. Calcium and phosphorus exert their effect chiefly by decreasing magnesium absorption (Lall, 1978).

6.3 Iron

Iron is an essential dietary element involved in respiratory processes including oxidation-reduction activity and electron transport. Iron usually exists in the body in complex forms bound to protein. Iron deficiency causes characteristic hypochronic anaemia in brook trout, yellow tail, Red Sea bream and eels (Lall, 1978). Nose and Arai (1976) report the work of Arai *et al.* that the iron requirement of eels is 0.017 per cent of the diet. Quantitative data on other finfish are lacking.

　　In monogastric terrestrial animals iron absorption and availability is depressed by high levels of phosphates and calcium in the diet. Ferrous iron is also much more readily absorbed than ferric iron (Lall, 1978).

6.4 Cobalt

The only established biological function of cobalt relates to its role as a component of cyanocobalamin (B_{12}). Lall (1978) reports Russian research indicating that the addition of cobalt chloride for cobalt nitrate to the feed enhanced growth and haemoglobin formation in carp.

6.5 Zinc

Zinc is involved in nucleic acid synthesis as well as the activity of many important enzymes. Very few data, however, exist on the requirement of finfish for dietary zinc. Ketola (1978) found that addition of 0.015 per cent zinc (as $ZnSO_4$) to a cataractogenic diet for rainbow trout overcame the problem and markedly improved growth. As previously mentioned many plant proteins contain phytic acid which is known to bind dietary zinc rendering it unavailable. The effect of phytic acid on zinc availability is accentuated by high levels of dietary calcium (Lall, 1978). Ogino and Yang (1979) investigated the requirement of carp for dietary zinc. Dietary zinc levels not only affected appetite, growth rate and mortality, but also levels of zinc, iron and copper in the tissues. Fish fed a diet containing 1 ppm suffered from skin and fin erosion and the requirement of carp for dietary zinc was found to be 15-30 ppm.

6.6 Iodine

Iodine deficiency has been noted as causing goitre and thyroid hyperplasia in salmonids (Lall, 1978). Plant proteins probably contain sufficient iodine to meet dietary requirements. Animal proteins, other than fishmeal, contain insignificant quantities of iodine.

6.7 Other Trace Elements

Copper, sulphur, fluorine, manganese and molybdenum deficiencies cause abnormalities in mammals and birds although few or no data exist on their requirements in fish (Lall, 1978).

　　In conclusion the mineral mixtures generally fed to fish are derived from those formulated for warm-blooded animals. These have been generally shown to be unsatisfactory as the mineral requirements of fish differ from those of warm-blooded animals. It is only when the mineral requirements of fish have been more extensively investigated that formulation of specific mineral mixtures will become possible and until then empirical formulation is all that is possible.

7. Other Additives to the Feed

Korneyev (1969) and Gribanov *et al.* (1966) refer to the unpublished results of Korneyeva (1961, 1965) on the growth-promoting effects of the antibiotic terramycin in carp diets. Terramycin appeared to act as a growth stimulant possibly by increasing appetite as well as a prophylactic and Korneyeva found that adding it to carp diets at a level of 6,000 to 10,000 units per kg of dry diet ensured an additional production of 5-15 per cent. The growth-promoting effect was found to be especially marked when terramycin was added to a vegetarian diet. These results are not in agreement with those of Hashimoto (1953) who found that the antibiotics aureomycin, penicillin, streptomycin and terramycin showed no remarkable growth stimulation in carp.

Korneyev (1969) also discusses Russian interest in adding to carp diets synthetic enzyme preparations to replace the exogenous ferments normally ingested with natural food. He used a Russian enzyme preparation called 'Avomarin' which had amylolytic, pectinolytic and proteolytic activity. This 'Avomarin' was fed at levels of 0.01, 0.02 and 0.10 per cent of the dry weight of the feed. The 0.1 per cent level caused a growth increase of 26 per cent and the 0.01 and 0.02 levels caused a 12-13 per cent increase compared to a standard. This, he says, demonstrates the possible use of enzymes to promote growth.

Other possible additives to the feed that may result in improved growth are hormones. Adelman (1977, 1978) studied the effects on growth and body composition of carp of injections of bovine growth hormone and also the relationship to temperature and photoperiod. Photoperiod did not influence the effects of bovine growth hormone whereas temperature had a marked effect. A much greater response was found to injections of bovine growth hormone at temperatures below the optimum (9.2 per cent and 19.5° C) and at temperatures above the optimum (35.8° C) than at the optimum.

The uses of sex steriods in the diets of fish, for their possible anabolic growth-promoting effects, have been investigated (Cowey *et al.*, 1973; Ghittino, 1970; McBridge and Fagerlund, 1973). However, no data on the effects of these substances on the growth of carp appear to have been published.

8. Temperature

The effect of temperature on the growth of mirror carp has been investigated by several authors and their results are summarised in Table 4.9.

The ways in which temperature affects the digestion and utilisation of feeds (and thus ultimately growth) include:

(1) The motor activity of the digestive tract is affected by environmental temperature. The rate of passage of food through the gastro-intestinal tract of fish is approximately doubled by a 10° C rise in temperature (Shcherbina and

Kazlauskene, 1971). The rate of gastic evacuation of sockeye salmon was quadrupled by a 10°C rise in temperature (Brett and Higgs, 1970).

(2) The activity of the digestive enzymes is increased by increasing temperature and there is evidence that the rate of absorption of nutrients from the fish intestine is also increased by increasing temperature (Shcherbina and Kazlauskene, 1971).

(3) Enzyme-substrate affinity may change significantly with temperature and this change may be in the direction to compensate for the lowering of enzyme activity with a fall in temperature (Cowey and Sargent, 1979).

Table 4.9: The Effect of Temperature on Mirror Carp

Temperature(s) $^\circ$C	Notes	Authors
32	final preferendum in graduated temperature apparatus	Pitt *et al.* (1956)
23-29	optimum for growth	Shpet & Kharitonaova (1963)
23-30	optimum for growth in cages	Gribanov *et al.* (1966)
33-34	growth greatly depressed	Korneyev (1969)
22	growth sub-optimal	Korneyev (1969)
29	preferred temperature in thermo-regulatory shuttlebox	Reynolds & Casterlin (1977)
28-30	optimum for growth	Adelman (1977)
29.6	optimum for growth	Adelman (1978)
28	optimum for growth	Aston & Brown (1978)
25-30	optimum for growth	Jauncey (1979)

These opposing effects of temperature combine to determine the degree of digestion and utilisation of dietary ingredients, but the effect of temperature on growth is further influenced by the metabolic rate of fish (and thus energy requirements) being raised by increasing environmental temperatures (Brett *et al.*, 1969).

Temperature has been observed to affect dietary protein requirements with the optimum dietary protein level increasing with increasing temperature (DeLong *et al.*, 1958). As temperature increases, up to the optimum, growth increases and thus so must the quantity of dietary protein required for protein synthesis. Metabolic rate also increases with increasing temperature accompanied by an increase in endogenous nitrogen excretion (ENE) and an increase in the protein requirement to cover these losses. The ENE of carp was found to be 7.2 mg of nitrogen per 100 g of fish per day at 20°C and 8.6 mg at 27°C (Ogino *et al.*, 1973).

Temperature also affects the carcass composition of the fish. Carcass lipid contents have been found to increase with moderate increases in temperature for channel catfish (Andrews and Stickney, 1972), rainbow trout (Papoutsoglou and Papoutsoglou, 1978) and carp (Jauncey, 1979). The effects of temperatures well above the optimum on carcass composition appear to have been reported only for carp (Jauncey, 1979) where a marked decrease in carcass lipid content occurred at a temperature of 35°C.

9. Ration Size

Ration size is a further important factor affecting feed utilisation and the requirements are influenced by fish age and size, diet composition and numerous other factors perhaps the most important of which is temperature. The combined effects of temperature and feeding rate on the growth and food utilisation of fish have been reported by several authors (Brett *et al.*, 1969; Andrews and Stickney, 1972; Huisman 1969, 1976).

Increasing the feeding rate at any one temperature has been found to decrease the carcass moisture content and increase the carcass lipid content in sockeye salmon (Brett *et al.*, 1969), channel catfish (Murray *et al.*, 1977) and carp (Nijkamp *et al.*, 1974; Jauncey, 1979) implying that dietary energy, excess to requirements, is stored as carcass lipid.

The efficiency of food utilisation at any one temperature has been found to decrease with increasing ration size. Brett *et al.* (1969) conclude, from their study of sockeye salmon, that as feeding rate is lowered the optimum temperature for growth is also lowered. This, they say, is because the decrease in maintenance metabolism at lower temperatures should permit better use of a restricted ration for growth if the temperature-dependent activities of digestive enzymes and growth processes were unaffected. However, results for carp (Jauncey, 1979) presented in Figure 4.1 show that the optimum temperature for growth was approximately the same at feeding levels of 3, 6 and 9 per cent (dry food) of the wet body weight per day. This may be due to insufficient data at lower feeding rates or a difference in the response of temperature-dependent activities. The specific growth rate (SGR) of carp in this experiment increased with increasing feeding rate at each temperature and did not appear to have reached a maximum even at the 9 per cent feeding rate. However, Figure 4.1 does show that increasing the feeding level resulted in proportionately smaller increases in SGR as the feeding level increases, suggesting that it was approaching a maximum.

Brett *et al.* (1969) found, with sockeye salmon, that increasing the feeding rate at any one temperature increased growth up to an asymptote beyond which it remained constant. Kausch and Ballion-Cusmano (1976) also found that increasing the ration size of carp increased their growth up to an asmyptote beyond which it remained fairly constant.

It should be emphasised at this stage that contrary to the results above, Huisman (1969), with carp, found that increasing the ration above 3 per cent at 17°C and 8 per cent at 23°C caused a fall in SGR. Such a fall in SGR at high levels of feeding was also reported for rainbow trout (Roberts, 1976) and it was postulated that a high feeding rate combined with a relatively high temperature resulted in poorer growth rates due to the high induced metabolic rate and consequent reduction in the amount of energy available for growth.

That the metabolic rate of fish increases 4-5 times with rations increasing from maintenance to satiation was shown by Paloheimo and Dickie (1966 a, b) who postulated that this may have been due to the energy required to metabolise excess nutrients, principally the deamination of excess amino acids.

However, this conclusion of the theoretical study of Paloheimo and Dickie (1966a, b) is not borne out by the recent experimental work of Smith *et al.* (1978) who found, by direct measurement, little heat increment in rainbow trout with increasing food intake.

Figure 4.1: Growth Responses of Carp at Three Feeding Levels (as Dry Food Per Cent Wet Body Weight Per Day) and Four Temperatures.

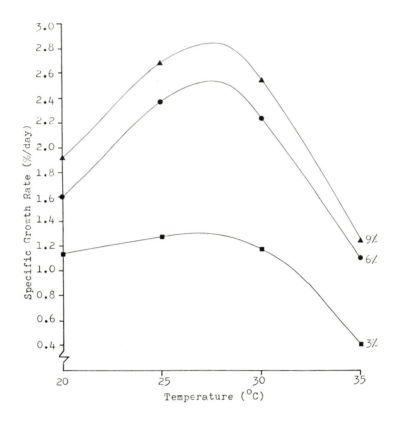

Huisman (1976) and Huisman *et al.* (1978) have extensively investigated the effects of ration size and environmental temperature in carp. The 'optimum' ration for 40 g carp at 23°C was reported to be 2.2 per cent of the body weight per day fed in nine equal feeds. This ration was optimum in terms of food utilisation, but maximum growth was obtained on a ration of 6.5 per cent per day. The commercially-viable ration size and feeding frequency is very much dependent on the cost of the diet in relation to the cost of the culture system.

It may be more economical to sacrifice optimum food conversion in order to obtain more rapid growth or *vice versa.*

10. Larval Nutrition

This section is necessarily brief as information is scarce on the particular requirements of carp larvae. The larvae of carp resorb their yolk sacs in approximately the first 24 hours after hatching at culture temperatures of 25° C. Carp larvae are extremely small on hatching (approximately 5-8 mm in length and weight from 3-5 mg) and the particle size of feed which they are capable of accepting is a problem. First feeding is usually performed with newly-hatched nauplii of *Artemia*, the brine shrimp, this usually being performed for seven days. A weaning stage then occurs during which the *Artemia* are gradually replaced with an inert dry feed such as salmon or trout starter feeds. At the end of this weaning phase (usually lasting about seven days) the fry are fed exclusively on dry foods. Recently, however, the demand for cysts of *Artemia* for raising fish larvae has exceeded supply and the quality of the cysts has been very variable. There is, therefore, an increasing incentive to find acceptable inert alternatives to *Artemia*.

Lakshmannan *et al.* (1966) evaluated a range of 19 possible feed ingredients, as larval feeds for carp, both singly and in combinations of 2, 3, 4 and 5 ingredients in varying proportions. Larvae reared on zooplankton achieved 94 per cent survival and a final length of 16.63 mm. The most promising of the alternatives was a mixture of notonectids, prawn (*Macrobrachium lamarrei*) and cow pea (*Vigna sinensis*) in a ratio of 5:3:2 which resulted in a mean length of 12.4 mm and a survival rate of 86 per cent. However, the diets were not analysed and there were no starved controls in the experiment making it impossible to determine what, if any, food was available from natural sources under the experimental conditions.

Appelbaum (1977) used a yeast grown on petrochemicals (*Candida lipolytica*) to rear carp larvae and compared this to feeding live *Artemia*. The yeast was found to be well digested with a gut passage time equivalent to that of *Artemia*. The slow sinking yeast had a particle size of 10-150 μm and after twelve days average lengths and survival of yeast fed larvae were as good as those of larvae fed *Artemia*. This study was continued by Appelbaum and Dor (1978) who fed carp larvae on two different rations. Group A received *Candida lipolytica* and group B received live *Artemia* and boiled egg yolk. For the first three days food of a particle size of 250 μm was fed which was then increased to 500 μ and finally 750 μ. Survival was high in both groups (over 95 per cent) and after ten days larvae fed diet A had attained a mean length of 10.37 mm and those fed diet B 9.02 mm. Grass carp larvae fed the yeast plus 4.5 per cent vitamin mixture and 5.5 per cent meat extract also had better growth and survival than larvae fed *Artemia* (Appelbaum and Uland, 1979). Dabrowski *et al.* (1978) performed a 25-day feeding trial on carp larvae using three basal diets. Two moist diets (31-37 per cent dry matter) consisted of mixtures of lyophilised spleen, lyophilised fish, fish protein, agar, starch, minerals, vitamins, fat and an amino acid supplement. The third basal diet was a dry diet (94.5 per cent dry matter) containing egg yolk, blood meal, fishmeal, starch, glucose, fat, vitamins and minerals, all three diets having protein levels of 73-83 per cent. These authors cite the hypothesis of Kainz (1976) that stomachless fish larvae, such as carp, cannot use artificial diets because they do not produce sufficient protease, relying on obtaining such enzymes from exogenous sources. Thus these diets were fed alone and in combination with bovine trypsin, neutralisation to pH 8.5 and additional feeding of cladocera, plankton and live zooplankton for various periods. Survival of larvae fed diets with the addition of bovine trypsin or neutralisation was improved compared to the basal rations: the moist diets showed no improvements in growth or survival over the dry diet. Survival ranged from 11 to 40 per cent but growth was negligible unless the diets were supplemented with zooplankton for the first three days of feeding.

Dabrowski *et al.* (1979) continued these studies of the effect of the addition of enzymes to diets for carp larvae. A basic diet of freeze-dried cattle spleen (40 per cent), fishmeal (20 per cent), lucerne meal (10 per cent), soya meal (10 per cent), milk (7 per cent), fish oil (2 per cent), yeast (5 per cent), minerals (2 per cent) and vitamins (2 per cent) was used. To this were added various levels of enzymes extracted from carp hepatopancreas and carp intestinal mucosa. The protein levels in the diets were fed for 16 days being compared to *Artemia*-fed and starved controls. The *Artemia*-fed fish reached a mean weight of 50 mg with 88 per cent survival during this period. The basal diet resulted in a final average weight of 6.68 mg compared to 4.03 mg for the starved controls and 23 per cent survival as compared to 8.6 per cent. Diets containing enzyme extracts produced very variable results although low levels of addition appeared beneficial with average weights up to 12.77 mg and survival up to 46 per cent.

Until such time as an adequate synthetic diet is developed for carp larvae it is impossible to investigate their specific nutrient requirements. Development of such a diet is regarded as a prime objective in the development of culture of those species·of fish having larval stages and much research is being performed on the requirements for taste, appearance, colour, particle size etc. for several species.

11. Conclusions

It may be seen from the preceding discussion that the understanding of the nutrition of fish, including carp, has progressed significantly over the last 10-15 years. However, a great deal still remains to be elucidated before fish nutritionists can be sure that the commercial rations formulated are optimal both in terms of fish health and growth and in terms of the economics of feeding.

Specifically major areas lacking in data are the mineral requirements of fish and, with fish such as carp, the possible contribution that intestinal microflora and exogenous enzymes may make to digestive processes. The enigma of the poor utilisation of amino acids and peptides discussed in Section 2.1 requires further investigation before it can be resolved. Interactions between dietary nutrients also require further study as does the possible satisfaction of dietary energy requirements by various forms of lipid and carbohydrate.

It can be concluded, therefore, that although substantial progress is being made in this field that much more investigation is required before the understanding of fish nutrition approaches that of the mammals.

References

Adelman, I.R. (1977). Effects of bovine growth hormone on growth of carp (*Cyprinus carpio*) and the influences of temperature and photoperiod. *J. Fish. Res. Bd Can., 3(4):* 509-15

——— (1978). Influence of temperature on growth promotion and body composition of carp (*Cyprinus carpio*) due to bovine growth hormone. *Trans. Am. Fish. Soc., 107(5):* 747-50

Adron, J.W., Blair, A., Cowey, C.B. and Shanks, A.M. (1976). Effects of dietary energy level and dietary energy source on growth, feed conversion and body composition of turbot (*Scopthalmus maximus* L.). *Aquaculture, 7:* 125-32

Andrews, J.W. and Murai, T. (1975). Studies on vitamin C requirements of channel catfish. *J. Nutr., 105:* 557-61

Andrews, J.W., Murray, M.W. and Davis, J.M. (1978). The influence of dietary fat levels and environmental temperature on digestibility and absorbability of animal fat in catfish diets. *J. Nutr., 108(5):* 749-52

Andrews, J.W. and Page, J.W. (1974). Growth factors in fish meal components of catfish diets. *J. Nutr., 104(8):* 1091-6

Andrews, J.W. and Stickney, R.R. (1972). Interaction of feeding rates and environmental temperature on growth, food conversion and body composition of channel catfish. *Trans. Am. Fish. Soc., 101:* 94-9

Anonymous (1973). What's happening to fishmeal? *Am. Fish. US Trout News, 18(6):* 6-9
——— (1978). The ingredient market. *Feedstuffs, 50(1):* 30

Aoe, H., Ikeda, I. and Saito, T. (1974). Nutrition of protein in young carp. II. Nutritive value of protein hydrolysates. *Bull. Jap. Soc. Sci. Fish., 40(4):* 375-9

Aoe, H., Masuda, I. and Takeda, T. (1967a). Water-soluble vitamin requirements of carp. I. Requirement for vitamin B_2. *Bull. Jap. Soc. Sci. Fish, 33(4):* 355-60
——— (1967b). Water-soluble vitamin requirements of carp. II. Requirements for p-aminobenzoic acid and inositol. *Bull. Jap. Soc. Sci. Fish., 33(7):* 674-80
——— (1967c). Water-soluble vitamin requirements of carp. III. Requirements for niacin. *Bull. Jap. Soc. Sci. Fish., 33(7):* 681-5
——— (1967d). Water-soluble vitamin requirements of carp. IV. Requirement for thiamine. *Bull. Jap. Soc. Sci. Fish., 33(10):* 970-4
———(1967e). Water-soluble vitamin requirements of carp. V. Requirement for folic acid. *Bull. Jap. Soc. Sci. Fish., 33(11):* 1968-71
——— (1968). Requirement of young carp for vitamin A. *Bull. Jap. Soc. Sci. Fish., 34(10):* 959-64
——— (1969). Water-soluble vitamin requirements of carp. VI. Requirement for thiamine and effects of antithiamines. *Bull. Jap. Soc. Sci. Fish., 35(5):* 456-65

Aoe, H., Masuda, I., Abe, I., Saito, T. and Togima, T. (1971). Water-soluble vitamin requirements of carp. VII. Some examinations on utility of reported minimum requirements. *Bull. Jap. Soc. Sci. Fish., 37(2):* 124-9

Aoe, H., Masuda, I., Abe, I., Saito, T., Toyoda, T. and Kitamura, S. (1970). Nutrition of protein in young carp. I. Nutritive value of free amino acids. *Bull. Jap. Soc. Sci. Fish., 36(4):* 407-13

Appelbaum, S. (1977). A possible substitute for live food of carp fry. *Arch. Fisch. Wiss., 28(1):* 31-44

Appelbaum, S. and Dor, U. (1978). Ten day experimental nursing of carp (*Cyprinus carpio*) larvae with dry food. *Bamidgeh, 30(3):* 85-8

Appelbaum, S. and Uland B. (1979). Intensive rearing of grass carp larvae (*Ctenopharyngodon idella*) under controlled conditions. *Aquaculture, 17:* 175-9

Aston, R.J. and Brown, D.J.A. (1978). Fish farming in heated effluents. In: Pastakia, C.M.R. (ed.), *Proceedings of the Conference on Fish Farming and Wastes*, University College, London, 4/5 January 1978. Institute of Fisheries Management, pp. 39-60

Atack, T.H., Jauncey, K. and Matty, A.J. (1979). The utilization of some single-celled proteins by fingerling mirror carp (*Cyprinus carpio* L.). *Aquaculture, 18:* 337-48

Atherton, W.D. and Aitken, A. (1970). Growth, nitrogen metabolism and fat metabolism in *Salmo gairdneri* Rich. *Comp. Biochem. Physiol., 36:* 719-47

Austreng, E. (1976). Fat and protein in diets for salmonid fishes. 2. Fat content in dry diets for rainbow trout (*Salmo gairdneri* Rich.). *Meld. Nord. Landsbrukshogsk, 55(6):* 1-14

Bender, A.E. and Miller, D.S. (1953). A new brief method for estimating net protein value. *Biochem. J., 53:* vii

Bondi, A., Spandorf, A. and Calmi, S.R. (1957). The nutritive value of various feeds for carp. *Bamidgeh, 9(1):* 13-22

Brett, J.R. and Higgs, D.A. (1970). Effect of temperature on the rate of gastric digestion in fingerling sockeye salmon *Oncorhynchus nerka. J. Fish. Res. Bd. Can., 27:* 1767-79

Brett, J.R., Shelbourne, J.E. and Shoop, C.T. (1969). Growth rate and body composition of fingerling sockeye salmon *Oncorhynchus nerka*, in relation to temperature and ration size. *J. Fish. Res. Bd Can., 26:* 2363-94

Brody, S. (1945). *Bioenergetics and Growth*. Reinhold, New York

Buhler, D.R. and Halver, J.E. (1961). Nutrition of salmonid fishes. IX. Carbohydrate requirements of chinook salmon. *J. Nutr., 74:* 307-18

Carpenter, J.K. and Ellinger, G.M. (1955). The estimation of 'available lysine' in protein concentrates. *Biochem. J., 61:* xi-xii

Castell, J.D. (1978). Fats – review of lipid requirements of finfish. *EIFAC Symp Finfish Nutr. and Feed Technol.*, Hamburg, June 1978. EIFAC/78/Symp.R/9.1

Castell, J.D., Lee, D.J. and Sinnhuber, R.O. (1972a). Essential fatty acids in the diet of rainbow trout (*Salmo gairdneri*): Lipid metabolism and fatty acid composition. *J. Nutr., 102:* 93-100

Castell, J.D., Sinnhuber, R.O., Lee, D.J. and Wales, J.H. (1972b). Essential fatty acids in the diet of rainbow trout (*Salmo gairdneri*): Physiological symptoms of essential fatty acid deficiency. *J. Nutr., 102:* 87-92

Castell, J.D., Sinnhuber, R.O., Wales, J.H. and Lee, D.J. (1972c). Essential fatty acids in the diet of rainbow trout (*Salmo gairdneri*): Growth, feed conversion and some gross deficiency symptoms. *J. Nutr., 102:* 77-86

Chiou, J.Y. and Ogino, C. (1975). Digestibility of starch in carp. *Bull. Jap. Soc. Sci. Fish, 41(4):* 465-6

Cowey, C.B. (1978). Protein and amino acid requirements of finfish. *EIFAC Symp. Finfish Nutr. and Feed Technol.*, Hamburg, June 1978. EIFAC/78/Symp.R/6

Cowey, C.B., Adron, J. and Blair, A. (1970). Studies on the nutrition of marine flatfish. The essential amino acid requirements of plaice and sole. *J. Mar. Biol. Assoc. UK, 50:* 87-95

Cowey, C.B., Adron, J.W., Blair, J.A. and Shanks, A.M. (1974). Studies on nutrition of marine flatfish. Utilization of various dietary proteins by plaice (*Pleuronectes platessa*). *Br. J. Nutr., 31(3):* 297-306

—— (1977). The production of renal calcinosis by magnesium deficiency in rainbow trout (*Salmo gairdneri*). *Br. J. Nutr., 38:* 127-35

Cowey, C.B., Adron, J.W., Brown, D.A. and Shanks, A.M. (1975). Studies on nutrition of marine flatfish. The metabolism of glucose by plaice (*Pleuronectes platessa*) and the effect of dietary energy source on protein utilization in plaice. *Br. J. Nutr., 33(2):* 219-31

Cowey, C.B., Pope, J.A., Adron, J.W. and Blair, J.A. (1971). Studies on nutrition of marine flatfish. Growth of the plaice (*Pleuronectes platessa*) on diets containing proteins derived from plants and other sources. *Mar. Biol., 10:* 45

—— (1972). Studies on nutrition of marine flatfish. The protein requirements of plaice (*Pleuronectes platessa*). *Br. J. Nutr., 28:* 447-56

—— (1973). Studies on nutrition of marine flatfish. The effect of oral administration of diethylstilbestrol and cyprohepadine on the growth of *Pleuronectes platessa. Mar. Biol., 19:* 1-6

Cowey, C.B. and Sargent, J.R. (1972). Fish nutrition. *Adv. Mar. Biol., 10:* 383-492

—— (1977). Lipid nutrition in fish. *Comp. Biochem. Physiol., 57(48):* 269-74

—— (1979). Fish Nutrition. In: Hoar, W.S. and Randall, D.J. (eds), *Fish Physiology*, Vol. VIII. Academic Press, New York and London, pp. 1-69

Csengari, I., Majoras, F. Oláh, J. and Farkas, T. (1978). Investigations on the essential fatty acid requirement of carp (*Cyprinus carpio* L.) *EIFAC Symp. Finfish Nutr. and Feed Technol.*, Hamburg, June 1978. EIFAC/78/Symp.E/11.

Dabrowski, H., Grudniewski, C. and Dabrowski, K. (1979). Artificial diets for common carp: Effect of the addition of enzyme extracts. *Prog. Fish. Cult., 41(4):* 196-200

Dabrowski, K. (1977). Protein requirements of grass carp fry (*Ctenopharyngodon idella* Val.). *Aquaculture, 12(1):* 63-73

—————— (1979). Feeding requirements of fish with particular attention to common carp. A review. *Pol. Arch. Hydrobiol., 26(1/2):* 135-58

Dabrowski, K., Dabrowska, H. and Grudniewski, C. (1978). A study of the feeding of common carp larvae with artificial food. *Aquaculture, 13(3):* 257-64

Dabrowski, K. and Wojno, T. (1978a). Use of non-protein nitrogen compounds for feeding of carp (*Cyprinus carpio* L.). 1. Feed characteristics, fish growth and feed utilization. *Zesz. nauk. ART Olszt., 7:* 83-100

—————— (1978b). Use of non-protein compounds for feeding of carp (*Cyprinus carpio* L.). 2. Digestibility of nutrients, feed protein utilization indices, absorption of amino acids. *Zesz. nauk. ART Olszt., 7:* 101-20

—————— (1978c). Use of non-protein nitrogen compounds for feeding of carp (*Cyprinus carpio* L.). 3. Chemical composition of fish body. *Zesz. nauk. ART Olszt., 7:* 121-31

—————— (1978d). Use of non-protein nitrogen compounds for feeding of carp (*Cyprinus carpio* L.). 4. Physiological indices. *Zesz. nauk. ART Olszt., 7:* 132-46

Davis, A.T. and Stickney, R.R. (1978). Growth responses of *Tilapia aurea* to dietary protein quality and quantity. *Trans. Am. Fish. Soc., 107(3):* 470-83

DeLong, D.C., Halver, J.E. and Mertz, E.T. (1958). Nutrition of salmonid fishes. VI. Protein requirements of chinook salmon at two water temperatures. *J. Nutr., 65:* 589-99

Dupree, H.K. (1969). Influence of corn oil and beef tallow on growth of channel catfish. *Tech. Pap. US Bur. Sport. Fish. Wildl., 27:* 13

Dupree, H.K. and Halver, J.E. (1970). Amino acids essential for the growth of channel catfish, *Ictalurus punctatus. Trans. Am. Fish. Soc., 99:* 90-2

Dupree, H.K. and Sneed, K.E. (1966). Responses of channel catfish fingerlings to different levels of major nutrients in purified diets. *Tech. Pap. US Bur. Sport. Fish. Wildl., 9:* 21

Edwards, D.J., Austreng, E., Risa, S. and Gjedrem, T. (1977). Carbohydrate in rainbow trout diets. 1. Growth of fish of different families fed diets containing different proportions of carbohydrate. *Aquaculture, 11(1):* 31-8

Ellinger, G.M. and Duncan, A. (1976). The determination of methionine in proteins by gas-liquid chromatography. *Biochem. J., 155:* 615-21

Erman, Ye.Z. (1969). The nitrogen saving effect of carbohydrates in the carp. *Probl. Icthyol., 9(4):* 615-17

Farkas, T. and Csengari, I. (1976). Biosynthesis of fatty acids by the carp, *Cyprinus carpio* L., in relation to environmental temperature. *Lipids, 11:* 401-7

Farkas, T., Csengari, I., Majoras, F. and Oláh, J. (1977). Metabolism of fatty acids in fish. I. Development of essential fatty acid deficiency in the carp (*Cyprinus carpio* L.). *Aquaculture, 11(2):* 147-58

—————— (1978). Metabolism of fatty acids in fish. II. Biosynthesis of fatty acids in relation to diet in the carp (*Cyprinus carpio* L.). *Aquaculture, 14(1):* 57-66

Garling, D.L. and Wilson, R.P. (1976). Optimum dietary protein to energy ratio for channel catfish fingerlings. *J. Nutr., 106(9):* 1368-75

—————— (1977). Effects of dietary protein to lipid ratios on growth and body composition of fingerling channel catfish. *Prog. Fish. Cult., 39(1):* 43-7

Ghittino, P. (1970). Riposte delle trotte d'allevamento al stilbestrols e metilburacile. *Riv. Ital. Pisciol. Ittiopatol., 5:* 9-11

Gribanov, L.V., Korneev, A.W. and Korneeva, L.A. (1966). Use of thermal waters for commercial production of carps in floats in the USSR. *FAO Fish. Rep., 44(5):* 218-26

Halver, J.E. (1957a). The nutrition of salmonid fishes. (III) Water-soluble vitamin requirements for chinook salmon. *J. Nutr., 62:* 225

—————— (1957b). The nutrition of salmonid fishes. (IV) An amino acid test diet for chinook salmon. *J. Nutr., 62:* 225-43

—————— (1969). Vitamin requirements. In: Neuhaus, O.W. and Halver, J.E. (eds), *Fish in Research*. Academic Press, New York and London, pp. 209-44

—————— (1971). Nutritional requirements of salmon and trout. *Proc. Georgia Nutr. Conf. Feed Manuf., 1971:* 128

—————— (ed.) (1972). *Fish nutrition*. Academic Press, New York and London, 713 pp.

—————— (1978). Vitamin requirements of finfish. *EIFAC Symp. Finfish Nutr. and Feed Technol.*, Hamburg, June 1978. EIFAC/78/Symp.R/8

Halver, J.E. and Coates, J.A. (1957). A vitamin test diet for long-term feeding studies. *Prog. Fish Cult., 19:* 112-18

Halver, J.E. and Shanks, W.E. (1960). Nutrition of salmonid fishes. (VIII) Indispensable amino acids for sockeye salmon. *J. Nutr., 72:* 340-6

Halver, J.E., Smith, R.R., Talbert, B.M. and Baker, E.M. (1975). Utilization of ascorbic acid in fish. *Ann. N.Y. Acad. Sci., 258:* 81-102

Hashimoto, Y. (1953). Effect of antibiotics and vitamin B_{12} supplement on carp growth. *Bull. Jap. Soc. Sci. Fish., 19(8):* 899-904

Hashimoto, Y. and Okaichi, T. (1969). Vitamins as nutrients for fish. Translation from *Nutrition of Fish and Feedstuffs for Fish Culture* (revised edition), pp. 52-85 (in Japanese). Roche Products Ltd, London.

Hepher, B., Chervinski, J. and Tagari, H. (1971). Studies on carp nutrition. 3. Experiments on the effect on fish-yields of dietary protein source and concentration. *Bamidgeh, 23(1):* 11-37

Hepher, B., Sandbank, E. and Shelef, G. (1978). Alternative protein sources for warm water fish diets. *EIFAC Symp. Finfish Nutr. and Feed Technol.*, Hamburg, June 1978. EIFAC/ 78/Symp.R/11.2

Higashi, H., Taneko, T., Ishii, S., Masuda, I. and Sugihashi, T. (1964). Effect of dietary lipid on fish under cultivation. I. Effect of large amounts of lipid on health and growth of rainbow trout. *Bull. Jap. Soc. Sci. Fish., 30:* 778-85

Higashi, H., Taneko, T., Ishii, S., Ushiyama, M. and Sugihashi, T. (1966). Effect of ethyl linoleate, ethyl linolenate and ethylesters of highly unsaturated fatty acids on essential fatty acid deficiency in rainbow trout. *J. Vitaminol., 12:* 74-9

Higuera, M. De La., Murillo, A., Varela, G. and Zamora, S. (1977). The influence of high dietary fat levels on protein utilization by trout (*Salmo gairdneri*). *Comp. Biochem. Physiol., 56A (1):* 37-41

Huisman, E.A. (1969). A study of the possibility of breeding carp in flow recirculation tanks (1968/9). Ann. Report of the OVB, Holland, 45 pp.

————— (1976). Food conversion efficiencies at maintenance and production levels for carp (*Cyprinus carpio* L.) and rainbow trout (*Salmo gairdneri* Rich.). *Aquaculture, 9:* 259-73

Huisman, E.A., Klein Breteler, J.G.P., Vismans, M.M. and Kanis, E. (1978). Retention of energy, protein, fat and ash in growing carp (*Cyprinus carpio* L.) under different feeding and temperature regimes. *EIFAC Symp. Finfish Nutr. and Feed Technol.*, Hamburg, June 1978. EIFAC/78/Symp.E/12

Ikeda, S. and Sata, M. (1964). Biochemical studies on L-ascorbic acid in aquatic animals. III. Biosynthesis of L-ascorbic acid by carp. *Bull. Jap. Soc. Sci. Fish., 30:* 365-71

Jauncey, K. (1979). Growth and nutrition of carp in heated effluents. PhD thesis, University of Aston in Birmingham, 202 pp.

Kainz, E. (1976). Weitere versuche zur aufzucht der brut des Karpfens (*Cyprinus carpio*) mit trockenfritter mitteln. *Osterr. Fisch., 29(4):* 58-62

Kaneko, T.P. (1969). Composition of food for carp and trout. *EIFAC Tech. Pap., 9:* 161-8

Kashiwada, K., Kanazawa, A. and Teshima, T. (1971). Studies on the production of B vitamins by intestinal bacteria. VI. Production of folic acid by the intestinal bacteria of carp. *Mem. Fac. Fish. Kagoshima Univ., 20:* 185-9

Kashiwada, K., Teshima, T. and Kanazawa, A. (1970). Studies on the production of B vitamins by intestinal bacteria. V. Evidence of the production of vitamin B_{12} by micro-organisms in the intestinal canal of carp (*Cyprinus carpio*). *Bull. Jap. Soc. Sci. Fish., 36:* 421-4

Kausch, H. and Ballion-Cusmano, M.F. (1976). Korperzusaro-mensetzung wachstun und naturungsausnutzung bei jungen Karpfen (*Cyprinus carpio* L.) unter Internsivhaltungs-bedingungen. *Arch. Hydrobiol., 48(2):* 141-80

Kayama, M. and Tsuchiya, Y. (1959). Fat metabolism in fish. 1. Intestinal absorption and distribution study of oil in the carp (*Cyprinus carpio* L.). *Tohoku J. Agric. Res., 10(2):* 229-36

Kerns, C.L. and Roelofs, E.W. (1977). Poultry wastes in the diet of Israeli carp. *Bamidgeh, 29(4):* 125

Ketola, H.G. (1978). Dietary zinc prevents cataract in trout. *Fed. Proc., 37:* 584 (Abs.)

Kim, Y.K. (1974). Determination of true digestibility of dietary proteins in carp with chromic oxide containing diets. *Bull. Jap. Soc. Sci. Fish., 40(7):* 651-3

Kitamikado, M., Morishita, T. and Tachino, S. (1964). Digestibility of dietary protein in rainbow trout. II. Effects of starch and oil contents in diets and size of fish. *Bull. Jap. Soc. Sci. Fish., 30(1):* 50-4

Kitamura, S., Suwa, T., Ohara, S. and Nakagawa, K. (1967). Studies on vitamin requirements of rainbow trout. *Bull. Jap. Soc. Sci. Fish., 33:* 1120-31

Koops, H., Tiews, K., Beck, H. and Gropp, J. (1976). The utilization of soybean protein by the rainbow trout (*Salmo gairdneri* Rich.). *Arch. Fischerewiss., 26(2-3):* 181-91

Korneyev, A.N. (1969). The biological requirements for warm industrial water used for fish breeding. In: *Fish Breeding in Warm Water in the USSR and Abroad.* Moscow, All-Union Scientific Research Institute of Maritime Fisheries and Oceanography, pp. 3-20

Krishnandhi, S. and Shell, W. (1967). Utilization of soybean protein by channel catfish (*Ictalurus punctatus* Raff.). *Proc. Ann. Conf. S.E. Assoc. Game and Fish. Commrs. 1965, 19:* 205-9

Lakshmanan, M.A.V., Murty, O.S., Pillai, K.K. and Banerjee, S.C. (1966). On a new artificial feed for carp fry. *FAO Fish. Rep. 44, 3:* 373-87

Lall, S.P. (1978). Minerals in finfish nutrition. *EIFAC Symp. Finfish Nutr. and Feed Technol.*, Hamburg, June 1978. EIFAC/78/Symp.R/9.2

Lee, D.J. and Putnam, G.B. (1973). The response of rainbow trout to varying protein/ energy ratios in a test diet. *J. Nutr., 103(6):* 916-22

Lee, D.J. and Sinnhuber, R.O. (1972). Lipid requirements. In: Halver, J.D. (ed.), *Fish Nutrition.* Academic Press, New York and London, pp. 145-80

Lee, D.J. and Wales, J.H. (1973). Observed liver changes in rainbow trout (*Salmo gairdneri*) fed varying levels of a casein-gelatin mixture and herring oil in experimental diets. *J. Fish. Res., 30(7):* 1017-20

Lesauskiene, L., Jankevicius, K. and Syvokiene, J. (1974). Role of microorganisms in the digestive canal in the nourishment of pond fish. 6. Amount of free amino acid in species of second year fish and ability of microorganisms to synthesise them. *Tr. Akad. Naut. Lit. SSR. Ser. B., 66(2):* 127-36

Lesauskiene, L., Syvokiene, J. and Grigorovic, G. (1975). Role of microorganisms in the digestive canal in the nourishment of pond fish. 8. Amount of free amino acid in some species of third year fish and ability of microorganisms to synthesise them. *Tr. Akad. Natu. Lit. SSR., Ser. B., 69(1):* 103-11

Lin, H., Romsos, D.R., Tack, P.I., and Leveille, E.A. (1977). Influence of diet on *in vitro* and *in vivo* rates of fatty acid synthesis in coho salmon (*Oncorhynchus kisutch*). *J. Nutr., 107(9):* 1677-82

Lovell, R.T. (1973). Essentiality of vitamin C in feeds for channel catfish. *J. Nutr., 103:* 134-8

Lukowicz, M. von (1978). Experiences with krill (*Euphasis superba* Dana) in the diet for young carp (*Cyprinus carpio* L.). *EIFAC Symp. Finfish Nutr. and Feed. Technol.*, Hamburg, June 1978. EIFAC/78/Symp.E/69

Mazid, U.A., Tanaka, Y., Katayama, T., Simpson, K.L. and Chichester, C.O. (1978). Metabolism of amino acids in aquatic animals. III. Indispensable amino acids for *Tilapia zillii. Bull. Jap. Soc. Sci. Fish, 44(7):* 739-42

McBridge, J.R. and Fagerlund, U. (1973). The use of 17-methyltestosterone for promoting weight gain in juvenile Pacific salmon. *J. Fish. Res. Bd Can., 30:* 1099-104

McLaren, B.A., Keller, E., O'Donnel, D.J. and Elvehjem, C.A. (1947a). The nutrition of rainbow trout. I. Studies on vitamin requirements. *Arch. Biochem., 15:* 169-78
—— (1947b). The nutrition of trout. II. Further studies with purified rations. *Arch. Biochem., 15(2):* 179-85

Mertz, E.T. (1969). Amino acid and protein requirements of fish. In: Newhaus, O.W. and Halver, J.E. (eds), *Fish in Research.* Academic Press, New York and London, pp. 233-44

Meske, Ch. and Pfeffer, E. (1977). Micro-algae, yeast or casein as components of fishmeal free dry feeds for carp. *Zeit für Tierphysiol. Tierernährung & Füttermittelkunde, 38(4-5):* 177-85
—— (1978). Influence of source and level of dietary protein on body composition of carp. *EIFAC Symp. Finfish Nutr. and Feed Technol.*, Hamburg, June 1978. EIFAC/78/ Symp.E/45

Meske, Ch. and Pruss, H.E. (1977). Mikroalgen als components von fischmehl freiem fischfutter. *Adv. An. Physiol. and An. Nutr., 8:* 71-81

Metailler, R., Febvre, A. and Alliot, E. (1973). Preliminary note of the amino acid require- ments of the sea bass, *Dicentrarchus labrax* (Linn.). *Stud. Rev. GFCM, 52:* 91-6

Miller, D.S. and Bender, A.E. (1955). The determination of the net utilization of proteins by a shortened method. *Br. J. Nutr., 9:* 382-8

Murray, M.W., Andrews, J.W. and DeLoach, H.L. (1977). Effects of dietary lipids, dietary protein and environmental temperatures on growth, feed conversion and body composition of channel catfish. *J. Nutr., 107(2):* 272-80

NAC (1973). Nutrient requirements of trout, salmon and catfish. National Academy of Sciences, Subcomm. on Fish Nutrition, No. 11, 57 pp.

—— (1977). Nutrient requirements of warmwater fishes. National Academy of Sciences, Subcomm. on Fish Nutrition, 78 pp.

Nagai, M. and Ikeda, S. (1971a). Carbohydrate metabolism in fish. I. Effects of starvation and dietary composition on the blood glucose, hepatopancreatic glycogen and lipid contents in carp. *Bull. Jap. Soc. Sci. Fish., 37(5):* 404-9

—— (1971b). Carbohydrate metabolism in fish. II. Effect of dietary composition on metabolism of glucose-6(^{14}C) in carp. *Nippon Suisan Gakkaishi, 37:* 410-14

—— (1973). Carbohydrate metabolism in fish. IV. Effect of dietary composition on metabolism of acetate-U^{14}C and L-alanine-U-^{14}C in carp. *Bull. Jap. Soc. Sci. Fish., 39(6):* 633-43

Nicolaides, N. and Woodall, A.N. (1962). Impaired pigmentation in chinook salmon fed diets deficient in essential fatty acids. *J. Nutr., 78:* 431-7

Nijkamp, J.H., Van Es, A.J.H. and Huisman, E.A. (1974). Retention of nitrogen, fat, ash, carbon and energy in growing chickens and carp. In: Menke, K.W., Lantzsch, H.J. and Reichl, J.R. (eds), *Energy Metabolism of Farm Animals*. University of Hohenheim, West Germany, pp. 277-80

Njaa, L.R. (1977). A method for determining the methionine sulfoxide content in protein concentrates. *Proc. 11th Meet. Fed. Eur. Biochem. Soc., A3-8:* 933

Nose, T. (1970). A preliminary report on some essential amino acids for the growth of the eel (*Anguilla japonica*). *Bull. Freshwat. Fish. Res. Lab., Tokyo, 19:* 31-6

—— (1978). Summary report on the requirements of essential amino acids for carp. *EIFAC Symp. Finfish Nutr. and Feed Technol.*, Hamburg, June 1978. EIFAC/78/Symp.E/8

Nose, T. and Arai, S. (1976). Recent advances in studies on mineral nutrition of fish in Japan. In: *FAO Conf. on Aquacult.*, Kyoto, 1976. FIR: AQ/Conf./76/E.25

Nose, T., Arai, S., Lee, D.L. and Hashimoto, Y. (1974). A note on amino acids essential for growth of young carp. *Bull. Jap. Soc. Sci. Fish., 40(9):* 903-8

Ogino, C. (1965). B vitamin requirements of carp (*Cyprinus carpio*). 1. Deficiency symptoms and requirements of B_6. *Bull Jap. Soc. Sci. Fish., 31(7):* 546-51

—— (1967). B vitamin requirements of carp (*Cyprinus carpio*). 2. Requirements of riboflavin and pantothenic acid. *Bull. Jap. Soc. Sci. Fish., 33(4):* 351-4

—— (1980). Requirement of carp and rainbow trout for essential amino acids. *Bull. Jap. Soc. Sci. Fish., 46(2):* 171-4

Ogino, C. and Chen, M.S. (1973a). Protein nutrition in fish. IV. Biological value of dietary protein in carp. *Bull. Jap. Soc. Sci. Fish., 39(7):* 797-800

—— (1973b). Protein nutrition in fish. V. Relation between biological value of dietary proteins and their utilization in carp. *Bull. Jap. Soc. Sci. Fish., 39(9):* 955-9

Ogino, C. and Chiou, J.Y. (1976). Mineral requirements in fish. II. Magnesium requirement of carp. *Nippon Suisan Gakkaishi, 41:* 71-5

Ogino, C., Chiou, J.Y. and Takeuchi, T. (1976). Protein nutrition in fish. VI. Effects of dietary energy sources on the utilization of proteins by rainbow trout and carp. *Bull. Jap. Soc. Sci. Fish., 42:* 213-18

Ogino, C., Cowey, C.B. and Chiou, J.Y. (1978). Leaf protein concentrates as a protein source in diets for carp and rainbow trout. *Bull. Jap. Soc. Sci. Fish., 44(1):* 49-52

Ogino, C., Kakino, J. and Chen, M.S. (1973). Protein nutrition in fish. II. Determination of metabolic fecal nitrogen and endogenous nitrogen excretions of carp. *Bull. Jap. Soc. Sci. Fish., 39(5):* 519-23

Ogino, C. and Kamizono, M. (1975). Mineral requirements of fish. I. Effects of dietary salt mixtures on growth, mortality and body composition in rainbow trout and carp. *Bull. Jap. Soc. Sci. Fish., 41(4):* 429-34

Ogino, C., Kawasaki, H. and Nanri, H. (1980). Method for determination of nitrogen retained in the fish body by the carcass analysis. *Bull. Jap. Soc. Sci. Fish., 46(1):* 105-8

Ogino, C. and Saito, K. (1970). Protein nutrition in fish. I. The utilization of dietary protein by carp. *Bull. Jap. Soc. Sci. Fish., 36(3):* 250-4

Ogino, C. and Takeda, H. (1974). Unpublished data cited by Nose and Arai (1976)

—— (1976). Mineral requirements of fish. III. Calcium and phosphorus requirements in carp. *Bull. Jap. Soc. Sci. Fish., 42:* 793-9

Ogino, C., Takeuchi, L., Takeda, H. and Watanabe, T. (1979). Availability of dietary phosphorus in carp and rainbow trout. *Bull. Jap. Soc. Sci. Fish., 45(12):* 1527-32

Ogino, C., Uki, N., Watanabe, T., Lida, Z. and Ando, K. (1970a). B vitamin requirement of carp. III. Requirements for biotin. *Bull. Jap. Soc. Sci. Fish., 36:* 734-40

—— (1970b). B vitamin requirements of carp. IV. Requirements for choline. *Bull. Jap. Soc. Sci. Fish., 36(11):* 1140-6

Ogino, C. and Yang, G.Y. (1979). Requirement of carp for dietary zinc. *Bull. Jap. Soc. Sci. Fish., 48(8):* 967-9

Osborne, T.B., Mendel, L.B. and Ferry, E.L. (1919). A method for expressing numerically the growth promoting value of proteins. *J. Biochem., 37:* 223-29

Page, J.W. (1974). Utilization of free amino acids by channel catfish. *Proc. Fish Feed and Nutr. Workshop, 3:* 9-13

Page, J.W. and Andrews, J.W. (1973). Interactions of dietary levels of protein and energy on channel catfish *(Ictalurus punctatus). J. Nutr., 103:* 1339-46

Paloheimo, J.E. and Dickie, L.M. (1966a). Food and growth of fishes. II. Effects of food and temperature on the relationship between metabolism and body size. *J. Fish Res. Bd Can., 23:* 869-908

—— (1966b). Food and growth of fishes. III. Relations among food, body size and growth efficiency. *J. Fish. Res. Bd Can., 23:* 1209-48

Papoutsoglou, S.E. and Papaparaskeva-Papoutsoglou, E.G. (1978). Comparative studies on body composition of rainbow trout *(Salmo gairdneri)* in relation to type of diet and growth rate. *Aquaculture, 13(3):* 235-44

Pfeffer, E. and Meske, Ch. (1979). Untersuchungen zür ermittlung des bedarfs an mineralischen mengelelementen von spiegelkarpfen *(Cyprinus carpio). Zeit. fur Tiephysiol., Tierernehrung und Futtermittelkunde, 42(5):* 225-31

Phillips, A.M. Jr. (1969). Nutrition, digestion and energy utilization. In: Hoar, W.S. and Randall, D.J. (eds), *Fish Physiology*, Vol. I. Academic Press, New York and London, pp. 391-423

Phillips, A.M. Jr. and Brockway, D.R. (1957). Nutrition of trout. IV. Vitamin requirements. *Prog. Fish Cult., 19:* 119

Phillips, A.M. Jr., Tunison, A.V. and Brockway, D.R. (1948). Utilization of carbohydrates by trout. *Fish. Res. Bull. No. 11*, New York Cons. Dept, Albany

Pieper, A. and Pfeffer, E. (1978). Carbohydrates as possible sources of dietary energy for rainbow trout *(Salmo gairdneri). EIFAC Symp. Finfish Nutr. and Feed. Technol.*, Hamburg, June 1978. EIFAC/78/Symp.E/21

Pitt, T.K., Garside, E.T. and Hepburn, R.L. (1956). Temperature selection of the carp *(Cyprinus carpio* L.). *Can. J. Zool., 34:* 555-7

Reinitz, G.L., Orme, L.E., Lemm, C.A. and Hitzel, F.N. (1978). Influence of varying lipid concentrations with two protein concentrations in diets for rainbow trout *(Salmo gairdneri). Trans. Am. Fish. Soc., 107(5):* 751-4

Reynolds, W.W. and Casterlin, M.E. (1977). Temperature preferences of four fish species in an electronic thermoregulatory shuttlebox. *Prog. Fish. Cult., 39(3):* 112-14

Roberts, J.K. (1976). The metabolism and growth of rainbow trout, *Salmo gairdneri*, in fresh and saline waters. PhD thesis, University of Aston in Birmingham, 329 pp.

Rozin, P. and Mayer, J. (1964). Some factors influencing short-term food intake of the goldfish. *Am. J. Physiol., 206:* 1430-6

Rumsey, G.L. and Ketola, H.G. (1975). Amino acid supplementation of casein in diets of Atlantic salmon *(Salmo salar)* fry and of soybean meal for rainbow trout *(Salmo gairdneri)* fingerlings. *J. Fish. Res. Bd Can., 32(3):* 422-6

Sakamoto, S. and Yone, Y. (1973). Effect of dietary calcium/phosphorus ratio upon growth, feed efficiency, and blood serum calcium and phosphorous level in Red Sea bream. *Nippon Suisan Gakkaishi, 39:* 343-8

Schlumberger, O. and Labat, R. (1978). Alimentation de la carpe *(Cyprinus carpio* L.) à partir de dechets industriels. *EIFAC Symp. Finfish Nutr. and Feed Technol.*, Hamburg, June 1978. EIFAC/78/Symp.E/35

Sen, P.R., Rao, N.G.S., Ghosh, S.R. and Rout, M. (1978). Observations on the protein and carbohydrate requirements of carps. *Aquaculture, 13(3):* 245-6

Shanks, W.E., Gahimer, G.D. and Halver, J.E. (1962). The indispensable amino acids for rainbow trout. *Prog. Fish. Cult., 24:* 68-73

Shcherbina, M.A. and Kazlauskene, O.P. (1971). Water temperature and the digestibility of nutrient substances by carp. *Hydrobiol. J., 7(3):* 40-4

Shpet, G.I. and Kharitonaova, N.N. (1963). Utilization of food by the goldfish (*Carassius auratus gibelio*, Bloch) and the carp (*Cyprinus carpio* L.). *Zool. Zn., 42(2):* 395-9

Sin, A.W. (1973a). The dietary protein requirements for growth of young carp (*Cyprinus carpio*). *Hong Kong Fish. Bull., 3:* 77-81

——— (1973b). The utilization of dietary protein for growth of young carp (*Cyprinus carpio*) in relation to variations in fat intake. *Hong Kong Fish. Bull., 3:* 83-8

Singh, R.P. and Nose, T. (1967). Digestibility of carbohydrate in young rainbow trout. *Bull. Freshwat. Fish. Res. Labl, Tokyo, 17(1):* 21-5

Sinnhuber, R.O. (1969). The role of fats. In: Neuhaus, O.W. and Halver, J.E. (eds), *Fish in Research*. Academic Press, New York and London, pp. 245-59

Sinnhuber, R.O., Wales, J.H., Ayres, J.L. and Engebrecht, R.H. (1968). Dietary factors and hepatoma in rainbow trout (*Salmo gairdneri*). II. Co-carcinogenesis by cyclopropanoid fatty acids and the effects of gossypol and altered lipids in aflatoxin-induced liver cancer. *J. Nat. Cancer Inst., 91(6):* 1293-301

Smith, P.R., Rumsey, G.L. and Scott, M.I. (1978). Heat increment associated with dietary protein, fat, carbohydrate and complete diets in salmonids. *J. Nutr., 108(6):* 1025-32

Stickney, R.R. and Andrews, J.W. (1972). The effects of dietary lipids on growth and food conversion of channel catfish. *J. Nutr., 102(2):* 249-57

Stickney, R.R. and Shumway, S.E. (1974). Occurrence of cellulase activity in the stomachs of fishes. *J. Fish. Biol., 6:* 779-90

Syvokiene, J., Jankevicius, K., Lesauskene, L. and Antanyniene, A. (1974). Role of micro-organisms of the digestive canal in the nourishment of pond fish. 5. Amount of free amino acids of first year fish. *Tr. Akad. Nauk. Lit. SSR, Ser. B., 65(1):* 143-7

Syvokiene, J., Jankevicius, K. and Lubianskene, L. (1975). Role of micro-organisms of the digestive canal in the nourishment of pond fish. 10. Synthesis of free amino acids by the digestive tract of three year old fish. *Tr. Akad. Nauk. Lit. SSR, Ser. B. 70(2):* 91-5

Takeuchi, L., Takeuchi, T. and Ogino, C. (1980). Riboflavin requirements in carp and rainbow trout. *Bull. Jap. Soc. Sci. Fish., 46(6):* 733-7

Takeuchi, T. and Watanabe, T. (1977). Effect of eicosapentaenoic and decosahexaenoic acid in pollock liver oil on growth and fatty acid composition of rainbow trout. *Bull. Jap. Soc. Sci. Fish., 43(8):* 947-53

——— (1978). Growth-enhancing effect of cuttlefish liver oil and short necked clam oil on rainbow trout and their effective components. *Bull. Jap. Soc. Sci. Fish., 44(7):* 733-8

Takeuchi, T., Watanbe, T. and Ogino, C. (1978a). Supplementary effect of lipids in high protein diets of rainbow trout. *Bull. Jap. Soc. Sci. Fish., 44(6):* 81

——— (1978b). Optimum ratio of protein to lipid in diets of rainbow trout. *Bull. Jap. Soc. Sci. Fish., 44(6):* 683-8

——— (1978c). Use of hydrogenated fish oil and beef tallow as a dietary energy source for carp and rainbow trout. *Bull. Jap. Soc. Sci. Fish., 44(8):* 875-81

——— (1979a). Digestibility of hydrogenated fish oils in carp and rainbow trout. *Bull. Jap. Soc. Sci. Fish., 45(12):* 1521

——— (1979b). Optimum ratio of dietary energy to protein for carp. *Bull. Jap. Soc. Sci. Fish., 45(8):* 983-7

Takeuchi, T., Yokoyama, M., Watanabe, T. and Ogino, C. (1978d). Studies on nutritive value of dietary lipids in fish. 13. Optimum ratio of dietary energy to protein for rainbow trout. *Bull. Jap. Soc. Sci. Fish., 44(2):* 729-32

Teshima, S. and Kashiwada, K. (1967). Studies on the production of B vitamins by intestinal bacteria of fish. III. Isolation of vitamin B_{12} synthesising bacteria and their bacteriological properties. *Bull. Jap. Soc. Sci. Fish., 33:* 979-83

——— (1969). Studies on the production of B vitamins by intestinal bacteria of fish. IV. Production of nicotinic acid by intestinal bacteria of carp. *Mem. Fac. Fish Kagoshima, 18:* 87-91

Tiemeier, D.W., Deyoe, C.W. and Weardon, S. (1965). Effects on growth of fingerling channel catfish of diets containing two energy and two protein levels. *Trans. Kansas. Acad. Sci., 68(4):* 180-6

Vallet, F. (1970). Alimentation artificielle et elevage de *Mugil* sp. et de morone labrax. Thèse de doctorat de specialité oceanographie biologique, Centre Universitaire de Luming (Universite d'Aix-Marseille), 95 pp.

Viola, S. (1975). Experiments on nutrition of carp growing in cages. 2. Partial substitution of fishmeal. *Bamidgeh, 27(2):* 40-8

Viola, S. and Rappaport, U. (1978). Acidulated soapstocks in intensive carp diets, their effect on growth and composition. *EIFAC Symp. Finfish Nutr. and Food Technol.,* Hamburg, June 1978. EIFAC/78/Symp.E/1

—— (1979). The 'extra-caloric' effect of oil in the nutrition of carp. *Bamidgeh, 31(3):* 51-68

Watanabe, T. and Hashimoto, Y. (1968). Toxic components of oxidised saury oil inducing muscular dystrophy in carp. *Bull. Jap. Soc. Sci. Fish, 34:* 1131-40

Watanabe, T., Kobayashi, I., Utsue, O. and Ogino, C. (1974a). Effect of dietary methyl linoleate on fatty acid composition of lipids in rainbow trout. *Nippon Suisan Gakkaishi, 40:* 387-92

Watanabe, T., Metsui, M., Kawabota, T. and Ogino, C. (1977a). Effect of α-tocopherol deficiency on carp. V. The compositions of triglycerides and cholesteryl esters in lipids of young carp. *Bull. Jap. Soc. Sci. Fish., 43(7):* 813-17

Watanabe, T., Ogino, C., Koshiishi, Y. and Matsunaga, T. (1974b). Requirements of rainbow trout for essential fatty acids. *Bull. Jap. Soc. Sci. Fish., 40:* 493-9

Watanabe, T. and Takashima, F. (1977). Effect of α-tocopherol deficiency on carp. VI. Deficiency symptoms and changes of fatty acid and triglyceride distributions in adult carp. *Bull. Jap. Soc. Sci. Fish., 43(7):*819-30

Watanabe, T., Takashima, F. and Ogino, C. (1970a). Requirement of a young carp for α-tocopherol. *Bull. Jap. Soc. Sci. Fish., 36(9):* 972-6

—— (1970b). Effect of α-tocopherol deficiency on carp. *Bull. Jap. Soc. Sci. Fish., 36:* 623-30

—— (1974c). Effect of dietary methyl linolenate on growth of rainbow trout. *Nippon Suisan Gakkaishi, 40:* 181-8

Watanabe, T., Takeuchi, T., Matsui, M., Ogino, C. and Kawabata, T. (1977b). Effect of α-tocopherol deficiency in carp. VII. The relationship between dietary levels of linolenate and α-tocopherol requirement. *Bull. Jap. Soc. Sci. Fish., 43(8):* 935-46

Watanabe, T., Takeuchi, T. and Ogino, C. (1975a). Effect of dietary methyl linoleate and linolenate on growth of carp. II. *Bull. Jap. Soc. Sci. Fish., 41:* 263-9

Watanabe, T., Utsue, O., Kobayashi, I. and Ogino, C. (1975b). Effect of dietary methyl linoleate and linolenate on growth of carp. I. *Bull. Jap. Soc. Sci. Fish., 41:* 257-62

Wolfe, L.E. (1951). Diet experiments with trout. *Prog. Fish Cult., 13:* 17-24

Yu, T.C. and Sinnhuber, R.O. (1972). Effect of dietary linolenic acid and docosahexaenoic acid on growth and fatty acid composition of rainbow trout (*Salmo gairdneri*). *Lipids, 7:* 450-4

Zeitoun, I.H., Ullrey, D.E. and Tack, P.I. (1974). Effects of water salinity and dietary protein levels on total serum protein and haematocrit of rainbow trout (*Salmo gairdneri*) fingerlings. *J. Fish. Res. Bd Can., 31:* 1131-4

5 THE INTENSIVE CULTURE OF TILAPIA IN TANKS, RACEWAYS AND CAGES

J.D. Balarin and R.D. Haller

1. INTRODUCTION
   1.1   The Different Concepts of Tilapia Production

2. THE SUITABILITY OF TILAPIA FOR INTENSIVE CULTURE
   2.1   Availability
   2.2   Food Habits
   2.3   The Parameters Critical to Intensive Culture
   2.4   Reproduction and Aggressive Behaviour

3. INTENSIVE FRY PRODUCTION
   3.1   Advantages of Controlled Fry Production
   3.2   Broodstock and Control of Reproduction
   3.3   Intensive Tank Production of Fry
   3.4   Special Techniques to Produce All-male Fry

4. TANK AND RACEWAY CULTURE
   4.1   The Characteristics of Intensive Tank Culture Units
   4.2   Stocking Practices and Production
   4.3   Grading and Harvesting
   4.4   Trends in the Intensification of Tank Systems

5. CAGE CULTURE
   5.1   Design and Construction
   5.2   Feeding
   5.3   Stocking Rate and Productivity
   5.4   Carrying Capacity

6. NUTRITION AND FEEDING IN INTENSIVE CULTURE
   6.1   Food Composition
   6.2   Diet Formulation and Pellet Size
   6.3   Feeding Rates

7. ECONOMIC EVALUATION AND FUTURE PROSPECTS

REFERENCES

1. Introduction

Tilapia culture is believed to have originated some 4,000 years ago, 1,000 years before carp culture was first initiated in China (Balarin and Hatton, 1979). However, other than a biblical reference to its culture, very little of this early work has been documented. The first recorded, scientifically orientated culture of tilapia was conducted in Kenya as recently as 1924 and soon spread throughout Africa (Meschkat, 1967). Tilapia were later transplanted and became established as a potential farmed species by the late 1940s in the Far East (Ling, 1977) and also by the early 1950s in North America (Iversen, 1976). Such has been the success of the tilapia that in the past 35 years it has risen from obscurity to become one of the most popular of farmed warm-water fish, contributing significantly to protein production in Africa and the Far East and becoming rapidly established as a food fish in Latin America. Tilapia are not only considered important for their food value, but have effectively been used as biological agents in weed and mosquito control, play a minor role as a bait fish (especially in the tuna-fishing industry), as a sport fish in angling dams and as an occasional addition to tropical aquaria.

Such has been the keen interest in tilapia that it was widely introduced into tropical areas other than its native Africa and can now be found in over 100 countries (Balarin and Hatton, 1979). Considerable information has therefore been accumulated on the biology and culture of the group and Balarin (in preparation) lists some 3,300 references. Much of the work pertaining to the extensive to semi-extensive pond culture of tilapia has been discussed in Balarin and Hatton (1979) and Caulton (1979). Although tank culture was initiated in Hawaii in 1959 (Uchida and King, 1962) and cage culture in 1969 (Pagan, 1969), little interest was shown in these techniques until recently.

The last five years have seen a considerable upsurge in intensification of husbandry methods and has led to the establishment of several tilapia research centres the world over, including: The International Centre for Aquaculture, Auburn, Alabama, USA; The Inland Fisheries Project, Central Luzon State University, Philippines; and The Institute of Aquaculture, Stirling, Scotland. So rapid has been the increase in research, that it has warranted the staging of a symposium devoted entirely to the tilapias, held in Bellagio, Italy in September 1980.

In view of the growing interest in the intensive culture of tilapia, this chapter is therefore mainly devoted to tanks and cages which are considered to fall under the category of intensive systems (see Figure 5.1). The literature has been reviewed and, where necessary, has been illustrated by previously unpublished data from the authors' own work on one of the more successful tilapia farms in Kenya. Space does not allow the inclusion of intensive pond culture and the interested reader is referred to Balarin and Hatton (1979). However, where it is necessary to illustrate a point or where there is an overlap between pond and tank systems (e.g. in water quality) reference has been made to work conducted in ponds.

267

Figure 5.1: Characteristics of Tilapia Production Systems.

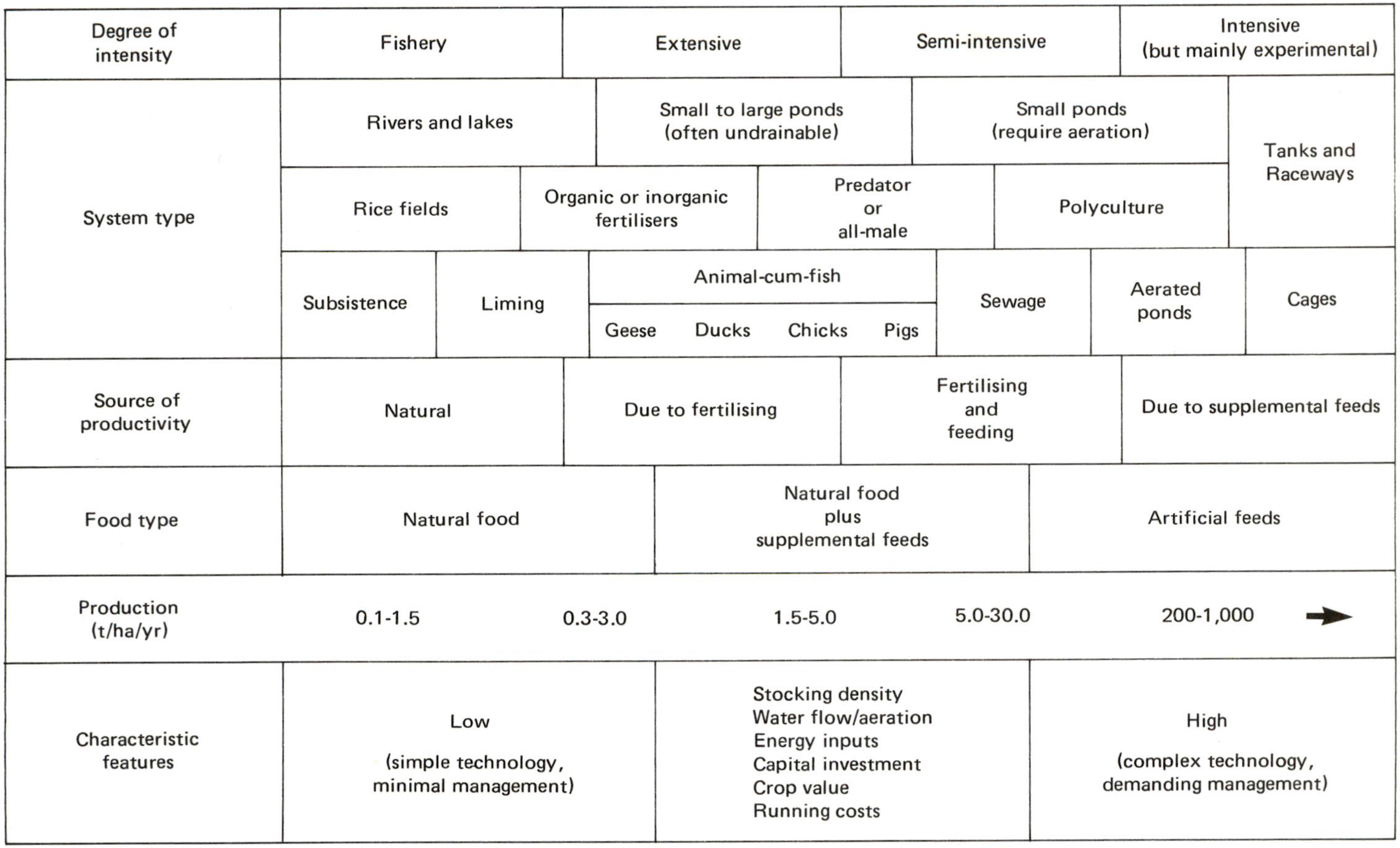

Characteristic	Fishery	Extensive	Semi-intensive	Intensive (but mainly experimental)
Degree of intensity	Fishery	Extensive	Semi-intensive	Intensive (but mainly experimental)
System type	Rivers and lakes; Rice fields (Subsistence); Organic or inorganic fertilisers, Liming (Natural)	Small to large ponds (often undrainable); Predator or all-male; Animal-cum-fish (Geese, Ducks, Chicks, Pigs)	Small ponds (require aeration); Polyculture; Sewage; Aerated ponds	Tanks and Raceways; Cages
Source of productivity	Natural	Due to fertilising	Fertilising and feeding	Due to supplemental feeds
Food type	Natural food	Natural food plus supplemental feeds		Artificial feeds
Production (t/ha/yr)	0.1-1.5	0.3-3.0	1.5-5.0	5.0-30.0	200-1,000
Characteristic features	Low (simple technology, minimal management)			High (complex technology, demanding management)

1.1 The Different Concepts of Tilapia Production

In tilapia culture, concepts of production may differ according to objectives and can be divided into three categories based on the definitions of Huet (1972):

(1) *Quantity production* produces the greatest weight with little consideration for quality, e.g. mixed age group stocking yielding small fish for food or restocking.
(2) *Economic production* produces as great a quantity as possible of fish of a high consumption value or for resale of fish of a high market value, e.g. bait fish, all-male fingerlings or table-size fish for restaurant or export.
(3) *Quality production* does not necessarily achieve maximum yields but aims at fish of uniform size and weight, e.g. by use of separate age groups and grading to produce fish for food or restocking.

The division is based on output per hectare per year (ha/yr) and the method of raising fish in each category may vary with food, water, labour and capital but generally is considered to be dependent on the feeding practice employed:

(1) *Extensive farming* is the production of fish usually in ponds, without artificial feeding. The sole food source is from natural productivity.
(2) *Intensive farming* seeks to produce a maximum quantity of individuals (or weight) in a minimum of water by means of intensive (or exclusive) feeding, often requiring some form of aeration or flow and needs a complex integrated infrastructure, requiring a relatively high capital investment and a large labour requirement.
(3) *Semi-intensive farming* is intermediate and often involves the use of fertilisers.

These features have been summarised in Figure 5.1 in relation to productivity. It may be argued that the intensity of production of a species is relative to the intensification of the methods most often employed. However, we have adopted the more classic definition that intensive culture is dependent almost exclusively on artificial feeding. By this definition tilapia culture is still very much in its infancy. A yield of 5 t/ha/yr is often given as the maximum pond productivity achieved in Africa (Shehadeh, 1976) and approximates that achieved in ponds in South East Asia (Ling, 1977). Experimental intensive pond culture can often achieve up to a fourfold increase while tank and cage culture practices can yield in excess of 50 times this productivity (i.e. 250 t/ha/yr), but necessarily demand higher levels of input. These techniques indicate the potential of tilapia culture but remain to be used on a large scale. Israel has in recent years embarked on an intensification programme and by implementing research findings, pond productivity has been raised from 3 to between 8 and 10 t/ha/yr (Tal and Ziv, 1978). This was founded on an increase in the use of artificial methods of aeration as an alternative to water circulation, increased stocking density, balanced polyculture and the supply of a balanced diet using automated feeders, as well as the use of all-male hybrids. These are all biological and mechanical means of increasing productivity, but as can be seen from Figure 5.2, socio-economic factors also decided the intensity of production. Intensification, in particular the use of tanks or cages, requires a definite infrastructure for successful culture, marketing and research. Although returns are relatively high, capital investments are also high and the absence of investment loans is believed by Tal (1979) to be one of the major reasons why intensive tilapia culture has not yet developed to any large scale in developing countries. Added to this is the shortage of trained personnel and competition with already established agricultural practices for feeds and fertilisers.

Figure 5.2: Factors Affecting Choice of System.

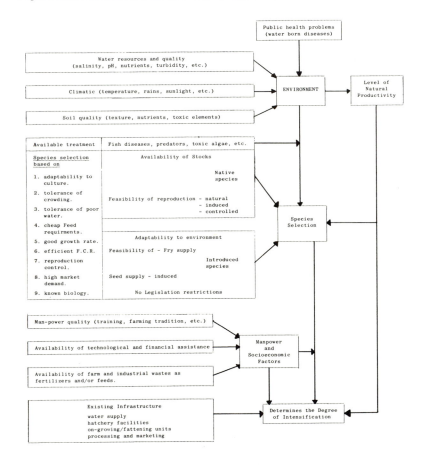

General outlay of the logistics which determine the degree of intensification of a tilapia production unit both at a national and private investment level.

In this chapter, therefore, we have dealt with intensive tank and cage culture to show that sufficient technological information is available to see the successful implementation of intensive tilapia culture. Problems in formulating a balanced diet and methods of intensive fry production are considered and the merits and economies of various techniques are discussed in the light of the socio-economic climate of the developing world.

2. The Suitability of Tilapia for Intensive Culture

Ease of cultivation, resistance to poor water quality and diseases, tolerance of a wide range of environmental conditions, ability to convert efficiently organic domestic and agricultural wastes into high quality protein, good growth rates and an amenability to intensification are some of the basic characteristics of tilapia which make them an ideal candidate for fish culture. However, these features have been offset by one major drawback, i.e. the prolific precocious breeding of the group which if left unhindered leads to overcrowding. The resultant competition for food results in stunting. Failures in tilapia culture have also resulted from a lack of understanding of the biology of the group.

Methods of reproduction control are discussed in section 3.2, while of specific consideration in this section are those values which repay careful consideration when selecting a species. Of particular importance in intensive systems are water quality and environmental requirements. By virtue of the nature of the system, intensive culture requires a well-balanced diet and the high densities used require that a large number of fry be readily available for stocking. Feeding and fry production are therefore considered in depth in separate sections of this chapter.

2.1 Availability

The natural distribution range of a species serves to indicate climatic areas to which the organism has adapted and within which it is likely to survive transfer for culture. Further, it also serves to indicate areas where wild-caught fry are plentiful, should there be no hatchery facilities available.

Generally the *Cichlidae* are confined to within the 20°C winter isotherm. The tilapia, in particular *S. mossambicus*, have been successfully distributed throughout this range and have become established in areas in which they were previously non-endemic. Outside of this zone, however, heated overwintering units are often necessary as the fish cannot survive the winter temperatures, for example in Auburn, Alabama (Avault *et al.*, 1968), Germany (Denzer, 1967) and Israel (Sarig, 1969).

The ease of transport and potential productivity in culture has led to such a widespread transplantation of tilapia that since its first transfer in 1939, with the probable exception of common carp, no other fish species can be considered to be as widely cultured. Thus, originally endemic to Africa, tilapia are now available in over 105 countries in the world (Balarin and Hatton, 1979).

2.2 Food Habits

The natural food preference of a species indicates the food type that it will take and which can then be used to determine the type of supplemental feeds to be used during intensive culture. Generally tilapia are herbivorous but have a tendency towards being opportunistic feeders. If the food habits of a particular species are not known it can be predicted from the dentition and distance between gill-rakers, characteristics which also serve to categorise the tilapia into two generic groups (after Trewavas, 1973 i.e.:

(1) *Planktophagous* — numerous long, thin, closely spaced gill rakers with fine raking dentition, e.g. *Sarotherodon* spp.
(2) *Macrophagous* — fewer, more robust, gill rakers with chisel shaped and chewing dentition, e.g. *Tilapia* spp.

(Trewavas (1981) has recently revised this grouping and erected three generic groups: *Sarotherodon, Oreochromis* and *Tilapia*.)

From the food habits listed in Table 5.1 it can be seen that tilapia of both feeding types readily adapt to artificial feeds, a feature which lends the group to intensification of production. Diet formulation often can be achieved economically, for tilapia can utilise such organic wastes as oil seed cakes, milling wastes, beer wastes and domestic wastes more efficiently than most domestically reared animals. The formulation of diets for intensive culture will be discussed more fully in Section 6.2.

2.3 The Parameters Critical to Intensive Culture

The following section is a brief outline of the environmental requirements of tilapia with particular emphasis on their tolerance limits, to indicate optimum ranges for maximum survival of stocks and within which growth remains unaffected.

Salinity Tolerance. Tilapia are believed to have evolved from a marine ancestor as they generally tend to be euryhaline and tolerant of a wide range of salinities. From Table 5.1 it can be seen that *T. zillii* is perhaps the most tolerant of the tilapia and has been reported to thrive at salinities of up to 40-45‰ (Bayoumi, 1969; Chervinski and Hering, 1973; El Zarka, 1956). *S. mossambicus* has been reported by Chen (1976) to be tolerant of even higher salinities, up to 75‰, however reproduction is affected and the fish fail to spawn at between 30-35‰ (Brock, 1954; Canagaratnam, 1966; Vaas and Hofstede, 1952). Generally, high salinities prevent or reduce the breeding of tilapia species and prolonged exposure to sea water may result in gonadal atrophy (Chervinski and Yashouv, 1971), a feature possibly worth exploiting as a mechanism of reproduction control.

At near isotonic concentrations, growth is enhanced, e.g. in *S. mossambicus* maximum growth occurs at 8.5-17‰ (Bashamohideen and Parvatheswararao, 1972), 17‰ (Canagaratnam, 1966) and 12.5‰ (Job, 1969). This has been attributed to the lower energy demand required for osmoregulation, leaving more for growth. Thus, as is often practised in intensive trout culture, tilapia too could be economically reared in estuarine areas. However, caution must be exercised where the species is near the limits of its range of temperature tolerance and where marked diurnal temperature fluctuations are likely, as the salinity tolerance of tilapia may vary with temperature (Whitfield and Blaber, 1976). However, increased salinity acclimation may increase the lower thermal tolerance by 2 to 3°C such that *S. mossambicus* has been able to extend its southern limit by colonising the coastal waters of South Africa (Caulton, 1979). Freshwater strains would require a longer acclimation period when being considered for culture in sea water and it must be noted that the double stress of transfer to sea water and lowering of temperatures can lead to mortality (Al Amoudi, in preparation).

Table 5.1: Summary of the Characteristics and Environmental Requirements of the More Important Tilapia Species Used in Fish Culture

Species	Food Habits	Upper Limits of Salinity Tolerance	Temperature and Oxygen Tolerance Limits	Growth	Reproduction and Temperament
S. andersonii	Adults tend to be omnivorous; fry below 5 cm feed mainly on zooplankton turning more to phytoplankton at 7 cm. Also take benthic organisms	22-33‰ if by gradual acclimation, though do survive a direct transfer to fresh water	Lower than that of *S. mossambicus*, can tolerate temperatures below 8°C	Maximum size is 0.7 kg (36 cm); generally grow 200-225 g/yr in ponds but may grow as much as 120 g in two months. Males show growth superiority	Male constructs a 18-60 cm saucer-shaped nest in 30-45 cm water. In lacustrine conditions fish often mature at 12-15 months (15-21 cm). Females lay 300-700 eggs, spawning only at temperatures above 21°C. Brood the eggs for 2-3 weeks and care for young for 5-6 weeks
S. aureus	Adults generally omnivorous. Fry below 5 cm feed on cladocera but fish larger than 2 cm eat filamentous algae, rotifers, etc. and take supplementary foods	Euryhaline species, normally survives and grows well at a salinity of up to 36-44‰ but cannot reproduce. Reproduces at 19‰. Can be acclimated to a salinity of 54‰	Lower lethal temperature about 8°C. Can tolerate 40°C for 3-4 hours. Larger fish are more tolerant but all are affected if held at <13°C. Temperature preference = 31-37°C	Male grows faster than female attaining a maximum size of 31.5 cm	Mature at 13.5-16 cm in second year in wild, at 7.6 cm in ponds. Eggs are laid in batches, 300-2,000 can be orally incubated at one time; hatch in 7-8 days. The young are cared for, for 8-10 days. Spawning can occur every 4-9 weeks (aggressive)
S. galilaeus (Galilee cichlid)	Mainly plankton-ophagic. Filter-feeds on algae including epilithic algae and make use of supplementary foods	Generally tolerates 13-29‰. Found to reproduce in wild at 29‰ and grows well at 19‰	Surviving at 8°C for 3-4 hours though lose equilibrium at 8.7°C. Are able to feed and grow between 11 and 16°C	Grow between 0.5-3 g/day reaching a maximum of 0.8 kg (40 cm). Male growth distinctly superior to female	Mature at 12-16 cm (80 g). Both sexes make nests and brood young. About 5,000 eggs/yr, 150-1,100 incubated at a time. Hatched fry are carried for 10-15 days. Not very territorial nor aggressive, though show display behaviour

Table 5.1: continued

Species	Food Habits	Upper Limits of Salinity Tolerance	Temperature and Oxygen Tolerance Limits	Growth	Reproduction and Temperament
S. macrochir (Green headed bream)	Young feed on zooplanton and phytoplankton. Adults generally phytoplankto-phagous, feeding almost entirely on planktonic and epiphytic algae. Fish above 4 cm readily take artificial feeds. At 12 cm can take 70 per cent of diet as artificial food	Fresh water species but some populations tolerate salinities of between 13 to 20‰ in Lake Mwerawa N'tipa in Zambia	Grows well between 23 and 24°C, with a lower lethal limit of below 11-13°C. Will only spawn at temperatures above 21-23°C. Die in wild at <1.2 mg O_2/l, for limited periods	Males grow 1.4 times faster than females, reaching 6-7 cm in 6 weeks and 14 cm in 6 months. Generally gain 150-200 g/yr in ponds. Maximum size 2.5 kg (40 cm)	Lake fish mature at 10-20 cm (8-12 months). Produce 1,000-5,000 of which they can incubate 1,000-1,500 eggs at a time with a 75 per cent breeding efficiency, spawning at 4-6 week intervals and 6-11 times/yr. Nests 50-300 cm in diameter, raised centre, in 30-150 cm water. Eggs laid in batches, fertilised and bucally incubated for five days and cared for, for 2-3 weeks. (Not very aggressive.) Become territorial when courting and tend to show more display behaviour than aggression
S. mossambicus (Java tilapia)	Adults omnivorous though mainly feed on plankton, vegetation debris and bottom algae. First feed at 6-9 mm. Juveniles (below 5 cm) feed on zooplankton. 5-7 cm fish are planktonic feeders: diatoms, green algae, small crustaceans and readily take artificial foods	Euryhaline and can reproduce at salinities of up to 35‰ and grow well in ponds at salinities of 35-40‰. Though they can survive transfer to a salinity of up to 69‰ and have been reported to tolerate levels of between 75 to 117‰ in brackish water ponds	Optimum temperature for spawning and growth is between 20 and 35°C. Need to be overwintered in heated tanks outside tropics, below 8-10°C being lethal. Because of this does not occur wild above 1000 m. Upper lethal temperature of between 39-42°C. Stops feeding at temperatures below 15.5°C. Temperature preference = 36°C. Metabolism is impaired at below 2-3 mg O_2/l, stop feeding at <1.5 mg O_2/l and die in ponds at 0.1 mg O_2/l	Males grow faster than females; growth 150-350 g/yr in freshwater and up to 450 g/yr in brackish water. Maximum size in wild 2.9 kg but stunting occurs in ponds	Monogamous mouth brooder. Matures at 2-3 months in ponds, 6-9 months in lacustrine conditions. Male constructs 30-35 cm saucer-shaped nest in 30-90 cm water. Females brood eggs at 75-1000/brood. Eggs hatch 2-5 days, emerge 5-8 days, staying with parent for 2-3 weeks. Spawn 6-11 times/ year every 22-40 days (somewhat aggressive)

Table 5.1: continued

Species	Food Habits	Upper Limits of Salinity Tolerance	Temperature and Oxygen Tolerance Limits	Growth	Reproduction and Temperament
S. niloticus (Nile tilapia)	Omnivorous but mainly feeds on phytoplankton; can use blue-green algae. Feeds also on benthic fauna and soft deposits. Fish of below 6 cm generally have a more diverse food spectrum. Adults readily take pelleted food	Thrive and reproduce at salinities of between 13.5-29 ‰ but less well adapted than *T. zillii*. Has survived salinities up to 35 ‰	Temperature below 12° C is lethal but tolerates 8° C for 3-4 hours. Survives prolonged periods at 15° C but does not feed or grow. Spawning is induced by 22-24° C. Upper lethal temperature 42° C. Temperature preference = 31-36° C. Metabolism is lowered below 2.5 to 3 mg O_2/l though fish survive 1.2 mg O_2/l in ponds, die at 0.7 mg O_2/l if prolonged	Males grow 2-5 times faster than females. Achieve 18-20 cm/yr in Sudan; 120-200 g in four months in cages in Ivory Coast. Maximum size in wild 50 cm (3.0 kg)	Mature in ponds at 4-5 months (10-17 cm), and in wild at 20-39 cm. Two or three hollow nests are made and female produces 300-3500 eggs at a time, spawning at least three times/yr, yet can only brood a maximum of 700 at any one time. (Tends to be aggressive towards other species.)
S. spilurus niger	Omnivorous grazer, feeds on epiphytic, epilithic and filamentous algae and benthic organisms but not on higher plants. Fry and adults readily accept artificial feeds	Have survived direct transfer into seawater at 33 ‰	Do not survive winters of below 8° C. Have been seen to breed young actively at temperatures of 40-43° C. Survive in tank culture at 1.2 mg O_2/l	Grows rapidly in the early stages (4-5 cm/month for 3-4 months). Males grow faster than females and at 4-5 months (10 cm) are 2 cm larger than females. Maximum size attained is 1 kg (38 cm)	Spawn at four months (11-16 cm) and at three-monthly intervals thereafter if conditions are suitable (highly aggressive).

Table 5.1: continued

Species	Food Habits	Upper Limits of Salinity Tolerance	Temperature and Oxygen Tolerance Limits	Growth	Reproduction and Temperament
T. rendalli (Red breasted tilapia)	Adults feed mainly on higher plants. Fry (below 5 cm) feed on zooplankton and phytoplankton, eating more plant material at 5-7 cm and thereafter 90 per cent of diet is filamentous algae and higher plants. Fry above 4 cm take artificial feeds	Cannot tolerate a salinity of above 13 to 19‰	Least cold tolerant of the *Tilapia*. Dies at 11-13°C therefore restricted to altitudes below 1000 m. Upper lethal limit 41°C. Has a preferred temperature of between 31 to 39°C. Spawns at temperatures above 21 to 23°C	Reaches 5-6 cm in 6 weeks and grows 100-150 g/yr. Maximum size attained in the wild is 2.5 kg (40 cm). Male growth is superior to females	Monogamous substrate spawner. Mature in seven months (12 cm), spawning at 6-7 week intervals (4-8 times/yr) provided temperature is above 21°C. Both sexes may make a nest of 5-10 holes of 10 cm diameter in 10-100 cm of water. 1000-6000 eggs are laid at one time in one hole and moved by the female from hole to hole. Hatch in 15-20 days, cared for for 2-3 weeks. (Aggressive.)
T. sparmanii (Banded tilapia)	Juveniles (below 6 cm) feed on zooplankton, gradually changing to benthic deposit and vegetation. By 10 cm are mainly phytophagous	Cannot tolerate a salinity of >18‰	One of the most hardy of *Tilapia* and can survive at temperatures as low as 7°C. Spawns at temperatures above 16-17°C	Seldom exceed 100 g though a maximum size of 300 g (27 cm) has been recorded	Mature at 8 cm. Lay 3,300 eggs at a time in one or more saucer-shaped nests. Both parents guard the eggs (highly aggressive)

Table 5.1: continued

Species	Food Habits	Upper Limits of Salinity Tolerance	Temperature and Oxygen Tolerance Limits	Growth	Reproduction and Temperament
T. zillii	Strictly phytophagous, feeding on leaves and stems of rooted aquatic vegetation and their associated epiphytic algae. Readily adapts to artificial feeding	One of the more saline tolerant of all tilapia, found in the Red Sea at a salinity of 45‰, but does not reproduce at salinities of 39-44‰, though grow well and reproduce at between 11 to 29‰. Survive direct transfer from freshwater to 24 to 27‰	Tolerant of low temperatures (6-8°C) but does not do well below 14°C. Grow well at a preferred range of between 20-31°C	Males generally grow 35 per cent faster than females. Can grow 10 cm in 3 months and have reached 26-29 cm (300-450 g) in a year. Maximum size 3 kg	Mature in second year (25-27 cm) in wild but in 3-5 months (20-50 g) in ponds. Saucer-shaped nest 20-25 cm diameter, 5-8 cm deep. 7,000 eggs laid per spawning; 6 spawnings/year. Eggs are attached to substrate guarded by female. Larvae are cared for by the female. (A highly aggressive, mouth fighter)

Temperature Tolerance. The influence of temperature on tilapia is summarised in Figure 5.3. Generally thermophilic, tilapia can tolerate temperatures as low as 8 to 10°C for short periods (Chimits, 1957; Sarig, 1969; Yashouv, 1960) though growth and reproduction are severely impaired at temperatures of below 20°C (Bishai, 1965; Fryer and Iles, 1972; Huet, 1972). Feeding is therefore reduced or stopped completely during the winter months and, where temperatures often drop to below lethal tolerance limits, overwintering is necessary either in covered ponds or heated tanks.

Figure 5.3: The Influence of Temperature on Tilapia

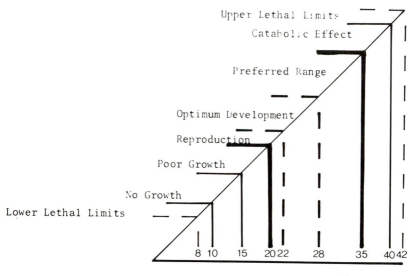

Temperature °C.

Temperature limits also account for changes in tilapia productivity with altitude (Kiener, 1963), and in highland areas deeper ponds are essential (Maar, 1956) to provide a winter retreat to warmer waters.

Physiological changes to kidney tissue may occur at lower temperatures (Allanson *et al.*, 1971) and handling stress at below 15°C can lead to severe stock losses (Cross, 1959). For this reason disturbance and handling is minimised during the winter months, unless harvesting, and extreme caution is exercised if transfer is necessary.

Tilapia tolerate temperatures of up to 35°C, but this varies with species type. Some, for example, *S. grahamii* (Reite *et al.*, 1973) and *S.s. chilwae* (Morgan, 1972) are found to thrive at temperatures of between 40 to 42°C under natural conditions. *S. niloticus* (Denzer, 1967) and *S. mossambicus* (Li Kuang Cong *et al.*, 1961) have an experimentally determined upper temperature limit of 42°C. This upper lethal temperature will increase, the higher the sedimentation temperature, and *S.s. niger* has been known (Haller, 1974) to be able to thrive and breed at 40 to 43°C after a slow seasonal acclimation. Tilapia, therefore, may be transferred over a wide range of temperatures, though fry generally are more tolerant — so the larger the fish, the more gradual is the temperature adjustments, e.g.:

Fry — (3-9 cm) can be acclimated in steps of between 10 and 13°C (Bishai, 1965).

Adults — (> 10 cm) may be acclimated in steps of between 6 and 8°C (Maar *et al.*, 1966); though Allanson and Noble (1964) recommend a much slower change of 1°C per two hours.

By careful acclimation in this way tilapia can be successfully transferred to areas outside of the extremes of the usual tolerance range, e.g. Tunisia (George, 1975a), though growth may still be affected.

The range for growth is 20 to 35°C with a reported optimum between 28 and 30°C, e.g. *T. rendalli* (Caulton, 1976), *S. niloticus* (Denzer, 1967) *S. mossambicus* (Jog, 1969) and *S. mossambicus/S. hornorum* hybrid (Suffern *et al.*, 1978). Productivity can, therefore, be assumed to be at a maximum within this temperature range. In addition, however, it would appear that tilapia have a behavioural mechanism exploiting diurnal fluctuations in temperature to enhance growth (Caulton, 1978). This theory was put forward to explain the diurnal migrations observed by Bruton and Boltt (1975), Caulton (1975), Maruyama (1958), Moriarty and Moriarty (1975) and Welcomme (1967) and could be utilised to improve productivity by regulated temperature adjustments.

Water Quality. From Table 5.2 it can be seen that tilapia can tolerate more adverse water quality conditions than most other cultured fish species even to the extent of surviving under dessicated conditions (Donnelly, 1978). Some of these tolerance limits are considered below.

Oxygen Tolerance Limits. Although the fish are tolerant of very low levels of oxygen saturation (see Table 5.1) in the wild, tilapia 'kills' are often associated with turnover in stratified water causing sudden deoxygenation and possible H_2S toxicity (Tait, 1965; Morgan, 1972). In pond systems, prolonged oxygen depletion is often the cause of mortality. Tilapia are able to survive long periods (up to six to eight hours in ponds, Sarig, 1971) of low oxygen concentration by 'gulping' at the air-water interface (Lowe-McConnell, 1959; Maruyama, 1958). In tanks fish can maintain oxygen levels at 1.2 mg/1 by this action, surviving up to 36 hours provided water quality remains good (Balarin and Haller, 1979). Further, survival is facilitated by the ability of tilapia haemoglobin to load oxygen at these low tensions, i.e. the blood of *S. macrochir* is 90 per cent oxygen saturated at 0.26 mg/1 (Dusart, 1963, also Figure 5.4), *S. esculentus* at 0.37 mg/1 (Fish, 1956) and *S. mossambicus* at 1.66 mg/1 (Perez and Maclean, 1975). It has also been suggested that tilapia are able to respire by anaerobic means (Kutty, 1972), a feature which would appear to be characteristic of the *Cichlidae*. Magid and Babiker (1975) believe this to be the possible reason for the decline in the rate of oxygen consumption of *S. niloticus* at levels below 2-3 mg/1 (see Figure 5.5). Tilapia are only able to sustain this process for a limited period, but in intensive systems this may allow sufficient leeway to engage emergency back up systems, should a systems failure or deteriorating pond conditions lead to deoxygenation.

Table 5.2: Water Quality Requirements for Survival and Growth Under Culture Conditions for Tilapia Compared with Carp, Channel Catfish and Trout

Parameter	Tilapia	Carp	Channel Catfish	Trout	Maximum Tolerable Level
Temperature (°C)	8-42	6-40	1.0-34.0	2-23	
Salinity tolerance limits (%)	<20-35	<12.5		<15.0	
Critical oxygen level (mg/l) (below which growth is affected)	0.1-3.0	3.0	3.0	4.0	5.0
pH	4.0-11	4.5-12	6.5-8.5	4.6-9.5	6.85
Lethal ammonia levels					
Total (mg/l)	>20.0	10-13		>2	
NH_3-N (mg/l)	2.3 (0.5)[a]		(0.13)[a]	0.35 (0.1)[a]	0.1
Turbidity tolerance (mg/l)	13000	>190		>15.0	15.0
Lethal CO_2 concentration (mg/l)	>73.0	—	>25.0	(15-20)[a]	22.0
Nitrite tolerance limits (D_{50})	2.1		<7.55	0.19 (0.015)[a]	0.1

Notes: a. Level at which growth is affected. An attempt has been made here to delineate environmental tolerance limits but these are by no means fixed and can vary and interact with other water quality parameters, e.g. total ammonia toxicity varies with pH. (See text.)

Figure 5.4: Oxygen Dissociation Curves for Blood Haemoglobin Solutions of Various Fish to Show the Bohr Effect.

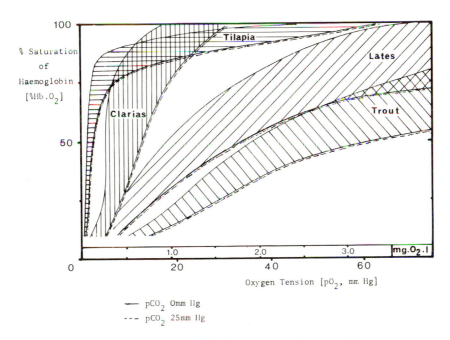

Source: *Clarias mossambicus* (Fish, 1956); *Lates albertianus* (Fish, 1956); *Sarotherodon macrochir* (Dusart, 1963); *Salmo gairdneri* (Beaumont, 1968, in Randall, 1971).

In semi-intensive culture practices, Prowse (1963) recommends stocking according to the species' oxygen demand but this would mean that fish are often held at low oxygen levels. Although tilapia can tolerate this, feeding (Balarin and Haller, 1979; Gruber, 1962; Payne, 1970) and subsequent growth (Allison *et al.*, 1976; Andren, 1976; Coche, 1976) are adversely affected. In more intensive systems, therefore, to maximise productivity, a continual water exchange rate of tilapia relative to that of trout (Denzer, 1967) will mean that the system's energy requirements can be expected to be lower. The numerous reports detailing tilapia oxygen consumption rates are currently under review (Balarin, in preparation), though the following is a tentative equation for the determination of flow rates required to supply the complete oxygen requirement under Bamburi tank conditions (Balarin and Haller, 1979):

$Q = 0.3459 \times dW^{0.7786}$ (at a temperature of 25.23°C)
where Q = optimum flow rate (1/min), W = x weight of fish (kg) and d = total number of fish stocked/tank.

Dependent on the size of fish, flow rates may vary between 0.5-1.0 l/min/kg.

Melard and Philippart (1980) however express the oxygen consumption of *S. niloticus* by the equation:

Oxygen consumption (O_2 mg/kg/hr) = $2.115W^{-0.61}$
where W = weight in g, and derive the following optimum loading formula

Loading (kg/1/min) = $0.112W^{0.614}$.

Figure 5.5: Pattern of Oxygen Consumption of *S. niloticus* in a Sealed Respirometer with Observations of Behaviour Under Deoxygenated Tank Conditions.

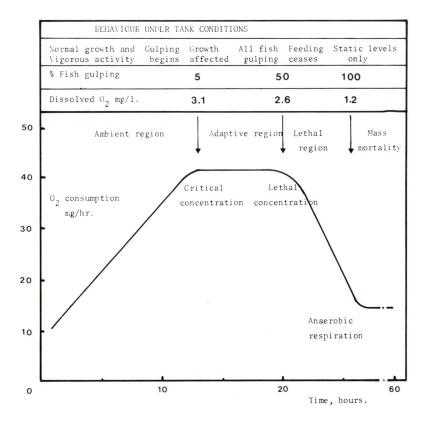

BEHAVIOUR UNDER TANK CONDITIONS					
Normal growth and Vigorous activity	Gulping begins	Growth affected	All fish gulping	Feeding ceases	Static levels only
% Fish gulping		5		50	100
Dissolved O_2 mg/1.		3.1		2.6	1.2

Source: Modified after Magid and Babiker (1975).

The Bamburi respiration values are higher (Figure 5.6); they more closely approach the actual situation by including the oxygen required for feeding and swimming as well as the BOD of the system. Therefore optimal loading is likely to be lower than that indicated by Melard and Philippart.

Carbon Dioxide. Carbon dioxide concentrations of between 50 and 100 mg/1 are considered to cause distress and are lethal to fish if prolonged (Doudoroff and Shumway, 1970); however, *S. macrochir* can tolerate concentrations of up to

Figure 5.6: Calculated Curve Showing the Relationship between Weight, Oxygen Consumption Measured before Feeding, and Optimal Loading for Male *S. niloticus* Reared at High Density in Tanks.

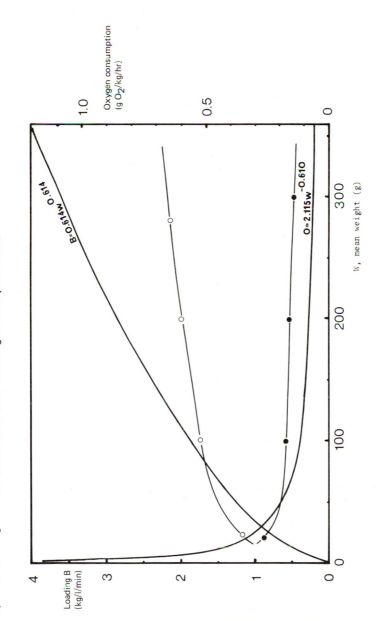

Source: Melard and Philippart (1980), temperature 30-32°C; Balarin and Haller (unpublished), temperature 25-29°C.

73 mg/1 in the wild (Dusart, 1963). Dusart and Perez and Maclean (1975) attribute this tolerance to tilapia haemoglobin having a smaller Bohr effect (see Figure 5.4) than that in other fish. Further, the latter authors were able to show that in *S. mossambicus* certain blood haemoglobins are not pH sensitive. Thus tilapia would appear to have evolved an adaptation enabling vigorous activity when blood pH is sufficiently low that most species, with a significantly larger Bohr effect, would have been unable to load oxygen and would have died of suffocation. In intensive systems, therefore, carbon dioxide levels are unlikely to have an adverse effect unless allowed to rise above 100 mg/1 (pCO_2 = 50 mm hg), a level which is normally believed to cause irreparable haemoglobin damage (Fish, 1956). However, as levels rarely reach this lethal limit, perhaps of more important consideration is that at prolonged concentrations of 7 to 12 mg/1, in slightly saline conditions and feeding a dry pelletised diet, tilapia though apparently not adversely affected may exhibit severe nephrocalcinosis (Balarin and Haller, unpublished). Severe lesions of this type are typical of intensive culture (Roberts, 1978) but little is known of their significance and they may well become a problem in tilapia culture. Severe cases could affect growth and lead to a loss in production as has been reported to be the case in trout (Harrison, personal communication).

pH and Alkalinity. As previously mentioned, low pH affects blood affinity for oxygen so that at a pH of less than 5, growth is affected (Bardach *et al.*, 1972; Pruginin, 1975) and this may well be due to a loss in appetite (Mabaye, 1971). High pH in pond conditions tends to stimulate feeding (Gruber, 1962) and the species can tolerate pH as high as 11 to 12 (George, 1975b; Reite *et al.*, 1973).

Tilapia are unlikely to be physically affected by alkalinity and tolerate levels as high as 700 to 3000 mg/1 $CaCO_3$ (Morgan, 1972), though at the higher levels, corneal damage may result in an opacity of the eyes. This tolerance of alkaline conditions is demonstrated by *S.s. grahamii* which lives in Lake Magadi, a soda lake (Reite *et al.*, 1973).

Turbidity. Although unlikely to become a problem in intensive systems turbidity in semi-intensive ponds, if due to suspended colloid particles, would reduce light intensity and adversely affect phytoplankton growth. Buck *et al.*, 1970 demonstrated that in a pond all other things being equal then the effect of turbidity on the mean fish weight was as follows:

Clear ponds (less than 25 mg/1) -- growth was 1.7 times that of
Intermediate ponds (25-100 mg/1) — where growth was 5.5 times that of
Muddy ponds (more than 100 mg/1).

Tilapia can tolerate highly murky waters and gill hyperplasia has only been noted at turbidities as high as 13,000 mg/1 (Morgan, 1972), however in view of the effect on pond productivity, where turbidity is a problem, settling ponds are used (Shehadeh, 1976). For more intensive systems, e.g. tanks, clear water is necessary for visual monitoring of stock performance. It is essential to be able to see the bottom of intensive systems though in tanks and cages a turbidity of < 100 mg/1 would be acceptable provided this is due to phytoplankton density or a fine colloidal suspension. Turbidity due to larger particle size, e.g. during floods, is responsible for gill damage which subsequently affects ventilation efficiency (Roberts, 1978).

Excretory Products. Systems design and good management should ensure that excretory wastes do not build-up in intensive culture systems. Faecal wastes will break down in ponds, unless overcrowded, and in tanks these can be readily flushed out. However, soluble metabolic by-products such as ammonia or by-products of the decomposition of organic waters, such as nitrites, are not as easily detected and may accumulate to toxic levels.

The lethal limits of unionised ammonia for *S. aureus* has been determined to be 2.35 mg/1 but it would appear that by prolonged exposure to lower concentrations tilapia can tolerate levels of up to 3.4 mg/1 (Redner and Stickney, 1979). Tilapia would, therefore, be ideal candidates for recirculation systems using minimal water replacement but caution must be exercised since resultant gill damage at levels of unionised ammonia above 0.5 mg/1 frequently results in mortality when fish are further stressed by, e.g. low oxygen, handling etc. (Balarin and Haller, 1979; Muir and McAndrew, personal communication).

In ponds and closed systems, biological breakdown of wastes may also produce toxic compounds such as nitrites, and although tilapia can tolerate levels up to 0.45 mg/1 (Balarin and Haller, unpublished), there would appear to be an as yet unidentified growth-inhibitor. At high densities in experimental ponds in Israel, carp and tilapia growth is inhibited 50 to 70 days after stocking (Rappaport and Sarig, 1975). This growth-inhibiting effect is lost if water is passed along a shallow 100 m channel, suggesting that the factor can be removed. But the problem requires further investigation if intensive stocking is to succeed at the highest levels.

Growth Rates. Growth varies with species and conditions of culture, and it is thus impossible to present fully accurate expected growth rates. Listed in Table 5.1, giving an indication of growth potential, are reported growth rates and maximum size attainable for wild stocks. It is generally accepted that males grow faster than females, the latter undergoing a reduction in growth once mature (Fryer and Iles, 1972). Males are, therefore, selected for intensive culture.

2.4 Reproduction and Aggressive Behaviour

The environmental parameters mentioned above may also influence the size of maturation, and in the wild, fish generally mature at a larger size than when cultured (see Table 5.1). This early maturation or precocious breeding, characteristic of reared fish, is believed to be an inherent, homeostatic, control mechanism which evolved to ensure the maintenance of populations under adverse conditions (see fry production section). Features peculiar to the breeding biology of various species are also considered in Table 5.1.

Tilapia males tend to be very territorial, especially during breeding, and are therefore very aggressive. This is particularly true of the *Tilapia spp.* Thus in defence or establishment of territory, dominant males severely attack other fish, damaging fins, scales and eyes, often leading to death. High density stocking or limited space reduces this aggressive behaviour (Haller, 1974; Balarin and Haller, 1979). It is thus important for stock well-being to maintain conditions where aggression is minimal and to choose compatible individuals as well as species when selecting stocks. Aggressive tendencies of various species, as observed by the authors, have therefore been given in Table 5.1. Aggression can also cause problems in attempting hybrid crosses and is of importance when initiating an intermediate harvesting programme in intensive systems. The reduction in numbers and disruption of a previously stable hierarchy can trigger off renewed fighting to re-establish territory or dominance.

3. Intensive Fry Production

Intensive culture requires a regular supply of large numbers of fry, preferably of a graded size group, of all-male fish. However, it would seem ironic that whereas in extensive pond culture excessive breeding leads to an overproduction of fry, fry production under more controlled hatchery conditions is not yet as productive. An inadequate supply of fry has often been blamed for intensive tilapia culture still being in its infancy, especially in Africa (Balarin and Haller, 1979). As fry production itself may be an important aspect of intensive fish production, in this section some of the more successful methods employed to produce large numbers of fry will be considered, with the physiological and environmental parameters which have been reported to affect the reproduction potential of tilapia.

Fry production can be divided into two categories:

(1) *Extensive* — fry netted from ponds or lakes with little reproduction control.
(2) *Intensive* — controlled fry production in ponds or tanks, often involving induced spawning.

The latter will form the main part of the discussion.

3.1 Advantages of Controlled Fry Production

Other than being able to produce large numbers of fry, as and when they are required, controlled breeding may allow for genetic selection of broodstocks, not only to produce offspring of a particular calibre (i.e. faster growth, deeper bodied, lighter colour, etc.), but also to enable hybrid crossing to produce all-male offspring. Otherwise, the controlled rearing conditions allow for hormonal or similar treatment to induce sex-reversal or sterilisation, resulting in monosex fry production. Further, as the fry are raised under intensive conditions, growth is dependent predominantly on the feeding intensity, and by selective feeding it is possible to hold certain stocks back and speed-up the growth of others until the required number of fry of the same size are obtained, enabling uniform size stocking of a system. Subsequent growth does not appear to be affected but benefits from the reduced competition and manageability of uniform sized stocks.

3.2 Broodstock and Control of Reproduction

Under natural lacustrine conditions, tilapia mature in the second or third year at a length of 17 to 20 cm. However, this may vary with the size of the water body concerned, e.g. *S. niloticus* matures at 17 cm in Lake Edward, but in the larger, deeper Lake Turkana, matures at 39 cm (Fryer and Iles, 1972). This would also appear to apply to pond size and often tilapia mature at under six months of age after stocking at a size of > 10 cm (Bardach *et al.*, 1972; Chimits, 1955, 1957; Hyder, 1970; Huet, 1972).

Tilapia can be considered as being shallow water dwellers, and thus are prone to predation and stranding as water levels fall suddenly in the dry seasons. In response to this the group appear to have evolved a mechanism whereby, when conditions are analogous to those of a pool undergoing desiccation, fish mature early and produce numerous numbers of smaller fry. Thus as larger fish die out or are predated upon, there will always remain a reserve of young potential breeders, capable of propagating the species when conditions are favourable again. Typical fishpond conditions therefore would appear to act as a stimulus, triggering precocious breeding. However, the actual physiological process is not fully understood and general weather conditions such as rain, wind, humidity and dark cover conditions are also known to have a seasonal effect on breeding. The following sections, therefore, consider those conditions which have been reported to affect spawning and are therefore of important consideration when attempting to control fry production.

Temperature. Tilapia only show secondary sexual characteristics and spawn at temperatures above 20-23°C. Thus, in equatorial regions the fish will spawn all year round but in subtropical and in high-altitude tropical areas, reproduction will be restricted during the cold months. Further, prolonged exposure of *S. niloticus* to temperatures below 20°C has been reported to result in a failure of immatures to mature and, in mature fish, gonad reabsorption took place (Mishrigi and Kubo, 1978). High temperatures have been shown to stimulate an increase in spawning under laboratory conditions (Hyder, 1970; Mironova, 1977) and Ray (1978) reports temperature control alone to be successful in increasing reproduction during the winter months even though the photoperiod was shorter. By lowering temperatures from 25 to 18°C for two weeks and then raising it back again it has been possible to induce over 50 per cent of female *S. niloticus* to spawn (McAndrew, personal communication). A similar technique is in operation in Israel, indicating that temperature plays a major role in inducing spawning and that it is therefore possible through a thermal shock to control the breeding of broodstocks.

Light. Bright sunlight would appear to stimulate breeding in pond fish (Chimits, 1955; Hyder, 1970; Legner, 1978; Ray, 1978) though Cridland (1962) believes strong illumination of *T. zillii* in aquaria delays sexual maturation and suggests that prolonged photoperiods enhance breeding activity. Strong illumination has also been reported to affect aquarium tilapia (e.g. Goldstein, 1970; Lanzing, 1978), though it is also felt that a light background may be equally stressful and affect reproductive behaviour (Balarin, unpublished).

Light intensity and photoperiod would therefore appear to influence the physiological processes leading to maturation. However, although high light intensity and long photoperiod can be used to enhance brood fish performance, the same conditions may be deleterious to fry growth, e.g. by triggering early maturation. Light manipulation therefore needs careful control or else fry must be removed as soon as they are released from the mouth of the parent.

Salinity. Fecundity decreases at salinities above 20°/oo and maximum fry production is often associated with near isotonic concentration, e.g. 4.6°/oo for *S. aureus* (Perry and Avault, 1972) and 10-13 °/oo for *S. mossambicus* (Uchida and King, 1962). Thus it may be an advantage to maintain broodstock in near iso-osmotic salinities to enhance fry production. Further, it is worth mentioning here that in those tilapia tolerant of sea water, such as *S. mossambicus* and *S. spilurus*, reproduction is considerably reduced, or inhibited altogether in sea water, a feature which might be worth exploiting in brackish water pond culture or sea cages.

Food: Quality and Quantity. A good supply of 'digestible' food stimulates an increase in fry production in natural waters (Lowe-McConnell, 1955), ponds (Guerrero, 1976a) and in tank stocks (Uchida and King, 1962). This may well be a mechanism adapted to exploit favourable conditions; however, it is of interest to note that poor quality rations and starvation induced early maturation in *S. niloticus* (Balarin, unpublished) and is said to be typical of all tilapia (Caulton, 1979). This could be a further response to the species survival mechanism ensuring spawning under adverse conditions, producing smaller fish which could more efficiently exploit the limited food source. In view of the likely adverse effects of a nutritional imbalance, starving would not be a favourable stimulus to induce early maturation of broodstocks. Furthermore, prolonged breeding would require that broodstocks receive a balanced diet. Uchida and King (1962) indicate that a high protein diet of 35 to 40 per cent protein results in maximum fry production but it is uncertain what the long-term effects would be. It is suspected that liver degeneration could occur as has been reported for trout and an acute anaphylactoid reaction in *S. mossambicus* fed on high protein pellets has been reported (Roberts, personal communication). Stocks would therefore require regular replacement, a practice which could be beneficial under any conditions in view of the energy drain to the fish incurred by very frequent breeding.

Broodstock Density and Sex Ratio. High stocking densities of up to 100,000/ha in ponds in Israel have been reported to depress spawning activity (Tal and Ziv, 1978), while in tanks tilapia did not spawn at a biomass of greater than 16 kg/m³ though they reproduce prolifically in ponds at 4 kg/m³ (Lauenstein, 1978). Similarly, in Kenya, no successful spawning was observed in tanks stocked at above 10 kg/m³ (Balarin, unpublished). This would suggest that the effects of crowding (i.e. increased social interactions and a breakdown in territorial behaviour) adversely affect reproduction. High density stocking is therefore a feature which favours the use of tilapia in intensive systems, as unwanted reproduction may be largely halted, but it creates problems when attempting intensive fry production. Numerous reports recommend an optimum broodstock density where fry production is maximised (Legner, 1978; Planquette and Petel, 1977; Rothbard, 1979; Uchida and King, 1962) but also indicate that the sex ratio is important. The picture is by no means fully understood though Table 5.3 summarises weights, sex ratios, density and expected production. From Table 5.4 the intensity of fry production of various systems is compared and only the most intensive unit is described below (after Uchida and King, 1962).

Table 5.3: Table of Sex Ratios and Brood Densities Used in Tilapia Fry Production

Species	Country	Stock Rate (n/m^2)	♂ : ♀	x̄ Weight (g)	Fry Production (n/m^2/month)	System and Comment	Reference
S. aureus	US – Alabama	10	1:3	(13-18 cm T.L.)	200	18 m^2 concrete tanks. Fry seined out daily	Shelton *et al.* (1978)
S. hornorum ♂ x *S. niloticus* ♀	Brazil	5.8	1:5	<100		350 m^2 pond results very variable. 1:5 ratio is equal to a 1:2 ratio	Lovshin (1977)
S. aureus x *S. vulcani*	Israel – Gan Schmuel; Ein Hamifraz		1:1 1:1.5		(167-250/♀/mth) (125/♀/mth) (62/♀/mth)	Use 1:1 ratio to ensure there is always a male ready; increased ratio lowers fry production. Lower productivities with hybrids	Muir (unpublished)
S. aureus x *S. niloticus*	Israel – Dor	(0.05 m^3/fish)	1:8-10	50-80	(83-167/♀/mth)	500 l tank with 3-5 cm gravel. Eggs may be removed to incubator	Mires (1977)
S. mossambicus	Hawaii	12.7 10.2	1:3 1:3	♂ = 154 ♀ = 134 ♂ = 450 ♀ = 226	400-750 600-770	78.6 m^2 trough with adjacent 90 l fry rearing tanks	Hida *et al.* (1962)
S. mossambicus	Hawaii	11.2	1:3	♂ = 142 ♀ = 113	1,688 (142-1,633)	Considered an optimum density maximising production in 4.48 m^2 tanks	Uchida & King (1962)
Malacca hybrid	Malaysia	0.04-0.05	1:3	♂ = 160-200 ♀ = 200-300	(300-600/♀)	Using a 0.04 ha pond	Pruginin (1965)

Table 5.3: continued

Species	Country	Stock Rate (n/m²)	♂ : ♀	x̄ Weight (g)	Fry Production (n/m²/month)	System and Comment	Reference
S. niloticus ♂ × S. hornorum ♀	Brazil	0.04-0.15 0.03-0.9	1:2 1:2	♂ = 436 ♀ = 617	(216/♀) (331/♀)	350 m² ponds and 100 m² ponds had no significant difference in fry production.	Lovshin & Da Silva (1975)
S. hornorum	Brazil	0.034	1:3	♂ = 90 ♀ = 45	(223-404/♀)	350 m² ponds from which ♀s are seined out once spawned	Da Silva et al. (1978)
T. zillii	US – California	0.43	1:1	♂ = 24 ♀ = 22	36	Considered an optimum density	Legner (1978)
S. niloticus	Ivory Coast	0.2	1:3			400 m² ponds — optimum density for maximum fry production	Planquette & Petel (1977)
S. mossambicus	Philippines	13	1:3.3	♂ = 72 ♀ = 54	275	In 1 m³ 'Hapas' = cages of 1/16″ mesh nylon screen suspended in ponds	IFP (1975)
		13	1:3:3	♂ = 39.5 ♀ = 25	104		
S.s. niger Tilapia	Saudi Arabia	0.56 22	1:2 1:2		62.5	Recommended for spawning ponds	Peacock (pers. comm.) Bardach et al. (1972)
Tilapia	South Africa	6.0	1:3	300-500	37-75	Using elevated pond-system with flow of 1 m³/hr in 18 m³ unit.	Caulton (1979)
Tilapia	South Africa	1-1.5	1:2.5-3	300-400	240	Intensive breeding units at 0.5-1.0 m³/hr flow in 6 m³ unit	Caulton (1979)

Table 5.3: continued

Species	Country	Stock Rate (n/m²)	♂ : ♀	x̄ Weight (g)	Fry Production (n/m²/month)	System and Comment	Reference
S. aureus × *S. niloticus*	Kenya	1.7-2	1:6	♂ = 250-500 ♀ = 80-200	180-200	3 ringed, circular arena is used, designed to minimise handling.	Haller & Parker (1981)
♀*S. niloticus* × ♂*S. hornorum*	N.E. Brazil	0.17 0.21	1:5 2:1	60-100	2.96 2.31	350 m² ponds	Lovshin (1980)
S. niloticus	Belgium	0.2-0.29	1:27-44	300	60-93 (257-467/♀/mth)	150 m² ponds = 200 m³	Melard & Philippart (1980)
S. niloticus	Scotland	8.4	1:3.75	300	2000-2700 (350-400/♀/mth)	2.25 m² (1.35 m³) tanks removing fry every two weeks	McAndrew (pers. comm.)

Table 5.4: Area Required to Produce One Million Fry Per Annum for Six System Types

System	Total Area (ha)	Number of Broodstock	Source	Country
1. Ponds — extensive	50	±1,000	Morissens (1977)	Israel
2. Special 2-tier pond design = 350 m^2 units	19.5	550	Lovshin and Da Silva (1975)	Brazil
3. Small pond units — intensive	0.23	490	Legner (1978)	California
4. Circular area tanks — 154 m^2 units	0.046	780	Haller and Parker (1981)	Kenya
5. Brood tank — 80 m^2 units	0.016	1,200-2,000	Hida *et al.* (1962)	Hawaii
6. Tanks — 4.5 m^2 units	0.005	550	Uchida and King (1962)	Hawaii

Note: One million fry = 250 t @ 250 g each.

3.3 Intensive Tank Production of Fry

Plywood rectangular tank units of 4.48 m^2 and 0.85 m deep (3.80 m^3) were stocked at various densities and sex ratios using males of 150 g and female fish of 100 g. From Figure 5.7 it would appear that a 3:1 (female:male) ratio produced the maximum number of fry at a density of 10 to 14 fish/m^2 (or 0.3 to 0.4 m^2/male). The breeding potential of the females remains relatively unaffected by sex ratio at the lower densities (6-7/m^2), but increasing density considerably reduces the number of fry produced. Crowding increases aggression between territorial males, reducing the period of time spent in courtship. Increased sex ratios, however, means a higher number of non-breeding fish which either predate on the young fry or disrupt the spawning of a breeding pair cycle by rushing in to feed on the eggs as they are spawned. This is a feature often seen in crowded tanks but at least the losses from cannibalism can be reduced by netting out young fry for ongrowing or removing them from the mouth of the parent (MacAndrew, personal communication).

At sex ratios below 3:1, there are insufficient females to carry on a continuous breeding run with the available males, i.e. the reproductive capacity of the female is unaffected but at any one time all females are likely to be brooding eggs and so there are periods when no spawning activity is taking place, thus reducing the total fry numbers produced.

Questions which as yet remain unanswered and in need of further investigation relate to the optimum size of brood stocks, and the length of time the same broodstock can be used without a reduction in fry production or before genetic aberrations occur. As can be seen from Table 5.3, generally smaller females are used and most workers have preferred fish of less than 200 g. Recently, Lovshin (1980) has shown that after three spawning periods (each of 2.5 months), fry production decreases. However, further work is still needed on this subject.

3.4 Special Techniques to Produce All-male Fry

Hybrid crosses and hormonal sex-reversal are not recent developments; the first successful cross was undertaken in the late 1950s by Hickling (1960) and hormone treatment was first pioneered in the late 1960s by Clemens and Inslee (1968). However, it is only recently that renewed interest has been shown in these techniques, and efforts are being made to make them applicable to commercial practice. The initial success was soon thwarted by a lack of technological understanding and by impurity of stocks, but it is now felt that either technique or a combination of both holds promise for increasing production through curbing reproduction in extensive and semi-intensive pond systems. In more intensive systems, reproduction is anyway adversely affected by the high stock density and competition for any suitable nesting surfaces, but the techniques are nevertheless important as intensive fry production can make use of these techniques to save extensive ongrowing of females and to enhance growth either as a result of hybrid vigour or the anabolic effects of the hormone.

Hormonal Sex Reversal. Two major distinctions can be made in the use of this technique:

Figure 5.7: Effect of Sex Ratio and Brood Density on (upper) Fry Production by Each Female and (lower) Overall Fry Production per Unit Area of *S. mossambicus* Reared in 3,800 l Tanks.

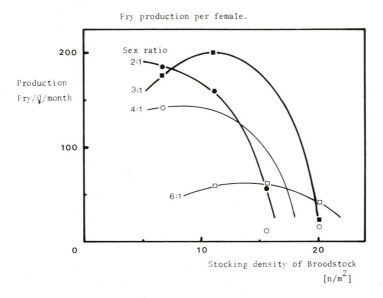

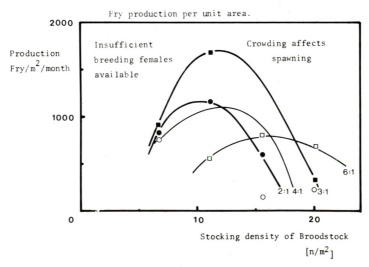

Source: adapted from results of Uchida and King (1962).

(1) *Androgens* — genetic females are induced to develop as functional males but genotypic males remain apparently unaffected, resulting in an all-male population.

(2) *Oestrogens* — genetic males are induced to develop as functional females which, if homogametic for sex determination, can then be bred to produce all-male progeny.

The desired hormonal action depends on the efficacy of the steroid, dosage rate, method of administration and time and duration of treatment, all of which may vary with species. Ideally a suitably potent level of hormone must be introduced either orally, in solution or injected, at a time when the gonad is still undifferentiated, and treatment maintained throughout the period of gonadal differentiation. Androgen treatment has been successful but oestrogen sex-reversal has met with little success and is not considered here. Interested readers are referred to Jensen (1976) and Jensen and Shelton (1979).

Androgen Treatment. 17, α-methyltestosterone (MT) and 17, α-ethynyltestosterone (ET), and other synthetic equivalents, have proven highly effective in all-male production, when administered orally to juvenile tilapia prior to gonadal differentiation. Gonad differentiation takes place between day 16 and 20 in *S. mossambicus* (Nakamura and Takahashi, 1973), though Clemens and Inslee (1968) believe this to be day 30 to 50; in *S. aureus* it occurs between day 30 to 60 (Eckstein and Spira, 1975) and before day 60 in *S. niloticus* (Jalabert *et al.*, 1974). Generally, therefore, emergent fry, 3 to 7 days after hatching (= 5 to 11 mm), are treated for a 20-40 day period at the most effective dosage rate of 30-60 mg/kg of feed. Variations to this practice and resultant successes are outlined in Guerrero (1975a, b, 1976 a, b), Jalabert *et al.* (1974), Shelton *et al.* (1978), Balarin and Hatton (1979) and MacIntosh (1980).

Androgens have also been administered in solution but have met with very little success and are certainly not as effective as when fed to the fish (Eckstein and Spira, 1975; Hackman, 1974; Katz *et al.*, 1976).

Practical Application of Hormonal Sex Reversal to Intensive Tilapia Culture. The anabolic effect of androgens, although evident in certain treatments, is inconsistent and there have been reported cases of growth being adversely affected (IFP, 1976a; Anderson and Smitherman, 1978). Prolonged crowding of fry during treatment is believed to inhibit growth and the fish suffer an initial setback. However, recent studies have shown that although crowding affected the overall growth rate, after treatment male fish can be as much as 70 per cent larger than controls (MacIntosh, personal communication). Optimum stocking density for maximisation of growth is currently being investigated.

A further drawback of this method is that although genotypic females are functional or sterile phenotypic males, growth is not enhanced as females are genetically slower growers (Anderson and Smitherman, 1978). The benefits of using sex-reversed fish in extensive systems is somewhat doubtful and though its use is likely to be of importance at lower levels of intensification in ponds, research is still needed into determining whether a certain percentage of females is genetically non-responsive to sex-reversal, potency of the drug in water, the

Table 5.5: Percentage of Male Hybrids Resulting from Known Parental Crosses, Fitted on to a Theoretical Gametry Derived from the Gametic Make-up as Proposed in the Text

Species (Female Parent)	Proposed gametes	S. hor.	S. mac.	S. aur.	S. var.	S. leu.	S. mos.	S. nil.	S. nig.	S. vul.	S. and.	T. zill.
Proposed gametes (Male Parent)		zz[1]	zz[2]	zz[3]	zz	zz**	xy[1]	xy	xy	xy	xy**	xy**
Sarotherodon hornorum	wz[1]		75				75-77	75				
S. macrochir	wz						<100	75			<100	
S. aureus	wz[3,4]	90-100					70-75	75-95			(100)	
S. variabilis	wz									(100)		
S. leucostictus	wz**											
S. mossambicus	xx[1,5]	98-100	+	+	+			50-89*	<100		<100	
S. niloticus	xx[2]	98-100	100*	50-100*	100	94	56-100		85			

S. spilurus niger	xx	100	+	+	95	43
S. vulcani	xx	98-100*	+	+	80-100	<100
S. andersonii	xx**					(100)
Tilapia zillii	xx**					

Key:
1 indicates gametes, proposed by Chen (1966).
2 homogametic, proposed by Jalabert *et al.* (1971).
3 homogametic, proposed by Liu (1977).
4 indicates gametes, proposed by Guerrero (1975a).
(From these proposed gametes and results of known parental crosses, this table was derived, e.g. male *S. hornorum* is suspected to be homogametic (zz), therefore, as the cross with female *S. niloticus* results in 98-100 per cent male offspring (Pruginin, 1967), it suggests that the female *S. niloticus* must also be homogametic (xx) to produce this result. The male *S. niloticus* is therefore heterogametic (xy), based on the proposal by Chen (1969) that z = ♂ > x = ♀ and y = ♂ > w = ♀.
5 homogametic, proposed by Clemens and Inslee (1968).
* hybrid reported to show hybrid vigour for faster growth rates.
** indicates insufficient information is available to enable a valid prediction of gametic make-up, tentative proposal.
<100 denotes not an 'all-male' cross; actual percentage male unknown, probably 75 per cent.
(100) denotes a suspect report.
+ indicates a cross which in theory is likely to yield nearly 'all-male' (>95 per cent) hybrid offspring.
N.B. Avtalion and Hammerman (1978) propose that there are only three chromosomes (w, x, y), therefore, zz = yy, where ♂ = y > x > w = ♀, wz = wy, but also include two autosomes (A, a) such that: A > a.
Thus, Genotype I = ♀♀   ♂♂
                   AAxx   AAxy   (e.g. *S. mossambicus, S. niloticus*)
      Genotype II = aaww   aayy   (e.g. *S. hornorum, S. macrochir, S. aureus*).

chemical pathway of hormone activity and excretion time and improved
techniques of administration.

All-male Hybrid Crosses.   Results of known hybrid crosses are summarised in
Table 5.5. Cases of heterosis have often been described from such crosses, e.g.
Chen, T.P. (1969), Hickling (1967), Mires (1977) and Pruginin *et al.* (1975). The
success of this technique in producing 100 per cent male hybrids is dependent
on the genetic purity of stocks. Transplantation or the mixing of non-endemic
tilapia in natural water bodies as well as contamination of farm stocks has
resulted in uncontrolled interspecific hybridisation. Stocks thus mixed do not
produce all-male offspring on further crossings, and accidental contamination
is believed to be the reason for the marked decline in the previous success of this
technique in Israel (Tal and Ziv, 1978).

The mechanism of sex determination is not fully understood. The generally
accepted theory was proposed by Chen (1969) that homogametic parents
produce all-male offspring as illustrated in Table 5.5. This would imply that
there are 16 possible combinations available which produce all-male hybrids.
More recently, however, Hammerman and Avtalion (1979) in a critical analysis
of the results of Chen (1969) proposed that there were only three gonosomal
types (w, x, y) and postulated that two autosomes (A, a) were involved in sex
determination, giving only four possible genetic types which could result in
all-male crosses (Avtalion and Hammerman, 1978). This hypothesis, however, is
not always valid and observed anomalies are explained as being due to a
'difference in chromosome strength' or due to questionable purity of stocks used.

It is hoped that through the refinement and use of electrophoretic techniques,
as have been successfully applied to blood sera (Badawi, 1971; Chen, 1969; Chen
and Tsayuki, 1970; Hines *et al.*, 1971), muscle proteins (Avtalion *et al.*, 1975,
1976; MacAndrew, 1981; Seki, 1976) or muscle enzymes (Basasidwaki, 1975)
and more recently to esterase patterns of surface mucosa (Herzberg, 1978),
that certified pure strains can be obtained and stocks selected for heterosis.

Techniques Employed to Obtain Hybrid Crosses.   Techniques have already been
described in Balarin and Hatton (1979) and therefore only a brief evaluation is
considered here. Generally, the aim of systems used is to restrict the broodstock
in a fenced off pond section (Chen, 1969; Chervinski, 1967; Lovshin, 1977) or
the pond design includes an elevated, drainable broodstock section (Lovshin and
Da Silva (1975). Fry produced are ongrown in the greater area of the ponds and
brood fish can be netted out before back-crossing can occur, i.e. three months
after stocking. Such handling of stocks would reduce their reproductive capacity
and often the design is such that stocks can be sealed off or fry netted out.
Haller and Parker (1981) have, however, devised a system which requires
minimal handling of all stocks and reduces male aggression and parental
cannibalism of fry. A three-ring arena, separated by entrances barred by grills
of selective width, ensures that males (250-500 g) are confined to the controlled
'breeding' arena. Smaller females (120-200 g) are able to escape to the second or
'nursery' ring but are free to enter the breeding arena whenever gravid and ready
to breed. Fry are then released into the shallow outer ring, escape predation and
are regularly flushed out into ongrowing tank units. Larger and therefore more
fecund fish can be used in this system. In brood tanks, it is necessary often to

stock fish of no more than 50 to 80 g, to avoid mortality due to aggression (Pruginin *et al.*, 1975). Stocking densities and fry productivity are also illustrated in Table 5.3.

Problems Encountered in the Commercial Production of Hybrids. Purity of stocks has already been mentioned as the major drawback to hybrid production. Selection of pure strains is time-consuming, purchase of certified stocks could be expensive and fry production is further aggravated by the lower fecundity of hybrid crosses, i.e. two-thirds of normal crosses (Mires, 1977). Mass production of all male progeny is therefore dependent on large numbers of certified broodstock, requiring lockable hatchery facilities employing specially designed systems which are often highly labour intensive and can raise the expense of fry production to 35 per cent of market value (Tal and Ziv, 1978).

The added expense of producing hybrids would certainly outweigh the losses in production of good quality fish were reproduction not controlled; however, in intensive systems where uncontrolled breeding is less of a problem, the resultant benefit from heterosis is questionable. Hybrid vigour remains to be verified under more controlled conditions. Perhaps a more important aspect of hybridisation is the production of coloured hybrids which are more readily acceptable to the consumer, e.g. the red-orange tilapia hybrid or 'cherry-snapper', which has proven commercially viable for sale to Japan (Fitzgerald, 1979). Sipe (1979) now offers six golden to red hybrid crosses as available for sale to an international market. Selection for such features as greater body depth and width to increase fillet size, faster growth, reduced aggression and other favourable characteristics are possible with intensive fry production system. It is therefore likely that the future trend may be towards producing a fish specific to the market requirements.

4. Tank and Raceway Culture

Tank units have only recently been used for the intensive rearing of tilapia and so are still very much at an experimental stage. Few commercial enterprises are operating on the same scale as that practised for more temperate species, such as salmonids. Perhaps one of the earliest reported systems involved the use of trough-like tanks (Uchida and King, 1962) and raceway-cum-square tank units (Hida *et al.*, 1962) used to rear *S. mossambicus* fry as bait fish for the Hawaii tuna fishery. About 15 countries in the world now report some form of experimental or active rearing of tilapia in tanks (Tables 5.6 and 5.7) and a few countries report the use of raceways (Table 5.8). The information currently available is very limited, however, so that the characteristics, features and design approaches to various units given below are based on general reports of tank systems and where details are available, special reference is made to tilapia.

Generally for tilapia the use of tanks and raceways can be classified as follows:

Raceways. Mainly used for the rearing of fry either to a size saleable as bait fish (Uchida and King, 1962) or for impoundment stocking (Ray, 1978) or prior to fattening in tanks (Balarin and Haller, 1979; Lauenstein, 1978) or cages (Konikoff, 1975).

Table 5.6: Countries in Which Tilapia are Reared Commercially or Experimentally in Tanks

Country	Species	Result	Comments	Reference
China	Tilapia Carp	= 50 kg/m^3/mth = 25 kg/m^3/mth	Experimental mix culture in concrete tanks and use double-doughnut tanks for fry production	Nash & Mayo (1979)
Colombia	*T. rendalli*	Enhanced predator *P. kraussii* and *C. occellaris* but reduced growth of *P. reticulatus* and trout	Tank experiments to determine compatable predator/prey combination	Henao & Corredor (1978); Popma (1978)
Ethiopia	*T. zillii/S. niloticus* and *C. carpio*		Experimental culture in concrete tanks	Meskal (1975)
Germany	*S. mossambicus*	= 20.0 kg/m^3 plus 20.0 kg/m^3 carp	Experimental recirculation system	Naegel (1977)
Fiji	*S. mossambicus*		Mention tank culture	Fiji Dept. Agric. (1959)
Hawaii	*S. mossambicus*		Fry production for tuna bait in concrete tanks	Hida *et al.* (1962)
Hawaii	*S. mossambicus*		Fry rearing for tuna bait in metal tanks	Uchida & King (1962)
Israel	*S. aureus*	Held at 100 kg/m^3	Experiments in brackish water culture in concrete tanks. High density stock for overwintering in tanks	Chervinski (1961) Muir (unpub.)
Kenya	Tilapia species	10.0 kg/m^3/mth in 6 m circular tanks	Experimental and commercial culture of nine species in concrete tanks and raceways	Haller (1974)

Table 5.6: continued

Country	Species	Result	Comments	Reference
Mexico	*S. niloticus* *S. mossambicus*		Intensive culture in water tanks has shown to be profitable	Gonzales (1976)
Nigeria	Tilapia/*Clarias* and *Chrysichthys sp.*		Indoor tanks used for research	Dada (1976)
Puerto Rico	*S. aureus*		Tanks used to acclimate fish to sea water	Miller & Ballantine (1974)
Sudan	*S. niloticus*		Cement tanks used to rear fry	George (1975a)
US – Colorado	*S. aureus*	20-60 kg/m^3/year	Use of geothermally heated water produces 50 t/yr in steel tanks, cement raceways and earthen ponds	Lauenstein (1978)
US – California	*S. mossambicus*	2.0 kg/m^3/month	At a ten per cent water change/month in a closed system heated by solar energy in a solar aquadome system	Serfling & Mendola (1978)
US – Texas	*S. aureus* and carp		330 l tanks in plastic greenhouse heated with power station effluent. Fish fed manure or algae slurry	Maddox *et al.* (1978)
Belgium	*S. niloticus*	25.7 kg/m^3/month	2 m^3 tanks using a semi-closed recirculation system under a greenhouse	Melard & Philippart (1980)
Japan	*S. niloticus*		Experimental in heated effluent from factories but was not thought economical	Chiba (1980)

Table 5.7: A Summary of Available Data on Tank Culture of Tilapia

Location	Species	Initial Stocking x̄wt (g)	n/m³	kg/m³	Description	Duration (days)	Harvest x̄wt (g)	kg/m³	Production kg/m³/month	Food Type	Rate	FCR	Reference
Kenya	S. niloticus	25-50	200-250	7-10	6 m circular concrete tank 0.7 m deep = 20 m³ with siphon drain. Pumped (brackish water) at 200-600 l/min/tank	150	250	50	10.0	22-40 per cent protein pellet	3-1.5% (3-6/dy)	2:1	Balarin & Haller (1979)
US – Colorado	S. aureus	25	(500)	(12.5)	2-7 m circular steel tanks 1 m deep	125	150	16-24	1 g/day (3.8-15.4)	36 per cent protein pellet	2%	1.5:1	Lauenstein (1978)
Hawaii	S. mossambicus	0.03-0.07	260-1600	0.01-0.1	1.2 x 3.7 by 0.85 m deep plywood fry rearing tanks of 3.8 m³ at 2.45 l/min	84	0.6-2.4	0.6-0.95	0.2-0.5	out feed — carter and fingerling grade	(2/dy)		Uchida & King (1962)
Germany	S. mossambicus + C. carpio	5.0	200	1.0	250 l tank with a recirculation system at 1 l/kg/min	147	83	16.6 (Σ=106.6)	3.2 (Σ=22.4)	Trout feed	5%		Naegel (1977)

Table 5.7: continued

Location	Species	Initial Stocking			Description	Harvest			Production kg/m³/month	Food		FCR	Reference
		x̄wt (g)	n/m³	kg/m³		Duration (days)	x̄wt (g)	kg/m³		Type	Rate		
Belgium	*S. niloticus*	20-25	100-800	16.0	4 m² (2 m³) tanks under a greenhouse using a semi closed recircula- tion system using 85 m³ heated water to produce 1 kg of fish	170-200	300	>30	25.7	Trout feed 46 per cent protein	7-1.5%	2:1	Melard & Philippart (1980)
Scotland — Stirling University	*S. niloticus* x *S. aureus*	30.0	90	2.7	2.25 m² (1.35 m³) tanks in tropical house using recirculated water	150	420	32.4	6.5	Trout feed 49 per cent protein	ad lib	—	McAndrew (pers. comm.)

Table 5.8: Summary of Available Data on Experimental and Commercial Culture of Tilapia in Raceways (Figures in parentheses have been extrapolated from available data)

Location	Species	Initial Stocking			Description	Duration (days)	Harvest			Food			Reference
		x̄wt (g)	n/m³	kg/m³			x̄wt (g)	kg/m³	Production kg/m³/month	Type	Rate	FCR	
US – Colorado	S. aureus	1-2	(650-2,500)	(1.5-3.7)	30 x 1 by 0.4 m deep concrete raceways using geothermally heated water	(60 est)	25	16-64	0.4 g/day (8-32)	Finely ground grain and blood meal plus vitamins and mineral premix	3%		Lauenstein (1978)
		150	(100-400)		Earth raceway 35 x 4 by 1 m deep	(150 est)	300-400	16-64	1.0 g/day (3.2-12.8)	36 per cent protein pellet	1%	1.5:1	
US – Oregon	T. zillii (+ S. mossambicus and I. punctatus)	3 cm			29.3 x 3.0 by 1.2 m deep set in series of four. Rearation by 0.6 m drop at a flow of 5000 l/min using geothermally heated water		7-10 cm						Ray (1978b) Klopfenstein & Klopfenstein (1977)

Table 5.8: continued

Location	Species	Initial Stocking			Description	Duration (days)	Harvest			Food			Reference
		x̄wt (g)	n/m³	kg/m³			x̄wt (g)	kg/m³	Production kg/m³/month	Type	Rate	FCR	
Kenya	*S. niloticus*	1-5	1,650	5.0	10 x 1.5 by 0.4 m deep concrete raceway using water from main production tanks to rear fry at an exchange rate of 3-4/hr	40-50	10-20	22.5	9.5	Tilapia pellet 22 per cent	ad lib		Balarin & Haller (1979)
		10-20	1,000	14.0		40-50	25-50	35.6	11.7				
US – Arkansas	*S. aureus*	3.0	100	0.3	7.6 x 0.9 m by 0.69 m deep divided into four sections	94	37.0	3.51	1.08	No food, feed off sides and catfish wastes			Allen & Carter (1976)
	Polyculture with *I. punctatus*	3.0	300	0.9	7.6 x 0.9 by 0.23 m deep	94	53.0	8.90	2.71				
Hawaii	*S. mossambicus*	0.03	(1,000-2,800)	0.8	9 x 0.76 x 0.76 m divided into six sections. Flow = 2.3-4.5 l/min	90	1.5-2.5	(5.6)	(5.52)	Ground trout pellets and granules	variable (21 day)		Uchida & King (1962)

Tanks may be used for fry production (see section 3.) or for early treatment with methyltestosterone (Balarin and Hatton, 1979; Guerrero, 1976a), but greatest potential lies in use of fattening units to produce marketable table-size fish (Balarin and Hatton, 1979; Haller, 1974; Lauenstein, 1978; Nash and Mayo, 1979) which should be adaptable to a high water economy by using recirculation systems if necessary (Naegel, 1977; Serfling and Mendola, 1978).

4.1 The Characteristics of Intensive Tank Culture Units

Whereas ponds and occasionally raceways are large and of earthen construction, tanks are generally smaller and constructed of concrete, wood, fibreglass, metal or other suitable materials. Tilapia units are mainly of concrete though portable units of asbestos/concrete are receiving considerable interest in Kenya and plastic pools are used in South Africa (Caulton, 1979).

Whatever the material used, a smooth interior finish is desirable to provide an easily cleaned surface, which does not create too great a resistance to flow and will not damage fish which rub against it. This forms one of the major design characteristics of an intensive system. Other features which are considered important are given in Table 5.9 and are summarised below.

An intensive fish culture tank unit is required to have a low flow to velocity ratio, so that the most efficient use can be made of water to maintain a high water quality, supply the fish's oxygen requirements, as well as provide a back-current to remove solid wastes and aid in food and fish distribution. The cost and complexity of construction should be low and the system adaptable to water reuse.

Table 5.9: Characteristics of an Ideal Intensive Culture Unit

---

1. Smooth interior — prevents abrasive damage to fish and makes for ease of cleaning, offering little hold for algae, bacteria etc.

2. Self-cleaning — has high water velocities and, if design creates a more uniform pattern of flow, cleaning efficiency can be high relative to flow.

3. Water quality is high — ensures optimum conditions for the cultured species.

4. Made of highly durable, non-toxic and non-corrosive material, preferably of sufficient mechanical strength to withstand moving or other stresses.

5. Easily cleaned and/or sterilised with independent drainage system for use at a time of severe disease infection.

6. Low construction cost, with a long amortisation period.

7. Good feed and fish distribution.

8. Flow to exercise the fish — resulting in better survival through improved muscle tone and stamina.

9. Adaptable to various stages of culture, i.e. fry rearing, ongrowing and fattening, with provision for improving system efficiency through water recirculation.

Systems Design. The variety of tanks used for the culture of aquatic organisms is endless, ranging from aquaria to tanks and raceways of up to 30 m along the major axis. The size and shape is variable depending on the required function in fish culture. Size is often decided on the most economical dimensions, e.g. a 10 m circular tank (10 t unit) is considered the most economic size for tilapia production (Balarin and Haller, 1979). The shape can be classified either as circular, square, rectangular (distinguishable from raceways by flow pattern and drainage) or oval with a central dividing wall. The characteristics of these various designs are summarised in Table 5.10 with illustrations of typical examples of each.

Wheaton (1977) considers circular tanks to have several advantages. Normally, water velocities are higher than in rectangular tanks (and raceways), leading to better conditioned fish for wild planting or restocking, but this can create a greater metabolic demand which may reduce conversion efficiencies to no advantage in fish to be sold for human consumption. Circular tanks tend to have a better feed distribution than raceways, are more self-cleaning, and use a lower flow (unless oxygen demand is the flow-determining parameter because of fish load). Cost of construction and installation of circular tanks is lower relative to raceways but this depends on the materials of construction.

Rectangular tanks are widely used because they are easier to construct. However, several problems are associated with them, in particular relating to their use in tilapia culture (see Table 5.11, after Balarin and Haller, 1979). Wild-caught fish when placed in rectangular tanks may crowd into one corner and deplete the oxygen, or, in an attempt to escape, swim into tank corners or walls and physically injure or exhaust themselves. Fish raised in tanks as fry rarely display such behaviour and the problem can be overcome either by using circular units or, preferably, by using raceways, allowing fish to run the whole length of the system. Raceways also are larger and therefore male territoriality does not restrict fish movements as is evident in smaller tanks.

Circulation in small rectangular tanks is often characterised by 'dead areas' and short-circuiting. Local oxygen depletion can occur, or metabolic wastes and infective organisms can build up, causing fish stress, if not death, particularly after handling. In circular tanks, short-circuiting occurs along the bottom and provides the characteristic self-cleaning efficiency of the system (Larmoyeux and Piper, 1973). However, the torus shaped 'dead area' can also be a major problem especially in larger tanks, although this can be displaced by the use of sprinklers (i.e. multiple nozzles extending to the centre of the tank) or by changing the angle of inflow relative to the wall (Balarin and Haller, 1979). Burrows and Chenoweth (1955) comparing flow patterns in circular tanks, raceways and a Foster-Lucas pond (oval tank) show circulation in raceways to be very different and to be dependent on inlet and outlet design. These must cover the full width of the raceway and for the most effective cleaning efficiency, length:width:depth must be in the ratio of 30:3:1 (Westers and Pratt, 1977). Water velocities are relatively low and baffles added to control inlet turbulance, maintain laminar flow and approach 'plug' flow conditions (i.e. all elements of water move with the same horizontal velocity). This eliminates both 'dead' spots and short-circuiting, maintaining a high cleaning efficiency.

In a Foster-Lucas pond, turbulent areas of lower than average flow occur on the downstream side of the centre wall, after the water has turned the corner.

Table 5.10: Characteristics of Various Designs of Intensive Culture Systems

Diagram showing flow patterns	Type	Flow:Velocity	Cleaning Efficiency	Food Distribution	Fish Distribution	Cost	Other
	Circular tank (also doughnut or spiral)	Low Inflow creates a spiral current	Self-cleaning, efficiency decreases as size increases	Very good Can use automatic feeders	Very good Current ensures a uniform distribution but graded in size	Relatively inexpensive but depends on materials	High velocity better fish conditions but has a high metabolic demand
	Rectangular/ square tank	High Requires greater flow to remove wastes	Poor 'dead areas' and short-circuiting creates deoxygenated zones	Depends on size and flow	Poor Tend to crowd and territorial males dominate corners	12% lower in construction costs than circular tanks	Easy to construct and can be used for fry rearing

	Oval tank (Foster-Lucas) (Burrours)	Low Controlled by a paddle wheel and/or tank design	Good Design creates turbulence which assists cleaning	Good Can be problematic because of central wall	Good Greater area and current forces fish to swim	High Complex construction	Permits greater water recirculation
	Silo tank	Medium Requires very high flow rates	High Usually recycling systems	Good Layering of fish may cause problems of jostling for food	Good Layered, permits high density stocking	Low capital investment but high running costs	Select harvesting may be difficult
	Raceway	High Design important otherwise require high flow rates	Poor-Good Requires plug flow to facilitate cleaning	Good Not as good as circular tanks	Good Often tend to crowd at inlets	Inexpensive simple construction	Suited to fry rearing

Burrows and Chenoweth (1970) used these areas as outlet points, increasing the self-cleaning properties of the tank. The characteristics of this tank conform to those of an ideal culture unit (Table 5.9), however construction is complex and relatively expensive. The unit is not often recommended for use and there are few reports of it having been tested for tilapia culture. Similarly, silo tanks have been proposed for high density fish culture and although not tested for tilapia, the results reported for trout (Buss *et al.*, 1970) might favour the future use of this system.

Table 5.11: An Evaluation of Circular vs. Rectangular or Square Tanks as Used for Intensive Tilapia Culture (Based on Observations of the Performance of Tilapia under Baobab Farm Conditions)

Circular Tanks	Rectangular Tanks
1. *Carrying capacity:* The shape and flow patterns give rise to a more uniform distribution of fish allowing for higher stocking densities in excess of 100 kg/m^3	Tank corners tend to act as 'hides' for dominants which then disrupt the uniform distribution pattern. Combined with poor flow conditions, a maximum carrying capacity of 70 kg/m^3 has been attained, under a wide range of conditions
2. *Disease:* Good flow patterns and high expulsion rates minimise 'dead areas' in which a build-up of infective components would occur. The system is also able to flush out particles which have gone into resuspension	Poor flow patterns create 'dead areas' in which faecal material builds up and decomposes, possibly leading to the production of toxic substances and an increase in the suspended material, creating an adverse environment
3. *Food distribution:* Water velocity ensures a good distribution of food, though there is some loss from the action of the centrally-spiralling currents short-circuiting food into the central sump	As there are no currents, food accumulates on the bottom, and often is uneaten if within the territory of a dominant fish. Further, jostling in the confines of the tank leads to injury and an uneven food allocation
4. *Cleaning efficiency:* Faecal material and waste food is carried towards the central drain by the higher water velocity. This results in a rapid expulsion of wastes	Slower flow velocities, characteristic of this shape, are often not fast enought to remove all faecal wastes. The resultant build-up on the bottom not only creates adverse water quality conditions, but tilapia have been noted not to feed on food which falls into these accumulated wastes. Further, if waste/excess food is allowed to accumulate in the tank, it decays quickly and fish refuse to eat and often 'shy' away
5. *Viability:* The good environmental conditions and forced exercise produces fish of high stamina, eliminating weak/sick fish. The rounded tank sides also reduce injury during times of mass hysteria resulting from netting/cleaning activities	Poor environmental conditions tend to select for a hardier fish tolerant of more adverse conditions, eliminating less tolerant individuals. However, fish are under stress, and if stressed further, especially during handling, mortalities result

Drain Design (Tanks). The large volumes of water used in open systems necessitate the need for careful consideration of drain design. Water outflow must match inflow and is often catered for by using a pipe diameter large enough to cope with the maximum flow rate required at optimal carrying capacity. Often an overflow pipe is used so that in the event of a blockage of the main drain, the tank will not overfill and spill over, allowing fish to escape.

The two conventional drain units most often used are the outside overflow stand-pipe and the central constant overflow stand-pipe. Although highly suited to small tanks and juvenile tilapia rearing, these units have been replaced by a siphon drain which through the years of research at Baobab Farm, Kenya, have been shown to be better suited to the requirements of large fattening tanks (Balarin and Haller, unpublished). The siphon drain required two stages of development before the final design was arrived at, and the characteristics of these are compared against the more conventional systems in Table 5.12.

The external siphon creates a greater flushing action through its stop-start cycle, and although complex in design, and therefore more expensive to construct, it does have a distinct advantage in that drainage can be stopped at any time and does not require priming. This is particularly beneficial at times of feeding. Tilapia tend to be slow feeders and if overfed, excess food may well be flushed out of the system before the fish have time to get to it in a normal system. Further, the removal of all of the drain structure to the tank exterior renders harvesting easier and the design allows for the use of a grill mask over the drain outlet which can be used as a self-grading device.

Flow Rates and Water Quality. Faecal load and oxygen requirements could be considered as the major parameters deciding flow rates. In an open system, rates of 0.5-1.0 l/kg/min maintain oxygen levels and a good cleaning efficiency as well as preventing the build up of metabolic products which are likely to inhibit growth. As discussed below, in semi-closed or recirculating systems accumulation of toxic wastes requires more careful consideration.

In open systems, therefore, flow has two functions:

Aeration. If not already saturated, oxygen saturation of incoming water can be aerated prior to its delivery to the tank by the use of an aerator tower, bubble or airlift device, venturi, cascade aerator or sprinkler pipe. A system more often used is that of reaeration using a pressurised water jet. The force of the jet striking the water surface creates a turbulence which introduces oxygen into the system. A jet angle of 60° from the horizontal, located one-third of the distance from the centre provides the greatest efficiency (Chesness *et al.*, 1973).

Current Speed. The peripheral current speed in a tank is dependent on inflow rate, water depth and angle of inflow pipe (Larmoyeux *et al.*, 1973). The velocity of the current, however, not only determines the cleaning efficiency but also affects the fish's active metabolic rates. There is therefore a need to maintain a balance between cleaning and the effects of the current on growth and oxygen consumption. Excessive current speeds mean a lowering of the oxygen levels as well as increasing the energy requirement for swimming at the expense of growth.

Studies by Balarin and Haller (unpublished) set the following flow limits for tilapia: (See Figure 5.8):

Table 5.12: Characteristics of Various Outflow Drainage Systems as Used in Tilapia Tank Culture Systems

Feature	Design Number				
	I Outside Overflow Stand-pipe	II Central, Constant Overflow Stand-pipe	III Internal Siphon System	IV Modified Internal Siphon	V External Siphon System
Flow: faecal removal efficiency	Low-medium Good when external overflow lowered manually	Low Good when stand-pipe lowered manually	Medium	Medium	Low – medium – high (depends on tank diameter and depth)
Maintenance/labour requirements	Low	Medium	High – very high	High	Low – medium
Ease of harvesting	Very good	Good	Problematic	Problematic	Very good (self-grading)
Food losses when feeding	Variable	High	High	High	Negligible

Cost of construction	Low	Low	Medium – high	High	Medium
Adaptability to fry rearing	Very good, no modifications needed	Very good with minimal modification	Not possible to lower water to levels recommended without major modification		Good, minor modifications needed
Recommended use	Fry rearing and juvenile ongrowing	Juvenile ongrowing	Only for use at exceptionally high flow rates in fattening tanks		Fattening tanks
Diagrammatic representation					

Lowest current speed = 7.5-10 cm/s (i.e. the settling velocity of the largest faecal particles)
Upper current speed = 20-30 cm/s (i.e. a velocity just below that at which fish are forced to swim)

As can be seen from Figure 5.8(a), the current speed is proportional to inflow rate and increasing depth has a dampening effect. However, these two parameters are dependent on stock density and system design, i.e. for a good cleaning efficiency large tanks must have a 5-10:1, diameter:depth ratio (Larmoyeux *et al.*, 1973) and often designs do not allow for depth to be adjusted to obtain the required flow. The angular momentum of the incoming water is not only a product of flow rate, but also its mass density, the radial position vector of the inlet and the horizontal component of the flow (Larmoyeux *et al.*, 1973). Thus an angle from the horizontal of 25-50° is often recommended to improve the self-cleaning action of a pressurised inflow. Adjusting this angle would affect the current speed but a more practical application, so as not to affect the cleaning action, would be to alter the radial position of the inflow as shown in Figure 5.8(b). Angling the inflow to between one-third and one-half the distance from the tank wall to the centre has given the most effective cleaning action (Balarin and Haller, 1979).

In raceways, an exchange rate of four changes/hour is often recommended for trout to maintain good water quality, in particular for fry rearing (Westers and Pratt, 1977). It therefore follows that the higher the exchange rate, the smaller is the required rearing space, thus reducing capital expenditure and operational costs. A common practice in tilapia culture therefore is to reduce the working depth and this can be calculated from the flow rates:

$$d = \frac{F}{A \times R}$$

where d = working depth (m), F = flow rate (m^3/hr) calculated from oxygen requirements, A = area of raceway (m^2) and R = required exchange rate (no/hr) minimum = 4.

Though it can be argued that lowering the depth may affect growth no adverse effects have been noted. It would appear that growth may even be enhanced in shallower raceways (Allen and Carter, 1976) possibly due to more rapid temperature rises or the high light intensity favouring phytoplankton growth on the walls and thus increasing available food.

4.2 Stocking Practices and Production

By careful stock manipulation, Nash and Mayo (1979) in China, report achieving a production of 50 kg/m^3 month ($\equiv$ 6000 t/ha/yr at 1 m depth). This is an exceptionally high production and a more realistic value would be of 10 kg/m^3/ month (1200 t/ha/yr at 1 m depth) as achieved both in Colorado, USA (Lauenstein, 1978) and Kenya (Balarin and Haller, 1979). In the latter case, growth inhibition was often noted and is attributable to crowding. This stunting effect is not only related to the competition for availability of space due to too high a total biomass (density as kg/m^3) but also reflects the effects of social interactions, i.e. the high numerical density adversely affects individual

Figure 5.8: Diagrams to Show (a) the Effect of the Angle of Inflow on the Current Speed in a 6 m Circular Tank Stocked at Capacity and (b) the Relationship Between Tank Depth and Current Velocity at Various Inflow Rates in a 6 m Circular Tank and the Corresponding Behaviour of Tilapia Indicating Recommended Flow Limits.

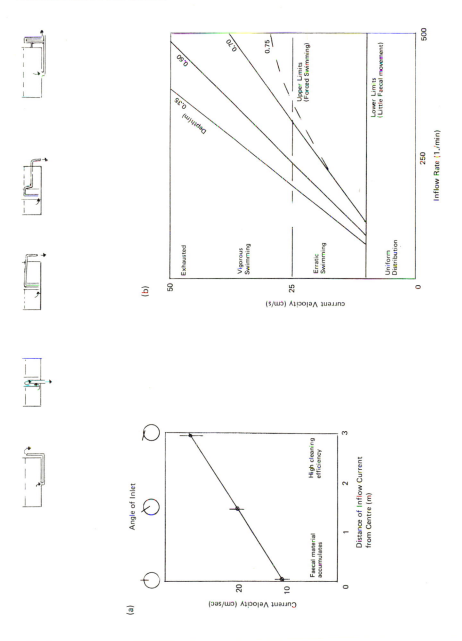

growth rates, though total production is increased. This substantiates the earlier findings of Uchida and King (1962) for juvenile (<5 g) *S. mossambicus* (Figure 5.9) where a maximum stocking rate of 320/m³ (1.5 kg/m³) was recommended. However for economy of space, Balarin and Haller (1979) recommend stocking fry (1-5 g) at 1600/m³ and thinning out by selective grading to 1000/m³ at 10-20 g (35 kg/m³) in fry tanks or raceways. Commercial fattening tanks are, however, only stocked at 200/m³ of 25-50 g fish (50 kg/m³) in order to take advantage of the enhanced individual growth rates to produce larger fish. Further, by this technique of selective grading only the upper 50 per cent of the initial stock (i.e. the faster growers) are ongrown and often these consist of up to 70 per cent males. This method therefore allows for two harvests a year and surplus stocks are sold as bait fish, for pond stocking, or can be used as an animal feed. A summary of the programme of operation is given in Figure 5.10. A similar practice is described by Lauenstein (1978), except that earthen raceways are preferred for the final fattening stages.

Figure 5.9: The Relationship between Fry (0.02-2.4 g) Growth Rate and Total Production at Various Stocking Rates in a 3.8 m³ Tank for Twelve Weeks.

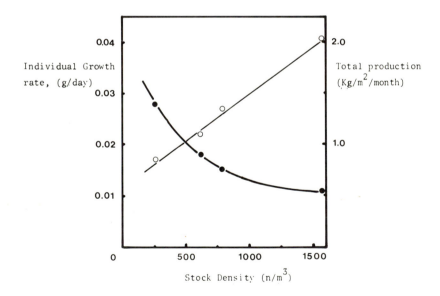

Source: Uchida and King (1962).

4.3 Grading and Harvesting

Tank and raceway systems are easily drained, facilitating the ease of stock removal. However, should stocks selected, or those remaining, be required for ongrowing, it is essential to minimise handling and disturbance to the fish. With

Figure 5.10: Suggested Model of Operation of an Intensive Tilapia Production Unit.

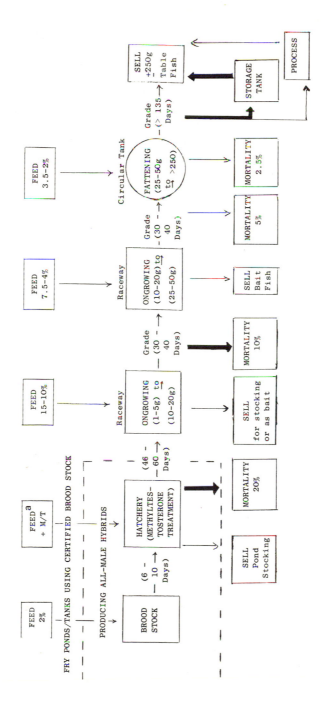

Note: a. M/T: Methyltesterone.
Source: Balarin and Haller (1979).

this aim in mind Balarin and Haller (unpublished) describe graders suitable for use in tilapia systems. The grader widths are given in Table 5.13 based on the body thickness of tank fish and vary with species and condition factor.

Table 5.13: Body Thickness of Three Species of Tilapia Cultured Under Tank Conditions with Recommended Dimensions for Grading

Mean weight (g)	*S. aureus*	*S. niloticus*	Pruginin and Shell (1962)	*S.s. niger*	Minimum grader width (cm)
>5	1.06	0.99	0.97	1.03	1.0
10	1.30	1.23	1.19	1.32	1.25
25	1.73	1.65	1.40	1.74	1.70
>250	3.60	3.45	–	3.76	3.50

Raceway graders are bag-shaped, with an entrance of equal dimension to the raceway. The grader is then pulled along the length of the raceway. A slotted cage at the end of the bag allows undersized individuals to escape, entrapping larger fish. These can then be further graded using a conventional box grader, before restocking or removing for sale.

Tank graders are of the conventional grading door design. Vertical slots and more recently, piano-wire, form the grills of the door. The door extends the radius of the tank and is attached to the outer wall and turns on a central pivot. A second door is then used to enclose the fish in a small segment of the tank. Undersized fish escape through the grills and entrapped fish are then scooped out for sale.

Both methods mean that undersized fish are not handled and have a further advantage over the conventional design of graders, where all size groups are fished out, in that small fish are not damaged or crushed by the violent activity of recently netted larger fish.

Suggested Production Cycle. The growing periods, feeding rates and expected mortalities given in Figure 5.10 are values obtained from tilapia grown in Kenya by the authors and therefore may apply only to equatorial conditions, but the basic principles are the same. Fry produced from the spawning units are ongrown in the hatchery tanks where they are fed a hormone treated diet. After the required treatment period, fingerlings are stocked into the first of two ongrowing units. By a process of selective grading, 50 per cent of those originally stocked (i.e. the faster growers) are then put into the fattening units where they are grown up to market size. Fish are sold as required, surplus fish being placed into a holding unit so as to be able to restart the cycle. During the fattening period, empty ongrowing units are used to rear fry for sale either for impoundment stocking or as bait fish.

4.4 Trends in the Intensification of Tank Systems

A market exists for tilapia as a luxury food item in some of the major developed nations, though often these are outside the natural distribution limits of the species. It is therefore necessary to overwinter the next season's stocks either in heated tanks as in Auburn, Alabama (Avault *et al.*, 1968) or in protected raceways as practised in Israel (Wohlfarth, 1977). Further, to reduce the cost of heating, such units often have recirculating water, a technique to which tilapia would appear well adapted and which has in recent years received more attention as a means of increasing production.

Warm Water Culture Using Heated Effluents. Heating costs are often prohibitive and thus cheaper sources of energy are constantly being sought. Perhaps the cheapest is the use of geothermally heated waters (Lauenstein, 1978; Ray, 1978b). A further economical source is the use of heated cooling waters from power stations or other industrial processes (Aston, 1980).

Because of their ease of culture, tilapia are considered by Hubert (1978) to have one of the strongest potentials as a cultured organism for use in such systems. A common practice is to stock the fish into cooling ponds and *S. mossambicus* was first used in ponds near London in the early 1960s (Iles, 1963). *Tilapia spp.* have also been used successfully to free such ponds of vegetation as well as to improve productivity in USSR and Germany (Aston *et al.*, 1976) and also in Texas (Noble *et al.*, 1975). More recently, in Germany, Otte and Rosenthal (1979) have used tank units supplied with heated effluent waters from power stations and in Tennessee, Suffern *et al.* (1978) have successfully reared tilapia hybrids in cages suspended in the water of a cooling reservoir fertilised with sewage.

Recirculation Systems. Uchida and King (1962) describe a simple system whereby water is reused after passing through a sand filter (this would remove solid wastes although waste metabolites are likely to accumulate). In a trial, water was reused up to seven times (86 per cent recycle), growth was not affected (Haller, personal communication), though prolonged recirculation within a tank system could result in poor water quality even if well aerated (Balarin and Haller, 1979). Biological filtration to remove toxic wastes (e.g. NH_3 and NO_2) is essential and Naegel (1977) has successfully used such a system to produce 3.2 kg/m^3/month (384 t/ha/yr at 1 m) tilapia in polyculture with carp at a total yield of 22.4 kg/m^3/month (2688 t/ha/yr at 1 m) with daily water losses as low as two to three per cent. The system incorporates separate nitrification and denitrification units but water can also be bypassed through hydroponic raft culture units. Here such plant crops as tomatoes and lettuces can be grown, acting as denitrifying agents. Mass cultivation of such algae as *Scenedesmus spp.* might be equally as effective; tilapia could then be stocked into these units to crop the algae.

Serfling and Mendola (1978) describe a more compact unit reliant on solar heating, called the 'Solar Aqua Dome System'. A central biological filter made up of nitrifying organisms, detritivorous invertebrates and aquatic plants removes nitrogenous wastes produced by the fish reared in an outer, doughnut-shaped raceway. As well as a supplemental diet, fish are fed on the plants and invertebrates flushed out from the central sump. A standing crop of 8.1 kg/m^3

yielding 85 t/ha/yr was achieved for only a ten per cent water change per month. Parker (personal communication) has designed a similar system for use in arid areas which can also be coupled to a hydroponic unit.

Melard and Philippart (1980) describe a more intensive semi-closed recirculation system capable of a production of 25.7 kg/m^3/month (3084 t/ha/yr at 1 m) which requires 85 m^3 of heated water to produce 1 kg of *S. niloticus*. The potential productivity and efficiency of water reuse using a combination of water treatment and ozonation is illustrated by the closed system employed by Otte and Rosenthal (1979). At a replacement rate of 0.7 per cent per day this system is capable of supporting a biomass of 217 kg/m^3 with no adverse effect on growth.

The use of a polyethylene dome or greenhouse to entrap solar energy, with the insulation of tanks, is a system currently being recommended for tilapia production in West Germany (Otte and Rosenthal, 1979) and in South Africa (Caulton, 1979). It would seem to have considerable potential and may become widespread as interest in tilapia increases throughout the developed countries.

5. Cage Culture

Cage culture permits the more intensive exploitation of a water system with a low capital expenditure. The advantages are numerous, particularly when considering that the method can be utilised with minimal infrastructural requirements and the ease of management lends the system to intensification (Table 5.14). The technique can be used in the sea, lakes and rivers as well as in water-filled excavations such as old quarries and mines, in irrigation canals, in outflow canals of power plants (i.e., heated effluents), undrainable ponds, where snags prevent seining and where the system has become invaded by undesirable fish species, as well as to make the most efficient use of sewage treatment ponds. Often cage culture can be the only practical and economically feasible method for exploiting an area and equally this method can be applied to rivers, lakes, lagoons, estuaries and coastal bays which may already have an existing fishery.

Perhaps one of the major advantages of tilapia cage culture is the prevention of successful breeding (Pagan, 1969; 1970) and the amenability of the species to intensification. Pagan's experiments are among the first reported cases of the use of tilapia in cages, but since then interest has increased and experiments have been undertaken in Lake Atitlan, Guatemala (Bardach *et al.*, 1972), in the research ponds of Auburn University, Alabama (Suwanasart, 1971; Pagan-Font, 1975) from where results were extended to cage culture in ponds in Puerto Rico (Jordan and Pagan, 1973), the Philippines (IFP, 1974b) and El Salvador (Bayne *et al.*, 1976). Cage culture has also been undertaken in Columbia (Patino, 1976), Brazil (FAO, 1977), in Taiwan (Maruyama and Ishida, 1976) and in old tin mines in Malaysia. In Africa a semi-intensive programme was carried out in Lake Victoria, Tanzania (Ibrahim *et al.*, 1976) and more intensively in Lake Kossou, Ivory Coast (Coche, 1975, 1976, 1977, 1978). Trials in Lake Kainji, Nigeria have not been too successful because of problems in obtaining sufficient numbers for stocking (Ita, 1976). Salt water cage culture has also been attempted in brackish water ponds in Louisiana (Perry and Avault, 1972)

Table 5.14: Advantages and Limitations of Cage Fish Culture Technique

Advantages	Limitations
Possibility of making maximum use with the greatest economy of all the available water resources	Difficult to apply when the water surface is very rough therefore location restricted to sheltered areas
Helps reduce the pressures on land resources	Back up food store hatchery and processing units necessary therefore requires strategic location
Possibilities of combining several types of culture within one water body, the treatments and harvests remaining independent	
Ease of movement and relocation	
Intensification of fish production (i.e. high densities, optimum feeding results in improved growth rates and reduces length of rearing period)	Need an adequate water exchange through the cages to remove metabolites and maintain high dissolved oxygen levels. Rapid fouling of cage walls requires frequent cleaning
Optimum utilisation of artificial food for growth, improves food conversion efficiencies	Absolute dependence on artificial feeding unless utilised in sewage ponds. High quality balanced rations essential. Feed losses possible through cage walls
Easy control of competitors and predators	Sometimes important interference from the natural fish population, i.e. small fish enter cages and compete for food
Ease of daily observation of stocks allows for better management and early detection of disease. Also economical treatment of parasites and diseases	Natural fish populations act as a potential reservoir of disease or parasites and the likelihood of spreading disease by introducing new cultured stocks is increased
Easy control of tilapia reproduction	
Reduces fish handling and mortalities	Increased difficulties of disease and parasite treatment
Fish harvest is easy and flexible, and can be complete and of a uniform product	Risks of theft are increased
Storage and transport of live fish is greatly facilitated	Amortisation of capital investment may be short
Initial investment is relatively small	Increased labour costs for handling, stocking, feeding and maintenance

Source: Modified after Coche, 1976.

and in Kenya (Haller, 1974) as well as off the coast of Dauphin Island, USA (Williams *et al.*, 1974) and in the coastal area around la Parquera, Puerto Rico (Miller and Ballantine, 1974). Tilapia have also been evaluated in the USA as a potential polyculture species for increasing productivity in *Ictalurus punctatus* cage culture, e.g. Alabama (Schmittou, 1969) and Louisiana (Perry and Avault, 1972) but at present there are very few reported commercial operations. Those documented are from Lake Calibato, Philippines, where some 170 cages are stocked with tilapia (Anon., 1978), Lake Kossou, Ivory Coast (de Kimpe, 1978), two private farms in Lake Ilopongo, El Salvador (Street, 1978) and the Peasant

Table 5.15: Summary of Available Data on Tilapia Cage Culture

Location	Species	Initial Stocking Rate x̄wts (g)	n/m³	kg/m³	Description of cage	Duration (days)	% Survival	Harvest x̄wt F (g)	kg/m³	Production (kg/m³/m)	Food Type	Rate	Food Conversion Ratio	Reference
Ivory Coast – Lake Kossou (exp. 1974)	S. niloticus (overall x̄)	9-55 (29)	215-488 (340)	2-21 (9.8)	Floating cages 1 m³, suspended in the lake with feeding rings	92 (125)	(94.1)	157-271 (197)	35-76 (61.1)	11.7-21.3 (12.9)	Pellets – 25% Protein	3-6% (1/day)	2.6-3.7 (3.1)	Shehadeh (1976); Coche (1975*, 1976*, 1977); Campbell (1976)
Puerto Rico (exp.)	S. aureus	10-15	300-500	3-5	Floating cages 1 m³ in a rock quarry pond	70		45-55	17-23	6-7.7				Jordan & Pagan (1973)
US – Alabama (exp.)	S. aureus	10-25.5	487	4.8-12.4	Floating cages 1 m³ in pond	87		128-150	66.5-85.5	21.3-25.2				Suwanasart (1971)
US – Alabama (exp.)	S. aureus	13.6	286-857	5.6-16.6	Floating cages 0.7 m³ in ponds	156		>200	35-94.3	7-18.8	Floating pellet		1.2-1.7	Pagan (1970)
Tanzania – Lake Victoria (exp.)	T. zillii	2.6	83	0.216	Floating cages made of bamboo and netting (a) 3.5x3.5x3.5 43 m³ (8 mm)	90		15.6	1.25	1.0	Brewery waste and fish meal (10:1), and green vegetable leaves	15-30%		Ibrahim et al. (1976)
	S. esculentus	19	19	0.36	(b) 5x5x5 125 m³ (20 mm)	180		46.6	0.84	0.48		50%		
	T. zillii and S. esculentus	16.3	22	0.36		150		83.1	1.74	1.38		(1/day)		
Kenya (exp.)	S. spilurus niger and T. zillii (♂ only)	32.0	71	2.28	Suspended in quarry pond 0.7 m³ made of wood frame and wire mesh covering (25 mm) with feeding tray	177	100	236.8	12.86	1.79	Minced tilapia, rice bran and vegetable leaves (1:1:1)	(2-3/day)	1.79	Haller (1974)

Table 5.15: continued

Location	Species	Initial Stocking Rate			Description of cage	Duration (days)	% Survival	Harvest			Food		Food Conversion Ratio	Reference
		x̄wts (g)	n/m³	kg/m³				x̄wt_F (g)	kg/m³	Production (kg/m³/m)	Food Type	Rate		
US – Louisiana (exp.)	S. aureus and I. punctatus	(12.5 cm) 11.3	200 200		Floating in BW ponds 1 m³ (12.5 mm mesh)	170	97	24.49	21.8 38.3	3.85 6.76	Purina trout chow	(1/day)	1.9	Perry & Avault (1972)
Columbia (farm unit)	T. rendalli	22.5	100	2.25	Floating or on legs in ponds, made of bamboo and netting 1 m³	150		165-250	26.0	5.7	Bore leaves and rice bran	To satiation (2/day)		McLarney (1978) Patino (1976)*
Nigeria – Lake Kainji (exp.)	S. galilaeus S. niloticus T. zillii	52.1	256	13.35	Suspended off a pier. Wooden framed, poultry mesh 1 m³ (12.5 mm)	171	70	85.4	17.43	0.72	Pelleted food		12.2	Ita (1976)* Konikoff (1975)
	S. galilaeus	33.9	328	11.12		164	65	70.6	16.95	1.07				
Philippines (exp.)	S. mossambicus (♂ only)	10.4	100	1.04	Floating cages, wood covered in wire mesh 1 m³ (6 mm) using feeding trays	60	100	25	2.50	1.27	Bulgur wheat (13 per cent protein)	5%/day (2/day)	4.9	IFP (1974b)* Guerrero (1975c)
		10.4	100	1.04		60	92	21.7	2.00	1.00		No food		
Saudi Arabia (Sharm Obhur) (exp.)	S.s. niger	0.9	331	1.04	Floating cages on a raft 1 m³ with 12 mm square galvanised wire mesh in sea water	150	71	46.2	10.87	2.17	Trout granules	Automatic feeder	–	Peacock (pers. comm.)
Sri Lanka (exp.)	S. niloticus	2.5 (3.8 cm)	50.0	0.2	Floating cages of 1.8x1.8 (7 m³) made of bamboo frame and mesh netting	±180 (est.)	good	0.9-1.36	56.77	9.46	Fish meal and rice bran + poonac, ox blood and kitchen refuse	(2x/day)		Anon. (1980)

Note: * Where several references appear, indicates the most descriptive.

Table 5.16: A List of Other Countries in Which Tilapia are Cultured or Experimentally Reared in Cages

Species	Country	Result	Comment	Reference
S. aureus	El Salvador	2-4 % optimum feeding rate. Profitable if use >60 g fish	Proliferation of *Hydrilla* and of the predator *C. manquense* is a problem	Sanchez (1978)
		1 m^3 cages, production = 1.25 kg/m^3/yr	Best results obtained by feeding a pellet based on coffee pulp	Ramirez (1977), Bayne et al. (1976)
	— Lake Ilopango	30-40 m^3 cages with good production	Estimate commercial production of 4.5 t/month	Hughes (1977), Street (1978)
	US — Alabama	0.3 cm mesh led to fry production but not at 0.6 cm	Experimental attempt to reduce reproduction in 0.7 m^3 cages	Pagan-Font (1975)
	US — Dauphin Island	Bacterial numbers were reduced	Attempt at sea water cage culture	Williams et al. (1974)
	US — Alabama		Describes polyculture with channel catfish started in 1968	Schmittou (1969)
	Puerto Rico — La Parqueva		Attempt at sea water cage culture	Miller & Ballantine (1974)

Table 5.16: continued

Species	Country	Result	Comment	Reference
S. mossambicus	Guatemala — Lake Atitlan		Experimental culture	Bardach *et al.* (1972)
	Philippines — Laguna de Bay		Experimental culture	Pantastico & Baldia (1979)
	Taiwan	Recommend best growth in cages >75 cm deep	Experiments on depth effects	Maruyama & Ishida (1976)
S. mossambicus x S. honorum (hybrid)	US — Tennessee	Estimated 25–50 t/ha/yr	Cage culture in sewage ponds heated by effluent from power station	Suffern *et al.* (1978)
S. niloticus	Ivory Coast — Lake Kossou	20 m^3 cages tried on commercial scale	Results are encouraging but costly	de Kimpe (1978)
	Philippines		Have been cultured experimentally since 1975	Guerrero (1975c)
Tilapia (?)	Brazil		Have been tried with some success	FAO (1977)
	Philippines — Lake Calibato	Maximum carrying capacity = 170 cages/40 ha lake	Overstocking of lake now leading to stunting	Anon. (1978)
T. rendalli	Columbia		Very economical feeding of fresh vegetable leaves	Popma (1978)

Fish Culture Unit (Unidad Piscicola Campesina) in use in Columbia (McLarney, 1978). All these reports are relatively recent and could therefore be an indication that the potential had not been realised previously, nor the technology been available to implement tilapia cage culture on a large scale. Results available from various experimental reports are given in some detail in Table 5.15 and information available from other culture practices is summarised in Table 5.16. These tables indicate the substantial amount of information now available. This, and the success of the comprehensive experiments by Coche (1975, 1976, 1977, 1978), should lead to an upsurge in interest, particularly when it is realised that tilapia culture can be as productive, if not more productive, than some of the more commercially favoured species (see Table 5.17).

Table 5.17: Summary of the Range of Recorded Total Yield (= Carrying Capacity) and Production Rates of Various Species as Achieved in Cage Culture Under Experimental and Commercial Conditions

Fish Species	Σ yield (kg/m^3)	Production $(t/ha/yr)$	Food Conversion Ratio
Cyprinus carpio	10-189	2.5-35.9 (300-4308)	1.6-2.5
Sarotherodon aureus	17-94	6.0-25.2 (720-3024)	1.2-3.07
Ictalurus punctatus	35-220	5.6-22.6 (672-2712)	1.0-1.7
Salmo gairdneri	4.5-118	1.2-19.5 (144-2340)	1.6-2.0
Sarotherodon niloticus	35-76	6-17.0 (720-2040)	1.2-3.7

Source: Modified after Coche (1976, 1978).

5.1 Design and Construction

The free-floating or walkway/pier type of cage is the system most often used. Cages have also been set on the bottom, either partially submerged in shallow bays or mounted on legs in ponds (Patino 1976). In this latter instance, cages are raised about 0.3 m from the benthic sediments and show favourable growth rates, though cages which rest on the bottom have been noted to have a significantly lower growth rate in experiments in Kenya (Haller, personal communication). This 'stunting' was attributed to a suspected growth inhibitor (either chemical in origin and/or due to micro-organisms) present in the sediments, but could also be due to deoxygenated conditions arising from the decomposition of the accumulated faecal wastes. Generally, therefore, it is recommended that where possible, tilapia cages should be suspended no less than 2 m from the bottom.

Size and Shape. Cage depth has also been shown to affect growth. *S. mossambicus* were stated to grow better in cages of a depth of 0.75 m (Maruyama and Ishida, 1976) though the change in density was not considered and this may have affected the results.

Tilapia however do grow better in deeper waters in tanks. In most reported cases of cage culture, depths of 0.5 to 1.0 m are often used (Table 5.15).

The most common type of cage used is of a box shape design which makes for easier handling and attachment to walkways. Experimental units are small in size, between 0.7 to 1.0 m^3, made up of a rigid wooden or metal frame, covered in netting or wire mesh. These rigid structures, though cumbersome and difficult to handle are suited to rough treatment, either through regular handling, when assessing stock performance or in rough waters and are easier to clean *in situ*. Commercial cages are much larger, up to 125 m^3, and are of the suspended bag type, where the netting, normally of nylon or polyester fibre, hangs from a rigid floating framework (Coche, 1976 and Ibrahim *et al.*, 1976). The use of bag nets reduces the need for regular maintenance required of wooden framed cages. These frames have a short amortisation period as they are prone to rotting and readily damaged by aquatic wood borers, e.g. *Povilla adusta*, a mayfly larva prevalent in African waters (Coche, 1976). This necessitates the use of hardwoods such as mahogany (e.g. *Khaya spp.*) or metal frames, thus increasing the capital cost of construction. Limnological conditions could also lead to corrosion of the metal parts, increasing the cost of maintenance as the frames would require regular painting and replacement.

Mesh Size and Material. Wire mesh has been recommended by Coche (1976) as suitable for African conditions as netting of artificial fibres was thought to perish quickly, where exposed to the sun. This does, however, increase the weight of the structure and a more recent report (CTFT, 1977) indicates plastic mesh to be better than either metal or nylon.

Except for cases where cages or small net enclosures, termed 'bitinan' (IFP, 1975) or 'hapas' (IFP, 1976d) are used for fry production and rearing and therefore made of a very fine nylon mesh (13 meshes per cm^2), mesh sizes range from 3 to 25 mm in diameter. At the smaller mesh size of 3 mm, eggs, which may be of a maximum diameter of 3 mm (Fryer and Iles, 1972), are retained long enough to be fertilised and picked up by the female so that successful breeding can occur (Pagan-Font, 1975). Fish are generally unable to reproduce at a mesh size of 6 mm but these small net diameters are prone to fouling, which allows *S. mossambicus* to breed successfully in cages of this mesh size (IFP, 1974b). On the other hand, too large a mesh diameter (25 mm) allows wild fry and smaller fish species freedom of movement, through the cages. This can lead to substantial feed losses (Haller, 1974) and can subject caged stocks to exposure to infections from external sources. Further, a large mesh diameter may lead to physical injury, particularly at first stocking, for tilapia appear at times to experience a 'mass hysteria' if frightened and dash blindly into the cage sides either gilling themselves or sustaining severe injuries and dying from secondary infections (Balarin and Haller, 1979). Cage mesh size would therefore be best set at between 6 to 20 mm, using the largest size of mesh that cannot gill the fish size stocked (section 4.3). Coche (1976) recommends a 20 mm mesh diameter as suitable for stocking 25 g tilapia, providing between 80 to 85 per cent open area of the total surface area, and allowing for maximum water exchange. Smaller meshes reduce the water flow and are more likely to become overgrown with algae. Tilapia have, however, been known to browse off the algae growing on the sides of cages and *S. niloticus* is considered to act as its own anti-fouling agent (Coche, 1976).

Flotation. Flotation is provided either by buoys made of styrofoam blocks (Coche, 1976), empty plastic or metal drums (Ibrahim *et al.*, 1976), or else cages are suspended separately on floating pontoons (CTFT, 1977) or are fixed to a landing pier (Ita, 1976).

Cage Cover. Covering may be optional. More rigid units often have lids either of an opaque wooden or metal structure (Coche, 1976; Konikoff, 1975) or of wire mesh (Haller, 1974) while larger cages often have a suspended net covering to keep fish from jumping out and to act as a deterrent to predatory birds. Opaque covers provide shade, which reduces algal fouling and may encourage 'shy' fish to feed. However, the value of this cover does not merit the added expense and increase in labour requirement at feeding. In certain cases a lid is essential where it is to be used as a lockable anti-theft precaution to protect valuable broodstocks. Generally overhead cover or protection is not necessary but in certain cases where large predators are present (e.g. otters), cages may be surrounded by a secondary stronger net (Coche, 1976) though more drastic measures are often more profitable, i.e. either predator removal or a relocation of cages.

5.2 Feeding

The value of feeding rings or trays to prevent feed losses is somewhat controversial. Where the tray extends the length of the cage, bottom food dispersion would be reduced but faecal wastes would accumulate; partial coverage on the other hand would only benefit the more aggressive feeders (Coche, 1976). Trays could also act as a surface for successful breeding (Ibrahim *et al.*, 1976). Reducing the size of tray also necessitates the use of feeding tubes for food administration (Konikoff, 1975) which could be both problematic, as they are prone to clogging, as well as time-consuming when feeding a commercial scale operation. This time would be better spent in more numerous applications of smaller amounts of feed as a remedy to reducing feed losses, through the cage sides. Feeding rings, however, may be used to better advantage during times of strong water currents and rough weather (Coche, 1976).

    As feed costs make up the greater portion of the running costs of a cage culture operation, further research into the prevention of food losses would be profitable. Feed trays and regular feeding may be suitable for small scale operations, whereas larger enterprises might find it more economical to use automatic feeders such as the demand feeding device used with some success in Lake Kossou, Nigeria (Coche, 1977) and the feeder triggered by wave action which was used by Peacock (pers. comm.) in experiments in Saudi Arabia. An important consideration here is that tilapia are not well adapted to triggering demand feeders, therefore such feeders must be reliant on mechanical or electrical trigger mechanisms.

5.3 Stocking Rate and Productivity

Site location is very important in determining the carrying capacity of a cage as the water quality is entirely dependent on exogenous conditions. Water currents, induced by the flow in a canal, or set up by wind action, or fish stock movement in a lake or pond, are essential to provide the fish's oxygen requirements and to remove metabolites. Thus during site selection, currents can be measured with the aid of a drogue or dye and the maximum stocking density calculated from the following equation:

$$w = \frac{A \times F \times (DO - 3.0)}{1000r}$$

where: w = final biomass (kg), A = cross-sectional area of submerged part of the cage (cm^2), F = water current (cm/min), DO = dissolved oxygen level (mg/1), 3.0 = lower lethal oxygen tolerance limit for tilapia and r = respiration rate (mg/kg/min).

As mentioned in Table 5.2, tilapia are tolerant of very low oxygen levels and although reported to survive in cages at 0.7 mg/1 in Lake Kossou, Nigeria, prolonged low oxygen conditions resulted in lower productivity (Coche, 1976) and 3.0 mg/1 is chosen as the critical level below which this loss will occur.

5.4 Carrying Capacity

The results of cage trials summarised in Table 5.14 are for differing conditions and although not comparable, indicate that of the reports available, a carrying capacity of up to 94 kg/m^3 is possible. This would appear to be considerably lower than values available for the cage culture of other fish species (Table 5.17) and as productivity is still high, could indicate that a maximum stocking density has not yet been achieved.

The growth rate of caged fish decreases with increasing stock biomass but production increases up to an initial stocking density of 36 to 50 kg/m^3 (Coche 1976). The apparent drop in productivity above this range could suggest some form of social interaction, due to crowding, as affecting growth. However, as results apply to Lake Kossou, it could also indicate that under the prevailing conditions, this is the maximum biomass that can be supported, i.e. oxygen availability or some other water quality parameter may have been affecting growth. A similarly high productivity has also been achieved in cages held in ponds (Pagan, 1970; Suwanasart, 1971; Table 5.15). As water quality is unlikely to be as good, it appears that it may well be crowding and not water quality that affects growth. If this were the case the problem would need to be identified and new husbandry techniques evaluated, e.g. selective cropping. Of particular significance here is the reported stunting and resultant deformities of cage stocks in Lake Calibato, Philippines (Anon., 1978). Overstocking of the lake due to an excessive number of cages is offered as an explanation, but the author does not specify that it is water quality that is to blame. The report does, however, illustrate that even a 40 ha lake does have limitations.

Stocking Rates. From the reports available (Table 5.15 and 5.16) the following stocking programme can be formulated: fingerlings (20 to 40 g) stocked at an initial biomass of 20 kg/m^3 (500 to 1000/m^3) would reach a harvestable size (> 200 g) in 120 to 150 days at a feeding rate of from 4 to 2.5 per cent body weight per day of a 25 per cent protein diet with an expected FCR of between 2 to 2.5. Two to three harvests a year would result in an annual production of 200 to 300 kg/m^3 (2000-3000 t/ha/yr at 1 m depth). This represents one of the highest levels of productivity of any cultured species but requires a considerable food input, is therefore costly and not likely to be viable on a large scale unless aimed at a select market. A practice more likely to favour the use of tilapia in cages is the exploitation of sewage ponds. With no feeding necessary, tilapia yields of between 25-505 t/ha/yr have been achieved from these ponds in Tennessee (Suffern, *et al.*, 1978). Although consumer reaction to the origins of the fish may present a problem in the Western world, this practice is likely to become a future trend in the use of tilapia cage culture.

6. Nutrition and Feeding in Intensive Culture

Intensive cultivation practices, by nature, require artificial feeding. When formulating a diet, the natural food preference of a species is an important consideration when deciding on the choice of raw materials, i.e. the proportion of animal:vegetable component, especially with reference to the protein type. Similarly, dentition will determine the mastication power of the pharyngeals or straining ability of the gill-rakers. Appetence or palatability are also a function of the species' feeding habits.

A well-balanced ration is of fundamental importance in intensive tilapia culture, as the fish are solely reliant on this diet to derive all essential elements necessary for healthy development and growth, and dietary imbalance can mean a loss in production, or mortality. Little is known, however, of the requirements of tilapia, though what information is available is discussed below with reference to establishing a commercial tilapia diet.

6.1 Food Composition

Much of the work on diets has been to determine the optimum protein levels required for growth.

Protein Component (Animal:Plant). Goldstein (1970), Gruber (1960) and Mathavan *et al.*, (1976) are of the opinion that for optimum growth in a herbivorous species such as *T. rendalli* and *S. mossambicus*, some form of animal protein (approximately ten per cent) is essential in the diet to meet the fish's metabolic requirements. The animal protein improves digestion and absorption efficiency from 56 to 73 per cent in a purely vegetable diet and to 94 per cent using a higher animal protein. Davis and Stickney (1978), in a series of diet trials, have arrived at an optimum value of 33 per cent animal protein for *S. aureus* feeds, diets of 100 per cent plant protein yielding the lowest growth rate. However, Fijan (1969) and Mironova (1975) both worked on *S. mossambicus*, and although in agreement that some animal protein is essential, showed that growth increased relative to the size of the algal component of the mix. Other results are similarly confusing. Hastings (1973) suggested that for *S. niloticus* fed on a 30 per cent protein diet, the animal protein may not be essential in intensive pond culture while Sitasit and Sitasit (1977) obtained the reverse. *S. niloticus* grew better in ponds fed entirely on a pellet diet made up

mainly of an animal protein component. Miller (1976) also working on
S. niloticus, obtained a higher production with a ration of high animal:plant
protein, but the Food Conversion Efficiency (FCE) was poorer. Cruz and
Laudencia (1976) also favour a high animal protein component (about 50 per
cent) for feeding juvenile (3-4 cm) *S. mossambicus*.

Though in need of further refinement, results would appear to suggest that
tilapia require a certain minimum level of animal protein for maximisation of
growth. This would raise the cost of production and might prohibit the use of
commercially prepared diets in less intensive systems. However, the successful
use of algae in feeds might reduce the costs without loss in growth performance.
The technique is, however, currently under review and is hoped to hold promise
for the formulation of an economical diet.

Per Cent Protein Requirement. In recent years numerous trials have been
conducted to obtain a better understanding of the protein requirements of
various size groups of tilapia. These have been summarised in Table 5.18.
Although different species were considered, the following generalisations can
be made:

below 1 g	require 35-50 per cent protein
1-5 g	require 30-40 per cent protein
5-25 g	require 25-30 per cent protein
above 25 g	require 20-25 per cent protein

Table 5.18: Protein Requirements of Tilapia for Maximisation of Growth of
Various Size Groups

Size Group (g)	Range	Species	Optimum % in Diet	Protein Range	Reference
<0.5		*S. mossambicus*	35-50		Ross & Muir (unpub.)
0.3-0.5		*S. aureus*	>36		Davis & Stickney (1978)
	<1			35-50	
0.3-0.8		*S. niloticus*	35-40		Cruz & Laudencia (1976)
<1.0		*S. mossambicus*	45-50		Jackson (pers. comm.)
1.65		*T. zillii*	30-35		Mazid *et al.* (1979)
1.0-3.0	1-5	*S. mossambicus*	29-38	30-38	Cruz & Laudencia (1976)
3-5		*T. zillii*	32-40		Hauser (1975)
1-10		*S. mossambicus*	37-42		Jauncey (in preparation)
9.0		*S. niloticus*	25-30		Cruz & Laudencia (1976)
14.5		*S. hornorum* x	25-30		Hughes (1977)
	5-25	*S. niloticus*		25-30	
21.0		*S. aureus*	26-36		Hughes (1977)
10-30		*S. mossambicus*	25-30		Ross (unpub.)
25-40		*S. mossambicus*	20-25		IFP (1976c)
50-70	>25	*S. niloticus*	20-25	20-25	IFP (1976c)
50-300		Tilapia	23		Marek (1975)

For the fattening of tilapia in ponds a diet of 20-25 per cent protein is required, but here natural productivity, though small, may offer a contribution to the overall food requirements. In more intensive tank culture systems, therefore, there is often a preference for use of a slightly higher protein component of between 25 to 35 per cent (Haller, 1974; Hughes, 1977; Melard and Philippart, 1980; Otte and Rosenthal, 1979). Fish feeding on such high protein diets often have a high mortality rate and poor FCE. Otte and Rosenthal (1979) attribute these losses to the partial destruction of liver tissue through fat degeneration, a result of increased fat deposition due to the high protein content. Viola and Amidan (1980) by feeding five per cent oil-coated pellets showed that almost one-third of the whole body fats were accumulated around the viscera (including the liver) whereas in carp this amounted to only one-sixth. It would appear, therefore, that tilapia use the visceral region as a fat reserve. Excessive deposits may be stressful and such fish more susceptible to secondary infections or death from liver malfunctions.

   It is of interest to note that diets in excess of 25 per cent lipid content (Lagler *et al.*, 1962) and fish of a body fat content of up to 44 per cent (Pandian and Ragharaman, 1972) and 54 per cent (Hauser, 1975) have been reported with no visible adverse effects. That these were all laboratory studies could suggest that it is only under the stresses of intensive culture (overcrowding, poor water quality, exposure to disease, etc.) that fish become weakened and die from excessive fat deposits. This emphasises not only a need for a balanced ration but that if cultured stocks are in a precarious balance with their environment, any upset, even dietary, can lead to mortality.

   Essential amino acids for *T. zillii* have recently been shown to be the same as those required for normal growth in other fish, e.g. carp, trout, etc. (Mazid *et al.*, 1978). The relative importance of each is indicated in Table 5.19.

Other Diet Components. Little to nothing is known of the fat and carbohydrate requirements of tilapia. Farinaceous matter ('mealy') used to bulk out diets often contains indigestible materials. In ponds these food items or their excreted residues act as a fertiliser (Chimits, 1955; Huet, 1972). However in tanks the increase in residue leads to fouling and deteriorating water conditions (Hauser, 1975; Iversen, 1976; Uchida and King, 1962). Digestibility of various raw materials is therefore of prime importance. Furthermore, the feeding of farinaceous feeds to *S. esculentus* in aquaria (Cridland, 1960) resulted in symptoms characteristic of tryptophane deficiency (Mertz, 1972). Pond fish reared on a similar diet showed no such abnormalities as they were able to derive essentials (in this case probably tryptophane) from the pond's natural productivity.

   Vitamin deficiency was believed to be the cause of the poor FCE of tilapia fed on a catfish diet (Otte and Rosenthal, 1979) but no qualitative information was given. Similarly an absence of trace elements such as Mn (Ishac and Dollar, 1968) and Ca, P (Scott, 1977) have been suggested as causing deformities and poor growth in tank fish, but no further information is known. Similarly, the eye cataracts observed in *S. mossambicus* fed in aquaria on a maize-based diet were believed to be due to a zinc deficiency and could be prevented by the addition of a vitamin and trace element supplement (Jauncey, personal communication). Excess calcium in the diet is also believed to give rise to nephrocalcinosis under certain conditions. Careful consideration must therefore be given to include vitamins and trace elements.

Table 5.19: Specific Growth Rate (%/day) of *T. zillii* Fingerlings (0.29-1.93 g) Fed Various Deficient Diets

Deficient Amino Acid	SGR	% Mortality
Alanine	0.55	
Arginine	-0.59^a	10
Aspartic acid	0.68	
Cystine	0.53	
Glutamic acid	0.65	
Glycine	0.44	
Histidine	-0.41^a	5
Isoleucine	-0.39^a	20
Leucine	-0.22^a	25
Lysine	-0.16^a	15
Methionine	-0.06^a	
Phenylalanine	-0.16^a	
Proline	0.65	
Serine	0.52	
Threonine	-0.10^a	5
Tryptophane	-0.17^a	5
Tyrosine	0.37	
Valine	-0.09^a	
Basal Control	0.87	
Casein Only	2.28-2.33	

Note: a. Essentials.
Source: Mazid *et al.* (1978).

6.2 Diet Formulation and Pellet Size

Numerous diets, pelletised or in other forms have been tested for tilapia culture. Balarin and Haller have listed those feeds used in semi-intensive systems and a detailed evaluation of some of the more common feeds is given in Table 5.20, after Miller (1976). There is as yet no universally accepted, standard tilapia diet. Different areas will tend to select the most economical diet based on locally available raw materials.

Fish meal and soybean meal make up the greater portion of the protein complement and have as yet failed to be replaced by cheaper substitutes such as rice, bran, copra meal or sorghum (Cruz and Laudencia, 1978), though the use of coffee pulp (Bayne *et al.*, 1976) and brewery wastes (Hastings, 1973) can be more economical. Intensive feeding in most countries, therefore, makes use of locally available industrial wastes, e.g. rum distillary waste (Puerto Rico), cotton seed cake (USA), rice husks (Far East), mill sweeping (Africa) and coffee pulp (El Salvador). Often, however, commercially available diets such as catfish pellets (32 per cent protein) (Legner, 1978), trout pellets (35 per cent protein), chicken broiler pellets (18 per cent protein) (Balarin and Haller, 1979; Kohler

Table 5.20: Per Cent Composition of Various Tilapia Diets Used Experimentally (E) and Commercially (C)

Size Group (g)	>25 C	>50 C	>25 E	>20 C	>20 C	7-40 E	3-5 E	2-6 E	0.3-12 E	>15 C	>20 C	>25 E
Country	Kenya	Israel	El Salvador	Central Afr. Rep.	Central Afr. Rep.	Stirling	Philip-pines	Philip-pines	Texas	El Salvador	Togo	Central Afr. Rep.
Fishmeal	10.00	10.00			5.00	15.00	12.50	36.60	17.90			
Meat/blood meal	10.00				5.00					2.50		5.00
Bone meal	1.00		1.00		2.00							2.00
Soybean meal	15.00	23.00				15.00	22.00		49.30			
Wheat bran	10.00	62.00	10.00	15.00	20.00					10.00		
Rice bran	20.00			15.00	15.00		51.50	63.40				
Maize screening	13.00										10	
Maize germ meal	10.00	5.00	24.00			70.00			10.70	2.50	15 spoils	15.00
Lucerne meal	5.00											
Molasses	5.00		20.00									
Coffee pulp			30.00							5.00		
Cottonseed meal/cake			14.00	45.00	30.00					32.50	30.00	25.00
Sesame cake				7.75	7.75							25.00
Brewery wastes				15.00	15.00						30.00	
Copra meal							10.00					
Ipil-ipil leaf							3.00					
Corn oil									5.00			
Shark liver oil										2.50		
Cellulose									4.10			

Table 5.20: continued

Size Group (g)	>25	>50	>25	>20	>20	7-40	3-5	2-6	0.3-12	>15	>20	>25
	C	C	E	C	C	E	E	E	E	C	C	E
Country	Kenya	Israel	El Salvador	Central Afr. Rep.	Central Afr. Rep.	Stirling	Philippines	Philippines	Texas	El Salvador	Togo	Central Afr. Rep.
Spoiled powder milk											15.00	
Peanut oil cake												28.00
Chicken manure									7.10	45.00		
Mineral mix						2.00						
Starch binder			1.00									
Urea												
Vitamin Supplement	1.00			0.25	0.25	2.00	1.00	0.50	6.00			
Energy (kcal/kg)		2682							4192	1166		
% protein	22	25	21	30	30	22	30	34	36	23	23	30
% fat	5.50					1.40		3.00				
Cost est. (£/kg/fish)	0.21-0.35		0.07	0.07	0.13			0.15		0	0.04-0.06	0.10
% feeding rate	2.00-2.50	2.0-3.0				2.0-4.0	1.50	2.50	10.00	3.00	3.00-4.00	4.00
Food Conversion Ratio (FCR)	2.30-2.60	2:1	2.30-3.70	1.20	1.60	1.5-2.0 / 1.2-1.8	1.7-2.5	1.24	1.43	1.40-1.70	4.70-6.70	1.42
Growth rate (g/day)	1.14	1.0-3.0		0.85	0.65			0.17	0.132	0.40-0.68	1.40-4.70	0.58
Production (t/ha/yr)	600[t]	3.0-25.0[p]	4.00-12.50[p]	5.10	6.20					3.70-7.80[p]		6.20[p]
Source	Haller	Marek	Bayne *et al.*	Hastings	Hastings	Ross and Muir	IFP	Cruz and Laudencia	Davis and Stickney	Davis and Hughes	Miller	Miller
	(1974)	(1975)	(1976)	(1973)	(1973)	(unpub.)	(1976d)	(1978)	(1978)	(1977)	(1976)	(1979)

Notes: p = pond; t = tanks.

and Pagan-Font, 1978) and rabbit pellets (16 per cent protein) (Uchida and King, 1962; Hauser, 1975) have been fed with marginally better success than the formulated tilapia pellet, though the rabbit pellet is not favoured because of residual (undigestible) material fouling the tanks.

The question of pelletising as opposed to powdered feeds is of little significance in pond feeding (Allison *et al.*, 1976; Miller, 1979) though it is felt (Sherry, personal communication) that pelleted feeds sink faster and can be lost to the bottom muds before tilapia have time to feed. Feeding smaller amounts of pelleted feeds would therefore be essential in pond systems. However pellet buoyancy can be controlled by the use of steam injection during manufacture. Although in small tanks (<2.5 m^3) pelleting has little effect on improvement of FCR (Balarin, unpublished) the disturbance when feeding commercial size units creates considerable wastage if in a powdered form (Balarin and Haller, unpublished). Floating feed-rings can reduce this loss but the benefits to be gained from pelleted feeds far outweigh the added expense of pelletising. Powdered diets (Bogdanova, 1970) or those of particle size less than 0.5 mm are therefore restricted in use to first feeding of 7-11 mm fry (3-7 days old) which can be gradually weaned on to a granular feed. A 3 mm pellet size is preferred for commercial stocks. Further pellet size increments, as used in trout feeds, is not felt necessary as tilapia do not have strong pharyngeal teeth and probably cannot chew larger pellets.

6.3 Feeding Rates

In subsistence and semi-intensive practices, tilapia are fed mainly farinaceous foods or fodder (i.e. leaves or grass). With leaves it is possible to determine the amount to feed by the quantity remaining. However, the bulk required makes this food type impracticable for intensive systems. With farinaceous feeds it is more difficult to determine what percentage is consumed and how much goes towards pond fertilisation. Pelleted feeds are more suited to intensification as they are more consolidated and therefore are readily consumed and easier to dispense from automatic feeders (Sarig and Marek, 1974).

The size of fish together with the particular tilapia species, protein component and quality, plus physical consistency of the feed, determine the intake and utilisation of a particular diet. For every size group, therefore, there will be an optimum feeding rate dependent on food type. To date there are a number of tilapia feeding formulae, equivalent to those in use for salmonids. These can be expressed as:

(a) *Pond fish* — fed a 25 per cent protein pellet (Sarig and Marek, 1974):
$$\log y = 1.343 - 0.459 \log x \quad (b = -0.429 \pm 0.0422)$$

(b) *Tank fish* — (i) fed a 35-40 per cent protein pellet (Balarin and Haller, unpublished):
$$\log y = 1.309 - 0.457 \log x \quad (b = -0.457 \pm 0.0388)$$

(ii) fed a 23-26 per cent protein pellet (Melard and Philippart, 1980):
$$\log y = 1.687 - 0.580 \log x \quad (b = -0.58 \pm 0.071)$$
where y = % body weight fed daily and x = $\bar{x}$ body weight (g)

(c) *Cage fish* — by extrapolation from natural feeds (Moriarty and Moriarty, 1973):

y = 271 – 13.3x

where y = dry weight of food (mg/day) and x = x̄ body weight (g)

Figure 5.11: The Variation of Daily Feeding Rate with Body Weight of Tilapia Fed Different Protein Rations.

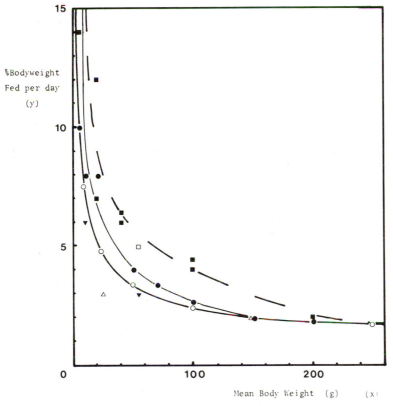

o   35-40% protein in tanks at 25-28°C (Balarin and Haller,1979)
     (Log y = 1.309 - 0.457 Log x ; b= -0.457 ±0.039)
•   25% protein in ponds at 20-28°C (Sarig and Marek, 1974)
     (Log y = 1.343 - 0.429 Log x ; b= -0.429 ±0.042)
■   23-26% protein in tanks at 27-31°C (Melard and Philippart 1980)
     (Log y = 1.687 - 0.580 Log x ; b= -0.580 ±0.071)
▼   46% protein in ponds (Swingle, 1967)
□   25% protein in cages at 27-30°C (Coche, 1976)
Δ   36% protein in tanks at 24-26°C (Lauenstein, 1978)

Feeding rates therefore vary with the protein content of the diet and, as can be seen from Figure 5.11, this may not simply be in relation to the amount of protein required. Accepting the above formulae, smaller fish have a higher protein requirement therefore one would expect a proportional change in required feeding rates. However, that the two equations (b)(i) and (b)(ii) above do not show this, suggests that either the protein in diet (b)(ii) is not completely digestible by smaller fish, hence proportionately more food is required or could be attributed to the temperature difference between the two systems, or both. Temperature alone can account for a 0.45 per cent difference in feed intake per 1°C, e.g. a 30 g tilapia has a maximum feed intake of 9 per cent at 30°C, 6.7 per cent at 25°C and 4.6 per cent at 20°C (Ross, 1981). These first two feeding rates have been adopted for the most efficient food conversion (i.e. 2:1). However, maximisation of growth often occurs at a slightly higher rate of feeding (Pandian and Ragharaman, 1972; Shell, 1967). That growth does not increase as the amount of food increases may be a result of the excess food not being properly digested while food absorbed would have a greater metabolic demand (SDA) requiring energy for protein deamination at the expense of growth. Excessive feeding is not only costly and inhibiting to growth but also increases fouling of tanks. Overfeeding may deplete oxygen levels and can clog gills, severely stressing the fish and reducing the availability of oxygen, and fish may die under deoxygenated conditions. Required feeding rates have also been shown to vary with species (Caulton, 1979) e.g.:

wt (g)	50	100	250	500
S. niloticus	1.8%	1.6%	1.4%	
T. rendalli		2.4%	1.5%	1.4%

Feeding Frequency.   Tilapia have no true stomach but rather a small intestinal bulb and it was felt by Shell (1967) and Coche (1976) that in view of this and of the slow digestion rate of the species (e.g. 20-30 hours for complete evacuation, Mishrigi and Kubo (1978) and 18-24 hours, Ross and Jauncey (1981)), it might be advisable to divide the daily ration into several portions. In Kenya, feeding tank fish three to six times daily not only improved food consumption but also water quality (Balarin and Haller, 1979). Further, in view of the diurnal feeding habits of the species (Maruyama, 1958; Harbott, 1975; Man and Hodgkiss, 1977) it may suggest that for maximum digestion efficiency fish should be fed as close to this cycle as is practicable. Moriarty and Moriarty (1973) for *S. niloticus* and Caulton (1976) for *T. rendalli* have shown that the fish when fed on natural foods is only in a 'physiologically prepared' state to digest this food efficiently some six to eight hours after first feeding. Excess food, fed before this state is achieved, would be only partially digested. Ross and Jauncey (1981) have, however, shown that larger fish can digest a meal faster than smaller fish and that there is negative correlation between stomach evacuation time and temperature. Therefore fish size and temperature would also affect the feeding frequency. Whether this applies to artificial feeds or specific diet components (i.e. of high fibre content) remains to be examined.

The development of a pelleted feed is of prime importance to the future of intensive tilapia production. This does, however, present problems. Careful storage is essential for in humid tropical conditions, characteristic of regions suited to tilapia culture, the shelf-life of the feeds will be rapidly reduced. Mouldiness or protein denaturation often leads to nutritional disorders. Well-ventilated stores are needed and quantities stored must be reduced to a minimum. This increases the cost of feed transportation as often farms are some distance from the commercial feed manufacturer.

7. Economic Evaluation and Future Prospects

Subsistence tilapia farming has been ranked alongside milkfish culture, as having one of the lowest energy requirements for protein production (inputs are often less per unit than protein produced) than that required of a low intensity inshore pelagic fishery (Edwardson, 1976). For this reason, tilapia have been recommended as suitable for culture in those parts of the world where there is a high level of malnutrition, low technological expertise and an economy too poor to develop an intensive fishery, or where the existing fishery is threatened or too small to supply the area's needs. Tilapia were, therefore, ranked alongside mullet, oysters and milkfish as having the highest potential for application to the developing world (Bell and Canterbury, 1976).

Tilapia are better food converters than broilers, pigs, sheep or cattle and have a proportionately larger dress-out weight. In terms of edible flesh production per unit area, tilapia are more productive than most farm animals at the same level of intensification (Haller, 1974; Payne, 1975). However, intensification requires a greater technology and a definite infrastructure to be successful. As indicated in this chapter, there is considerable documented information on tilapia biology and technology of culture, yet despite this there has not as yet been established a viable, thriving fish culture industry in the underdeveloped countries. The current practices are mainly of an extensive to semi-extensive nature, mainly due to poor dissemination of information, with little financial support. Early disillusionments, through the lack of knowledge of reproduction control resulting in the production of small unmarketable fish, has caused many a potential investor to become wary of tilapia culture. Further, tilapia are now common in almost all tropical freshwater fisheries, and competition therefore exists between farm-reared and wild-caught fish, often for a restricted market. Skin colour, and general consumer preference for a more traditional fish species has resulted in lower market prices, especially in the Far East, though in certain parts of the world tilapia commands a better price (see Table 5.21). Such can be the demand that tilapia is considered a luxury food item to the extent that fish are imported to satisfy the market, e.g. Japan (Fitzgerald, 1979) and Britain.

At present, integrated fry production and intensive ongrowing operations require a great deal of attention and capital if they are to be viable in those areas where there exists a high market price. However, capital investments and operational costs of those systems (semi-intensive and intensive) considered in Table 5.22 are comparably lower than, for example, trout rearing systems in the UK. (This may not be a valid comparison for other than the Kenyan tank system, as the other costings are mainly from family or communal run operations so that labour costs are minimal and seed supply is often free or subsidised by government assistance.) Food and/or fertilisers therefore make up the single most expensive component in intensive pond and tank culture, the incidence of cost increasing with intensification (Table 5.23). Rice paddy stocking is characterised by a high labour cost, but this is mainly due to rice being a labour-intensive practice. Cages have a short working life, and therefore require a great deal of attention, raising running costs although if used as a family unit, they can be highly profitable (McLarney, 1978).

Table 5.21: World Market Prices Per Live Weight of Tilapia

Country	Price (£/kg)	Year	Market	Source
Columbia	0.18	1975	Local sales	McLarney (1978)
El Salvador	0.4-0.55	1976	Local sales	Hughes (1977)
Brazil	0.32	1976	Local sales	Lovshin (1977)
Philippines	0.40	1978	Local sales	Anon. (1979)
Taiwan	0.15	1975	Local sales	Chen (1977)
	0.5-0.8	1977	Export to Japan	Fitzgerald (1979)
US — Alabama	0.44	1977	Local sales	Crawford *et al.* (1978)
	0.38-1.10	1977	Wholesale to supermarkets	
Israel	1.00-1.25	1978	Local sales	Anon. (1979)
Central African Republic	0.42	(1975)	Local sales	Miller (1976)
Kenya	0.77	1979	Local sales	Balarin & Haller
	1.29	1979	Wholesale to restaurants	(1979)
Nigeria	2.00-2.50	1978	Local sales	Roberts (personal communication)
South Africa	0.57	1979	Wholesale price	Caulton (1979)
Saudi Arabia	2.00-2.20	1979	Minimum fish price	Peacock (personal communication)
United Kingdom	2.00-3.00	1979	Wholesale to restaurants	Muir (personal communication)
Belgium	50-300 (BF/kg)	1979	Marketed in restaurants	Melard & Philippart (1980)

Intensification therefore increases the rate of returns per unit area although not necessarily per ton of fish produced (Table 5.22). However, this would depend on the site location, experience of operator and availability of cheap feeds. One particular option is combined fish and animal husbandry, in which, with necessary experience, a zero-cost food source is available, in the form of fresh manure. The profit index of such operations are given in Table 5.24 from which it can be seen that tilapia production offers a considerable return on production costs, representing a saving to normal farm operations.

From Table 5.22 it can therefore be seen that, apart from cage culture, the economics of intensive systems have three main sensitive areas: seed stock, food and labour costs. Each of these vary with local conditions and are therefore of importance when assessing the feasibility of introducing tilapia culture into a particular area:

Seed stock — Wild-caught seed either from lakes or ponds offers the cheapest means of stocking a system but as the farmer has no control over reproduction it is practically impossible to maintain a six month production cycle. Hormone treatment or hybrid crossing further raises the seed costs. It becomes therefore necessary for the farmer to become self-sufficient in producing his own fry using such hatchery systems as described in the section on intensive fry production.

Food — Tilapia feeds which include fish meal are expensive and compete with other livestock feeds. The limited supply of fish meal and seasonal fluctuations in availability mean that a farmer cannot expect to budget on a fixed feed price and is not always certain of its supply, especially in the developing world. Alternative sources of protein are constantly being sought and it is often the case that the cheapest locally available raw materials are used in feeds. A balance is therefore sought between the cheaper feed costs and the loss in production through using an inferior diet and the concomitant increase in running costs.

Labour — Automation can reduce labour costs, but in the developing world, labour is relatively inexpensive when compared with the capital cost of importing or constructing as well as the relatively short amortisation period, resulting from the harsh tropical climate. It can, however, be argued that cheap unskilled labour can be unreliable and in view of the careful attention necessary for the successful running of a system, losses through negligence or mismanagement could favour automation wherever possible.

The emergence of more economical forms of production of tilapia in developing countries opens up the prospect of export, where local prices in the developed countries do not favour intensification. Thus, by exploiting the naturally favourable conditions within the tropics, making use of the relatively cheaper labour, fish can be produced economically for an export market. This practice has led to an upsurge in tilapia production in Taiwan and other Far East countries, mainly for export to Japan. The benefits in terms of generating trade and creating employment make this practice worthy of consideration at government levels. The absence of an economical and commercially available diet as well as problems of obtaining a regular supply of fingerlings, still hampers such developments in Africa (Balarin and Hatton, 1979).

Tilapia are thus suitable as a food fish, to improve protein nutrition, only when at a low level of intensification, e.g. family/communal pond culture. To be successful this requires government assistance, both financial as well as through back up fry production units and distribution centres with an efficient extension service. More intensive commercial scale operations can only operate where tilapia as a food fish are locally high priced, where an export market exists or there is a market for a particular size group. In Kenya, for example, tank culture can yield up to 400 per cent returns on capital if put to fry rearing for resale (Balarin and Haller, 1979). Similarly intensive rearing of tilapia as a bait fish (Gopalakrishnan, 1979), an agent of aquatic weed control (Ray, 1978a) or as a fishmeal concentrate (Anderson *et al.*, 1978; Apandi *et al.*, 1974; Rowland *et al.*, 1977) have proven to be highly profitable.

Table 5.22: A Comparison of the Proportion of Cost and Benefits from Various Levels of Intensification of Tilapia Production Units

	Extensive		Semi-intensive										Intensive		Trout Farms	
System Type	Subsistence	Rice-cum fish	Fertiliser + Predator	Fertiliser	Fertiliser + Predator	Fertiliser	Animal-Fertiliser	Fertiliser + Feed	Fertiliser + Feed	Fertiliser + Feed	Fertiliser + Feed	Fertiliser + Feed	Cages	Tanks	Earth Ponds	Tanks
Country	Cent. Afr. Rep.	Philippines	El Salvador	El Salvador	El Salvador	S. Africa	Brazil	Brazil	Brazil	Thailand	Sudan	Sudan	Columbia	Kenya	UK	UK
Productivity (t/ha/yr)	2.2	0.6	2.66	3.00	3.86	6.4	4.69	4.26	3.81	6.3	10.0	10.0	250.0	695.0		300.0
Capital (£/t)	418.41	991.2	385.45	1051.36	268.3	218.84	46.67	89.76	267.35	253.78	2625	3675	23.50	1179.06	434.0	1284.0
% Variable costs:																
Seed stock	98.7	11.7	24.7	12.9	13.9				5.9					27.8	2.0	2.0
Food		1.6				0.6		25.0	48.8		21.8	18.0	10.8	31.6	36.0	29.0
Fertiliser		6.3	49.4	37.1	28.9	22.2	Combined husbandry	20.4	8.0		7.6	9.0				
Labour		51.3	12.8	23.3	28.1	9.9	33.3	18.2	4.0		34.9	28.7		8.4	28.0	15.0
Fuel/power		0.3				9.7			6.2			4.5		12.3		14.0
Marketing		0.4							1.3		5.4					3.0
Miscellaneous		8.5		2.7					4.9				19.5		4.0	
Total variable (£/t)	35.23	314.2	58.13	73.1	36.85	68.47	5.44	22.48	144.61	49.65	1920	4020	9.14	885.94	74.0	63.0
%	98.7	80.1	86.84	76.1	70.93	42.43	33.3	63.6	79.1	90.7	69.7	60.2	30.3	80.1		

Table 5.22: continued

	Extensive		Semi-intensive										Intensive		Trout Farms	
System Type	Subsistence	Rice-cum fish	Fertiliser + Predator	Fertiliser	Fertiliser + Predator	Fertiliser	Animal-Fertiliser	Fertiliser + Feed	Fertiliser + Feed	Fertiliser + Feed	Fertiliser + Feed	Fertiliser + Feed	Cages	Tanks	Earth Ponds	Tanks
Country	Cent. Afr. Rep.	Philippines	El Salvador	El Salvador	El Salvador	S. Africa	Brazil	Brazil	Brazil	Thailand	Sudan	Sudan	Columbia	Kenya	UK	UK
% Fixed costs: Regular labour		0.3				44.01	47.6	27.3	1.1			22.4		11.2	6.0	5.0
Maintenance			5.0	3.80			9.5	9.1	3.5		5.8	3.3	54.1	0.3	3.0	1.0
Depreciation – structure	1.3	0.4			24.7	13.56			15.1		6.7	5.0	15.6	5.4	4.0	9.0
– equipment		2.7	4.81	15.2	4.3		9.6		1.1		9.2	4.2		2.9		
Interest/rents		16.7	3.3	4.93							8.6	4.9			13.0	22.0
Total fixed (£/t)	0.45	77.8	8.80	22.80	15.10	92.91	10.89	12.85	38.14	5.08	833	2562	21.02	220.0		
%	1.3	19.9	13.1	23.9	29.1	57.6	66.7	36.4	20.9	9.3	30.3	39.8	69.7	19.9	26.0	37.0
Σ operational (£/t)	35.68	392.02	66.90	95.93	51.96	161.37	16.33	35.33	182.75	54.73	620.3	668.2	30.16	1106.83	820.0	1031.0
Profits – (£/ha/yr)	351.82	427.13	232.50	448.54	203.29	2092.22	629.64	1024.39	346.33		1994	818	26 405.33	152 159.44	580	110 700
– (£/t/yr)	159.92	711.88	87.41	149.51	52.67	326.91	134.25	240.47	90.90		199.4	81.8	105.62	218.93	133.6	369.0
% Return on investment	84.1	31.9	22.0	42.7	19.6	149.4	287.7	195.6	34.0		37.9	22.3	449.5	18.6		28.7
Source of reference	de Kimpe (1978)	Sevilleja and McCoy (1979)	McCoy (1974)	Parkman and McCoy (1977)	Bayne (1974)	Caulton (1979)	Lovshin (1977)	Lovshin (1977)	Jensen (1976)	Kloxe and Potaros (1975)	Bassa (1979)	Bassa (1979)	McLarney (1978)	Balarin and Haller (1979)	Varley (1977)	Varley (1977)

Table 5.23: A Comparison of Various Fertilisers and Feeds Including Possible Fish Productions, Incidence of Cost/kg of Fish Produced and Profit Index

Fertiliser or Feed	Application Rate (kg/ha) or Feeding Rate (% body weight)	Frequency of Application	Expected Production (kg/ha/yr)	Incidence of Cost/kg of Fish (£ sterling)	Profit Index
Inorganic fertilisers					
Triple superphosphate	25	monthly	1,400	0.04	10.98
Triple superphosphate	60	monthly	1,600	0.08	5.20
Double superphosphate	56	monthly	1,100	0.39	1.03
Lime	1,680	initially			
	280	monthly			
Organic fertilisers					
Chicken manure	8,000	monthly	2,300	0.07	6.06
					$\bar{x} = 5.82$
Feeds					
Rice Bran	10%	daily	300	0.09	4.40
Corn meal	10%	daily	350	0.15	2.59
Cassava flour	10%	daily	325	0.16	2.50
Cooked corn	10%	daily	430	0.11	3.46
Cooked cassava	10%	daily	300	0.15	2.72
Ration A — vegetable } 30% protein	4%	daily	5,000	0.07	5.30
Ration B — animal	4%	daily	6,000	0.13	3.18
Beer waste + spoiled feeds of 20% protein	33%	daily	5,000	0.10	3.93
Ground cottonseed (coarse)	10-15%	daily	4,200	0.08	4.98
					$\bar{x} = 3.67$
Feed and Fertiliser					
Beer waste and chicken manure	33%	monthly	4,300	0.10	4.07
	500%	monthly			

Profit index = $\dfrac{\text{Value of fish crop}}{\text{Cost of fertiliser or feeding}}$

Incidence costs = $\dfrac{\text{Cost of fertiliser or feed}}{\text{Kg of fish produced}}$

Table 5.24: Comparison of Various Animal-cum-Tilapia Culture Practices Given Expected Production and Profit Index

Description	Food Type (Animal Feed)	Fish Production (t/ha/yr)	Animal Production (t/ha/yr)	Profit Index Food	Σ Cost	% Tilapia Contribution
Ducks and Fish						
1. 100/ha ducks; *S. niloticus* and *Clarias*	Commercial duck feed	3.8-4.5	5.8-8.0	3.22	2.17	62.5
2. 750/ha ducks; *S. niloticus* and Carp	Duck food	1.8-2.2	2.0-2.5	2.5	1.29	23.1
Pigs and Fish						
1. 150/ha piglets and tilapia	Flour meal	9.0	25.0	2.95	2.27	55.8
2. 50/ha porkers and *S. niloticus*	Greens and fodder	8.0	5.4	10.85	1.73	58.6
Chicken and Fish						
1. 1000-3000/ha chicks and *S. niloticus*	Chicken feed	3.5-5.0	3.6-8.0 (+ 1,200 eggs)	2.55	1.55	17.5

$$\% \text{ Tilapia contribution} = \frac{\text{Value of fish crop}}{\text{Total production costs}}$$

Source: Modified after Vincke (1976).

It would therefore seem to be only a matter of time, given the rate at which fishery reserves are being exhausted and competition for agricultural land necessitating intensification of crop production, that the potential of tilapia farming is realised and intensification becomes a fully economical proposition.

References

Al-Dahan, N.K. (1970). The use of chemosterilents, sex hormones, radiation and hybridisation for controlling reproduction in *Tilapia* species. PhD thesis, Auburn University, 176 pp.

Allanson, B.R., Bok, A. and Van Wyk, N.I. (1971). The influence of exposure to low temperature on *Tilapia mossambica* (Peters) (Cichlidae). II. Changes in serum osmolarity, sodium and chloride ion concentrations. *J. Fish Biol., 3 (2):* 181-5

Allanson, B.R. and Noble, R.G. (1964). The tolerance of *Tilapia mossambica* (Peters) to high temperature. *Trans. Am. Fish Soc., 93 (4):* 323-32

Allen, K.O. and Carter, R.R. (1976). Polyculture of channel catfish, Tilapia and hybrid buffalo in divided raceways. *Progve. Fish Cult., 38 (4):* 188-91, ASFA, *7*, 9213, BRI, *13*, 052082

Allison, R., Smitherman, R.O. and Cabrero, J. (1976). Effects of high density culture on reproduction and yield of *Tilapia aurea. FAO Tech. Conf. on Aquaculture, Kyoto, Japan:* AQ/Conf/76/E.47, 3 pp.

Anderson, C.E. and Smitherman, R.O. (1978). Production of normal male and androgen sex-reversed *T. aurea* and *T. nilotica* fed a commercial catfish diet in ponds. In: Smitherman, R.O., Shelton, W.L. and Grover, J.H. (eds.), *Culture of Exotic Fishes, Symp. Proc.* Fish Culture Section, American Fisheries Soc., Auburn, Alabama; 34-42

Anderson, R.G., Griffin, W.L., Stickney, R.R. and Whitson, R.E. (1978). Bioeconomic assessment of a poultry and tilapia aquaculture system. In: *Proc. 3rd Ann. Trop. and Subtrop. Fish. Tech. Conf. of the Americas*, Nickleson, R. (Comp), Texas A and M., Univ. College St., College Station, Texas: 126-41, ASFA, *9,* 11177

Andren, L.E. (1976). Pollution and degradation of environment affecting aquaculture in Africa. *CIFA Tech. Pap., 4* (Suppl. 1): 728-44

Anon. (1978). Lake Tilapia threatened by too many cages. *Fish Farming Int., 5 (3):* 48

Anon. (1979).Tilapia farmers cut prices to push sales. *Fish Farming Int., 6 (2):* 4

Anon. (1980). Cage farming can be simple and cheap. *Fish Farming Int., 7 (1):* 46

Apandi, M., Atmadilago, D. and Bird, H.R. (1974). Indonesian fish meals as poultry feed ingredients. Effects of species and spoilage. *World's Poult. Sci. J., 30 (3):* 176-82

Aston, R.J. (1980). The availability and quality of power station cooling water for aquaculture. *EIFAC Symp. on New Dev. in Util. of Heated Effluents and Recirc. Systems for Ints. Aquacult.* (EIFAC/80/Symp.-AR/3), 29 pp.

Aston, R.J., Brown, D.J.A. and Milner, A.G.P. (1976). Heated water farms at inland power stations. *Fish Farming Int., 3 (2):* 41-4

Avault, J.W., Shell, E.W. and Smitherman, R.O. (1968). Procedures for overwintering tilapia. *FAO Fish. Rep., 44 (4):* 343-5 (V/E-3)

Avtalion, R.R., Duczyminer, M., Wojdani, A. and Pruginin, Y. (1976). Determination of allogeneic and xenogeneic markers in the genus of Tilapia. II. Identification of *T. aurea, T. vulcani* and *T. nilotica* by electrophoretic analysis of their serum proteins. *Aquaculture, 7 (3):* 255-65

Avtalion, R.R. and Hammerman, I.S. (1978). Sex determination in *Sarotherodon (Tilapia)* 1. Introduction to a theory of autosomal influence. *Bamidgeh, 30 (4):* 110-15

Avtalion, R.R., Pruginin, Y. and Rothbard, S. (1975). Determination of allogeneic and xenogeneic markers in the genus of Tilapia. I. Identification of sex and hybrids in Tilapia by electrophoretic analysis of serum proteins. *Bamidgeh, 27 (7):* 8-13

Badawi, H.K. (1971). Electrophoretic studies of serum proteins of four *Tilapia* species (Pisces). *Mar. Biol., 8 (2):* 96-8, ASFA, *1,* 280

Balarin, J.D. and Haller, R.D. (1979). Africa tilapia farm shows the profit potential. *Fish Farming Int., 6 (2):* 16-18

Balarin, J.D. and Hatton, J.D. (1979). *Tilapia: A Guide to Their Biology and Culture in Africa*. Unit of Aquatic Pathobiology, Stirling University, 174 pp.

Bardach, J.E., Ryther, J.H. and McLarney, W.D. (1972). *Aquaculture: The Farming and Husbandry of Freshwater and Marine Organisms*. Wiley-Interscience, New York and London, 868 pp.

Basasidwaki, P. (1975). Comparative electrophoretic patterns of lactate dehydrogenase and malate dehydrogenase in five Lake Victorian cichlid fish. *Afr. J. Trop. Hydrobiol. Fish., 4 (1):* 21-6

Bashamohideen, M. and Parvatheswararao, V. (1972). Adaptions of osmotic stress in the freshwater euryhaline teleost *Tilapia mossambica*. 4. Changes in blood glucose, liver glycogen and muscle glycogen levels. *Mar. Biol. (Berl.), 16 (1):* 69-74

Bassa, G.K. (1979). Some technological and economic aspects of investment appraisal in aquaculture in a developing country. MSc thesis, University of Stirling, Scotland, 77 pp.

Bayne, D.R. (1974). Progress report on fisheries development in El Salvador. *Res. Rpt. No. 7*. Int. Centre for Aquaculture, Auburn University, Auburn, USA, 11 pp.

Bayne, D.R., Dunseth, D. and Rumirios, C.G. (1976). Supplemental feeds containing coffee pulp for rearing *Tilapia* in Central America. *Aquaculture, 7 (2):* 133-46

Bayoumi, A.R. (1969). Notes on the occurrence of *Tilapia zillii* (Pisces) in Suez Bay. *Mar. Biol. (Berl.), 4 (3):* 255-6

Bell, F.W. and Canterbery, E.R. (1976). *Aquaculture Countries: A Feasibility Study*. Ballinger Publishing Company, Cambridge, Mass. 266 pp.

Bishai, H.M. (1965). Resistance of *Tilapia nilotica*. L. to high temperatures. *Hydrobiologia, 25:* 473-88

Bogdanova, L.S. (1970). The transition of *Tilapia mossambica* (Peters) larvae to active feeding. *J. Ichthyol., 10 (3):* 427-30, ASFA, *1,* 2355

Brock, V.E. (1954). A note on the spawning of *Tilapia mossambica* in sea water. *Copeia, 1:* 72

Bruton, M.N. and Boltt, R.E. (1975). Aspects of the biology of *Tilapia mossambica* (Peters) (Pisces: *Cichlidae*) in a natural freshwater lake (Lake Sibaya, South Africa). *J. Fish Biol., 7 (4):* 423-7

Buck, D.H., Charts, C.F. and Rose, C.R. (1970). Variation in production in duplicate ponds. *Trans Ameri. Fish Soc., 99:* 77-9

Burrows, R.E. and Chenoweth, H.H. (1955). *Evaluation of Three Types of Fish Rearing Ponds.* USDI Fish and Wildlife Service Research Dept.

Buss, K., Graff, D.R. and Miller, E.R. (1970). Trout culture in vertical units. *Prog. Fish Cult., 32 (10):* 187-91

Campbell, D. (1976). Lake Kossou fishery development project, Ivory Coast. *FAO Aquacult. Bull., 8 (1):* 22-3

Canagaratnam, P. (1966). Growth of *Tilapia mossambica* (Peters) at different salinities. *Bull. Fish. Res. Stn Ceylon, 19:* 47-50

Carey, T.G. (1965). Breeding behaviour of *Tilapia macrochir. Fish. Res. Bull. Zambia,* 1962/3: 12-13

Caulton, M.S. (1975). Diurnal movement and temperature selection by juvenile and sub-adult *Tilapia rendalli* (Boulenger) *(Cichlidae). Proc. Trans. Rhod. Scient. Ass., 56 (4):* 51-6

—— (1976). The energetics of metabolism, feeding and growth of sub-adult *Tilapia rendalli* (Boulenger). PhD thesis, University of Rhodesia

—— (1978). The importance of habitat temperatures for growth in the tropical cichlid *Tilapia rendalli* Boulenger. *J. Fish Biol., 13 (1):* 99-112

—— (1979). *The Biology and Farming of Tilapia in Southern Africa.* Fish. Dev. Corp., Box 19, Gingindlovu, 380, South Africa, 115 pp.

Chen, F.Y. (1966). Preliminary studies on the sex-determining mechanisms of *Tilapia mossambica* (Peters) and *Tilapia hornorum* (Trewavas). *Rep. Trop. Fish. Cult. Res. Inst. Malacca,* (1967): 25-9, ABA, *1,* 914

—— (1969). Preliminary studies on the sex-determining mechanisms of *Tilapia mossambica* Peters and *T. hornorum* Trewavas. *Verth. Internat. Verein Limnol., 17:* 719-24

Chen, F.Y., and Tsayuki, H. (1970). Zone Electrophoretic studies on the proteins of *Tilapia mossambica* and *T. hornorum* and their F. hybrids, *T. zillii* and *T. melanopleura. J. Fish. Res. Bd Can., 27:* 2167-77

Chen, H.H. (1977). Taiwan. In: Brown, E.E. (ed.), *World Fish Farming: Cultivation and Economics:* AVI Publ. Co. Inc., Westport, Conn., Ch. 23: 345-58

Chen, T.P. (1969). Hybridization and culture of hybrids. *FAO Fish. Cult. Bull. (33):* 6

—— (1976). *Aquaculture Practices in Taiwan,* Fishing News Books, Farnham, UK, 161 pp.

Chervinski, J. (1961). Study of the growth of *Tilapia galilaea* (Antedi) in various saline concentrations. *Bamidgeh, 13 (3-4):* 71-4

—— (1967). Polymorphic characters of *Tilapia zillii* (Gervais). *Hydrobiologia, 30:* 138-44

Chervinski, J., and Hering, E. (1973). *Tilapia zillii* (Gervais) (Pisces, *Cichlidae*) and its adaptability to various saline conditions. *Aquaculture, 2 (1):* 23-9

Chervinski, J., and Yashouv, A. (1971). Preliminary experiments on the growth of *Tilapia aurea* (Steindachner) (Pisces, *Cichlidae*) in seawater ponds. *Bamidgeh, 23 (4):* 125-9

Chesness, J.L., Fussel, J. and Hill, T.K. (1973). Mechanical efficiency of a nozzle aerator. *Trans. Amer. Soc. Agric. Eng., 16 (1):* 67-8, 71

Chiba, K. (1980). Present status of flow-through and recirculation systems and their problems in Japan. *(EIFAC/80/Symp.-R/16),* II, 16 pp.

Chimits, P. (1955). Tilapia and its culture: a preliminary bibliography. *FAO Fish. Bull., 8 (1):* 1-33

—— (1957). The *Tilapia* and their culture: a second review and bibliography. *FAO Fish. Bull., 10 (1):* 1-24

Clemens, H.P., and Inslee, T. (1968). The production of unisexual broods *Tilapia mossambica,* sex reversed with methyltestosterone. *Trans. Am. Fish. Soc., 97 (1):* 18-21

Coche, A.G. (1975). Fish culture in cages, in particular *Tilapia nilotica* (L) in Lake Kossou, Ivory Coast, *FAO/CIFA Symp. Aquaculture in Africa, Accra,* 30 September 1975. FAO/CIFA-75/SE-13: 46 pp.

—— (1976). A general review of cage culture and its application in Africa. *FAO Tech. Conf. on Aquaculture, Kyoto, Japan,* FIR: AQ/CONF/76/E.72: 33 pp.

—— (1977). Preliminary results of cage rearing *Tilapia nilotica* (1) in Lake Kossou, Ivory Coast. *Aquaculture, 10 (2):* 109-40 (in French, English summary)

—— (1978). A review of fish cage culture as practised in inland waters. *Aquaculture, 13:* 157-89

Crass, R. S. (1959). Tilapia in Natal. In: *Proc. 1st Fish. Day in S. Rhodesia, Aug. 1957.* Govn. Printer, Salisbury: 29-31

Crawford, K.W., Dunseth, D.R., Engle, C.R., Hopkins, M.L., McCoy, E.W. and Smitherman, R.O. (1978). Marketing tilapia and Chinese carps. In: Smitherman, R.O., Shelton, W.L. and Grover, J.H. (eds), *Culture of Exotic Fishes,* Symp. Proc. Fish Culture Section, American Fisheries Soc., Auburn, Alabama: 240-57

Cridland, C.C. (1960). Laboratory experiments on the growth of *Tilapia* species. 1. The value of various foods. *Hydrobiologia, 15:* 135-60

Cruz, E.M. and Laudencia, I.L. (1976). Preliminary study on the protein requirements of Nile Tilapia (*Tilapia nilotica*) fingerlings. *IFP Tech. Rept. No. 10, 2nd half Cy 1976,* UP, Diliman, Quezon City, Philippines: 117-20

CTFT (1977). Window on aquaculture techniques world-wide. *Fish Farming Int., 4 (4):* 43-4

Dada, B.F. (1976). Present status and prospects for aquaculture in Nigeria, Africa. *CIFA Tech. Pap.,* 4 (Suppl. 1): 79-85

Da Silva, A.B., Carneiro-Sobrinho, A., Melo, F.R. and Lovshin, L.L. (1978). Mono e Policultivo Intensive de Tembaqui, *Colossomo macropomum,* e do Pirapitinga, *Colossoma bidens,* com o Hibrido Macho dos Tilapias. *Sarotherodon niloticus* Female e *Sarotherodon hornorum* male. *Second Symp. Latin Amer. Aquacult. Assoc.,* Mexico City, Mexico

Davis, A.T. and Stickney, R.R. (1978). Growth responses of *Tilapia aurea* to dietary protein quality and quantity. *Trans. Am. Fish. Soc., 107 (3):* 479-83

de Kimpe, P. (1956). Végétation phanérogamique et le plancton. *Bull. Agr. Congo B., 47 (4):* 1148-56

——— (1978). First fish farming experiment in floating cages in lagoons (Ivory Coast) and prospect of this technique development in tropical Africa. *Notes Doc. Peche Piscic. (Nouv. Ser.), 17:* 27-35 (in French)

Denzer, H.W. (1967). Studies on the physiology of young Tilapia. *FAO Fish. Rep., 44 (4):* 358-66

Donnelly, B.G. (1978). Evidence of fish survival during habitat desiccation in Rhodesia. *J. Limnol. Soc. South Afr., 4 (1):* 75-6, BA, *67,* 020893

Doudoroff, P., and Shumway, D.L. (1970). Dissolved oxygen requirements of freshwater fishes. *FAO Fish. Tech. Pap. (86),* 291 pp.

Dusart, J. (1963). Contribution a l'étude de l'adaption des Tilapia (Pisces, *Cichlidae*) a la vie en milieu mal oxygène. (Contribution to the study of the adaption of Tilapia (Pisces, *Cichlidae*) to oxygen depletion) (in French, English summary). *Hydrobiologia, 21:* 323-41

Eckstein, B. and Spira, M. (1975). Effects of sex hormones on gonadal differentiation in a cichlid, *Tilapia aurea. Biol. Bull., 129:* 482-9

Edwardson, W. (1976). Energy demands of aquaculture: a worldwide survey. *Fish Farming Int., 3 (4):* 10-13

El Zarka, S. (1956). Breeding behaviour – of the Egyptian cichlid fish, *Tilapia zillii. Copeia, 1956 (2):* 112-13

Fijan, N. (1969). Fish feeds containing Chlorella. *FAO Fish. Cult. Bull., 2 (2):* 6

Fiji Colony. Department of Agriculture (1959). *Annual Report for the Year 1958.* Suva, Fiji, Gvt. Press, 16 pp.

Fish, G.R. (1956). Some aspects of the respiration of six species of fish from Uganda. *J. Exp. Biol. 33 (1):* 186-95

Fitzgerald, W.J. (1979). The red-orange Tilapia: a hybrid that could become a world favourite. *Fish Farming Int., 5 (5):* 26-7

Fryer, G. and Iles, T.D. (1972). *The Cichlid Fishes of the Great Lakes of Africa: Their Biology and Evolution.* Oliver and Boyd, Edinburgh, 641 pp.

George, T.T. (1975a). Introduction and transplantation of cultivable species in Africa. *Proc. FAO/CIFA Symp. on Aquaculture in Africa, Accra, Ghana.* CIFA/75/SR. 7: 25 pp.

——— (1975b). Observations on the growth of *Tilapia nilotica* (L) in tropical fish ponds treated with different fertilizers. *Proc. FAO/CIFA Symp. on Aquaculture in Africa, Accra, Ghana.* CIFA/75/SE.11: 16 pp.

Goldstein, R.J. (1970). *Cichlids.* TFH Publications (H-939), 254 pp.

Gonzales, E.P. (1976). A study on the food habits of *Tilapia nilotica.* Undergraduate thesis,

Central Luzon State University, Munoz, Nueva Ecija, Philippines, 16 pp.
Gopalakrishnan, V. (1979). Status and problems of culture of baitfish for skipjack fishery in the Pacific region. In: Pillay, T.V.R. and Dill, Wm. A. (eds), *Advances in Aquaculture.* Fishing News (Books) Ltd, Farnham, Surrey, England: 58-62
Gruber, R. (1960). Considérations sur l'amélioration des rendements en pisciculture congolaise. *Bull. Agr. Congo Belge, 51 (1):* 139-57
—— (1962). Etude de deux facteurs inhibant la production de *Tilapia melanopleura* (Dum). Study of two factors. *Hydrobiologia, 19 (2):* 129-45 (in French, English summary)
Guerrero, R.D. (1975a). Sex reversal of Tilapia. *FAO Aquaculture Bull., 7 (3-4):* 7
—— (1975b). Use of androgens for the production of all-male *Tilapia aurea* (Stein-dachner). *Trans. Am. Fish. Soc., 104 (2):* 342-8
—— (1975c). Cage culture of male and female *Tilapia mossambica* with and without supplementary feeding in a fertilized pond. *Central Luzon State Univ. Scient. J., 9 (2):* 18-20
—— (1976a). Culture of male *Tilapia mossambica* produced through artificial sex reversal. *Proc. FAO Tech. Conf. on Aquaculture, Kyoto, Japan.* FAO: AQ./Conf./76/ E.15: 3 pp.
—— (1976b). Sex reversal of tilapia. *FAO Aquaculture Bull., 8 (1):* 5-6
Hackman, E. (1974). The influence of androgens on the gonad differentiation of various cichlids (Teleostei). *Gen. Comp. Endocrinol., 24 (1):* 44-52 (German edn) ASFA, *5:* 3904
Haller, R.D. (1974). *Rehabilitation of a Limestone Quarry. Report of an Environmental Experiment.* Publication by Bamburi Portland Cement Co. Ltd, Mombasa, Kenya, 32 pp.
Haller, R.D. and Parker, I.S.C. (1981). New Tilapia breeding system tested on Kenya farm. *Fish Farm. Int., 8 (1):* 14-18
Hammerman, I.S. and Avtalion, R.R. (1979). Sex determination in *Sarotherodon* (Tilapia) II. The sex ratio as a tool for the determination of genotype – a model of autosomal and gonosomal influence. *Theor. Appl.Genet., 54 (0-00):* 243-54
Harbott, B.J. (1975). Preliminary observations on the feeding of *Tilapia nilotica* (Linn.) in Lake Rudolf. *Afr. J. Trop. Hydrobiol. Fish., 4 (1):* 27-37, ASFA 7, 5732
Hastings, W.H. (1973). Regional project on research and fisheries development (Cameroon-Central African Republic – Gabon – Congo, Peoples Rep.). Experience related to the preparation of fish feed and their feeding. Report prepared for the regional project. *FAO Project Rep.* FAO-FI-DP/RAF-66/054/1): 24 pp. (Fr.) ASFA, *4,* 3900
Hauser, W.J. (1975). Influence on diet on growth of juvenile *Tilapia zillii. Progve. Fish Cult., 37 (1):* 33-5
Henao, R.A. and Corredor, G.G. (1978). Development and production of *Prochilodus reliculatus* associated with herbivorous Tilapia (*Tilapia rendalli*, Boulenger). *Inf. Tec. Cent. Piscic. Exp. Univ. Caldas., 2:* 73-8 (Sp.), ASFA 9, 10020
Henao, R.A. and Popma, T.J. (1978). Comparative study of the effects of three organic fertilizers on the ponderal growth and production of *Tilapia rendalli* (Boulenger). *Inf. Tec. Cent. Piscic. Exp. Univ. Caldas., 2:* 51-5 (Sp.), ASFA, *9,* 10016
Herzberg, A. (1978). Electrophoretic esterase patterns of the surface mucus for the identification of Tilapia species. *Aquaculture, 13 (1):* 81-3
Hickling, C.F. (1960). The Malacca *Tilapia* hybrids. *Journ. Genet., 57 (1):* 1-10
—— (1967). Fish hybridization. *FAO Fish. Rep., 44 (4):* 1-11 (IV/R-1)
Hida, T.S., Harada, J.R. and Kind, J.E. (1962). Rearing *Tilapia mossambica* for tuna bait. *Fish Bull. U.S., 62 (198):* 20
Hines, R., Yashouv, A. and Wilamovski, A. (1971). Differences in electrophoretic mobility of the haemoglobins of *T. aurea, T. vulcani* and their hybrid cross. *Bamidgeh, 23 (2):* 53-5, ABA, *2,* 5626
Hubert, W. (1978). Waste Heat Projects – State of the art aquacultural uses. *TVA/EPRI Workshop on Factors Affecting Power Plant Waste Heat Utilization,* 28 November 1978, A. Hanta, GAAD, 4; 10; 101
Huet, M. (1972). *Textbook of Fish Culture: Breeding and Cultivation of Fish,* translated by H. Kohn. Fishing News (Books) Ltd, Farnham, Surrey, England, 436 pp.
Hughes, D.G. (1977). Progress Report on Fisheries Development in El Salvador. *Int. Centre for Aquacult., Auburn Univ., Auburn, Alabama, Res and Dev., 15,* 16 pp.
Hyder, M. (1970). Histological studies on the testes of pond specimens of *Tilapia niger*

(Gunther) (Pisces *Cichlidae*) and their implications for the pituitary-testis relationship. *Gen. Comp. Endocr., 14 (1):* 198-211

Ibrahim, K.H., Nozawa, T. and Lema, R. (1976). Preliminary observations on cage culture of *Tilapia esculenta* (Graham) and *Tilapia zillii* (Gervais) in Lake Victoria waters, at the Freshwater Fisheries Institute, Nyegezi, Tanzania. *Afr. J. Trop. Hydrobiol. Fish., 4 (1):* 121-7

IFP (1974a) *Technical Report* (1 January 1974-30 June 1974). Inland Fisheries Project, T.R. 5, Second Half

—— (1974b) *Technical Report* (1 July 1974-31 December 1974). *Inland Fisheries Project, TR 6, First Half CY 75,* UPCF, Diliman, QC, Philippines: 100 pp. *Inland Fisheries Project IFP (1975).* Tilapia Fry rearing in Nylon Net Enclosures (Bitinan). *Inland Fisheries Project-* TR 6, *Second Half FY 75,* UPCF, Diliman, QC, Philippines: 91-2

—— (1975). Tilapia fry rearing in nylon net enclosures (Bitinan). *Inland Fisheries Project, TR6, Second Half FY 75.* UPCF, Diliman, QC, Philippines: 91-2

—— (1976a). *Inland Fisheries Project − TR9., First Half CY76,* UPCF, Diliman, QC, Philippines: 109-12. Pond evaluation of Tilapia hybrids treated for sex reversal

—— (1976b). *Inland Fisheries Project − TR9, First Half CY 76,*UPCF, Diliman, QC, Philippines: 115-18. Pond evaluation of *Tilapia zillii* treated with ethynyltestosterone for sex reversal

—— (1976c) *Technical Report* (1 January 1976-30 June 1976). *Inland Fisheries Project, TR 9, First Half CY 76.* UPCF, Diliman, QC, Philippines: 181 pp.

—— (1976d). *Technical Report* (1 July 1976-31 December 1976). *Inland Fisheries Project, TR 10, Second Half FY 76,* UPCF, Diliman, QC, Philippines: 168 pp.

Iles, R.B. (1963). Cultivating fish for food and sport in power station water. *New Scient., 17 (324):* 227-9

Ishac, M.M. and Dollar, A.M. (1968). Studies on manganese uptake in *Tilapia mossambica* and *Salmo gairdneri* in response to manganese. *Hydrobiologia, 31:* 534-72

Ita, E.O. (1976). Approaches to the evaluation of fishery reserves in the development and managements of Inland Fisheries. Rome, FAO, CIFA/72/514, 18 pp.

Iversen, E.S. (1976). *Farming the Edge of the Sea.* Fishing News Books Ltd./Whitefriars Press Ltd, London, 436 pp.

Jalabert, B., Kammacho, R. and Lessent, T. (1971). Sex determination in *Tilapia macrochir* x *Tilapia nilotica* hybrids. *Ann. Bio. Anim. Biochem. Biophys., 11(1):* 155-65

Jalabert, B., Moreau, J., Planquette, P. and Billard, R. (1974). Sex determination in *Tilapia macrochir* and *Tilapia nilotica*: effect of methyltestosterone administered in fry food on sex differentiation; sex-ratio of the offspring produced by sex reversed males. *Ann. Biol. Anim. Biochem. Biophys., 14 (4B):* 729-39 (Fr. edn), ASFA, *5,* 9673

Jensen, G.L. and Shelton, W.L. (1979). Effects of estrogens on *Tilapia aurea:* implications for production of monosex genetic male Tilapia. *Aquaculture, 16:* 223-42

Jensen, J.W. (1976). Progress report on fisheries development in Northeast Brazil. *Int. Centre for Aquacult., Auburn Univ., Auburn, Alabama, Res. and Dev., 10:* 7 pp (AID-1152, TO2)

Job, S.V. (1969). The respiratory metabolism of *Tilapia mossambica* (Teleostei). I. The effect of size, temperature and salinity. *Mar. Biol.* (Berl.), *2 (2):* 121-6

Jordan, D.T. and Pagan, F.A. (1973). Developments in cage culture of *Tilapia aurea* in a rock-quarry pond in Puerto Rico. *Comm. 10th Ann. Meet. Ass. Islands Mar. Lab. Carib.:* 59 pp. (resume)

Kantor-Perikanon, Darat (1951). *Pedoman Perikanan Air Tawar.* Dhakarta, Indonesia: 36 pp. On *T. mossambica* (not seen; quoted from Chimit)

Katz, Y., Abraham, M. and Eckstein, B. (1976). Effects of adrenosterone on gonads and body growth in *Tilopia nilotica.* (Teleostei, *Cichlidae*). *Gen. Comp. Endocrinol., 29:* 414-18

Katz, Y., Eckstein, B., Ikan, R. and Gottlieb, R. (1971). Estrone and estradiol-17B in the ovaries of *Tilapia aurea* (Teleostei, *Cichlidae*). *Comp. Biochem. Physiol (B)., 40 (4):* 1005-9 ASFA, *2,* 1759

Kiener, A. (1963). Poissons, pêche et pisciculture à Madagascar. *Publ. CTFT,* 244 pp. Mentioning *T. melanopleura* (= *T. rendalli*), *T. zillii, T. macrochir, T. nilotica, T. mossambica* and *T. nigra*

Kloke, C.W. and Potaras, M. (1975). Aquaculture as an integral part of the agricultural farming system − a case study in the N.E. of Thailand. *Indo Pac. Fisheries Circuit*

Occasional Paper, Bangkok, Thailand

Klopfenstein, P. and Klopfenstein, I. (1977). Tilapia farm in Oregon. *Fish Farming Int., 4 (3):* 6-8

Kohler, C.C. and Pagan-Font, F.A. (1978). Evaluations of rum distillation wastes, pharmaceutical wastes and chicken feed for rearing *Tilapia aurea* in Puerto Rico. *Aquaculture, 14:* 339-47

Konikoff, M. (1975). Nigeria – feasibility of cage culture and other aquaculture schemes at Kainji Lake. FAO Publ. FINIR/66/524/18 10 pp.

Kutty, M.N. (1972). Respiratory quotient and ammonia excretion in *Tilapia mossambica. Mar. Biol.* (Berl.), *16 (2):* 126-33

Lagler, K.F., Bardach, J.E. and Miller, R.R. (1952). *Ichthyology.* Wiley and Sons, New York and London, 545 pp.

Lanzing, W.J.R. (1978). Effect of methallibure on gonad development and carotenoid content of the fins of *Sarotherodon mossambicus (Tilapia mossambica). J. Fish. Biol., 12 (2):* 181-5, ASFA, *8,* 9038, BA, *66,* 030948

Larmoyeux, J.D. and Piper, R.G. (1973). Effects of water reuse on rainbow trout in hatcheries. *Prog. Fish Cult., 35 (1):* 2-8

Lauenstein, P.C. (1978). Intensive culture of Tilapia with geothermally heated water. In: Smitherman, R.O., Shelton, W.L. and Grover, J.H. (eds), *Culture of Exotic Fishes,* Symp. Proc. Fish Culture Section, American Fisheries Soc., Auburn, Alabama: 82-5

Legner, E.F. (1978). Mass culture of *Tilapia zillii (Cichlidae)* in pond ecosystems. *Entomophaga, 23 (1):* 51-5, ASFA, *9,* 7591

Li, Kuang Cong, Nguen, Din Zau and Nguen, Kuang Ving (1961). First results from a study of the effects of temperature on *Tilapia mossambica* (Peters) acclimatized in Vietnam since 1951. *ZH. Obshch. Biol., 22:* 444-51 (Russ.)

Ling, S.W. (1977). *Aquaculture in Southeast Asia: A Historical Overview.* Univ. of Washington Press, Seattle, Washington, 120 pp.

Liu, C. (1977). Aspects of reproduction and progeny testing in *Sarotherodon aureus* (Steindachner). MSc thesis, Auburn Univ., Auburn, Alabama: 42 pp.

Lovshin, L.L. (1977). Survey of the fish culture potential in San Julian and Alto Beni colonization projects of Bolivia. *Int. Centre for Aquacult., Auburn Univ., Auburn, Alabama,* (CSD-2730211d): 27 pp, ASFA, *9,* 3408

—— (1980). Progress report on fisheries development in Northeast Brazil. *Int. Centre for Aquacult., Auburn Univ. Alabama, Res. and Dev., 26* (AID 1152T. 02): 15 pp.

Lovshin, L.L. and Da Silva, A.B. (1975). Culture of monosex and hybrid tilapias. *FAO/ CIFA Symp. on Aquaculture in Africa, Accra, Ghana.* CIFA/75/SR.9. 16 pp.

Lowe-McConnell, R.H. (1955). New species of Tilapia (Pisces, *Cichlidae)* from Lake Jipe and the Pangoni River, East Africa. *Bull. Brit. Mus. Nat. Hist. Zool., 2 (12)* 347-68

—— (1959). Observations on the biology of *Tilapia nilotica* L. in East African waters. *Rev. Zool. Bot. Afr., 57 (1-2):* 129-70

—— (1975). *Fish Communities in Tropical Freshwaters: Their Distribution, Ecology and Evolution.* Longmans, London and New York, 337 pp.

Maar, A. (1956). Tilapia culture in farm dams in Southern Rhodesia. *Rhodesia Agric. J., 53 (5):* 667-87

Maar, A., Mortimer, M.A.E. and Van der Lingen, I. (1966). *Fish Culture in Central East Africa.* FAO Publ. 53608-66/E, 158 pp.

Mabaye, A.B.E. (1971). Observation on the growth of *Tilapia mossambica* fed on artificial diets. *Fish. Res. Bull. Zambia, 5:* 379-96

MacAndrew, B.J. (1981). Project report: Tilapia genetics (unpublished). Institute of Aquaculture, University of Stirling, Scotland

MacIntosh, D.J. (1980). Manual of sex reversal of tilapia using methyl testosterone. *Information Note* (Unpublished). Institute of Aquaculture, University of Stirling, Scotland, 2 pp.

Maddox, J.J., Behrends, L.L., Madewell, C.E. and Pile, R.S. (1978). Alga-swine manure system for production of Silver carp, Bighead carp and Tilapia. In: Smitherman, R.O., Sherton, W.L. and Grover, J.H. (eds), *Culture of Exotic Fishes,* Symp. Proc. Fish Culture Section, American Fisheries Soc. Auburn, Alabama: 109-20

Magid, A. and Babiker, M.M. (1975). Oxygen consumption and respiratory behaviour in three Nile fishes. *Hydrobiologia, 46:* 359-67

Man, H.S.H. and Hodgkiss, I.J. (1977). Studies on the ichthyo fauna in Plover Cove reservoir, Hong Kong. II: feeding and food relations. *J. Fish. Biol., 11 (1):* 1-4, ASFA, *8,* 4104, BA, *64,* 061727

Marek, M. (1975). Revision of supplementary feeding tables for pond fish. *Bamidgeh, 27 (3):* 57-64

Maruyama, T. (1958). An observation on *Tilapia mossambica* in ponds referring to the diurnal movement with temperature change. *Bull. Freshwat. Fish. Res. Lab., Tokyo, 8 (1):* 25-32 (in Japanese, English summary and figure headings)

Maruyama, T. and Ishida, R. (1976). Effect of water depth in net cages on growth and body shape of *Tilapia mossambica. Bull. Freshwat. Fish. Res. Lab., Tokyo, 26 (1):* 11-19 (in Japanese, English summary)

Mathavan, S.V. and Paudian, T.J. (1976). Food utilization in the fish *Tilapia mossambica* fed on plant and animal foods. *Helgolander wiss. Meereswiters, 28 (1):* 66-70

Mazid, M.A., Tanaka, Y., Katayama, T., Simpson, K.L. and Chichester, C.O. (1978). Metabolism of amino acids in aquatic animals Part 3. Indispensable amino acids for *Tilapia zillii. Bull. Jap. Soc. Sci. Fish., 44 (7):* 739-42

McCoy, E.W. (1974). Economic analysis of the Inland Fisheries Project in El Salvador. *Int. Center for Aquacult., Auburn Univ., Auburn, Alabama, Res. and Dev., 6* (AID/La-688): 15 pp.

McLarney, B. (1978). Family scale aquaculture in Colombia. Professor develops the Peasant Fish Culture Unit. *Commer. Fish Farmer Aquacult. News, 4 (6):* 28-9, ASFA, *9,* 7593

Melard, Ch. and Philippart, J.C. (1980). Intensive culture of *Sarotherodon niloticus* in Belgium. *EIFAC Symp. on New Dev. in Util. of Heated Effluents and of Recirc. Systems for Ints. Aquacult.* (EIFAC/80/Symp.-E/11), 11: 28 pp (Fr. edn)

Mertz, E.T. (1972). The protein and amino acid needs. In: Halver, J.E. (ed.), *Fish Nutrition.* Academic Press, New York and London: 105-43

Meschkat, A. (1967). The status of warm-water fish culture in Africa. *FAO Fish Rep., 44, (2):* 88-122 (I/RR-6)

Meskal, F.H. (1975). Preliminary studies on the possibility of starting aquaculture of fish in Ethiopia. *FAO/CIFA Symp. on Aquaculture in Africa, Accra; Ghana.* CIFA/75/SE 6: 3 pp.

Miller, J.W. (1976). Fertilization and feeding practices in warm-water pond fish culture in Africa. *CIFA Tech. Paper, 4* (Suppl. 1): 512-41, BRI, *15,* 000258

────── (1979). A preliminary study of feeding pelleted versus non-pelleted feeds to *Tilapia nilotica* L. in ponds. *Proc. World Symp. Finfish Nutrition and Fish feed Technology, Hamburg, 20-23 June 1978, Vol. 1, Berlin:* 371-7

Miller, J.W. and Ballantine, D.I. (1974). Opercular algal growth on the cichlid fish *Tilapia aurea,* cultured in sea water. *Aquaculture, 4 (1):* 93-5

Mires, D. (1977). Theoretical and practical aspects of the production of all-male Tilapia hybrids. *Bamidgeh, 29 (3):* 94-101, AB, *66,* 001579

Mironova, N.V. (1975). Food value of algae for *Tilapia mossambica* (Peters) (English summary). *Vopr. Ikhtiol., 15 (3):* 567-71

────── (1977). Energy expenditures and egg production in young specimens of the Tilapia (*Tilapia mossambica*) and effects of environmental conditions on reproductive intensity. *Vopr. Ikhtiol., 17 (4):* 708-14 (Russ.), ASFA, *8,* 1509, BA, *65,* 051517

Mishrigi, S.Y. and Kubo, T. (1978). The energy metabolism of *Tilapia nilotica.* II. Active metabolism at 20° and 26°C. *Bull. Fac. Fish. Hokkaido Univ., 29 (4):* 313-21

Morgan, P.R. (1972). Causes of mortality in the endemic Tilapia of Lake Chilwa (Malawi). *Hydrobiologia, 40:* 101-19

Moriarty, C.D. and Moriarty, D.J.W. (1973). Quantitative estimation of the daily ingestion of phytoplankton by *Tilapia nilotica* and *Haplochromis nigripinnis* in Lake George, Uganda. *J. Zool. Lond., 171 (1):* 15-23

Morissens, P. (1977). Production of male monosex hybrid of Tilapia in Israel. *Piscic. Fr., 50:* 39-46 (Fr. edn), ASFA, *8,* 6143

Naegel, L.C.A. (1977). Combined production of fish and plants in recirculating water. *Aquaculture, 10 (1):* 17-24, ASFA, *8,* 7213, BA, *64,* 001612

Nakamura, M. and Takahashi, H. (1973). Gonadal sex differentiation in *Tilapia mossambica,* with special regard to the time of estrogen treatment effective in inducing complete feminization of genetic males. *Bull. Fac. Fish., Hokkaido Univ., 24 (1):* 1-13, ASFA, *4,* 6929

Nash, C. and Mayo, R. (1979). Double-doughnut tanks for carp-rearing. *Fish Farming Int.,*

6 (3): 31

Noble, R.L., Germany, R.D. and Hall, C.R. (1975). Interactions of Blue Tilapia and large mouth Bass in a power plant cooling reservoir. *Proc. Annu. Conf. Southeast Assoc. Game Fish Comm., 29:* 247-51, BA, *77,* 031713

Otte, G. and Rosenthal, A. (1979). Tilapia grow on carp feed plus vitamins (report by FFI). *Fish Farming Int., 6 (2):* 47

Pagan, F.A. (1969). Cage culture of Tilapia. *FAO Fish Cult. Bull., 2 (1):* 6

——— (1970). Cage culture of Tilapia. *FAO Fish Cult. Bull., 3 (1):* 6

Pagan-Font, F.A. (1975). Cage culture as a mechanical method for controlling reproduction of *Tilapia aurea* (Steindachner). *Aquaculture, 6 (3):* 243-7

Pandian, T.J. and Ragharaman, R. (1972). Effects of feeding rate on conversion efficiency and chemical composition of the fish *Tilapia mossambica. Mar. Biol.* (Berl.), *12 (2):* 129-36

Pantastico, J.B. and Baldia, J.P. (1979). Supplementary Feeding of *Tilapia mossambica. Proc. World Symp. Finfish Nutr. and Fishfeed Technology, Hamburg, 20-23 June 1978, Berlin, 1:* 587-93

Parkman, R.W. and McCoy, E.W. (1977). Fish marketing in El Salvador. *Project, AID/csd-2780* (ICARD -D. No. 12). Int. Centre for Aquacult., Auburn Univ. Auburn, Alabama: 19 pp.

Patino, R.A. (1976). Cultivo experimental de peces en Estanques (in English). *J. New Alchemists, 3:* 86-90

Payne, A.I. (1970). An experiment on the culture of *Tilapia esculenta* (Graham) and *T. zillii* (Gervais) (Cichlidae) in fish ponds. *J. Fish Biol., 3 (3):* 325-40

——— (1975). Tilapia — a fish of culture. *New Scient., 67 (960):* 256-8

Perez, J.E. and Maclean, N. (1975). The haemoglobins of the fish *Sarotherodon mossambicus* (Peters). Functional significance and ontogenetic changes. *J. Fish Biol., 9 (5):* 447-55

Perry, W.G. and Avault, J.W. (1972). Comparisons of striped mullet and tilapia for added production in caged catfish studies. *Progve. Fish Cult., 34 (4):* 229-32

Planquette, P. and Tetel, C. (1977). Data on mass production of *Tilapia nilotica* fry. *Notes Doc. Peche Piscic., (Nouv. Sér.), 14:* 1-6 (Fr.)

Popma, T.J. (1978). Experiment on the growth of *Tilapia rendalli* (Boulenger) cultured in cage and fed with fresh leaves of *Alocasia macrorhiza. Inf. Tec. Cent. Piscic. Exp. Univ. Caldas, 2:* 43-50 (Sp. edn), ASFA, *9,* 10010

Prowse, G.A. (1963). Introduction of exotic fish. *Nature, Lond., 197 (4872):* 1123

Pruginin, Y. (1965). Mono-sex culture of tilapia through hybridization. *STRC Symp. on Fish Farming, Nairobi. 21-24 Sept. 1965 Pap., (65):* 3

——— (1967). Culture of carp and Tilapia hybrids in Uganda. *FAO Fish Rep., 44 (4):* 223-9

——— (1975). Species combination and stock densities in aquaculture in Africa. *FAO/CIFA Symp. on Aquaculture in Africa, Accra, Ghana.* CIFA/75/SR 6, 5 pp.

Pruginin, Y., Rothbard, S., Wohlfarth, G., Halevy, A., Moav, R. and Hulata, G. (1975). All male broods of *Tilapia nilotica* x *T. aurea* hybrids. *Aquaculture, 6 (1):* 11-21

Pruginin, Y. and Shell, E.W. (1962). Separation of the sexes of *Tilapia nilotica* with a mechanical grader. *Progve Fish Cult., 24 (1):* 37-40

Ramirez, G.R. (1977). Background and perspective of aquaculture in Mexico and its role in the international trade of fishery products. *Proc. Symp. on Aquacult. in Latin America, 3, National Reports, Montevideo, Uruguay, (26 Nov.-2 Dec. 1974), FAO Fish Rep., 159 (3):* 6-13 (FAO-FIRR/R.159-3) (Sp. edn), ASFA, *8,* 13821

Rappaport, U. and Sarig, S. (1975). The results of tests in intensive growth of fish at the Genosar (Israel) station ponds in 1974. *Bamidgeh, 27 (3):* 75-82, ASFA, *7,* 2011

Ray, L.E. (1978a). Water quality: The single most important factor in fish production. *Commer. Fish Farmer Aquacult. News, 4 (4):* 8-9, 31, ASFA, *10,* 679

——— (1978b). Production of tilapia in catfish raceways using geothermal water. In: Smitherman, R.O., Shelton, W.L. and Grover, J.H. (eds), *Culture of Exotic Fishes,* Symp. Proc. Fish Culture Section, American Fisheries Soc., Auburn, Alabama: 86-9

Redner, B.D. and Stickney, R.R. (1979). Acclimation to Ammonia by *Tilapia aurea. Trans. Am. Fish. Soc., 108:* 383-8

Reite, O.B., Maloiy, G.M.O. and Aasehaug, B. (1973). pH, salinity and temperature tolerance of Lake Magadi Tilapia. *Nature, Lond., (5439), 247:* 315

Roberts, R.J. (ed.) (1978). *Fish Pathology.* Baillière Tindall, London

Ross, B. and Jauncey, K. (1981). A radiographic estimation of the effect of temperature on gastric emptying time in *Sarotherodon* hybrids. *J. Fish Biol., 19:* 331-44

Rothbard, S. (1979). Observations on the reproductive behaviour of *Tilapia zillii* and several *Sarotheroden spp.* under aquarium conditions. *Bamidgeh, 31 (2):* 33-43

Rowland, L.O., Hooge, D.M. and Stickney, R.E. (1977). Evaluation of Tilapia meat as a protein source for broilers. *Poultry Science, 56 (5):* 1952 (Abstr.)

Sanchez, C. (1978). Cage culture of Tilapia. *Pesca Mar. Barco Pesq., 30 (2):* 19-20 (Sp. edn), ASFA, *9,* 3431

Sarig, S. (1967). A review of diseases and parasites of fishes in warm water ponds in the Near East and Africa. *FAO Fish Cult. Bull., 44 (5):* 278-89 (II/R -2)

―――― (1969). Winter storage of Tilapia. *FAO Fish Cult. Bull., 2 (2):* 8-9

―――― (1971). Diseases of fishes. The prevention and treatment of diseases of warmwater fishes under subtropical conditions, with special emphasis on intensive fish farming. In: Snieszko, S.F. and Axelrod, H.R. (eds), *Diseases of Fish. Book 3.* TFH Publ., Reigate, Surrey, England, 127 pp.

Sarig, S. and Marek, M. (1974). Results of intensive and semi-intensive fish breeding techniques in Israel in 1971-1973. *Bamidgeh, 26 (2):* 28-48, ASFA, *5,* 3218

Schmittou, H.R. (1969). The culture of channel catfish *Ictalurus punctatus* (Rafinesque) in cages suspended in ponds. *Proc. 23rd Ann. Conf. South-eastern Ass. Game Fish Commissioners, Alabama, USA:* 226-44

Scott, P.W. (1977). Preliminary studies on disease in intensively farmed Tilapia in Kenya. MSc thesis, Stirling University, Scotland, 159 pp.

Seki, N. (1976). Identification of fish species by SDS-polyacrylamide gel electrophoresis of the myofibrillar proteins. *Bull. Jap. Soc. Sci. Fish., 42 (10):* 1169-76 (Jap. edn), ASFA, *7,* 5467, BA, *64,* 031225

Serfling, S. and Mendola, A. (1978). Aquaculture using controlled environment ecosystems. *Commer. Fish Farmer and Aquacult. News. 4 (6):* 15-18

Sevilleja, R.C. and McCoy, E.W. (1979). Fish marketing in Central Luzon, Philippines. *Research and Development Series No. 21* (AIDea/180) Int. Center for Aquacult., Auburn Univ., Auburn, Alabama: 23 pp.

Shehadeh, Z.H. (ed.) (1976). *Report of the Symposium on Aquaculture in Africa, Accra, Ghana.* CIFA Tech. Pap., *4:* 36 pp.

Shell, F.W. (1967). Mono-sex culture of male *Tilapia nilotica* (Linnaeus) in ponds stocked at three rates. *FAO Fish. Rep., 44 (4):* 253-8 (V/E -5)

Shelton, W.L., Hopkins, K.D. and Jensen, G.L. (1978). Use of hormones to produce monosex Tilapia for aquaculture. In: Smitherman, R.O., Shelton, W.L. and Grover, J.H. (eds), *Culture of Exotic Fishes.* Symp. Proc. Fish Culture Section, American Fisheries Soc., Auburn, Alabama: 10-33

Sipe, M. (1979). Announcing: 6 New Golden Tilapia Hybrids. *Fish Farming Int., 6 (4):* 29

Sitasit, D. and Sitasit, V. (1977). Comparison of the production of *Tilapia nilotica* (Linn.) fed with protein from different sources. In: *Indo-Pacific Fisheries Council, Proc. 17th session, Colombo, Sri Lanka 27 Oct.-5 Nov. 1976: Section 3-Symp. on Dev. and Util. of Inld. Fish. Res.* FAO Regional Office for Asia and the Far East, Bangkok, Thailand: 400-3, ASFA, *9,* 1594

Street, D.R. (1978). The socioeconomic impact of fisheries programs in El Salvador. *Int. Center for Aquacult., Auburn Univ., Auburn, Alabama, Res. and Dev., 17:* 14 pp.

Suffern, J.S., Adams, S.M., Blaylock, B.G., Coutant, C.C. and Guthrie, C.A. (1978). Growth of monosex hybrid Tilapia in the laboratory and sewage oxidation ponds. In: Smitherman, R.O., Shelton, W.L. and Grover, J.H. (eds), *Culture of Exotic Fishes.* Symp. Proc. Fish Culture Section, American Fisheries Soc., Auburn, Alabama: 65-81

Suwanasart, P. (1971). Effects of feeding mesh size and stocking size on the growth of *Tilapia aurea* in cages. *Ann. Rep. Int. Centr. Aquacult. Agric. Exp. St., 1971, Auburn Univ., Alabama:* 71-9

Swingle, H.S. (1967). Biological means of increasing productivity in ponds. *FAO Fish. Rep., 44 (4):* 243-57

Tait, C.C. (1965). Mass fish mortalities. *Zambia Fisheries Res. Bull., 3:* 28-30

Tal, S. (1979). The concept of aquaculture. *Bamidgeh, 31 (2):* 23-5

Tal, S. and Ziv, I. (1978). Cultivation of exotic (fish) species in Israel. *Bamidgeh, 30 (1):* 3-11

Trewavas, E. (1973). On the cichlid fish of the genus *Pelmatochromis*, on the relationship between *Pelmatochromis* and *Tilapia* and the recognition of *Sarotherodon* as a distinct genus. *Bull. Brit. Mus. (Nat. Hist.) Zool., 25:* 1-26

Uchida, R.N. and King, J.E. (1962). Tank culture of tilapia. *US Fish Wildl. Serv., Fish. Bull., 62:* 21-52

Vaas, K.F. and Hofstede, A.E. (1952). Studies on *Tilapia mossambica* Peters (ikan mudjair) in Indonesia. *Contrib. Inl. Fish. Res. St., Djakarta, Bogor (Indonesia), 1:* 1-88

Varley, R.L. (1977). Economics of fish farming in the UK. *Fish Farming Int., 4 (1):* 17-19

Vincke, M.M.J. (1976). La rizipisciculture et les élevages associés en Afrique. *CIFA Tech. Pap., 4* (Suppl. 1): 659-707

Viola, S. and Amidan, G. (1980). Observations on the accumulation of fat in carp and *Sarotherodon* (Tilapia) fed oil-coated pellets. *Bamidgeh, 32 (2):* 33-40

Welcomme, R.L. (1967). Observations on the biology of the introduced species of Tilapia in L. Victoria. *Rev. Zool. Bot. Afr., 76:* 249-79

Westers, H. and Pratt, K.M. (1977). Rational design of hatcheries for intensive Salmonid culture, based on metabolic characteristics. *Prog. Fish Cult., 36 (2):* 86-9

Wheaton, F.W. (1977). *Aquaculture Engineering.* Wiley, New York, 708 pp.

Whitfield, A.K. and Blaber, S.J.M. (1976). The effects of temperature and salinity on *Tilapia rendalli* (Boulenger 1896). *J. Fish Biol., 9 (2):* 99-104

Williams, E.M., Phelps, R.R., Gaines, J.L. and Buldey, L.K. (1974). Gram negative pathogenic bacteria of some fishes before and after cage culture. *Proc. 5th Ann. Workshop. World Mariculture Soc., S.C. Jan. 1974*

Yashouv, A. (1960). Effect of low temperatures on *Tilapia nilotica* and *Tilapia galilaea. Bamidgeh, 12 (3):* 62-6

6 RECIRCULATED WATER SYSTEMS IN AQUACULTURE

James F. Muir

1. Introduction

A major input and constraint in aquaculture is the supply of water of suitable quality and quantity, and thus the methods for conserving and reusing water within recycled systems are of considerable interest. A number of recycled systems have been developed to date, and descriptions and design outlines have been given (e.g. Speece, 1973; Meade, 1974; Liao and Mayo, 1974). Reviews and bibliographies of recycled water use have been provided (e.g. Parker and Simco, 1974; Pettigrew *et al.*, 1978) and a recent conference (EIFAC, 1980) has dealt particularly with the subject. As yet, there are few commercially viable recycled systems in operation, though a wide variety of designs is employed (Muir, 1978). It is the aim of this chapter to discuss existing techniques for recycling water, to examine in closer detail some of their more important components, and to consider the prospects for water recycling within the constraints identified.

Within the widest sense of water reuse the cycles of evapotranspiration and biological nutrient transfer ensure the replenishment of water used for fish production. We are concerned here, however, with the short-circuiting of this process by the maximising of water use within a culture system and the immediate reconditioning and return of used water. An approximate division may also be made in that used water may be treated in specifically designed units, outside the culture area (technological), or treated within the culture area, using controlled or intensified biological processes (ecological). The distinctions between these approaches are shown in Table 6.1; technological methods typical of intensive aquaculture are the main subject of this review, though an understanding of the processes of nutrient transfer and maintenance of water quality in the 'static' systems supplying the bulk of the world's aquaculture production is of great importance for the future.

Table 6.1: Approaches to Water Reuse

Technological	Ecological
water moved outside culture area	Water retained within culture area
specific treatment	maximise water use
pumping energy	photosynthetic energy
controlled and designed treatment	limited control over biological processes
intensive culture	extensive culture
monoculture	polyculture
capital intensive	low capital input
high technology	low technology
high energy use	low energy use

1.1   Basic Concepts of Water Reuse

The relationships between water flows, recycle rates, production levels and
system efficiencies are shown in Figures 6.1 and 6.2. The expressions in Figure
6.1 are derived from a simple static mass-balance and describe flow and the
concentrations of water quality parameters which may limit the acceptability
of the water. In this way the use of defined treatment efficiencies, waste outputs,
flows and quality limits may provide recycle rates, or existing recycle systems
may be evaluated for operating efficiency. It is thus important to identify
expected water quality effects and system efficiencies, and to examine relationships between them. From this point the more complex relationships between
stock weight and waste output, waste concentration and system efficiency, and
those between component efficiencies and system performance can be considered.

Figure 6.1: Open and Recycle Flow Mass Balances.

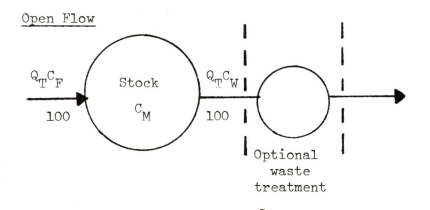

Open Flow

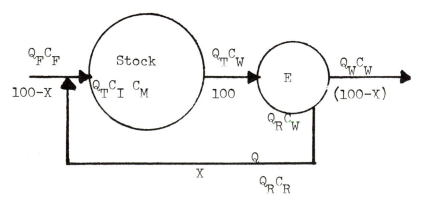

Recycle Flow

X = recycle ratio = Q_R/Q_T; A = water use advantage = 100/(100-X); Q = flow;
C = concentration; E = efficiency = $1-C_R/C_W$. Subscripts: T = total; R = recycle;
W = waste; F = fresh; I = inlet; M = added by stock. Where $C_F = 0$:

$$C_I = XC_M(1-E)/1+EX-X, \quad C_W = C_M/1+EX-X, \quad C_R = C_M(1-E)/1+EX-X.$$

Where RE, recirculation efficiency, $= (1+EX-X)$:

$C_I = XC_M(1-E)/RE$, $C_W = C_M/(1-E)$, $C_R = C_M(1-E)/RE$.

Recycle Definitions

%Recycle $= Q_R/Q_T$ 100, or 100X, or $(1-Q_F/Q_T)$ 100.

%per day $= Q_F/V$, where V = system volume, $= Q_F/\theta Q_T$,

where θ = system residence time.

Recycle residence time $= \theta Q_T/Q_F$.

Figure 6.2: Water Reuse Advantage. A: Advantage in water use relative to % recycle. B: Advantage for given % recycle relative to treatment efficiency, E.

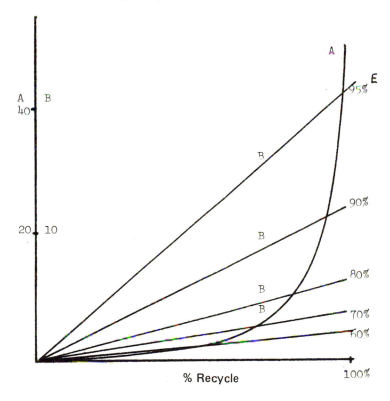

2. Operating Conditions

The flow and quality of water required in a culture area will be set both by the environmental requirements of the stock held and by the rate of water quality change due to the metabolic processes of the stock. This can be expressed simply by:

$$C_o = C_i + \frac{dC}{dt}\triangle t$$

where C_o = outlet concentration, C_i = inlet concentration, $\frac{dC}{dt}$ = rate of change of concentration, $\triangle t$ = time in culture system. Where C_o = limiting concentration, the maximum use is made of the water with respect to an individual water quality factor.

Thus the main limiting factors, and hence the priorities for water treatment, may be defined by the order in which the factors reach the limits of environmental requirements. The most important factors in high-density fish culture are normally those affected directly by the processes of metabolism: oxygen, ammonia, CO_2 and suspended solids. Other significant factors include temperature, light, water velocity and variation of environmental quality.

2.1 Environmental Requirements

The most critical environmental requirements are those associated directly with respiration, metabolism and ionic regulation, and the transfer of gases and ions across gill surfaces. Environmental requirements based on threshold values in short-term toxicity tests must be interpreted with care for application in the longer-term exposures typical of the recycle systems, and the effects of combinations of environmental factors have received little attention. Reviews or summaries have been compiled for salmonids and general use (Brockway, 1950; Doudoroff and Katz, 1950; Liao, 1970; Alabaster, 1974; Muir, 1975; Munro, 1978; Forster *et al.*, 1977), channel catfish (Colt *et al.*, 1975), tilapia (Balarin and Hatton, 1979), penaeids (Wickins, 1976), oysters (Harman and Maurer, 1971), and in general terms by Wedemeyer *et al.* (1976) and Chiba (1980). Colt *et al.* (1979) have produced a useful abstract collection on environmental requirements for North American species and Wickins (1980) has provided a review on water quality in intensive aquaculture. Burrows (1964), Liao and Mayo (1973, 1974), Larmoyeux and Piper (1973), Wickins (1976) and Muir (1978) describe effects in recycled systems.

The use of an EC_{50} relationship (Colt and Tchobanoglous, 1978), measuring concentrations at which growth is depressed to 50 per cent of 'normal' (control) growth over a specified period, is particularly useful in this context, though distinction must be made between EC_{50} figures on length and those on weight bases. These might be extrapolated if necessary to set levels at EC_{90} if an economic assessment must be made of the relationship between reduced growth and the lower cost associated with poorer treatment. However, few results are currently available on this basis. As a full exposition of environmental requirements and the nature of physiological effects is outside the scope of this chapter, a number of critical or representative figures are summarised for the most important parameters in Tables 6.2-6.6. Table 6.7 shows typical acceptable limits in aquaculture use and the main mediating effects on these limits. As few of these limits are obtained experimentally, the determination of the combined effects typical of intensive or recycled water culture is difficult, though the earlier experiments by Brockway (1950), Burrows, (1964), Larmoyeux and Piper (1973) and Mayer and Kramer (1973) showed the overall effects of sequential or recycled water use, typically showing progressive decrease in stamina and growth, and greater susceptibility to disease under combined oxygen deficit, CO_2, ammonia and solids loading. As suggested by Alabaster (1974), a time/concentration distribution approach over the period of exposure may be more appropriate to describe environmental conditions.

Table 6.2: Oxygen Limitations and Requirements

Concentration mg/1	Species	Effect	Reference
0-0.2	*Clarias*	growth in ponds	Muir (unpublished)
2-3	*Tilapia*	lowest levels for growth	Balarin & Hatton (1979)
2.5	*S. salar* smolts	72 hr LC_{50} 30-80% seawater	Alabaster *et al.* (1979b)
2.8	*Cyprinus*	possible mortalities, long exposure	Downing & Merkens (1957)
<3	*Cyprinus*	growth decrease	Itazawa (1971)
<3-4	*Anguilla japonica*	growth, feeding, conversion effect	Chiba (1980)
3	*S. salar* smolts	72 hr LC_{50} fresh water	Alabaster *et al.* (1979b)
4	*Penaeus japonicus*	feeding reduced	Liao (1969)
4-6	*O. kisutch*	minimal level	Davies (1975)
5	Salmonids	minimum sustained level	Forster *et al.* (1977)
>6	*Penaeus japonicus*	optimum	Liao (1969)
7	*Cyprinus*	maximum growth rate	Huisman (1974)
$\triangle 0 \sim 0.5$	*Cyprinus*	sudden drop causes stress, surface breathing	Albrecht (1977)
$\triangle 0 > 3$	*M. salmoides*	max. fluctuation for no effect	
$\triangle 0 > 6$	*M. salmoides*	reduced growth	Stewart *et al.* (1967)

Table 6.3: Ammonia Limitations

Concentration (mg/l NH_3-N)	Species	Effect	Reference
0.15 (freshwater); 0.30 (seawater)	*O. tshawytscha*	stimulates growth	Robinson-Wilson & Seim (1975)
0.06	*I. punctatus*	stimulates growth	Robinette (1976)
~0.044	*A. japonicus*	growth decrease	Chiba (1980)
~0.011	*Plecoglossus altivelis* (Ayu)	growth decrease	Chiba (1980)
~0.14 +	*I. punctatus*	growth decrease	Knepp & Arkin (1973)
>0.01	Salmonids	growth, health impaired	Burrows (1964) Smith & Piper (1975)
0.01, 0.06	*I. punctatus*	no effect, 4 wks	Robinette (1976)
0.066	*Solea solea*	threshold, no effect	Alderson (1979)
<0.07	*S. gairdneri*	no growth effect, 126 days	Forster & Smart (1979)
0.05-1.0	*I. punctatus*	linear growth effect	Colt & Tchobanoglous (1978)

Table 6.3: continued

Concentration (mg/1 NH_3–N)	Species	Effect	Reference
0.103	*S. gairdneri*	growth decrease first 2 weeks, thereafter adaptation	Forster & Smart (1979)
0.11	*C. carpio*	mortalities, 35 days	Flis (1968)
0.11	*Scophthalamus maximus*	threshold, no effect	Alderson (1979)
0.12, 0.13	*I. punctatus*	growth decrease after 4 weeks	Robinette (1976)
0.15	*Salmo salar* smolts	24 hrs LC_{50} full O_2	Alabaster *et al.* (1979a)
0.30	*Salmo salar* smolts	24 hrs LC_{50} full O_2	Alabaster *et al.* (1979a)
~0.20	*Clarias batrachus*	levels in culture conditions	Muir (unpublished)

Table 6.4: CO_2 Limitations

Concentration (mg/l)	Species	Effect	Reference
12-55	*S. gairdneri*	increasing nephrocalcinosis	Smart *et al.* (1978)
70	Salmonids	no adverse respiratory effect	Albrecht (1977)
300	*C. carpio*	no adverse respiratory effect	Albrecht (1977)

Table 6.5: NO_2 Limitations

Concentration (mg/l)	Species	Effect	Reference
0.012	Salmonids	stress shown	Weston (1974)
0.015-0.06	*S. gairdneri*	minimum level for methaemo-globinaemia	Smith & Russo (1975) Wedemeyer & Yasutake (1978)
0.15	*S. gairdneri*	methaemo-globinaemia, 48 hrs	Smith & Williams (1974)
0.19-0.39	*S. gairdneri*	96 hrs LC_{50} dep. fish size	Russo *et al.* (1974)
0.55	*S. gairdneri*	methaemo-globinaemia, 24 hrs, 50% mortality, fingerlings	Smith & Williams (1974)
1.6	*S. gairdneri*	50% mortality, yearlings, 24 hrs	Smith & Williams (1974)
1.8	*Macrobrachium rosenbergii*	larval growth reduced, 8 days	Armstrong *et al.* (1976)
2.4	*O. tschawytscha*	mortality, 7 days	Smith & Williams (1974)
6.4	*Penaeus indicus*	growth reduced, 3 weeks	Wickins (1976)

Table 6.5: continued

Concentration (mg/1)	Species	Effect	Reference
7.55	*Ictalurus punctatus*	96 hrs LC_{50}	Konikoff (1975)
15.4	*Macrobrachium rosenbergii*	3-4 wk LC_{50}	Wickins (1976)
24.8	*Ictalurus punctatus*	96 hrs LC_{50}	Collins *et al.* (1975)
29.8	*O. kisutch*	48 hrs, zero mortality, high Cl^-	Perrone & Meade (1977)

Table 6.6: NO_3 Limitations

Concentration (mg/1)	Species	Effect	Reference
90	*I. punctatus*	grow normally	Colt & Tchobanoglous (1976)
275	*S. gairdneri*	no adverse effect	Bohl (1977)
>400	*I. punctatus*	tolerated	Knepp & Arkin (1973)
>400	*S. gairdneri*	tolerated	Westin (1974)
>800	*S. gairdneri*	stops growth	Berka *et al.* (1980)
1,300	*S. gairdneri*	lethal level	Westin (1974)
1,400	*I. punctatus*	lethal level (osmoregulation)	Colt & Tchobanoglous (1976)
2,400	*C. carpio*	successful growth	Kaüsche (1973)
4,400	*O. tshawytscha*	96 hrs LC_{50} sea water	Westin (1974)
5,800	*O. tshawytscha*	96 hrs LC_{50} fresh water	Westin (1974)

Physiologically, the common action of many of these environmental factors on the gills and the respiratory processes suggests an accumulative effect, and the evidence of Forster and Smart (1978) and Alabaster *et al.* (1979), showing high oxygen levels to alleviate ammonia toxicity, confirms this. Hampson (1976) notes that higher local levels of CO_2 in water produced by respiration might alleviate the effects of high stock loadings by accelerating transfer of NH_3 to NH_4+ outside the gills, though this effect should be dependent more on the overall carbonate system.

To assess or allow for combined effects on an empirical basis Muir (unpub.) has used an index of the form

$$M = (O_s - O)^a U^b N^c S^d C^e, \text{etc.}$$

where M = multiple index, $(O_s - O)$ = oxygen deficit, U = unionised ammonia (e.g. mg/l), N = nitrite (mg/l), S = solids (mg/l), C = CO_2 (mg/l). While this does not reflect physiological interactions, it may provide a basis for further development.

It is clear that there are few complete assessments of sub-optimal conditions and, as recycled systems can potentially be designed to maintain any combination of environmental qualities, there is no possibility of defining the effects of every

Table 6.7: Summary of Effects, Adaptations, Recommended Limits and Mediating Factors of Significant Environmental Parameters

Factor	Effects	Adaptation	Mediation	Limits	References
O_2	activity, growth, stamina; lactic acid build-up, pH decrease in blood; (1) urine changes; (2) mucus production	amplitude and frequency of gill movement	CO_2 depresses uptake, NO_2 reduces Hb capacity, temperature affects O_2 affinity of Hb	2 mg/1, *Clarias, Ophiocephalus*; 2-3 mg/1, *Anguilla*, carps, tilapia; 5-6 mg/l salmonids, marine Flatfish	(1) Kirk (1974); (2) Hunn (1969); Warren *et al.* (1973)
NH_3	gill hyperplasia and reduced efficiency activity, growth, stamina; blood (1), liver, kidney changes (3); brain cholinesterase inhibited; diuretic effect (6); increased O_2 consumption (7)		high O_2 (2), salinity (4), CO_2 (5)	0.01-0.05 mg/l	(1) Smith & Piper (1975); (2) Alabaster *et al.* (1979); (3) Mukherjee & Bhattachanya (1974); (4) Herbert & Sherben (1965); (5) Lloyd & Herbert (1960); (6) Lloyd & Orr (1969); (7) Smart (1976)
NO_2	methaemoglobinaemia reduces O_2 uptake		NaC1 at 16:1, $Cl^-:NO_2^-$ (1), hardness, NaC1, $CaC1_2$ (2)	0.05-0.20 mg/1	(1) Tomasso *et al.* (1979); (2) Wedemeyer & Yasutake (1978)
CO_2	respiratory activity; nephrocalcinosis (2) implication	adaptation shown (1)	high O_2	<10 mg/1	(1) Eddy & Morgan (1969); (2) Smart *et al.* (1978)

Factor	Effects	Adaptation	Mediation	Limits	References
Suspended solids	mucus production, bacterial gill disease implicated	'coughing'	—	20-40 mg/1	
Cl				0.01 mg/l	
H_2S	neurotoxic effect, erratic swimming behaviour, reduced growth	—	poss. pH effect	0.01 mg/1 -0.4 mg/l	
O_3			0.01 mg/l		
NO_3	osmotic effect	normal hyperosmotic adaptation?	—	400 mg/1	Weston (1974)
N_2	gas bubble disease, swimming ability decrease	adaptation observed	—	105-120% sat'n	Fickeisen & Schneider (1976)

circumstance. Thus it may be necessary to evaluate conditions on a case-by-case basis, or if the cost of maintaining specific environmental levels can be determined more exactly, effects on growth could be used to optimise design. The comment by Sowerbutts and Forster (1980), concerning the danger of setting single tolerated values as design levels in fluctuating conditions, should be noted.

Other environmental factors which are of possible significance in water reuse are heavy metals, where metallic components are in long-term contact with moving water, supersaturated gases, and disinfectant or treatment agents, though these can effectively be eliminated by good design and management.

The presence of pheromones or other organic compounds acting intra- or inter-specifically to control growth or reproduction has been suggested to affect fish kept in enclosed areas (Yu and Perlmutter, 1970; Pfuderer *et al.*, 1974; Henderson-Arzapalo *et al.*, 1980). The need for such compounds to be effective in dilute concentrations for possibly prolonged lengths of time in biologically productive natural waters suggests a refractory nature and possibly an accumulating effect in recycled systems.

The speed of flow may be controlled specifically or be determined as a consequence of water exchange and stock requirements, together with the configuration of the holding facility. In addition to controlling energy utilisation, and in some circumstances stimulating growth, feeding opportunity and behavioural structures can be influenced. In conventional designs, velocities may be determined by the need to scour and self-clean the holding areas, normally at least 1 cm/sec (Kerr, 1980) to 3 cm/sec (Burrows and Chenoweth, 1970). The relationship between these velocities and those maintained by fish stocks is shown in Figure 6.3. Table 6.8 shows the normal range of stock densities employed, based on volume and flow requirements.

The effects of water temperature are relatively well understood and described (EIFAC, 1969), typical effects for a number of species being shown in Figure 6.4. Optimum temperatures for growth have been reviewed by McAuley and Casselman (1980). Within the acceptable range for growth, the greatest effect will be that on feeding and growth rate and as a behavioural trigger, though in water reuse these effects are complicated by the effect of temperature on the process water and on the treatments employed.

Intensity and periodicity of light are known to have effects on feeding, growth and reproductive activity in a number of species (e.g. Poston, 1978), particularly those indigenous to temperate areas, where seasonal changes are associated with changing light patterns. Similar reproductive triggers may be found in water level and temperature changes in tropical species such as *Clarias* catfish and tilapia (Balarin and Hatton, 1979).

2.2 Metabolic Effects

Metabolic activity results in a range of solid and soluble wastes being produced, while respiratory activity removes oxygen and replaces carbon dioxide. The most significant effects on these are caused by temperature, feed consumption, feed type and activity, though a comprehensive relationship has yet to be determined. The processes of metabolism have received considerable attention in respect of energy balance and growth and temperature (e.g. Winberg, 1956; Paloheimo and Dickie, 1966; Brett, 1970; Elliott, 1976; Braaten, 1978). While relationships can

Table 6.8: Range of Stock Densities

Stock Density Effects (kg/m^3)

Density	Species	Effect	Reference
4-15	*Cyprinus carpio*	highest stock rate, recirculated systems	Chiba (1980)
16	*O. kisutch* smolts; *S. gairdneri* (4-5 inches)	physiological stress	Wedemeyer (1976)
25	*Pleuronectes platessa; Scopthalmus maximus*	conventional production	Kerr (1976)
20	*S. salar*	normal ongrowing	Edwards (1978)
25	*S. salar* fry	normal maximum	Edwards (1978)
	I. punctatus		Allen (1974)
30-95	*Cyprinus carpio*	highest stock rate, running water	Chiba (1980)
up to 50	*S. gairdneri*	conventional production	Shepherd (1973)
⩾50	Tilapia	normal intensive production	Balarin (unpub.)
100	*A. japonicus*	acceptable growth at 1.20 lpm/kg	Chiba (1980)
⩽250	*C. carpio*	little growth suppression	Meske (1973)

Water Supply Effects (Flow Density, lpm/kg)

Rate	Species	Effect	Reference
0.16-0.66	general use		Kerr (1980)
0.2-2.8	*S. salar*	normal requirements, first feeding	Edwards (1978)
0.05-0.7	*S. salar*	normal requirements, ongrowing	Edwards (1978)
<0.84	*S. gairdneri*	growth decrease	Chiba (1980)
1.20	*A. japonicus*	acceptable at 100 kg/m^3	Chiba (1980)

be derived on the basis of laboratory studies, it is difficult to quantify energy balance terms with populations which are of disparate size, have different feeding opportunities and have different activities, and bulk measurements may be more appropriate for aquaculture conditions.

A number of early measurements of this form have been correlated with fish size or temperature, and thus Liao (1971) relates oxygen consumption as:

$$O_2 = KT^m W^n$$

where T = temp ($^\circ$C), W = individual fish weight (g), with O_2 in g/day, and K values are derived for a specific feeding regime. Details of weigh-related outputs are given in Table 6.9a. However, as feeds and feeding rates become more established and convenient for use, a number of relationships have been based on the simplified assumption of metabolic effects being proportional to feed input

Figure 6.3: Fish Weight and Water Velocities.

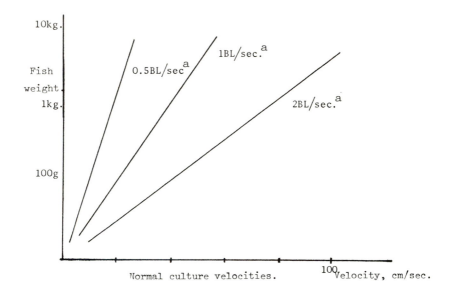

Note: a. BL = Bodylengths

Figure 6.4: Typical Temperature Effects

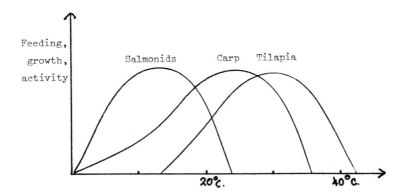

within the normal ranges of food demand:

$$(X) = kF$$

where (X) = metabolic effect, F = food input. Typical values are shown in Table 6.9b.

This approach appears to be generally valid for overall effects, but neglects diurnal variations, activity and the specific dynamic action (SDA) effects of increased metabolic activity after feeding. Moreover the relationships assume correct matching of feed to fish size and temperature, and by inference an accurate stock weight.

However, in terms of ease of assessment the accuracy of this method is probably acceptable; a more exact approach, such as the utilisation equation for ammonia production by Meade (1974), is:

$$NH_3N \text{ production} = \text{feed rate x biomass x protein nitrogen in feed x protein}$$
$$\text{utilisation x N excreted as ammonia}$$

However, this requires more information than can normally be provided in routine use. Hambrey (pers. comm.) has noted that in ammonia production growth was a better index of protein used and metabolised and would explain output more accurately, but although growth may be assessed as food rate times conversion, growth and food conversion are poorly predictable and so the method may only have *post facto* use.

The output of nitrate, phosphate, urea, creatine and other nitrogenous compounds is generally insignificant, although the first two may increase as solids break down. Levels of urea are generally low, and most cultivated species do not have a functional ornithine cycle for urea production, but may produce quantities through dietary arginine or the uricolytic pathway, allowing increased tolerance to high ammonia levels (Colt *et al.*, 1979). Burrows (1964) noted that salmonids in higher stock densities produced lower amounts of urea, while Muir (1978) found that urea levels were generally unaffected by stocking density of juvenile rainbow trout, though levels appeared to increase during periods of low ammonia concentration.

2.3 Variation in Output

Diurnal variations in metabolic effect have been recorded in a number of instances, apparently related to feeding and photoperiod-based activity. Thus Brett and Zala (1975) noted a peak of 35 mg NH_3-N/kg/hr about 4.5 hours after feeding in *O. nerka*, from a baseline of 8.2 mg/kg/hr and Muir (1978) found a variation of output directly related to feeding events in a marine recycling system. In intensive tilapia production, oxygen consumption rose within 2-3 hours post-feed (Muir, unpub.), while in carp production, Huisman (1969) noted similar effects. Studying fasted and fed fish, Rychly and Marina (1977) noted a single morning peak with the former and a double peak with the latter. The increased activity of fish at sunrise and sunset is commonly noted.

The effect of cleaning and other husbandry procedures has been described by Elliott (1969) and others. As depicted in Figure 6.5 levels of 2-3 times baseline levels may be produced.

Table 6.9a: Metabolic Effects — Stock Weight Related (kg/ton/day)

	Feed	O_2	NH_3-N	NO_3-N	Tot-N	PO_4-P	Tot-P	BOD[a]	Solids		Reference
Rainbow trout	trash fish	—	—	—	0.5-0.7	—	0.06-0.10	2.4-3.6	—	pond and	Warrer-Hansen (1979)
	dry food	—	—	—	0.12-0.24	—	0.03-0.05	0.9-1.5	—	tank	
	moist pellets	—	—	—	0.30-0.45	—	0.04-0.06	1.0-1.7	—	effluent	
O. nerka	baseline		0.20								Brett and Zala (1975)
	post-feed		0.84								
Rainbow trout	trash fish	—	0.54	0.06	0.81	0.09	0.18	4.88	8.50	pond and	Warrer-Hansen (1979)
	dry food	—	0.13	0.06	0.38	0.05	0.10	1.88	5.00	tank	
	moist pellets	—	0.25	0.05	0.50	0.04	0.10	2.12	—	effluent	
Salmonids	pellets	—	0.59								Meade (1974)
Rainbow trout		—	0.41							growing fish	Shirahata (1964)
Rainbow trout (200 g)		2.6 (6°C) 6.8 (16°C)									Forster *et al.* (1977)
Rainbow trout (130 g, 10°C)		0.31									Kaushik (1980)

Note: a. Biochemical Oxygen Demand.

Table 6.9b: Metabolic Effects — Food-related

	Feed % BW	Temp	ρF^a (pm/kg)	ρV^a (kg/m3)	O_2	NH_3-N	NO_3-N	PO_4-P	BOD	Solids	Other	Reference
Rainbow trout (1.5-15 cm)	1-5	15	0.85	27.8	—	.0289F[b]	.024F	.0162F	.60F	.52F	COD[c]1.89F	Liao & Mayo (1974)
Rainbow trout	—	—	—	—	.25F	.032F	.087F	.005F	.34F	.30F		Willoughby et al. (1972)
Channel catfish (1.25-2.5 kg)	2	25	0.16	4.3	—	.0035F	.017F	—	.25F	—	CO_2 .305F	Murphy & Lipper (1970)
Salmonids — average	1.14 -7.15	10- 12.5	2.35	5.6	.21F	.0365F	.112F	.0094F	.376F	.85F	50% solids settleable	Liao (1970)
(Range)			(.384) (-19.2)	(1.1) (-18.3)	—	(.0014) (-.1790)	(.0043) (-.313)	(.0022) (-.0248)	(.192) (-1.145)	—	—	
Trout (1)	—	8-15	0.94	16	—	.1025F	—	—	—	—	—	Brockway (1950)
(2)	—	8-15	0.94	32	—	.0405F	—	—	—	—	—	
Goldfish (2.5 kg)	—	20-29	—	—	—	.0229F	—	—	—	.308F	—	Kawamoto (1961)
Sunfish (29.7 kg)	—	—	—	—	—	.0242F	—	—	—	—	—	Gerking (1955)
Salmonids	—	—	—	—	—	.076F- .280F	.006F .057F	.001F .025F	.470- 1.64F	.58F	NO_2 .002F -.039F	BSFW (Liao & Mayo, 1974)
Chinook fingerling	—	8.13	0.12 -0.61	4.23 -42.3	—		—	—	—	—	Urea-N .012-.176F	Burrows (1964)
Various	Unfed	12	—	—	—	.0025F -.018F	—	—	—	—	Decrease at high F, ρF	Wood (1958)

Notes: a. ρF = flow-based stock density; ρV = volume-based stock density.
   b. F = weight of food supplied.
   c. COD = Chemical Oxygen Demand.

Figure 6.5: Typical Husbandry Effects on Water Quality.

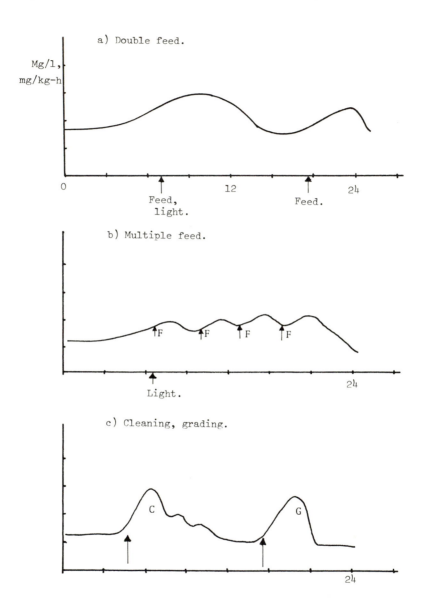

Source: After Elliott (1969) and Muir (1978).

2.4 Uneaten Food

The quantification of uneaten food in aquaculture conditions is difficult, and little allowance is made for either unconsumed food or the effects of its breakdown. Warrer-Hansen (1979) estimated wastes as 10-30 per cent when trash fish were fed, 5-10 per cent for moist pellets and 1-5 per cent with dry pellets, though it was not stated whether these differences were also due to poor water stability of the food, poor distribution or inflexibility of supply (i.e. dry pellets being normally more easy to regulate in quantity and distribution).

The effects of uneaten food in a recycled system will depend on its stability and the amount of time it spends in the system. If the solid wastes are removed effectively, uneaten food will appear to reduce overall waste output based on food fed. If it is held within the system, the oxygen consumed to metabolise it should be approximately similar to that used if the fish had consumed the food, though nitrogen accumulation could be greater.

The effects of environmental conditions on waste output are poorly described, though it would be expected that mechanisms could act to alter the fish's ability to metabolise feeds ingested, and thus affect output levels. Thus, Muir (1978) found rainbow trout waste output and oxygen consumption to be reduced at higher stock densities, and Kawamoto (1961) noted that carp held in water at NH_3-N 0.0037 mg/l showed reduced oxygen consumption.

Kaushik (1980) examined the effects of temperature rise (10-18°C) on rainbow trout, finding an increase from 16 mg N/100 mg food consumed to 29 mg N/100 mg on temperature increase, stabilising to approximately 20 mg N/100 mg. Lloyd and Orr (1969) noted increased urine flow in rainbow trout exposed to sublethal ammonia concentrations. However, Olsen and Fromm (1971) did not find different nitrogen excretion patterns for rainbow trout in similar conditions, but did find goldfish to alter nitrogen excretion. In terms of total waste nitrogen and ammonia nitrogen however, Fromm and Gillette (1968) found decreases with increased ammonia levels.

Perhaps more importantly in fish culture conditions, impaired feeding and growth or diseases related to environmental factors are likely to result in poorer food utilisation and possible excess of uneaten food. Problems in levels of wastes produced in freshwater open-flow systems have recently encouraged the development of high-digestibility foods, with reduced solid waste output. The effect of wastes on water quality is shown in Table 6.10, which is adapted to show the normal contribution to overall waste levels.

Table 6.10: Normal Contribution to Overall Waste Levels (mg/l)

Typical Effluent	Muir (1978)	Warrer-Hansen (1979)
BOD	5	5.0-20
Tot-N	2	1.0-4.0
NH_3-N	0.5	0.3-0.5
Tot-P	0.10	0.05-0.15
SS[a]	7	5-50

Note: a. Suspended solids.

2.5 Water as a Medium

The properties of water in supplying environmental requirements or mediating metabolic effects affect the need for its replenishment or treatment. Table 6.11 shows some of the important properties of gases or ions in water; pH, temperature and, to some extent, salinity have the greatest effects; Figure 6.6 and Table 6.12 describe the most significant equilibrium relationships.

Figure 6.6a: Oxygen Concentration in Water.

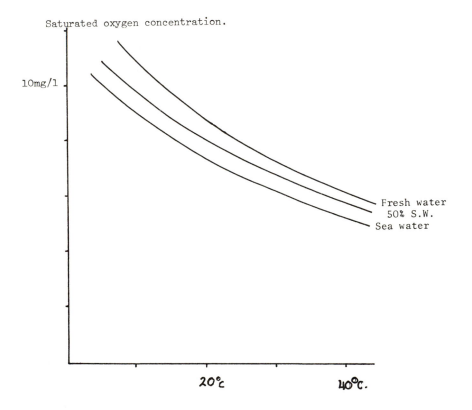

When comparing environmental requirements with waste outputs to determine design limits, a simple procedure of 'critical factors' as shown in Table 6.13 may be used to identify priority of treatment; the lower the factor the more immediate the treatment required. Thus it can be seen that oxygen is normally the most critical factor, followed by ammonia, carbon dioxide, or solids levels, depending on pH, temperature or other variables.

3. Overall System Design

The equations in Figure 6.1 can be used to calculate the amount or type of treatment required at a particular rate of recycling, or conversely the recycling

Figure 6.6b: Ammonia Dissociation in Water.

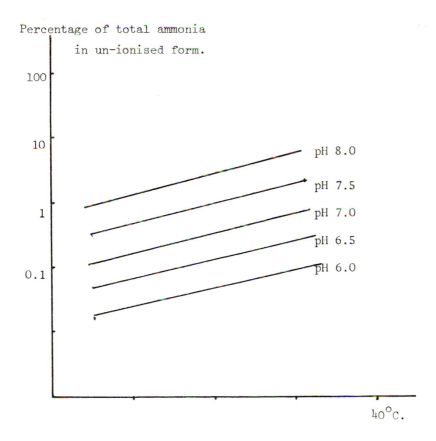

Percentage of total ammonia in un-ionised form.

Table 6.11: Dissolved Gas Characteristics

Gas	Absorption Coefficient at STP (ml/1)			Density at STP (g/1)	Diffusion Coefficient $(cm^2/hr \times 10^{-2})$	Transfer Coefficient (cm/hr) $[T^{\circ}C]$	
	$10^{\circ}C$	$20^{\circ}C$	$30^{\circ}C$				
O_2	38.4	31.4	26.7	1.429	9.4	0.41	$[25^{\circ}C]$
CO_2	1,190	878	665	1.977	7.4	1.46	$[25^{\circ}C]$
N_2	18.5	15.5	13.6	1.251	6.8	[Unstirred condition]	
NH_3	910	711		0.771	6.4		
H_2S	3,520	2,670		1.539	5.8		
O_3	520	368	233	2.144			
Cl_2	3,100	2,260	1,770	3.214	5.1		

Table 6.12: Equilibrium Relationships

Low pH				High pH
Carbonate System				
CO_2+H_2O $\rightleftharpoons$	H_2CO_3 $\rightleftharpoons$	$H^++HCO_3^-$ $\rightleftharpoons$		$2H^++CO_3^{2-}$
Bicarbonate Replenishment				
$CaCO_3+CO_2+H_2O$		$\rightleftharpoons$		$Ca^{2+}+2HCO_3^-$
Ammonia				
$NH_4^++OH^-$		$\rightleftharpoons$		NH_3+H_2O
Nitrite				
HNO_2		$\rightleftharpoons$		$H^++NO_2^-$
Hydrogen Sulphide				
H_2S $\rightleftharpoons$	HS^-+H^+		$\rightleftharpoons$	$S^{2-}+2H^+$
Phosphorus				
H_3PO_4 $\rightleftharpoons$	$H^++H_2PO_4^-$ $\rightleftharpoons$	$2H^++HPO_4^{2-}$ $\rightleftharpoons$		$3H^++PO_4^{3-}$

Table 6.13: 'Treatment Factors'

Water Quality Parameter	Typical Conc'n (mg/1)			Typical Rate of Change $R = \frac{dC}{dt}$	Treatment Factor $\triangle C/R$
	C_i	C_l	$\triangle C$		
Oxygen*	0	5	5	.09W	55.5
Ammonia	0	1	1	.009W	111.1
Nitrate	5	100	95	.001W	95,000
Solids	0	10	10	.15W	66.6

*$(C_{SAT}-C)$.
W = stock weight.

rate available with a given type and efficiency of treatment. It can also be shown that up to specified recycle rates, fresh inflow water is sufficient to maintain particular parameters. The limits before treatment is required can be defined by

$$C_W = C_M/(1 + EX-X)$$

where C_W = waste (outlet) concentration, C_M = metabolic output, E = treatment efficiency, and X = recycle ratio. If E = O (i.e. no treatment)

$$C_W = C_M/(1 - X)$$

Hence, where C_W is set at its maximum, defined by environmental requirements (i.e. = C_1), and C_M, based on normal metabolic outputs, is specified, X, the recycle ratio, can be determined.

In this manner, we can show (Figure 6.7) the levels of recycling possible before treatment. Figure 6.8 shows the typical order of complexity of treatment. It can be noted for example for maximum unionised ammonia of 0.05 mg/l, at recycle levels of up to 98% at pH6, and 65% at pH8 (15° C), ammonia removal will not be required, while at a limit of 100 mg/l NO_3 renewal will not be required until 90% recycle.

Figure 6.7: Recycling Ratios and Treatment Levels, Oxygenation Only. The figure shows the amount that water can be recycled (without treatment) before specified water quality limits are reached.

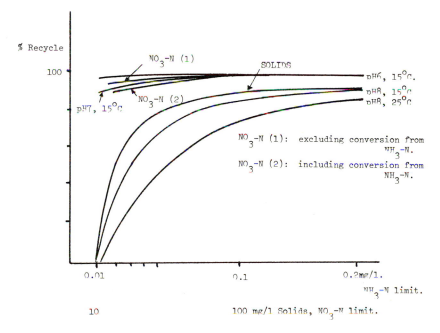

Where oxygenation is the sole form of treatment, the C_W levels for other factors can be calculated on the basis of their relationship to feed levels. Thus where oxygen demand is 5 mg/l at 0.3F, ammonia addition will be 0.5 mg/l at 0.03F, and corresponding recycle ratios can be calculated.

3.1   Treatment Methods

Table 6.14 illustrates the role of types of treatment; individual treatment stages may affect more than one water quality factor, though in the following sections, treatments are grouped according to their effect on individual factors. The acceptable operating levels will in turn affect the efficiency of treatment of individual parameters, particularly if treatment is concentration-dependent, as are most of the biological methods. Thus if oxygen levels are increased, producing higher ammonia levels, biological treatments for ammonia removal will tend to be relatively more efficient.

Individual treatments are discussed in the following sections according to their approximate order of importance as water reuse increases.

Figure 6.8: Recycle Ratio and Treatment Required.

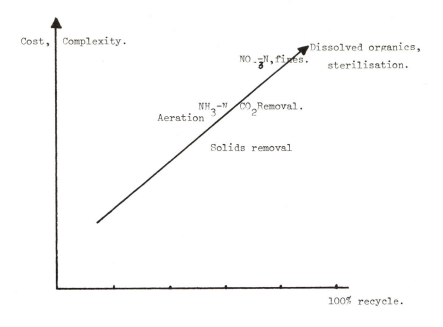

Table 6.14: The Roles of Types of Treatment

Water Quality Factor	Treatment Methods
Oxygen	Aerate or oxygenate water, mechanically (weirs, paddles, jets) or chemically; grow photosynthetic organisms.
Ammonia (NH_3)	Convert to nitrites and nitrates with nitrifying bacteria in filters or aeration tanks; adsorb onto ion exchange resin; transfer to atmosphere by air bubbling through water; react chemically with chlorine.
CO_2	Lime dosing; desorption by air stripping.
pH	Use limestone filter rock or clam, oyster shells to neutralise; pH dosing.
Suspended solids	Settle in ponds or tanks, filter mechanically, coagulate; use particulate feeding organisms.
Biochemical Oxygen Demand (BOD)	Remove solid wastes and ammonia.
Nitrites (NO_2)	Convert to nitrate with nitrifying bacteria.
Nitrates (NO_3)	Convert to nitrogen with denitrifying bacteria; nutrient for other aquatic organisms — e.g. algae.
Phosphate	Supply other aquatic organisms; form insoluble salts — e.g. calcium phosphate.
Colour	Activated carbon, ozone.
Bacteria or other pathogens.	Ultra-violet irradiation, chlorination, ozone treatment.
Cl^-	Activated carbon, UV, sodium sulphite.
N_2	Desorption by airstripping.

4. Oxygen Supply

Oxygen may be supplied to water from atmospheric air or by using pure oxygen. The rate of transfer of oxygen from gas to liquid may be determined according to:

$$dD/dt \text{ (mg/l/hr)} = K_\triangle D$$

where D = concentration, K = mass transfer co-efficient, $\triangle D$ = difference between actual concentration and saturation concentration.

This relationship can also be used with the appropriate 'K' value for transfer of N_2, NH_3 or CO_2. While there is advantage in removing the latter two, the undesirable introduction of N_2 during oxygen transfer needs to be avoided and deserves careful attention in design. The 'K' term (or K_La, as transfer is controlled by the liquid film) is dependent on the opportunity of contact between gas and liquid phase, and is normally expressed as

$$K_La = K_L(A/V)$$

where K_L = transfer coefficient (cm/hr), A = area of gas/liquid interface (cm^2), and V = water volume (cm^3). The term (A/V) represents the turbulence and renewable contact opportunity with the water volume, and K_L is controlled by gas diffusivity and bubble size. Thus comparative transfer rates for the gases can be obtained normally by development from experimental determination of transfer rates at standard pressure and temperature (usually $20°C$), using maximum $\triangle 0$. Petit (1980) notes that in standard conditions oxygenation using pure oxygen as the gas phase could give up to five times the oxygen transfer rate of conventional aeration, though in practice this advantage is normally about four times.

In the context of fish culture it is particularly important to note the effects of temperature, level of organic wastes and surfactants on transfer rates. Thus Chesness and Stephens (1971) found K_La to increase by 1.56 per cent per $°C$ temperature rise, using catfish culture water, and Ogden *et al.* (1959) described a reduction of up to 60 per cent in aeration rate at levels of up to 3 mg/1 of surfactive detergent, and up to 50 per cent reduction as a linear relationship with salinity from 0 to 36 per cent.

Comparisons between aerators have been provided by Mitchell and Lev (1971) and Shell and Cassady (1973) for waste treatment and by Liao (1971), Busch *et al.* (1974), Mitchell and Kirby (1976), Rapaport *et al.* (1976), Colt and Tchobanoglous (1979) and Sowerbutts and Forster (1980) for aquaculture use. Descriptions of design aspects in aquaculture have been given by Liao and Mayo (1974) and Wheaton (1977) for aeration, and by Petit (1980) and Sowerbutts and Forster (1980) for oxygenation. Evaluation is normally in the form of oxygenation rate (kg/hr) and efficiency (kg O_2/kWhr) at a standard oxygen deficit ($\triangle 0$). A description and comparison of the efficiencies of aerator and oxygenator types are shown in Table 6.15. It should be noted in particular that the efficiencies recorded for fish culture conditions are considerably less than those obtained in standard conditions (Mitchell and Kirby, 1976). The selection of appropriate aeration devices may also depend on characteristics such as capital cost, space available, additional water-mixing effect desired and

Table 6.15: Oxygen Transfer Efficiencies (η)

Type	η (kg O_2/kW)	Conditions
Surface aerators (water spray)	1.2-2.4	standard
	0.48	25°C, 4.2 mg/1, ponds
	0.25-0.5	20°C, 6 mg/1, ponds
	1.9-2.3	near standard
Surface agitators	1.2-2.4	standard
Brush agitators	1.3-3.5	sewage (low O_2)
Venturis	0.6-2.4	standard
Paddlewheel	up to 2.4	standard
Waterjets	1.3-2.6	fish culture
Cascades	1.2-2.3	standard
	1.3-1.7	raceway water
	1.9-2.3	lattice, raceway water
Coarse bubble (to 10 mm)	~1	standard
Coarse bubble airlift	to 1.8	
Fine bubble, 2-5 mm	1.5-6.0	standard
	0.96	25°C, 4.2 mg/1, ponds
	0.25-0.42	20°C, 6 mg/1, ponds
U-tube	2.5-4.5	standard
	1.36	lake oxygenation
Oxygen chamber	0.25	5.7 mg/1

Sources: Petit (1980), DOE (1973), Wheaton (1977), Rapaport *et al.* (1976), Liao (1974).

control of oxygen input.

In comparing oxygenation with aeration, Sowerbutts and Forster (1980) suggest that oxygenation is only likely to be cost-effective in large-scale systems, but has advantages in controllability, freedom from nitrogen supersaturating effects, lower power, distribution and maintenance requirements, and additional supply safety if reservoirs are used. Furthermore, while aeration is limited in efficiency to maintaining levels somewhat below full saturation, oxygenation could maintain fully saturated conditions. The use of a 'sidestream' operation, using oxygen supersaturated water to mix with the main process water offers particularly controllable conditions if activated by sensing electrodes. At present, however, the distribution and storage costs of oxygen present difficulties compared with the relatively simple and cheap aeration equipment available, and until developments in molecular sieve (pressure swing arm) systems are made, on-site oxygen generation will not be economically feasible.

Petit (1980) considered dual aeration/oxygenation systems using aeration for the first stages and oxygen-only near full saturation levels, to compare the advantages of each system. Although overall power requirements were reduced, capital costs of the dual system were greater than those of the system for oxygenation alone.

5. Nitrogen Removal

The removal of ammonia from water has been reviewed by Short (1973) and
Spotte (1979): the main systems are described in Table 6.14. Much of the
available literature concerns the removal from waste water at ammonia
concentrations higher than those normal in fish culture, and note must be made
of the dependence of removal rate on concentration. Physico-chemical methods
in contrast are generally independent of concentration.

Short (1973), in comparing methods for ammonia removal from river wastes,
identified biological treatment as the most cost-effective and Burrows and
Combs (1968) found biological treatment to be the most practical and
economical treatment method for aquaculture use. The conventional nitrification
process is based on filtration, though other fixed media equipment, such as
biosedimentation (Short, 1973; Liao, 1980), biodiscs (Steels, 1974; Lewis and
Buynak, 1976), or suspended-media systems such as activated sludge (Meske,
1976; Rosenthal, 1980b) may be employed.

Nitrification. Nitrification has received considerable attention for its role in
water treatment practice. Painter (1970) reviewed nitrifying processes and work
on pure cultures of nitrifying bacteria, and recent descriptions have included
treatment of low ammonia levels in river waters (Short, 1973), nitrification in
sewage treatment (Haug and McCarty, 1971; Wylde *et al.*, 1972) and nitrification
of fish culture wastes (Burrows and Combes, 1968; Liao *et al.*, 1972; Liao and
Mayo, 1972; Muir, 1978). Design procedures for nitrification in fish culture have
been developed by Speece (1973), Liao and Mayo (1974) and more recently
by Hess (1979).

The principal organisms responsible for nitrification are the autotrophic
(independent of organic carbon sources) bacteria *Nitrosomonas* and *Nitrobacter*,
of which a number of freshwater and marine species have been identified. The
overall equations for cell synthesis and oxidation are described by Haug and
McCarty (1971) as

Nitrosomonas $55\ NH_4^+ + 5\ CO_2 + 76\ O_2 \rightarrow C_5H_7O_2N + 54\ NO_2^- + 52\ H_2O$
$\qquad\qquad + 109\ H^+$

Nitrobacter $\quad 400\ NO_2^- + 5\ CO_2 + NH_4^+ + 195\ O_2 + 2\ H_2O \rightarrow C_5H_7O_2 +$
$\qquad\qquad 400\ NO_3^- + H^+$

Similar relationships for synthesis and metabolism have been described by Meade
(1974).

By stoichiometry, a number of relationships can be derived:

$1\ kg\ NH_4-N \rightarrow 147\ g$ *Nitrosomonas*
$\qquad\qquad\qquad 20\ g$ *Nitrobacter* $\rightarrow NO_3-N$
$1\ kg\ NH_4-N \rightarrow 3.02\ kg\ O_2 \qquad \rightarrow 1\ kg\ NO_2-N$
$1\ kg\ NO_2-N \rightarrow 1.02\ kg\ O_2 \qquad \rightarrow 1\ kg\ NO_3-N$

In practice, cell yields and oxygen consumption may vary depending on growth
conditions, inhibition and cell age (Haug and McCarty, 1971).

The addition of hydrogen ions and, to a lesser extent, the removal of CO_2

during nitrification may in turn affect the carbonate system or the pH of process water; on the basis of neutralisation by bicarbonate ions, recycled systems should incorporate some form of buffering to supply carbonate or bicarbonate ions. The rate of nitrification is controlled by a number of factors affecting the population, activity and nutrient supply of the nitrifying bacteria. The principal factors are shown in Table 6.16.

Table 6.16: Factors Influencing Nitrification

Effects	Factor(s)
Growth	Mg, Fe, Cu, Ca, Zn, Mo, PO_4; optimum K:N 1.35[a].
Inhibition	Thiourea, allylthiourea, 8-hydroxyquinoline, peptone, some lipid-soluble compounds, NO_2[a], NH_4[b], compounds with S and N on same C atom, metals.
Activity	O_2 above 0.2 mg/1, 3-4 mg/1 in bulk concentration; 5°C min, c. 35°C optimum; pH min. 5.5, adapt 6.0, optimum 6-9

Notes: a. *Nitrosomonas*.
b. Nitrobacter.
Sources: Painter (1970), Haug and McCarty (1971).

There is evidence from specifically inhibited waste treatment systems that nitrification may also depend on heterotrophic species (Painter, 1970) but the difficulty of examining pure cultures of autotrophic nitrifiers does not assist in assessing their significance. For this reason particularly, evidence from the performance of treatment systems tends to be of greater value than results based on pure cultures.

5.1 Nitrification in Treatment Systems

Nitrification is usually associated with the film or floc bacterial or mixed-community materials present in the treatment system, and the means by which the nitrification factors shown in Table 6.16 are provided are determined by the process variables shown in Table 6.17. There are clearly a number of options in system design and it is preferable to establish at the outset the nature of the duty required, e.g. type and concentration of waste loading, cleaning cycle, etc. In fish culture terms, the stability of operation is particularly important, especially where a risk of shock loading, possible toxicants and process changes may occur.

At the micro-level, the processes of nutrient transfer and biological growth are similar in most biological treatment systems. Fair *et al.* (1971) note three successive processes: transfer from water to film or floc, preservation of contact quality (i.e. available surface area) by cell growth, enabling continuous transfer by absorption and adsorption of nutrients and metabolites, and conversion of biomass into removable form. In extended processes, a fourth stage, stabilisation of removable biomass into inert material, may also occur.

A number of models have been proposed for describing the rate of transfer through nitrifying films and floc; Bruce and Boon (1970) showed that in plug-flow

Table 6.17: Design Variables for Biological Treatment[a]

Type of Treatment:	FIXED	SUSPENDED	HYBRID
(1) Waste loading	Contact surface, structure, mixing/distribution, microbial conditions	Size, no. and structure of suspended particles, mixing, microbial condition	Contact surface and suspended particle quality, solids loading
(2) Waste removal	Contact time, microbial condition, mixing/distribution dilution (recycle), nature of waste	Hydraulic loading, dilution (recycle), microbial condition, nature of waste	Contact time (disc rotation in biodisc)
(3) Continuity/stability	Accumulated solids, void volume, load patterns, ecology	Particle concentration, ecology, settling behaviour load patterns	Accumulated solids, particle concentration, ecology, load patterns
(4) Metabolic requirements	Ventilation, nutrient distribution, film thickness	Aeration, recycle	Rotation speed, air supply

Note: a. Criteria for fish culture use include simplicity and economy of construction and management, and adaptability and stability with respect to nutrient level and other environmental factors.

or completely mixed filters modified Monod equations such as those for first order reaction rates in bacterial growth could describe nutrient removal and cell growth. Thus reaction rate would vary directly with nutrient concentration:

$$dC/dt = -k(C_L - C)$$

Haug and McCarty (1971), developing the model of Atkinson *et al.* (1967) used a numerical integration method to describe concentration gradients and removal rates within the film of submerged filters (Figure 6.9). For nitrification, these authors found the active film depth (L_a) to be temperature-dependent at 700 microns (25°C) to 1,300 microns (5°C), with higher NH_3–N concentrations also increasing L_a. Over a range of 0.5-40 mg/l, movement of nutrient was proportional to the square root of its diffusion coefficient. Within the liquid area, with the horizontal mixing normally present, diffusion rates within the film are affected by particle charge, and hence a counter-flow of oppositely charged ions is required, the process being temperature-dependent. A typical overall transport scheme is shown in Figure 6.10, though in the case of nitrification, the transport of major species is unlikely to affect diffusion coefficients. A similar approach is applicable to floc structures in suspended-medium equipment, though the model developed was not tested in this respect.

Figure 6.9: Biological Film Zones and Removal Rates.

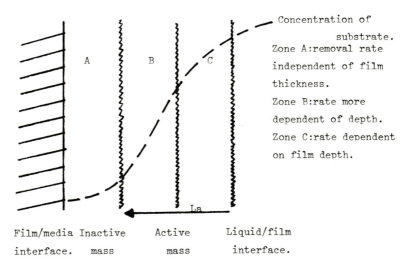

Concentration of substrate.

Zone A: removal rate independent of film thickness.

Zone B: rate more dependent of depth.

Zone C: rate dependent on film depth.

Film/media interface.   Inactive mass   Active mass   Liquid/film interface.

Source: Modified after Haug and McCarty (1971).

5.2 Film and Floc Ecology

The ecology of films and flocs in conventional waste treatment has been studied extensively by Hawkes (1963), and a considerable and changing diversity of organisms has been demonstrated to occur. Thus it is unlikely that single-species models are completely applicable to waste treatment; this is supported by the many observations of change in performance with change in ecology. This aspect is particularly important in terms of system stability, either at constant loading or under the external shock loadings easily produced in aquaculture conditions.

Figure 6.10: Ion Transport and Nitrifying Film Surfaces.

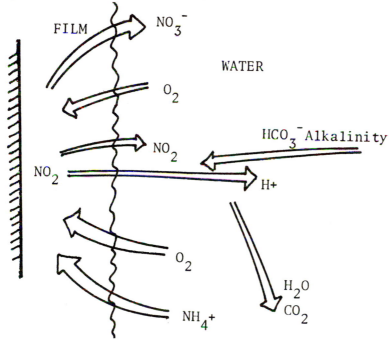

Source: Modified after Haug and McCarty (1971).

In submerged nitrifying filters Haug and McCarty (1971) observed a granular bacterial floc, with a variety of cell forms, though few free-swimming or filamentous organisms were noted. Higher organisms included ciliated protozoa, amoebae, rotifers and a large number of nematodes. Seasonal or temperature effects were not described. In a typical activated sludge system, where a more uniform environment is normally available (Hawkes, 1963), Pike and Curds (1971) noted a predominance of heterotrophic bacteria, with 228 protozoans, 160 ciliates, 25 amoebae, 17 phytoflagellates, 16 zooflagellates and 6 actinopod species. Changes with load were also observed by Curds and Cockburn (1970).

Bulking of activated sludge, where normally-settlable material rises and contaminates the outflow, is usually associated with organisms such as *Sphaerotilus, Nocardia, Beggiatoa, Geotrichium* and *Zoophagus* growing in conditions of low oxygen level and high C:N and C:P ratios (Pike and Curds, 1971). Studies of the ecology of fish-waste treatment are more limited, and may reflect more than mere nitrification activity. Thus Kawai *et al.* (1965) monitored populations during start-up of a sea-water aquarium system and found populations (per gram filter sand) as shown in Table 6.18. Though the primary role may be ammonia removal, these illustrate the mixed activity normally present and particularly the occupancy of treatment volume by non-nitrifying functions.

Table 6.18: Active Populations at Start-up of Aquarium

Total aerobes	2×10^7	Total anaerobes	3×10^5
Aerobic starch hydrolysis	1×10^6	Anaerobic starch hydrolysis	2×10^5
Gelatine liquefaction	2×10^4	Sulphate reduction	1×10^2
Aerobic nitrate reduction	2×10^6	Cellulose decomposition	2×10^2
Nitrification (N'as) SW	3×10^5	Urea decomposition	1×10^6
Nitrification (N'as) FW	2×10^2	Denitrification	1×10^5
Nitrification (N'er) SW	1×10^5	Ammonification	2×10^5
Nitrification (N'er) FW	1×10^2	Anaerobic nitrate red'n	1×10^5

5.3 Nitrification in Filters

The use of filters has been described by Haug and McCarty (1971), Muir (1978), Wheaton (1977) and Spotte (1979). Research into their performance in fish culture has been described by Speece (1969), Gigger and Speece (1970), Liao and Mayo (1974), Meade (1974), Forster (1974) and Muir (1978) for a number of different system types (Figure 6.11). Table 6.19 shows application rates and the relative advantages of particular filter types. The major operating parameters to be identified are filter configuration and flow distribution, available surface area, waste and hydraulic loading, duty cycle, and environmental variables such as temperature, pH, oxygen supply and organic loading.

Table 6.19: Filter Types and Comparative Advantages

	Advantages	Disadvantages
Trickling filter	Good oxygenation; grazing and external loss of accumulated biomass; lighter and possibly simpler (e.g. cages of simple media)	Lower loadings, poor dispersion, hold-up control; possible fly problem; head loss
Downflow submerged filter	High loadings, good dispersion, hold-up control; head retained; backwash in opposite direction to flow	Poor oxygenation; generally heavier construction
Upflow submerged filter	As above, save backwash; can operate in 'expanded media' mode	As above; backwash in same direction as flow
Recycled filter	Improved efficiency at high waste loadings	Return water flows required; more complex design and control
Alternate double filter	Alternate heavy load and endogenous respiration, high loadings	Additional valves and pipework; changing performance

Where flow is evenly and perfectly distributed over the internal filter surface, the rate of ammonia removal can be directly related to surface area, and so upflow or downflow regimes (Figure 6.11) are effectively equivalent. The

Figure 6.11: Filter Configurations.

Trickling Filter

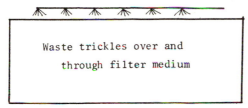

Flooded Filter

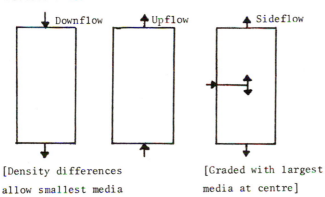

[Density differences
allow smallest media
    to the base]

[Graded with largest
media at centre]

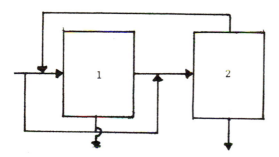

Filter order
changed to control
biological film.

Alternate Double Filter

percolating filter used for conventional waste treatment is more dependent on the thickness and distribution of the biological film in holding up solid and liquid wastes, and in directing flow through the medium. Eden *et al.* (1966) found flow-rates to have little effect on hold-up within these filters over a range of 0.3-1.2 m³/m³/day though surface tension forces were found to be significant.

The reactor dispersion number E/uH describes the range of mixing in filters from plug flow (0) to perfect mixing (∞). This is defined (Levenspiel, 1967), using flow tracer tests, as:

$$\sigma^2 = 2(E/uH) - (E/uH)^2 \, (1 - e^{-uH/E})$$

where σ^2 = variance of pulse tracer output curve in dimensionless time units (t/t_0), t_0 = theoretical detention time (V/Q), u = flow velocity, H = reactor length, E = dispersion coefficient.

Table 6.20: Ammonia Removal vs Hydraulic Flow-rate

	Removal (g/m³/day)	Flowrate (m³/m³/day)	Type	Other
Liao & Mayo (1974)	216	86.4	Upflow filter	Koch rings 10 cm, FW, trout
	123	49.7	Trickling filter	,,
	162	65.5	Activated sludge	,,
	41	16.5	Extended aeration	,,
Meske (1973)	33	16.1	Activated sludge	Carp, FW, 25°C
Risa & Skjervold (1975)				Salmon fry and fingerling
Goldizen (1970)	214	360.0	Filter	SW, lobster
Hall (1974)	300	68.2	Submerged filter	Limestone rock, catfish, FW, 25-30°C
Meade (1974)	665	101.1	Surfpac, submerged filter	Prior solids removal, Chinook salmon, FW, 15°C
Forster (pers. comm.)	501	20.5	Submerged filter	12-25 mm gravel, $(NH_4)_2\,SO_4$ Feed, 26±2°C SW
	1,112	82.0		
	2,178	246.0		
Short (1973)	14	18.5	Submerged filter	River waters, ammonia removal, media 20-40 mm, FW, 11-19°C, 3 mg/l
	13-19	17-25.5		
	245	12.0		
	438	30.0		
	334	22.8-57.6	Airstripping at ph 11.0	FW, 11-19°C, 3 mg/l
	1,000	63.6	Biosedimentation	FW, 11-19°C, 3 mg/l
Duddles *et al.* (1974)	48	13.7	Filtration	1520 g BOD/m³/day, air 401.1 m³/m³/d
	138	8.4	Filtration	331 g BOD/m³/day

Figure 6.12: Ammonia Removal and Hydraulic Loading.

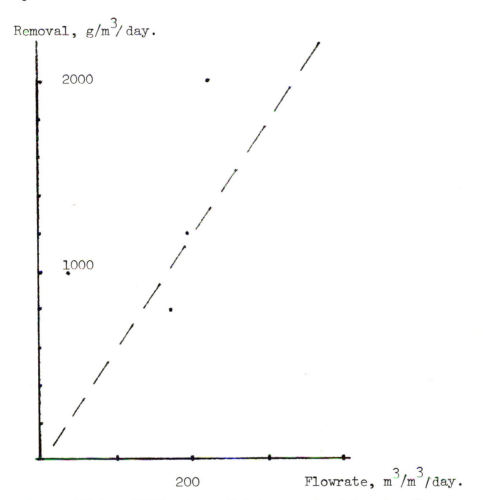

Removal, $g/m^3/$ day.

2000

1000

200

Flowrate, $m^3/m^3/$ day.

Haug and McCarty (1971), using an alkalinity tracer, found dispersion values to vary from 0.5 at t_o = 15 mins to 0.04 at t_o = 60 minutes with no bio-film, while clogging filters had dispersion numbers of 0.2-0.4, and normally developed filters had values of approximately 0.1.

The effect of hydraulic loading on nitrification has been examined by Forster (1974) and Muir (1978) using simulated fish-farm wastes, where NH_3 removal rates $(g/m^3/day)$ were found to vary directly with loading $(m^3/m^3/day)$ over a range of 0-100 $m^3/m^3/day$ (i.e. efficiency of removal remained approximately constant). This confirms the observation by Liao *et al.* (1972) that from 86.4 to 146.9 $m^3/m^3/day$, removal efficiency was not affected, and that retention time was the primary factor affecting removal. Other recorded results (Table 6.20) have been plotted in Figure 6.12 to illustrate the relationship, though it should be noted that the performance at higher hydraulic loadings was obtained with relatively tall cylindrical filters. The interrelated effects of turbulence and ionic transfer on the available film area suggests a more subtle relationship, and a more complete description would involve local velocities and film thickness and condition. It must be assumed that an ultimate limit is set by the available surface of actively metabolising films, and by the effect of high fluid velocities on the species, structure and performance of the film. Thus a maximum removal rate $(g/m^3/day)$ must be reached regardless of hydraulic loading. In practice, however, the pumping energy required at the higher velocities may impose limits well below this theoretical maximum.

Table 6.21: Filter Media

Type		Specific Surface Area (m^2/m^3)	% Void	'B'[a]	Weight (kg/m^3)	Assembly/Other
Rock	½ "	420	50	8.4	1,300	random, low cost (rock shape)
	1 "	210	50	4.2	''	''
	2 "	105	50	2.1	''	''
Ceramic Raschig	½ "	364	63	5.8	864	random or stacked, high cost
Ring	1 "	190	73	2.6	640	''
	2 "	92	74	1.2	608	''
Carbon	½ "	374	74	5.1	432	''
Raschig	1 "	187	74	2.5	432	''
Ring	2 "	93	74	1.3	432	''
Berl	½ "	466	63	7.4	864	''
Saddle	1 "	249	69	3.6	720	''
	2 "	105	72	1.5	640	''
Pall Ring[b]	1 "	217	93	2.3	528	''
	2 "	120	94	1.3	440	''
'Surfpac'						
Standard		82	94	0.9	64	1-4 cm space stacked
Close		187	94	2.0	48	modules, moderate cost (4 rock for area)
'Flocor'	PVC	85	98	0.9	37	
	RC	330	95	3.5	70	
'Cloisonyle'						
	PVC	220	94	2.3	80	single channel module

Notes: a. B = Index of 'Blockability' = Specific surface area/% Void: hence theoretical film thickness on single surface before 100% blockage (0% void) = 1/B.
b. PVC materials available for lower weight.

Filter Media. A range of filter media is available for use, including sand, gravel and rock material, random plastic shapes and structured plastic modules. Miscellaneous waste materials such as brushwood, plastic offcuts and net mesh have also been used. Table 6.21 illustrates typical characteristics, together with an 'Index of Blockability' determining the relative ease with which a particular material will block as film thickness increases. Bruce (1972) has noted that waste removal efficiency normally correlates with surface area over a number of media at high hydraulic loadings (6 $m^3/m^3/day$), though angled-channel media showed lower efficacies, presumably due to different liquid retention characteristics. Average film thickness was considerably greater on aggregate media, and higher relative nitrification rates were observed on smaller (30 mm) cracked rock than on larger (60 mm) rounded material. The relative effects at lower loadings were not noted, though reduced turbulence would tend to accentuate the liquid retention effect. Liao *et al.* (1972) and Speece (1973) used surface-area-related removal rates (i.e. $g/m^2/day$), which may be related to volumetric rates by the specific surface area of the media employed.

Backwashing. The reaction equations and yield coefficients for nitrification predict cell growth up to 0.7 times the mass of substrate metabolised. Assuming the cell mass forms a film of a similar water content (95 per cent to that of typical sludge floc (Fair *et al.*, 1971), at a representative NH_3–N removal rate of 400 $g/m^3/day$, Table 6.22 shows the effect of cell growth on typical media.

Table 6.22: Effect of Cell Growth on Typical Media

Removal rate 400 $g/m^3/day$; cell growth @ 0.2 x substrate = 80 $g/m^3/day$.

Film growth $= \ 80 \ \times \ \dfrac{100}{5} \ = 1,600 \ g/m^3/day$

Medium	m^2/m^3	Mean Film Growth (@ 1 g/cm^3)		Blocking Depth (mm)	Theoretical Blocking Time (days)
		$g/m^2/day$	mm/day		
½ " rock	420	3.80	0.0038	1.19	314
1 " ring	187	8.60	0.0086	3.96	460
Surfpac	187	8.60	0.0086	5.03	585

Bruce *et al.* (1970) found that in trickling filters with 2" media efficiency decreased sharply when loading reached 7.7 kg wet film/m^3, though this was not found with 1" media at even higher levels.

In practice, the trapping of suspended solid particles by the developing film tends to accelerate film growth, while a continuous sloughing of unstable film, dependent on water velocity and biological control by grazing organisms, reduces the blocking effect. Most filter systems are however arranged to provide a backwash cycle in which air or water is allowed to scour excess material into the waste — though Haug and McCarty (1971) note the ease by which complete stripping of film may occur, leaving a temporarily inadequate biomass for water treatment duty. For nitrification systems, these authors found a velocity of 0.25-0.75 cm/sec at fortnightly intervals of unspecified duration to be most effective in controlling growth. This compared with higher rates of 10-20 cm/sec used to expand solids filter beds (Short, 1973).

Csavas and Varadi (1980) describe the use of a backwash in a 3-5 cm rock filter for aquaculture use; the initial designed flow of 4.4 m/hr (0.12 cm/sec) was replaced by a short duration (10 minutes) flow of 16.6 m/hr (0.46 cm/sec) for greater effect. It was noted that normal operating power requirements of 10 kW rose to 50 kW during backflush cycle. The use of additional air scouring, as an adjunct to water backwash may reduce the power requirement; Burrows and Combs (1968) note the successful use of a weekly air scour at 37.6 l/min/ft^2 (6.74 cm/sec) in a crushed rock/oyster shell filter.

The multiple of flow and duration will determine the volume of water used in the backwash cycle, and depending on backwash efficiency, the solids removed. If water is to be conserved, this supply may be treated for solids removal prior to reuse, or kept as a self-contained stock. Water of poorer or different quality (e.g. sea water in fresh-water systems) may also be considered, subject to limits of toxic effect on the film organisms.

Operating Conditions. The performance of nitrifying filters is affected by a number of operating conditions, such as temperature, oxygen supply, presence of other nutrients and of inhibiting substances. The effect of temperature is mainly metabolic and may be defined by conventional equations. Additionally, temperature is found to affect film thickness in trickling filters, with winter conditions producing thicker films and lower BOD removal efficiency (Bruce *et al.*, 1970) though Haug and McCarty (1971) considered increased thickness and hence cell mass to balance out the reduced activity at lower temperatures.

Stoichiometric requirements for oxygen are up to 4.56 mg/mg N oxidised, though in younger cultures particularly, the oxygen contributed during CO_2 fixation reduces this demand to about 4.3 mg/mg N (Haug and McCarty, 1971). These authors found oxygen to be rate-limiting unless supplied sufficiently for these requirements and Painter (1970) suggests minimum absolute levels of 1-2 mg/l O_2. Higher figures have been suggested for rapid growth though these may actually be 'bulk' concentrations rather than those at the cell level. Gigger and Speece (1970) suggest an allowance of 1.5 times the stoichiometric requirements for filter design, presumably to ensure that oxygen levels remain sufficiently high to avoid limitation.

It is thus essential to provide adequate oxygen throughout the filter body. Where a fish culture system imposes lower limits of say 3-5 mg/l, the ammonia removal required may necessitate additional submerged oxygenation within the filter. Muir (1978) noted that aerated model-scale filters treating simulated fish-farm wastes showed considerable improvements in performance, particularly at NH_3-N concentrations of 2 mg/l. Birkbeck and Walden (1976) obtained removal rates of 48 g, 93 g and 135 g NH_3-N/m^3/day without aeration, with aeration, and with oxygenation, respectively, in filters loaded at 137.4 m^3/m^2/day.

The effect of pH on nitrification can be considerable, with marked inhibition outside pH 6.0-9.0 and an optimum of 8.0-8.5 (Painter, 1970). Haug and McCarty (1971) note that the production of H$^+$ during nitrification reduces the pH, requiring an alkalinity equivalent of approximately 3 mg/mg NH_3-N oxidised to buffer the reaction.

The effects of specific inhibitors and poisons must be extrapolated from single-culture tests with care, as there is considerable evidence of protection within the bacterial film and by cell adaptation (Painter, 1970). Of particular interest to fish culturists is the effect of disease treatment agents, particularly those with generally toxic or bacterial effects. Table 6.23 shows some of the recorded effects.

Table 6.23: Recorded Toxic Effects in Nitrification

Chemical	Contact	Effect
Terramycin (Oxytetracycline)	constant 10 mg/l	negligible
Tribrissen (Thioprin & Sulphadiazine)	wash 10 mg/l	negligible
	constant 2 mg/l	negligible
	constant 10 mg/l	slight disruption at 3 days
Chloramine T. (Tosylchloramide)	constant 10 mg/l	negligible
Malachite green	constant 0.1 mg/l	slight disruption at 6 days
	3 hrs, alternate days	negligible
Formaldehyde	constant 0.1 mg/l	slight disruption
	constant 5 ml/l, 45 mins exposure	disruption and recovery within 3 days
	constant 5 ml/l, 12 hrs exposure	complete disruption
	10 mg/l, 3 times on alternate days	negligible
Formaldehyde & malachite green	10 mg/l + 0.1 mg/l, 3 times alternate days	negligible
Copper sulphate	1 mg/l, 3 times on alternate days	negligible
Methylene blue	single dose, 5 mg/l	disruption over 16 days
Potassium permanganate	4 mg/l, single dose	negligible

Sources: Collins *et al.* (1975), Muir (1978).

Muir (1978) examined a number of disruptive effects such as load and salinity changes and concluded that within normal fish culture operating conditions, nitrifying filters were easily able to withstand variations. Knowles *et al.* (1965), Laudelout *et al.* (1968) and Haug and McCarty (1971) studied the effects of substrate starvation, showing little effect over 24 hours starvation, up to 50 per cent less activity after 7 days starvation, but with full restoration of activity within 4 days.

Effects of Organic Loading. The presence of carbonaceous wastes in the form of dispersed solids can be a major operating variable in fish culture use and competition for the filter surface by carbonaceous oxidisers can have a significant effect on filter performance. Muir (1978), studying sequential nitrifying filters, showed an approximately inverse relationship between solids loading and ammonia removal at solids loadings of approximately 4-150 g/m^3/day with ammonia removal decreasing by as much as 50 per cent. Duddles *et al.* (1974) noted similar effects in terms of BOD loading (Table 6.20), where at 1,520 g BOD/m^3/day, ammonia removal was 48 g/m^3/day, and at lower hydraulic loading, 331 g BOD/m^3/day allowed removal of 138 g/m^3/day NH$_3$-N.

The effects of organic loading have also been noted by Gigger and Speece (1970), Liao *et al.* (1972) and Meade (1974). In fish culture use, Berka (1980) quotes a ratio of 100:5:1 of BOD:N:P as ideal for filter operation, and notes the suitability of typical untreated fish culture water (ratio 100:6:2). For this type of operation, Otte and Rosenthal (1976) consider a trickling filter to be preferable in offering better zonation of effect (carbonaceous oxidation in the upper surfaces, nitrification below). Because of such organic loading, Liao *et al.* (1972) recommend a maximum loading of 0.98 g NH$_3$-N/m^2 surface area per day, though non-nitrifying loads were not specified. At a typical media surface area of 200 m^2/m^3, this is the equivalent of 196 g NH$_3$-N/m^3/day.

The presence or survival of other organisms, particularly pathogens, within nitrifying filters can be of considerable significance. In a freshwater recirculating system holding channel catfish, Collins *et al.* (1975) noted total bacterial counts to stabilise at 10^6 per ml, with about 10^4 per ml directly associated with the fishes' gut flora. Tomlinson *et al.* (1962) found trickling filters to reduce coliform counts by over 95 per cent, attributing removal to physical processes. Although potential pathogens may be retained, their survival and further transmission is not certain.

Starting Nitrification. Normal start-up time has been found to vary particularly with temperature, seed material and pH (Muir, 1978), though with suitable substrates full activity can be attained within 40 days, and indeed normally within considerably less time. Carmignani and Bennett (1977) found nitrification to commence in 10.7 days with seeding, compared with 22.2 days without. Meade (1974) describes the activation of nitrifying filters by isolating the filters, inoculating with a 'seed' (e.g. garden soil, stream sediment, or if contamination is to be avoided, pure cultures) and using a buffered culture medium for growth (Table 6.24). NH$_3$-N levels were maintained at 10-20 mg/l. Birkbeck and Walden (1976) however recommend a level of not more than 1.25 mg/l NH$_3$-N on starting. Similar seeding procedures are recommended by Spotte (1979), and Forster (1974) describes the use of shucked mussel in sea-water nitrification as an additional nitrogenous source.

Table 6.24: Nutrients for Start-up of Nitrification

Meade (1974)	Muir (1978)
40 mg/l (NH$_4$)$_2$ HPO$_4$	(a) 0.1 mg/l NH$_3$-N as NH$_4$OH
40 mg/l Na$_2$ HPO$_4$	(b) 50% wash of disturbed filter solids plus
40 mg/l 'Instant Ocean' solids	5 mg/l NH$_3$-N as (NH$_4$)2 PO$_4$
0.5 mg/l 'Instant Ocean' liquid	
250 mg/l CaCO$_3$	
Forster (pers. comm.)	
1 mg/l NH$_3$-N as NH$_4$ Cl	

Filter Models. Haug and McCarty (1971), studying nitrification in submerged filters for conventional nitrification, developed a series of relatively simple relationships of the form:

$$-dS/dt = a(S/10)^b$$

where $-dS/dt$ = ammonia removal rate, S = ammonia concentration and for various temperatures:

	5°C	10°C	15°C	25°C
a =	0.38	0.99	1.15	2.29
b =	1.10	1.25	0.93	2.4

so giving a first to second order rate relationship. Muir (1978) found that at the lower ammonia levels typical of fish culture, these rate equations tended to overestimate ammonia removal rates, though this was not determined over the complete temperature range.

Several models have been proposed for fish culture application. Speece (1973) developed a simple design method based on nitrification rate (assessed at 1 g NH_3–N/m^2 media surface area per day) at 20°C, together with Haug and McCarty's (1971) temperature function to plot a graph of nitrification rate and temperature. Using data on specific surface areas of filter media, overall filter volume for desired loading could be calculated. The method does not however allow for hydraulic loading or filter configurations, and may not cover operation at higher temperatures.

The method of Liao *et al.* (1972) is based on an observed relationship between 'removal efficiency'

$$E_A = \frac{\text{removal rate } (kg/m^2/day)}{\text{loading rate } (kg/m^2/day)}$$

and the residence time, t_m, based on actual filter volume, for trickle, upflow and downflow filters. For residence times of 0.206-0.46 hours, and temperatures of 10-15°C, the relationship $E_A = 0.96 t_m$ was developed, and in conjunction with nitrification/temperature relationships extended to:

$$E_A = (9.8T - 21.7)t_m$$

This assumed filters to operate above the recommended minimum loading and hydraulic rates, and is assumed to be independent of the (unspecified) solids or carbonaceous loading. Hess (1979) presented an equation for submerged filters comparing actual ammonia removal with expected removal (based on filter sizing and film area), based on actual recorded performance, to derive a function NXC:

$$NXC = \frac{1000\,(C_i - C_o)Q}{C_i\,T_c} \times \frac{V_F}{(D^{0.8})\,(A)\,(SSA)^{0.8}}$$

where C_i and C_o are inlet and outlet concentrations respectively, Q is flow (g/m), T_c is temperature correction (i.e. $0.0604T - 2.15$), V_F is voids fraction,

D is bed depth (ft), A is bed area (ft^2), SSA is specific surface area (ft^2/ft^3).

This function was found to be applicable to a wide range of filters, producing NXC values of approximately 4.8. The NXC value could be used either as an index of performance or set to allow design calculation. The method does not, however, account for solids or organic loading.

There is a need to develop a more complete design procedure to take account of rates intermediate between the typical 1 mg/l NH_3-N in aquaculture, and the 10 mg/l NH_3-N in secondary or tertiary waste treatment, and to include the effects of solid or carbonaceous loading. Thus:

$$E_A = K^a K^b(t_m)$$

where K^a = initial concentration and K^b = additional load effect.

In actual operation, filter removal efficiencies recorded are generally low. Mayo (1980) reports about 30 per cent efficiency in Utah hatcheries using styrofoam media. Otte and Rosenthal (1979) quote an average efficiency of 31 per cent, and Liao records removal efficiencies of 48 per cent at 86.4-319.7 m^3/m^2/day. Muir (1978) recorded similar efficiencies in a sea-water recycle system. It should be noted that a self-regulating effect may occur in that reduced efficiencies lead to increased ammonia levels, and in turn to an improved removal rate.

5.4 Activated Sludge Systems

A number of activated sludge treatment systems have been developed to treat fish culture wastes (Meske, 1973; Nagel, 1977; Otte and Rosenthal, 1979; Berka *et al.*, 1980). Typical configurations and loading of conventional waste treatment systems are shown in Figure 6.13 which also compares the system reported for fish culture and general waste treatment.

The advantages claimed for these systems in normal waste treatment include higher loading rates and more complete control of oxidation processes, and these have gained wide use for high BOD wastes. It is less certain whether the lower loadings and the greater demand for stability and ease of control required in fish culture, can be met as satisfactorily, though the 'Ahrensberg' system developed by Meske (1973) has been the basis of considerable research.

Operating Conditions. The maintenance of stable nitrifying populations is a particular problem in suspended medium systems in that their long mean generation time requires particular care to avoid 'washing out' and replacement by other, faster-growing populations (Downing *et al.*, 1964; Downing, 1966). This is a greater problem with mixed carbonaceous and nitrogenous loads, and may be a cause of considerable instability. Muir (1978) noted that a model-scale activated sludge system operating at equivalent loading rates of fish culture effluent was less stable than a submerged filter, and shock loads could cause particular problems.

In conditions of deoxygenation and high C:N and C:P ratios, or when denitrification and nitrogen gas production occurs, the normally settled material may rise to the surface, 'bulking' and contaminating the outflow. This is often associated with communities of *Sphaerotilus, Beggiatoa, Geotrichium, Nocardia* and *Zoophagus* (Pike and Curds, 1971).

Figure 6.13: Suspended Media Treatments.

	Stabilisation Pond	Oxidation Pond	Conventional Activated Sludge (CAS)	High-rate Activated Sludge	Contact Stabilisation	Extended Aeration	Step Aeration
		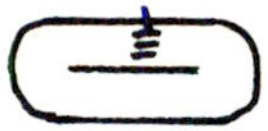	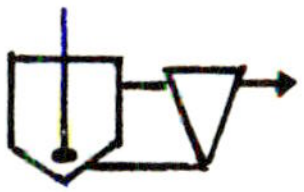	As CAS	As conventional with sludge re-aeration	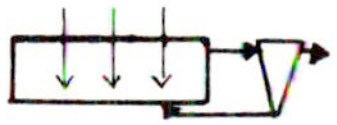	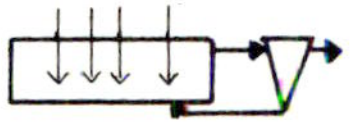
Size	50 m by 200 m or larger	50 m by 200 m by 1-1.6 m	30-120 m by 3-5 m channel, plus settling tank	As CAS	As CAS	As CAS or larger	As CAS
Contact	30 days	1-3 days	6-12 hrs, 25% return	1-4 hrs	30 min, then 3-5 hrs	30 min, then 40-60 hours	As CAS
Air	Mechanical or diffused air, photosynthesis	Mechanical, by rotor or brush photosynthesis	Mechanical or airlift, 6 m³ per m³ waste	As CAS 3 m³ per m³	As CAS	As CAS 12-18 m³ per m³	As CAS
Loading	50-150 kg BOD per hectare	100-500 kg BOD per hectare	0.5-0.6 kg BOD per m³ per day, 50 kg BOD per 100 kg MLVSS	1.6 kg BOD per m³ per day	As CAS	0.2-0.4 kg BOD per m³ per day	As CAS
Other	Long-term, poor control; possible inclusion of fish, plant crop	Denitrification if diurnal aeration	Typically 90% BOD reduction 2,000-3,000 mg/l MLVSS	High or low (650 mg/l) MLVSS, poor nitrification, less BOD removal	MLVSS to 20,000 in stabilisation section	Reduced waste sludge volume, nitrification	Higher initial air supply reduces overall demand

Note: MLVSS = mixed liquor volatile suspended solids (active mass).

The effect of activated sludge treatment on organisms pathogenic to fish is not well documented, though removal of human pathogenic indicators is frequently 90-99 per cent, by a combination of floc adsorption and ciliate grazing. A number of activated sludge systems used in aquaculture employ the sedimentation stage as a means of allowing denitrification, which can proceed effectively (Berka *et al.*, 1980) apparently without disturbing settled material by gas production. These authors cite Muller *et al.* in suggesting the addition of sugar as a carbon source to assist denitrification.

Mixing and Oxygenation. The age of the suspended floc, dependent on hydraulic loading and recycle rates within the activated sludge system, determines its relative activity (proportion of live cells in the floc material), and the ability to metabolise wastes, while floc size, controlled by recycle rate and mixing turbulence, affects rates of waste and metabolite diffusion. Wuhrmann (1963) described active constituents (per cent volatile solids – normally termed MLVSS) by:

$$MLVSS/_{TS} = exp. (-0.26t)$$

where t = residence time, TS = total solids.

Within the mixing areas, concentrations of 10,000 mg/1 MLSS (mixed liquor suspended solids) became difficult to suspend, and so give an upper limit to available floc (Bruce and Boon, 1970).

Oxygen demands vary with aeration time, typically ranging from 20 to 80 mg/1 per 1000 mg/1 MLVSS. For BOD removal, Downing (1966) described consumption as:

$$R = aL + bS$$

where $R = O_2$ (g), L = BOD removal (g), S = sludge mass (g), a = 0.5-0.6, b = 0.1, the latter term (bS) representing that oxygen required for endogenous respiration of floc material. Where there are low overall oxygen requirements, it is more energy-efficient to use baffles for the turbulence needed. Adams (1973) found oxygen levels of 3 mg/1 to be necessary for nitrification and Berka *et al.* (1980) quote a minimum of 2 mg/1 for successful treatment of fish culture waste. For air supplied systems, a range of 3.5-15 m^3 air was required per m^3 water treated (Liao, 1980). Downing (1966) quotes 3.12-12.42 m^3 air/m^3 treated in conventional practice.

Sludge Production. Solids, instead of accumulating on fixed media, are incorporated into particle material, resulting in a net gain of biomass and an excess sludge production. This is related to BOD production as follows (Heukelekian *et al.*, 1951):

$$VSS (kg/day) = 0.5 \times BOD (kg/day) - 0.055 MLVSS$$

and is related to residence time by a multiple of $(0.1 + 0.9t^{-\frac{1}{2}})$ (Downing *et al.*, 1964), where t = residence time.

System Design. Design in terms of operating conditions and limitations is similar in many respects to that of fixed medium equipment, save that little information has been developed on performance of specifically nitrifying systems. A number of models, based either on absorption rates (Katz and Rohlich, 1956) or Michaelis functions (Tischler and Eckenfelder, 1969; Boon and Burgess, 1972) have been proposed for BOD removal, though the latter authors note that most rate equations can be made to fit observed performance at the levels of accuracy obtainable.

Design approaches for use in aquaculture are limited, particularly as the NH_3-N levels are much less than those normally encountered in waste treatment. Systems are thus normally designed at loadings typical of extended aeration operation. Thus Berka *et al.* (1980) quote a limit of 500 g BOD/kg sludge dry matter per day, or less than 100 g if stabilised sludge is required. Muller *et al.*, quoted by these authors, suggest a loading of 130-150 g/kg sludge to ensure nitrification. In none of these systems, however, is the relationship between nitrification and BOD removal considered; in terms of the need to maintain sludge mass, and to avoid excessive return flow within the activated sludge system, additional BOD may be an advantage in encouraging cell growth.

On the basis of quoted BOD loadings, and average oxygenation efficiencies, the energy requirements of removing 20 mg/l BOD are approximately 20-20 Whr/m^3. Albertson *et al.* (1970) quote requirements of 13-23 Whr/m^3 using pure oxygen, though at these loadings turbulence may not be sufficient.

One interesting aspect which has not received particular attention is the concept of the complete recycle system as a type of activated-sludge unit. It has been observed in some recycle systems that considerable nitrification takes place in pipes, channels and presumably the holding tanks (Muir, 1978; Aldridge, pers. comm.). On the basis of observed film area on these sites, this removal cannot be accounted for by surface activity, and so it is suggested that a free-floating population of nitrifiers may be responsible. This idea has also been quoted by Mayo (1976) from Meade (pers. comm.), but has not been investigated further. Should this be a significant factor in ammonia removal, processes such as sedimentation and sterilisation may have to be reconsidered in terms of their effect on these populations.

5.5 Hybrid Equipment

This term is used to describe systems incorporating aspects of both fixed and suspended-media operation, chief of which are the 'biodisc' or 'biodrum' system (Figure 6.14) in which fixed media are rotated through a suspended-media bed, and 'biosedimentation' or fluidised bed systems, where solid-particle media are held in suspension by upward waste flows (Figure 6.15).

Biodiscs. The basic design procedures have been described by Stells (1974); the disc system rotates at 2-3 rpm through the suspended medium, both aerating and allowing contact with growth surfaces on the discs. Suspended solids are normally separated and recycled within the troughs below the discs; high and low recycle rates and rotation velocity approach suspended or fixed medium operation respectively. Solids sloughing from the biodisc, or running from the collecting trough, are normally more settlable (Lewis and Buynak, 1976); Holmberg (1978) quotes 50 per cent settling within 10 minutes and 100 per cent settlement within 30 minutes. A rotating system may thus combine many of the advantages of each type of operation.

Figure 6.14: Biodisc and Biodrum Systems. Above: two-stage biodisc system — disc holds fixed film, trough contains suspended film while aeration comes from revolving discs. Below: single-stage biodrum system — cage holds plastic or other small media, aeration as biodisc.

Two-stage Biodisc System

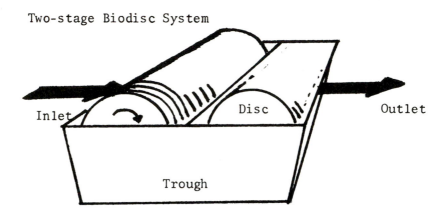

Single-stage Biodrum System

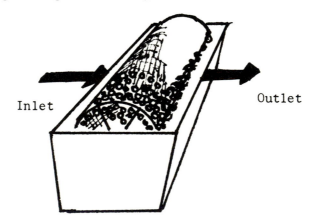

A similar system has been developed by Holmberg (1980) employing polystyrene spheres within a rotating cage, providing a greater surface area and aeration effect, though with greater energy cost. The system has recently been developed for fish-culture applications, where in a sense the fish tank is employed as the suspended medium area for nitrification, and a separate system is used for solids and BOD removal. Operation in fish culture effluents has been described by Lewis and Buynak (1976), Muir (1978), Holmberg (1978) and

Lewis *et al.* (1980). Muir (1978) using a 35 l model-scale two-stage system with 30 cm discs, found NH_3-N removal rates of up to 55 g/m^3/day, with the discs contributing up to 66 per cent of ammonia removal at hydraulic loadings of 20.6 m^3/m^3/day.

Figure 6.15: Biosedimentation System.

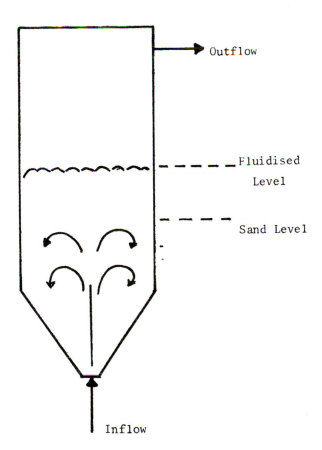

Lewis and Buynak (1976) note a requirement of 0.5 m^2 disc area per kg channel catfish. Liao (1980) quotes a removal of more than 95 per cent BOD and NH_3-N at a loading of 0.06-0.1 m^3/m^2/day. Joost (1969) quotes a 45-minute residence time (32 m^3/m^3/day) for 90 per cent BOD reduction.

On the basis of Steels' (1974) design criteria of 30 x 3 m diameter discs or 34 x 2 m discs per metre shaft length, a specific surface area of 60-88 m^2/m^3 is provided. Power consumption is approximately 10 W/m^3, or at typical loadings 7.5-12 Wh/m^3 treated.

The biodrum system described by Holmberg (1978) is normally loaded at

8-12 kg BOD/m^3/day, achieving 92-97 per cent removal. Loading rates are typically 10-50 m^3/m^3/day, and at 700 W/m^3 drum volume, the power required is 336-1,680 Wh/m^3. However, in the use described, the drum oxygenates the fish culture system served and hence the treatment power requirement is only a part of this total consumption.

Fluidised Bed Systems. Short (1973) described an upward flow sedimentation tank incorporating biological ammonia removal, as reported by Lewis (1966), Parker (1972) and Pugh (1945). Using a sand load of 280 kg/m^3, a flow rate of 60 m^3/m^3/day produced stable fluidisation, with sand occupying 1.5 m of a conical base 4 m tank, with almost 100 per cent removal of 2 mg/l NH_3-N. Further capacity was limited only by oxygen input.

Design requirements and head loss in operation may be determined by use of backwash equations for fine sand filters. Though the process was not assessed for fish culture use, Short (1973) considered it to have the greatest economic potential for treating river waters of similar characteristics. Mayo (1980) describes the proposed use of a biosedimentation system in a salmonid hatchery, operating at design rates of 864 m^3/m^3/day, though performance is not detailed.

In terms of controllability, high available surface area and effective solids removal (using a wash cycle or continuous recycling or solid medium), this system offers considerable potential limited only by the head losses incurred in fluidising the filter.

6. Other Biological Processes

The specific processes and design variables for algae and for plant- and animal-based reuse systems are outwith the scope of this chapter, though reported ratios of treatment size to fish biomass are provided in the text where appropriate. The microbiological processes of carbonaceous oxidation and denitrification are, however, more immediately significant. Both may be supported in systems described earlier.

6.1 Carbonaceous Oxidation

Of the wastes produced by the fish, a certain amount will be available for the range of carbonaceous oxidisers, regardless of the efficiency of solids removal; though this will affect significantly the competition for nitrifying medium. In freshwater self-purification processes a primary carbonaceous stage oxidises 70-80 per cent of organic carbon to CO_2, and a secondary nitrification stage oxidises nitrogen, using the remaining 20-30 per cent and forming relatively inert humic-acid type compounds. This is similarly reported in marine fish holding systems (Spotte, 1979).

As the mixture of carbon compounds is rarely definable, and oxidisers are of many species, removals are normally based on BOD reduction, as it is assumed that BOD measurements will normally exclude nitrogenous components of waste, and thus the conventional BOD rate equations can apply (Fair *et al.*, 1971), e.g.:

$$y/y_0 = 1 - (1 + \Sigma kt)^{-1}$$

where y = final conc., y_0 = initial conc., k = rate constant and t = residence time.

It can be assumed that unless the BOD removal requirements can be satisfied within designed systems, nitrification will be restricted, and where BOD removal must be accommodated, effective nitrification should be based on the calculated filter volume remaining. There remains scope at this stage for optimising the relationship between solids/BOD and nitrogen removal.

As nitrification and other biological processes proceed, a number of longer-chain carbon compounds are produced, either as extra-cellular products or as cell breakdown materials. These compounds give the typically brown colour to highly recycled water, and although apparently harmless to fish stocks are undesirable in marine display systems (Spotte, 1979) where they must be removed, normally by oxidation reactions splitting up the larger chains into biodegradable carbon compounds. The quantification of these compounds has been described by Rosenthal (1980a) and is discussed more fully later. Final concentrations of long-chain compounds are limited by dilution.

6.2 Denitrification

Anaerobic processes, normally employing species of the generae *Pseudomonas* and *Bacillus*, can be used to reduce nitrates to atmospheric nitrogen. St Amant and McCarty (1969) describe a two-step process:

$$3NO_3^- + CH_3OH \rightarrow 3NO_2^- + CO_2 + 2H_2O$$

$$2NO_2^- + CH_3OH \rightarrow N_2 + CO_2 + H_2O + 2OH^-$$

Although anaerobic conditions are normally desired, Meade (pers. comm.) notes that denitrification could occur in locally anaerobic filter areas while bulk oxygen concentrations were maintained at 1.5-10 mg/1. Wheatland *et al.* (1959) found nitrate to be reduced to nitrite only in activated sludge at a bulk O_2 concentration of 5 mg/1 or less.

The selection of a suitable carbon substrate requires some care: methanol is most frequently employed, though care has to be taken to avoid toxic effects to humans. Other possibilities include glucose, sugar, acetone and acetic acid, though St Amant and McCarty (1969) consider methanol to be the least expensive. Depending on the position in the recycle system, carbonaceous sources in waste may also assist. According to St Amant and McCarty (1969), the equation

$$C_m = 2.47 (NO_3-N) + 1.53(NO_2-N) + 0.87(O_2)$$

describes the requirement for methanol, C_m being in mg/1. Insufficient methanol may inhibit yields or may produce higher levels of NO_2 in the effluent. Eckenfelder and Balakrishnam (1968) and Smith *et al.* (1972) showed overall reduction rates to be proportional to COD up to stoichiometric limits.

At suitable dilution rates the normal limits of approximately 100 mg/1 NO_3-N may be maintained in recycled sytems without denitrification. The rate of denitrification depends primarily on pH and temperature. Wijker and Delwiche (1954) noted end-products to vary with pH, N_2 being produced at pH above 7.3, N_2O and NO appearing at pH above 7.0. Rate equations appear to follow Arrhenius relationships (Dawson and Murphy, 1972):

$$k = k_o e - E/RT$$

Reported rates of denitrification vary from 2 mg (Painter, 1970) to 17-120 mg N/mg dry cells per hour (Wuhrmann, 1963). Wheatland *et al.* (1959) found a Q_{10} value (rate increase per $10°C$ rise) of approximately 2.0, with *P. denitrificanis* between $5°C$ and $25°C$. On the basis of St Amant and McCarty's substrate requirements cell yields are approximately 0.57 mg/mg NO_3-N reduced.

In fish culture systems, Otte and Rosenthal (1979) described a laboratory-scale activated sludge denitrification unit of $1.06 m^3$, using glucose or methanol, achieving 50-98 per cent reduction of nitrate supplied. As approximately 3.57 mg alkalinity are produced per mg/N_2, the system improves pH stability. Berka *et al.* (1980) refer to an average designed retention time of three hours, and O_2 levels of less than 0.5 mg/l for successful operation. Meade (1974) quotes a surface area based removal rate of $2.32 g/m^2$ at $14°C$, which can be used for design purposes. In operation, Meade (1974) recommends the return of denitrification effluent through the nitrifying stage to avoid methanol contamination. This would also assist to remove any nitrite produced in inefficient operating systems.

6.3 Lagoons

Lagoons and/or settling ponds have been used in a number of experimental recycling systems (Hill *et al.*, 1973; Muir, 1977). Where heat losses are not critical, land costs are acceptable and performance is not subject to excess diurnal or seasonal variation, these can be effective, particularly if additional fish crops can be obtained within the lagoons themselves. Designs can be made on the basis of conventional sewage treatment (Cooper *et al.*, 1965), of 36 kg BOD/ha/day, on which basis, sizing will be approximately 4 ha/tonne stock. On the basis of 70 per cent removal of nutrients, up to 12 t/ha/yr dry weight of algae may be obtained. As plankton density is by intention relatively high, a filtration or settling stage will be required prior to reuse of water.

Lagoons are likely to have greatest applicability in areas of high photo-synthetic productivity, though in all areas the operation of simple lagoons is likely to be considerably less reliable than conventional treatment. Multi-stage lagoons, or more intensive aerated and recycled oxidation-ditch systems, might also be employed and may offer more predictable performance. On the basis of conventional practice (Fair *et al.*, 1971) oxidation ditches would be sized at approximately 0.1 ha/tonne.

An experimental algal/bacterial system operating on the activated sludge principle has been described (McGriff, 1972), recording removals of up to 97 per cent BOD, 87 per cent COD, 92 per cent N and 74 per cent phosphate. *Chlorella* was the dominant species.

6.4 Macrophytes

Aquatic vegetation such as water hyacinth (*E. crassipes*) and hydroponic vegetable crops have been used experimentally as treatments for fish wastes (Nagel, 1977; Lewis *et al.*, 1980; Jesperson and Hodal, 1980; van Toever and Mackay, 1980). Table 6.25 illustrates typical nutrient requirements and uptakes.

However, although theoretical removal rates may be favourable, practical problems of harvesting and competition by, e.g. *Cladophora*, may limit aquatic macrophytes (Jesperson and Hodal, 1980). Nagel (1977) found that higher plants did not significantly reduce nitrate in aquaculture wastes. In hydroponic systems, the control of water level and the need to avoid clogging by solids may create design or cost problems, and treatment for disease of crop plants may cause difficulties.

Table 6.25: Removal Rates: Macrophytes (kg/ha/yr)

	Hydroponic Culture[a]	E. crassipes	J. americana	A. philoxeroides	T. latifolia
N	3,600	1,980	2,293	1,779	2,630
P	1,200	322	136	198	403
S	—	248	204	180	250
Ca	3,600	750	1,022	322	1,709
Mg	1,100	788	465	322	307
K	7,200	3,188	3,723	3,224	4,570
Na	—	255	193	229	730
Fe	100	19	123	45	23
Mn	13	296	13	27	19
Zn	3	4	30	6	6
Cu	14	1	3	1	7

Note: a. Removal at 10 per cent of nutrient supplied.
Sources: Boyd (1970) and Harris (1974).

Yields and design relationships may be calculated on the basis of theoretical removal rates. Jesperson and Hodal (1980) quote a plot surface area of 22 m^2/ 36 kg fish (0.61 m^2/kg); Lewis *et al.* (1980) quote an area of 0.53 m^2/kg.

7. Solids Removal

The effective flushing and removal of solids from aquaculture process water can be the most critical aspect of water reuse. As earlier sections show, solids levels may become the first limiting parameter as recycle rates increase, and the solids may physically block filtration systems and may themselves break down and exert an oxidative load on treatment systems. Solids are normally removed either by physical interception or adsorption between particulate beds (filtration), or by settling out of the process stream (sedimentation). In aquaculture systems sedimentation is normally used for coarse solids and filtration for removal of finer material. For a fuller description of design parameters, the reader is referred to Fair *et al.* (1971) and Wheaton (1977). Other methods of solids removal, such as centrifugation, are discussed in Wheaton (1977), though the energy requirements are generally too high for use in aquaculture; Mayo (1976) quotes the use of a 3 m operating head in a centrifugal unit, although solids removals of 61 per cent were recorded.

7.1 Characterisation of Solid Wastes

Water-borne solids are conventionally described in a continuum from 0.01 μ to 1 cm (Figure 6.16). The movement of discrete solid particles in water is described approximately by the Navier-Stokes equations. Thus:

$$v^2 = 4g_d/3C_d/(s-1)$$

where v = settling velocity, s = specific gravity of particle, C_d = drag coefficient and g_D = gravitation force, and thus settling velocity, and hence ease of removal, depend largely on particle size and density. In practice, settler configuration, inlet and outlet design, currents, density layers and temperature may all influence performance.

In aquaculture, the type and size of solid particles depends primarily on type of food used and faeces produced, on the size of the animal fed, and on the amount of turbulence and resuspension in the culture system. Faecal material typically comprises undigested dietary materials bound together in a mucus coat, frequently produced in long strings. These strings are normally denser than water, though Haller (pers. comm.) has noted floating casts in tilapia culture, and the author has observed similar effects in supersaturated systems where gas bubbles come out of solution, surround the particles and make them buoyant.

The relationship between holding facilities and faecal material has been described by Burrows and Combs (1968), Chapman *et al.* (1971), Muir (1978), Warrer-Hansen (1979) and Balarin (unpublished), the critical feature being the creation of an adequate flow of water towards the drain area (normally a velocity greater than the settling rate), without resuspension, agitation and the production of smaller, less settlable particles.

Settling rates for particles produced in individual systems can be determined approximately by jar-settling tests in which samples are timed to determine the percentage of solids settling within a particular time. For trout culture wastes, Muir (1978) noted that approximately 66 per cent of solids settled within five minutes, up to 90 per cent within 15 minutes, and compaction of solids, producing a stable base, occurred within 90 minutes. Warrer-Hansen (1979) found solids from earth ponds to settle less well than those from circular tanks, in trout culture; 37 per cent and 73 per cent settled within 30 minutes, respectively, though differences could be due to a number of husbandry factors. Ellis *et al.* (1978) found 72-90 per cent of solids from channel catfish ponds to be settlable.

7.2 Design of Settling Areas

The main types of settler or sedimentation system are described in Fair *et al.* (1971) and Wheaton (1977). A simple settler used in fish culture systems can be formed from an assembly of standard pipe components (Muir, 1978). Diagrams of typical configurations are shown in Figure 6.17. The full design procedure of settling areas may be complex, as the differential settling rates of the different sized particles result in effects whereby some particles trap, filter or coalesce with others, and thus predictions from simple particle sizes and velocities do not fully correspond.

Figure 6.16: Solids in Water.

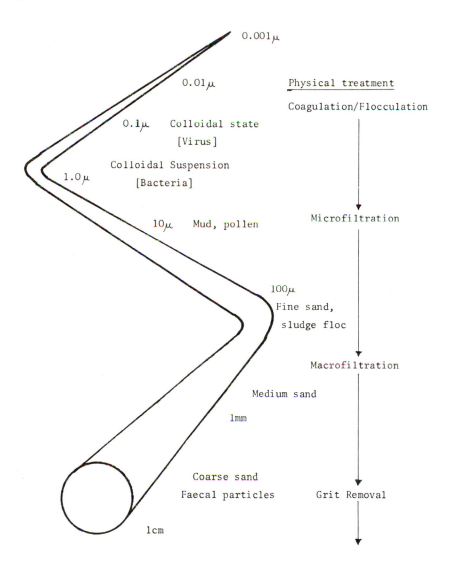

In conventional settling basins, Fair *et al.* (1971) describe four zones:

(1) Inlet zone — flow and suspended matter disperse out from the inlet point.
(2) Settling zone — suspended particles settle in the flowing water.
(3) Bottom zone — removed solids accumulate and may be continuously withdrawn.
(4) Outlet zone — remaining particles are drawn in and emit from the outlet.

Figure 6.17: Settling Systems.

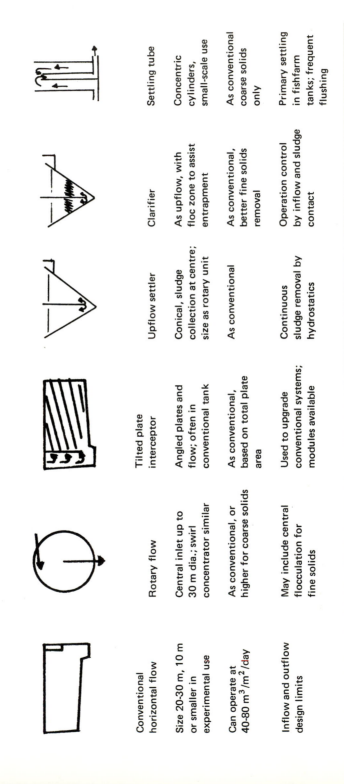

Conventional horizontal flow	Rotary flow	Tilted plate interceptor	Upflow settler	Clarifier	Settling tube
Size 20-30 m, 10 m or smaller in experimental use	Central inlet up to 30 m dia.; swirl concentrator similar	Angled plates and flow; often in conventional tank	Conical, sludge collection at centre; size as rotary unit	As upflow, with floc zone to assist entrapment	Concentric cylinders, small-scale use
Can operate at 40-80 m³/m²/day	As conventional, or higher for coarse solids	As conventional, based on total plate area	As conventional	As conventional, better fine solids removal	As conventional coarse solids only
Inflow and outflow design limits	May include central flocculation for fine solids	Used to upgrade conventional systems; modules available	Continuous sludge removal by hydrostatics	Operation control by inflow and sludge contact	Primary settling in fishfarm tanks; frequent flushing

Thus in Figure 6.18, the relationship between settling velocity, flow and settler area can be shown. Hence we have the relationship: $v_o = Q/A$ as the surface loading, or overflow rate where particles of v greater than v_o will settle out in vertical flow systems. In horizontal flow systems, where v is less than v_o, particles can be captured where depth of fall, $h = vt_o$.

Figure 6.18: Particle Settlement and Overflow Rate.

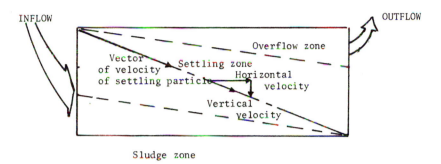

The design of inflows and outflows is particularly critical, and inflows may require baffles or distributors to provide dispersion and reduce kinetic energy. Outflows must ensure equally dispersed flow to eliminate short-circuiting and resuspension of settled particles. Currents in weir-type outflows are normally limited to 125 l/min/m length (180 m^2/day).

The actual performance of a sedimentation basin may be related to theoretical performance by a correction function representing the extent to which current formation and mixing occurs. This can be described as:

$$y/y_o = 1 - (1 + nv_o/(Q/A)^{-1/n})$$

where y/y_o is fraction settled if v exceeds v_o, v_o is settling velocity of particle settling h_o in relation to time t_o, Q is mass flow rate, A is surface area and n is performance coefficient.

The performance coefficient, n, varying from 0 to 1 from good to poor operation, has been noted by Thomas and Archibald (1952) to approximate the ratio:

$$\frac{t_{mean} - t_{mode}}{t_{mean}}$$

where the 't' time values are obtained by tracer studies.

Collected sludge may be removed by mechanical scraping towards an outlet, via a self-drawing sump, or by vacuum suction. In systems with limited settling depth, sludge removal is correspondingly more critical.

7.3 Settling Aquaculture Wastes

The recorded range of particle velocities in fish-farm effluents suggests an overflow rate of approximately 20-50 m^3/m^2/day. Muir (1978), studying a tube-flow settler, found considerable settlement at rates of 30-90 m^3/m^2/day, though correlation with particle sizes was limited by the poor classification of sizes between 5 and 75 μ.

Chapman *et al.* (1971), using catfish culture wastes, found observed removal rates to accord relatively well with conventionally predicted values in small model basins, though a larger model showed lower actual performance. Wheaton (1977) quotes a range of design residence times from 15-60 minutes in fish culture use, corresponding to an overflow rate of 24-96 m^3/m^2/day at a depth of 1 metre. Liao (1980) suggests loadings of 24.2-61.3 m^3/m^2/day in salmonid hatchery design, residence time of 2-6 hours and a length:width ratio of more than 3:1.

The use of surface-area type settlers such as the tilted plate separator offers particular advantages; based on conventional area loading rates, the greatly increased area provided gives a considerably improved volumetric efficiency. The author has found that any simple interceptor (e.g. plastic off-cuts) can improve performance. Even if randomly packed, where operation resembles that of a filter, placing the media in a net-bag provides simple cleaning of any solids not being removed at the sump of the tank.

7.4 Effects of Solids Removal

The removal of other factors in addition to solids may also be significant. Liao and Mayo (1974) observed that 70 per cent of NH$_3$-N output was associated with organic solids, and that a one-hour settling time reduced NH$_3$-N by 20 per cent. Murphy and Lipper (1970) note that 42 per cent of BOD from *Clarias* culture is in non-soluble form. Chapman *et al.* (1971) note, however, that settlement in catfish raceways was not alone sufficient to provide recycling of water. The rate of sludge removal from settling areas has not been studied to any extent, and the degradation of solid material and leaching out of soluble products may affect quite considerably the ability of sedimentation to remove other waste components from process water.

7.5 Removal of Fine Solids

Apart from the incidental removal of solids within biological filters, the removal of finer particulate material deserves particular attention in recycling systems (Spotte, 1979; Wheaton, 1977). This is normally achieved in fine-particle filter beds, though additional processes such as coagulation or precipitation may occur within the system, particularly if pH changes occur.

The design and operation of fine filter beds is described by Fair *et al.* (1971), Muir (1976), Wheaton (1977) and Spotte (1979), the principle modes of operation being upflow, downflow, pressurised or open. The critical factors in aquaculture use are the particle removal effectiveness, the cycle time before clogging and backwash, ease of backwash, and the pressure head required for operation.

Spotte (1979) quotes normal particle size removal to 30 μ, though size removed does not necessarily relate to filter bed size, as accumulated particles reduce the effective pore opening. Filter beds are therefore commonly graded in pore size from coarse to fine in the flow direction, to increase the capacity before blocking. Typical media sizes are 1-5 mm, with normally a coarser bed of gravel or rock, though materials may also be selected on a density basis, to allow media particles to settle correctly if the bed is backwashed.

Cycle time is related to filter particle size, solids loading and the pressure head allowed. Relationships are described more fully in Wheaton (1977); the pressure head in conventional practice reaches 1-2 m before backwash and may continue to rise until 'breakthrough' of poor-quality water occurs. Typical loadings are shown in Figure 6.19, though there has been little systematic work done on fish culture effluent. Mayo (1976) notes that using No. 30 sand, filters operating at 285 m^3/m^2/day remove 70-90 per cent of suspended solids and 15 per cent of the NH_3-N.

The effectiveness of backwash is particularly critical in fish culture use, as this can be a major source of wasted process water. Additionally, backwashing normally necessitates a regular and periodic maintenance schedule, whereas biological filters and settling tanks may operate more continuously. Although there are self-backwashing designs, operating when head losses reach set levels (Degremont, 1974), these do not appear to be widely used. Backwash flow-rates are typically 0.5-1 m^3/m^2/min (720-1,440 m^3/m^2/day), held for 0.5-10 minutes for removal of fish-culture wastes, though rates depend on water temperature and filter particle density, and must avoid washing out filter media.

A number of proprietary filters operate on a continuous or intermittent automatic backwash, the latter sensed by increasing pressure loss. The manufacture of plastic media can be controlled to grade pore space through the media body to maximise cycle time or reduce overall head loss.

For extremely fine particulate removal (0.1-5 μ), diatomaceous earth filters can be employed, in which a central core retains a continuous feed of fine diatomite particles to remove a feed of fine particulates from the water. Periodically, the combined material is washed to waste. Typical loading rates are 0.03-0.09 m^3/m^2/min (43.2-129.6 m^3/m^2/day) with a precoat of 1 kg/m^2 core area, and feed of 0.144 kg/m^2/day. Full procedures for use and maintenance are described in Spotte (1979). On a smaller scale, Strand *et al.* (1969) described an aquarium disc filtration unit removing up to 97 per cent of particulates from 5 to 15 μ at an equivalent flowrate of 60 m^3/m^3/day.

A continuous-contact system of process stream and medium may be of considerable use in aquaculture, as cycles can be avoided and the medium separated and returned to use. Meade (1979) has reported a thin layer sand-bed system incorporating a semi-continuous wash cycle, and fluidised-bed type operation might be adapted to bleed off mixed solids for separation and return of media solids. This approach has now been adopted commercially.

The selection of filtration to minimise pressure loss is a critical factor in reducing operating costs, though this may be offset against size removal and backwash use. Depending on amounts of solids present, fine filters may be used intermittently or on a sidestream to reduce operating costs.

Kujal (1980) describes a sand-bed clarifier for pretreatment of fish culture water. BOD and suspended solids removals were 30-75 per cent at loadings of 43.2-86.4 m^3/m^3/day, using 1-2.2 mm sand grains.

8. Sterilisation

One of the quoted advantages of recycled systems is the potential, by

Figure 6.19: Solids Removal Filters.

	Slow sand filter	Rapid sand filter	Trickle filter	Pressure/vacuum filter	Diatomaceous earth filter	Alternate double filter
Type	Slow sand filter	Rapid sand filter	Trickle filter	Pressure/vacuum filter	Diatomaceous earth filter	Alternate double filter
Size	Wide range	Wide range	To 20 m diam.	1-5 m³	1-2 m³	As rapid or trickle
Loading (m³/m³/day)	0.2	60-200	0.5	120-1,200	24-240	
Depth (m)	0.7	0.7-1.0	1.5-2.0	1.0-2.0		1.0
Medium	Ungraded fine sand	Sand, gravel, plastic, loose or formed	Gravel, plastic, loose or formed	Sand, charcoal, plastic, fibre, metal	Diatomaceous earth on plastic sleeves	As rapid or trickle
Particle Size	0.3-0.4 mm	0.5-1.5 mm, larger at base	to 5 mm, larger at base	0.3-1.0 mm, or similar pore size		
Particle Removal	To 30 mμ	To 30 mμ, usually larger		To 30 mμ	To 0.1 mμ	
Cleaning	Surface scraping	35-40 m³/m²/day		5 m³/m²/hour		
Uses	Final cleaning nitrification	Primary and secondary, some nitrification Also flooded filter	Primary and secondary, main biological	Secondary treatment	Final polishing	Primary filtration, heavy loadings
Other	Head loss 1 m, surface biological action	Head loss 0.1-3.0 m; grading of media reduces loss; considerable biological action possible	High rate plastics 20 m³/m³/day; open surface allows grazing to reduce solids; flooded filters operate at similar loadings	Head loss 2-20 m	Requires precoat each wash, plus feed during run	Double filters may be linked in series for 2-stage operation

incorporating a sterilisation stage, of reducing or eliminating disease organisms, and by implication, the risk of disease. While acceptable in theory, the means of transmission of disease, and the specific effectiveness of sterilisation procedures, must be examined closely. Moreover, the cost of treating significant quantities of water must be balanced against the actual advantage in disease control.

Sterilisation procedures normally involve ultraviolet light, ozone treatment and chlorination, though the latter is not normally used in aquaculture. The processes, described in further detail by Fair *et al.* (1971), Wheaton (1977) and Spotte (1979), are compared in Table 6.26. Studies on the effectiveness of sterilising in aquaculture use have been conducted by Herald *et al.* (1971), Hoffman (1974), Kimura *et al.* (1976) and Bullock and Stuckey (1977) for ultraviolet, by Burleson *et al.* (1975) and Conrad *et al.* (1975) for ozone, and by Wedemeyer and Nelson (1977) and Wedemeyer *et al.* (1978) for ozone and chlorine.

Table 6.26: Sterilising Treatments

	Ultraviolet	Ozone	Chlorine
Application	Lamps, direct contact or suspended	Electrical discharge or UV generator and O_3 gas diffuser	Dose-pump, drip feed Cl or hypochlorite
Exposure	35,000 μW sec/cm^2, 2600 A most effective	1-5 mins x 0.5-1 mg/1	30 min x 1-2 mg/1 residual
Effect	Disruption of DNA chains	Breaks long-chain organic modecules	Strongly oxidising, breaks organic molecules
Limitations	Turbidity and colour; lamp life	Oxidising other organic compounds hence reduced dose; heat removal; dry air or O_2 required	Chloramine formation with nitrogenous compounds alters dose effect, as different toxicities
Advantages	No residuals; supplies easily available	Quick reaction, toxic on contact; additional uses in recycle systems, residuals removed quickly	Relative cheapness, ease of measurement; residuals removable
Disadvantages	Expense-power and lamp replacement; dependent on water quality	Some residuals, control and measurement; toxicity to fish and humans; energy costs; can remove soluble minerals in seawater	Toxicity of residuals, unpredictability of chloramine formation, pH and temperature effect; handling problems (Cl gas)

The relatively high energy requirements involved in UV or ozone use, and the high degree of control required in ozone and chlorine treatment, necessitates a high operating cost, and use is often confined to sidestream operation, early rearing stages or quarantine. Spotte (1979) has evaluated the use of ultraviolet and ozone, concluding that apart from the recorded advantage of UV in combating whirling disease (*Myxosoma cerebralis*) in salmonids, there was little evidence of any effect on fish mortality rates, although several instances showed kill-off of potential pathogens. Moreover, there was evidence that both treatments might have mutagenic or toxic effects on marine larvae; and in marine systems in particular, routine use was of doubtful benefit.

In aquaculture use, a number of treatments have been described. Flatow (1980) reviews the basic requirements of UV systems and the factors affecting their performance. In salmonid production for release to the wild a ratio of 17:1 in returning numbers is quoted for UV-treated hatchery systems, compared with untreated water systems. Seegert and Brooks (1978) noted that in UV for dechlorination use, 2 x 1200 W sterilisers would cost $2,850 per year to treat 50 l/min including lamp and power costs. Eagleton and Herald (1968) quoted a power requirement of 20 W/m^3 treated in hatchery use, based on the standard dosage of 35 $mW/sec/cm^2$, though results of trials by Bullock and Stuckey (1977) indicate that lower dosages may be used. However, the interfering effect of colour, turbidity and variable lamp intensity (Flatow, 1980) suggest use of higher doses.

Use of ozone is discussed by Liao (1980), who recommends a concentration of 1-7 mg/l for effective sterilisation depending on oxygen demand of the treated water. Normally 1-3 mg/l at a contact time of two minutes is satisfactory for bacteria and virus control. Otte *et al.* (1977) suggest the use of 0.095 g O_3/day per kg stock held. At 2 mg/l and conventional loadings costs were £1.45/tonne stock per day.

Chlorine does not appear to be used as a regular procedure because of its residual-level toxicity, though the results of Seegert and Brooks (1978) suggest the possibility of relatively inexpensive and reliable dechlorination (£0.0023/m^3) and the relative cheapness of chlorination may have further potential. Dosages vary from 1 mg/l with good quality effluent and 3-9 mg/l for filter effluent, to 24 mg/l for raw waste water.

9. Chemical Oxidation Processes

The powerful oxidising ability of both ozone and chlorine has led to their consideration for use as general purpose treatments for oxidising nitrogenous and carbonaceous wastes, in addition to their sterilising effects (Spotte, 1979; Rosenthal, 1980a). Moreover, the more resistant residual colour-causing molecules may be broken down to more degradable forms for treatment by conventional processes.

While chlorine, because of its longer-lasting residuals and difficulty of control, has not been greatly considered, ozone has been studied extensively (Honn and Chavin, 1975; Otte *et al.*, 1977; Colberg and Lingg, 1978; Rosenthal *et al.*, 1978, 1980; Honn, 1979). Rosenthal (1980b) has provided a review of the production, chemistry and effects of ozone in aquaculture use. As a supplement to biological filtration, Honn and Chavin (1975) noted consistent reduction of total ammonia levels, though reports from the literature quoted by Rosenthal (1980b) suggest that at normal pH ammonia oxidation in fresh water is slow, first order reactions occurring only at pH 7 to 9, the rate increasing with pH. Colberg and Lingg (1978) found oxidation of ammonia as measured by nitrate production to be almost negligible at pH 7.2, but showed an approximate 8-fold and 16 fold increase at pH 8.2 and 9.3 respectively.

Nitrate oxidation was rapid at 1 mg/l O_3, but was not significantly pH-dependent; Rosenthal *et al.* (1978) showed 2-4 mg/l O_3 to produce approximately 50 per cent reduction in nitrite concentration, while higher doses oxidised nitrite almost completely.

Measurements of BOD, COD and TOC have confirmed the ability of ozone to break-down refractory compounds (Rosenthal, 1980a). Initial BOD was found to increase using 6.8-10.4 mg/l O_3, with best results being obtained using intermittent ozonation, presumably because of higher efficiency at higher organic concentrations. The effectiveness of the biological stage in removing these organic materials was not stated, though eventual BOD levels appeared to indicate that removal was effective. In trials using a glucose substrate, Colberg and Lingg (1978) noted increased oxidation rates at higher pH, supporting the suitability for ozone use in brackish or marine water systems.

The design equipment for ozone production and contact with water is described by Wheaton (1977), Spotte (1979) and Rosenthal (1980a). Efficiency of use depends both on ozone generation efficiency and effective gas/liquid transfer, usually in co-current or counter-current columns, where bubble size is normally the most critical factor. In addition to mass transfer the removal of organic molecules at the surface of bubbles and their entrapment in foam is a further requirement. Rosenthal (1980b) describes a spinning-disc mixing device offering advantages in controllable retention time, though mixing energy comparisons with conventional airlift or flow contactors were not given. The stability of foam depends on pH, temperature and bubble size, and is generally considerably better in sea water, possibly due to fractionated inorganic ions having a significant effect, in addition to the effects of differences in viscosity and surface tension (Rosenthal, 1980b).

On stoichiometric requirements, it is unlikely that ozone will be acceptable economically for complete oxidation of metabolic products, but its use as an adjunct to conventional treatments deserves further consideration, particularly in sea-water systems, though the constraints of atmospheric exposure and the production of bromates in sea water (Crecelius, 1979; Rosenthal, 1980a) must be considered. As a means for improving recycled-water supplies, compensating for fluctuations in treatment efficiency and removing colour from process water, ozone, if efficiently applied, will have considerable potential.

10. The Carbonate System and pH Control

The long-term addition of nitrates, carbonates and phosphates in a highly recycled system will tend to decrease pH, as the culture water buffering capacity is used up (Spotte, 1970). The carbonate system provides the main source of stability against pH change:

$$CO_2 + H_2O \rightleftarrows H_2CO_3 \rightleftarrows H^+ + HCO_3^- \rightleftarrows 2\,H^+ + CO_3^{2-}$$

the reaction moving to the left at low pH, and Co_3^{2-} being replenished from $CaCO_3$ where present.

The acidifying velocity in aquarium systems has been calculated by Saeki (1958) as 2.1 eq/day by nitrate accumulation, and 0.5 eq/day by combination of phosphates with calcium and magnesium, per tonne of fish held. Spotte (1970) considers that systems where nitrate levels are kept below 20 mg/l will experience little pH reduction.

In recycle systems incorporating a number of treatments, pH may be affected at a number of points (Figure 6.20). Thus denitrification, in addition to removing nitrate, will increase alkalinity; overall the balance may be different from that in a simple system. It should be noted also that the effect of pH on toxicity of metabolites, as mentioned earlier, may favour a reduced pH for operation, although the incidental removal of ammonia, for example, will be less significant.

Figure 6.20: pH Changes through a Recycling System.

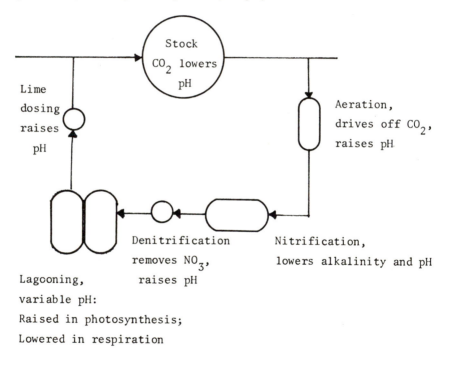

There are likely to be other reactions occurring in the carbonate system. Berener (1968) notes the combination of free fatty acids:

$$2\ RCOOH + CaCO_3 \leftrightarrows Ca(RCOO)_2 + CO_2 + H_2O$$

and suggests that organic materials may precipitate to reduce buffering ability and decrease pH. Depending on pH, reactions may also occur with ammonia (Berener, 1968):

at pH >9, $NH_3 + Ca^{2+} + HCO_3^- \leftrightarrows CaCO_3 + NH_4^+$

at pH <9, $CO_2 + NH_3 + H_2O \leftrightarrows NH_4^+ + HCO_3^-$

$$CO_2 + 2NH_3 + H_2O \rightleftharpoons 2NH_4^+ + CO_3^{2-}$$

The buffering of water using a range of mineral sources is discussed by Spotte (1979). On the basis of solubility and stability, dolomite ($CaMg(CO_3)_2$), oyster shell, coral rock and pure brimstone ($CaCO_3$) are suitable materials. Filter beds should be 100 per cent calcaraeous for sea water, at least 25 per cent for hard fresh water and more for soft water. Sacki (1958) suggests the neutralisation of aquarium water at 78 eq/tonne stock per month, corresponding to 4.4 kg CaO, 3.1 kg MgO, or 6.5 kg $NaHCO_3$. Chave (1965) showed that organic coating could completely inhibit the solubility of some calcareous materials and hence additional capacity should be provided.

For CO_2 removal, where low-pH regimes are being operated at levels of less than 10 mg/l, lime-dosing is normally the preferred method. Alternatively gas/liquid contact, using a high gas:liquid ratio may be employed (Fair *et al.*, 1971).

Apart from normal buffering, pH control, for more rapid results, can be achieved by dosing with acid or alkali. A typical installation consists of a metering pump, mixing tank and control system. This type of system may be useful in pretreating make-up water, particularly from ground water or sub-beach sources. Rates and designs depend greatly on individual circumstances and must be assessed accordingly.

Other Treatments. Treatments such as ion exchange, activated carbon and air stripping have been proposed for ammonia removal (e.g. Spotte, 1970). Designs are presented in Wheaton (1977) and Spotte (1979). Short (1973) has compared ion-exchange and airstripping with other methods of ammonia removal for river water use, concluding that although offering the potential of high efficiency, overall costs were considerably higher.

Ion exchange has received some attention in fish-culture use, and in improving water quality in transport tanks. Jorgensen *et al.* (1979) and Berka *et al.* (1980) discuss the capacity of clinoptilolite, a naturally occurring zeolite, in ammonia removal. Csavas and Varadi (1980) describe a submerged biofilter system using a zeolite filter-bed, though in terms of the loadings used, it is not certain whether the zeolite is effective, as some residual ammonia is present. Mayo (1976) quotes a 98 per cent removal at a loading of 2.59 m^3/m^3/day, using clinoptilolite. A brine solution is used for regenerating the exhausted exchanger.

Particular problems in ion exchange concern interference by other ions, combination with organic compounds and ineffective regeneration. Use in saline water is particularly difficult. The capacity of clinoptilolite is relatively low; hence regeneration cycles may be frequent at high loadings. Spotte (1970) considers the best application may be as an occasional adjunct to biological treatment, or as a final stage in removing a dilute concentration of ammonia.

Eliassen *et al.* (1965), using strongly basic anion exchangers, observed 92 per cent nitrate and 95 per cent phosphate reduction in sewage plant effluent, though application in fish culture treatment has not been considered.

Activated carbon has particular application for dissolved organic removal, though ammonia removal is generally insignificant. Waggott *et al.* (1972) note 75-99 per cent COD (10-72 mg/l COD) removal at loadings of 7.2-55 g

COD/100 g carbon, 7-30 m^3/m^2/hour. At a carbon consumption rate of 25-270 g/m^3 treated costs were £0.007-0.026/m^3 treated, though at the lower levels of COD probable in fish culture effluent costs may be proportionately less.

Miklosz (1971) notes the efficiency of activated carbon in removing methylene blue, malachite green, sulfathiozode and other medicaments. Copper and chlorine are also removed, and Seegert and Brooks (1978) have evaluated its use in dechlorination, finding it to be not completely effective. Problems in operation include physical clogging, particularly of finer-pored carbon, competition for pore space and difficulty of regeneration. Spotte (1979) considers regeneration only to be feasible at capacities of more than 1 million gals/day (4,540 m^3/day).

Air stripping of ammonia occurs as it is desorbed from water, ideally in conditions of high air/water contact, normally provided in spray towers, or by aeration. Short (1973), reviewing its operation, noted that a high gas/liquid ratio was required (3,000:1 for 95 per cent removal in sewage liquor, and 3,900-9,750:1 in lower ammonia concentrations in river water). An ammonia removal efficiency of 50 per cent occurs at normal pH levels, rising to 90 per cent at pH 11. At loading rates of 43,200 m^3/day a column of 6.5 m x 450 m^2 gave maximum efficiencies at water flows of 146 m^3/m^3/day. At high pH, CO_2 may also be removed, to cause scaling by $CaCO_3$ formation.

Although some incidental NH_3-N removal may occur during aeration processes, air stripping does not appear to be feasible in fish culture use as a major treatment, because of the high gas flows and the need to adjust the pH to achieve reasonable efficiency.

11. Energy Use in Recycled Systems

The significance of energy use in aquaculture production has been described by Edwardson (1976), who found most intensive flowing water systems to be less energy-efficient than deep-sea fishing. Recycle systems, particularly if involving heat as well as water circulation, are especially energy-intensive, and operating costs should dictate the most efficient design. Energy use in an air or oxygen supply (see earlier) must also be included.

11.1 Pumping Energy

The energy required for returning the water through the system constitutes the main demand for pumping, though waste-solids removal and chemical dosing may also employ pumps. Energy requirements are simply estimated as:

$$P = 9.80 \ Qh/\eta$$

where P = power (kW), Q = flow (m^3/sec), h = head (metres) and η = pump efficiency. For seawater, at a specific gravity of 1.025, power requirements are $P = 10.05 \ Qh/\eta$, actual energy demand being obtained by multiplying by a pump efficiency factor, typically 60-80 per cent.

The selection of pumps for the duty envisaged falls outside the scope of this chapter, although Table 6.27 outlines the major options. Sources such as Perry and Chilton (1973), Fair *et al.* (1971) or manufacturers' literature may be consulted for further details.

Table 6.27: Pump Selection

	Duty	Advantages	Disadvantages
Centrifugal	Low-medium head, most flow ranges, cont's or semi-cont's operation; fish pump	Simplicity, relative efficiency, can be submerged; most common and wide range available; can accept solids; easily controlled	Non-priming unless submerged or with foot-valve or priming tank
Axial	Very low head, high flow, e.g. pumping of ponds	Most efficient and cheapest for low-head systems; can accept solids	Limits in applicability; cavitation limits in propeller types
Mixed	Medium head, medium flow	Intermediate between centrifugal and axial; accept solids	
Positive Displacement	High head, low volume	Self-priming, high pressures	Sensitive to solids; seal maintenance
Peristaltic	Low, controlled flow; dosing chemicals	High controllability and accuracy	Relatively inefficient and expensive

As the flow is dictated by the stock requirements, pumping head is the most critical determinant of energy use. Head losses are caused chiefly by friction loss in pipes, valves, inlets and outlets, and by similar loss through filter media. Pressurised filters, where process water is forced through fine filter media, are particularly significant, and the blocking of filters by particulate matter also adds to losses. Friction losses in pipes and fittings depend on water velocity, material employed and suddenness of change (e.g. curves, constrictions), and the reduction of friction loss by using large diameters must be balanced against increased capital cost. Head losses can be calculated simply by using published data on friction factors and roughness coefficients (Fair *et al.*, 1971; Wheaton, 1977).

Pressure losses through filter media depend mainly on the water velocities, the nature of filter media and the degree of clogging, and as the last two may be difficult to quantify, empirical measurements are frequently used.

As an overall indication of the head losses to be expected, Table 6.28 gives approximate losses and typical totals for basic system units. On the basis of pumping efficiencies of 60 per cent, power required per tonne of stock is 55-100 W/metre head.

11.2 Airlift Pumps

Airlift pumps have been used in a number of instances to combine water movement with reaeration, and so attempt to reduce overall energy requirements. These are essentially similar to small aquarium systems in concept. Figure 6.21 illustrates typical configurations. A number of design equations have been presented relating to air flow, bubble size, pipe diameter, temperature and submergence depth to water flow (Wheaton, 1977; Murray *et al.*, 1980), and

Figure 6.21: Airlift Configurations.

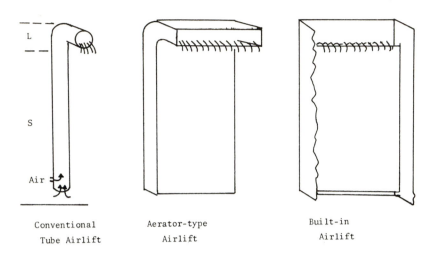

| Conventional | Aerator-type | Built-in |
| Tube Airlift | Airlift | Airlift |

S : Submergence
L : Lift

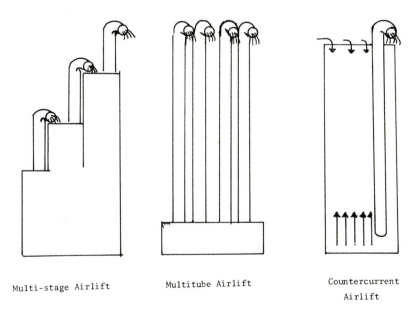

Multi-stage Airlift   Multitube Airlift   Countercurrent
                                                        Airlift

energy requirements may be assessed by determining the air flow required, although designs for low-submergence, large diameter pipes have not a present been fully investigated, and thus large flow rates are less accurately defined at the low heads typical of aquaculture use.

Table 6.28: Typical Component and System Head Losses[a]

Item	Head Loss (m)
90° bend	0.01-0.05
T-piece	0.02-0.10
Globe valve	0.10-0.50
Gate valve	0.02-0.10
Apertures	0.01-0.05
10 m pipe	0.05-0.10
Trickle filter	0.5-1.5[b]
Submerged filter	0.05-0.3
Sand filter	0.1-0.5
Sand filter (clogged)	0.3-1.0
Pressure filter	5.0-20.0
Cartridge filter	5.0-20.0
Settling tank	0.05-0.2

Notes: a. Assuming conventional design practice.
b. Height of filter bed.

According to Wheaton (1977) total efficiencies are normally considerably less than 50 per cent; on the basis of 35 per cent overall efficiency and 0.5 m system head, power requirements per tonne stock will be approximately 40-80 W (80-160 W/m head).

However, the addition of oxygen by airlifts may also be important, and the energy required for an equivalent oxygen transfer may be deducted from the total operating requirements. Thus if the airlift adds 1 mg/l to the water moved, this is the equivalent of 30-60 gO_2/tonne of stock per hour (0.5-1 m^3/min water flow), which at a transfer efficiency of 1 kg/kWhr represents 30-60 W, and so net pumping power used is 10-20 W at 0.5 metre.

11.3 Heat Supply

Heat can be supplied either directly (e.g. live steam) by electrical heat elements, or by heat exchange with other heated sources or with the atmosphere. In a recycled system, heating is required primarily for make-up water and system losses through conduction and convection. Heat lost through waste water could in theory be recovered, although in practice the temperature difference between inflow and outflow is often not sufficient to make this feasible. In fish culture, particular care has to be taken to ensure that heat exchangers are constructed of inert materials and that there is no risk of accidental cross-contamination. Other heat supplies must also be free of contamination and must be reliably controlled.

Heat requirements for make-up water can be simply assessed according to:

$$Q_H = Sp.Ht \times flow \times \triangle T$$

and thus relationships can be developed to determine the effects of stock weight and recycle rate.

Conduction, convection and air heat losses in enclosed areas or tanks may be more difficult to determine; in simple, circular, fully enclosed insulated tanks, assuming no convection transfer, conduction heat losses can be calculated according to total tank area, thickness of wall and heat transfer coefficient. For buildings a similar approach may be used, with the addition of a factor for heating replaced air. The heat supplied by pumps may also be considered; much of their mechanical inefficiency is in the form of heat, some of which is transferred to the process water. This can be estimated on the basis of:

$$\text{Heat supply} = \text{Rated pump power} \times (1 - \eta) \times H_W$$

where η = pump efficiency and H_W = proportion of heat transferred to water. H_W may reach 100 per cent in enclosed areas; a figure of 50 per cent can be used as an approximate guide.

Solar heating has been employed in experimental use as a back-up heat supply. Ayles *et al.* (1980) describe a system for trout production in Canada, using recycle or open flow depending on heat input. At an operating temperature of 12°C approximately 70 per cent of energy demand was supplied, though on overall production costs the use of heat appeared expensive, and the additional capital costs of solar heating were not assessed. A model study by Al-Shamma (1977) suggests that solar heating and recycling would be uneconomic for trout culture.

12. System Operation and Control

It is important to integrate the operation of individual stages adequately, to ensure the relatively efficient operation of each stage and of the system as a whole. By examining the process requirements for each stage, it is possible to outline a sequence for treatment and to establish whether particular conditions (e.g. flow-rates, concentrations) suggest full-stream, side-stream, continuous or intermittent operation. Table 6.29 illustrates typical process requirements and Figure 6.22 shows the layouts suggested.

In terms of process flows, Figure 6.23 shows the essential relationships involved. The use of a central sump has been suggested by a number of designers, having the advantage of allowing improved control over individual processes, providing a 'radial' as opposed to a conventional 'ring' system, in which each process is controlled by the process prior to it.

The use of activated sludge systems in particular can involve a number of different flow circuits. Thus Meske (1976) treated 50 per cent of the process water through the activated sludge unit, while Berka *et al.* (1980) describe a small-scale unit in which recirculation from sediment tank to contact tank proceeds at a greater flow (2.5 l/min) than the main flow (1.5 l/min).

Denitrification and sterilising stages are frequently used on sidestream, as determined by concentration limit or cost.

There have been no 'classical' systems models developed for aquaculture recycle systems in terms of dynamic responses, though these are being developed

(Mantle, pers. comm.). However, a number of observations have been made of fluctuations in conditions through recycling systems, as a result of fluctuating waste loadings. In practice, many systems are designed on a relatively conservative basis, though stock rates may be revised upwards if the capacity is shown to be available.

Table 6.29: Process Requirements for Ideal Operating Conditions

Process	Preconditions
Biological filtration	Oxygen supply, pH control
Nitrification	BOD removal, oxygen supply, pH control
Denitrification	Carbon source, low oxygen levels
Fine filtration	Coarse filtration or sedimentation
Activated carbon treatment	Fine filtration
Ultraviolet sterilisation	Solids removal (fine filtration)
Ion exchange treatment	Fine filtration
Air stripping	pH adjustment
Chlorination	Organic, NH_3 removal
Ozone treatment	Primary organic and solids removal

Srna and Baggaley (1975) noted the response of perturbed marine nitrification systems, and Rosenthal *et al.* (1980) described the response of recycled systems, to feeding of fish, noting incomplete adaptation, a gradual rise in metabolic concentration during the day and a return to baseline levels during the night. A change of more than 100 per cent in NO_2-N levels was observed, suggesting a poorer adaptation by the second nitrification stage. Muir (1978) noted similar effects in a marine recycled system, though the overall treatment efficiency was low.

It should be noted that as many processes are concentration-dependent, there is some degree of self-regulation. Thus increased oxygen deficit improves aerater efficiency, increased ammonia levels increase nitrification rate, and higher nitrate levels improve denitrification. The critical factor may be the degree of 'lag' present in the response — thus aeration acts instantaneously, as a physical process, whereas if biological treatment depends to any extent on cell growth there may be an appreciable delay.

The role of the fish stocks in cyclical patterns may also be significant, particularly if feeding drops as a result of poor water quality. Unless feeding is carefully controlled, the additional waste food load may further aggravate water quality conditions.

There is inevitably a compromise between full adaptation and control of processes to production conditions and the need to simplify routine operation. Thus while it is possible to control flow rates, filter operation or solids removal in conjunction with metabolic output changes, it is usually simpler to allow overcapacity to accommodate peaks in output.

Figure 6.22: Water Treatment Stages.

Simple Production Enhancement

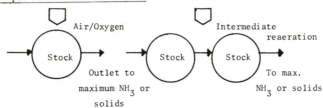

Further Enhancement at low pH   Solids and Ammonia Removal

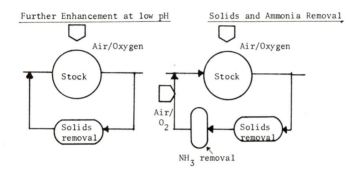

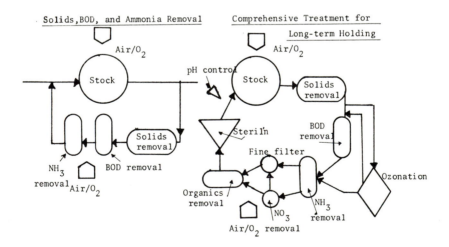

Figure 6.23: Recycle Layouts. Top — Ring System: individual treatments dependent on output from preceding treatment; limited control by by-pass. Middle — Radial System: individual treatments individually controlled; greater management required. Bottom — Radial System with Sump: individual treatments independently controlled; sump allows intermediate area before use, to optimise use of component treatments.

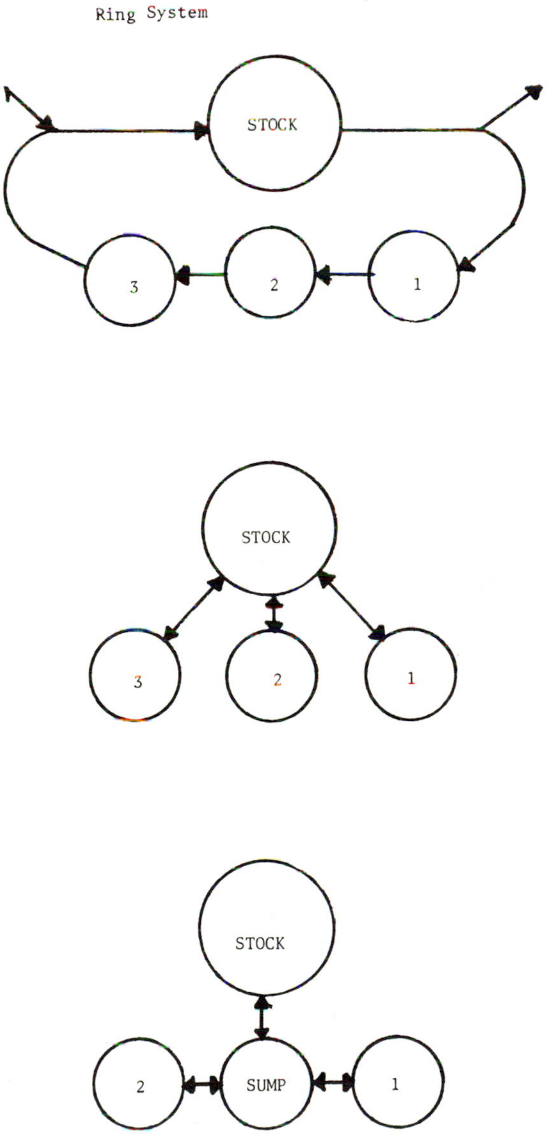

Table 6.30: Recirculation Systems

Reference	Type	Main Treatment	Media	Size (m³)	%	Flow (m³/m³/day)	Size/stock (l/kg) System	Size/stock (l/kg) Treatment	Other
Liao & Mayo (1973)	Pilot 20 m³ R. trout	Upflow filter Trickling filter Activated sludge Extended aeration	Koch Rings 10 cm — —	3.75 6.60 5.00 19.90	18.8 33.0 25.0 99.5	3.6 2.1 2.7 0.7	74.5 79.0 54.5 157.0	13.9 24.5 18.5 71.0	15°C, UV treatment, 90% recycle
Parisot (1967)	Full Chinook Salmon	Filtration	0.3 m oyster shell 1.3 m gravel (10-75 mm)	340	19.3	2.1	6.9- 13.8	1.3- 2.6	13-15°C, 90-95% recycle, UV treatment, aerated to 90% sat'n, Backwash: 0.4 m³/m² water 0.33 m³/m³
Risa & Skjervold (1975)	Salmon Lab 41 m³	Gravel filter	0.2 m rock (2-4 cm) 0.6 m limestone 0.6 m limestone	18.00 (0.8-1.5 cm) (0.3-0.8 cm)	43.9	0.013	146.0	90.0	9-10°C, 90% recycle, 9 m³ sedimentation
Hall (1974)	Catfish Prod'n 962 m³	Filter & pressure filter	Limestone (2-10 cm) Silica No. 8	600	62.4	0.28 58.5	32.1	12.0	28-35°C, 90-95% recycle, temperature control in insulated barn
Greer (pers. comm.)	" 4368 m³	Filter & pond	Limestone	600	13.7 42.9	—	84.2	20.0 — 62.4 —	filter pond
Meske (1973)	Carp Pilot 14 m³	Activated sludge	—	8.10	58.1	0.67	90.0- 172.0	50.0- 100.0	23°C, 99% recycle
Wohlfarth & Lahman (1971)	Carp 32.6 m³ Lab	Filter & settling tank	Gravel/shell —	6.36 6.00	19.5 18.4	5.82 6.17	700.0	136.6 128.9	Additional aeration
Hill et al. (1973)	Trout/Catfish 26,077 m³	Settling pond	—	26,000	99.7	0.005	3,040	3,031	Complete recycle, rotation of production

Reference	Type	Main Treatment	Media	Size (m³)	%	Flow (m³/m³/day)	Size/stock (l/kg) System	Treatment	Other
Meade (1974)	Salmon Pilot 37 m³	Filtration	PVC module 88.6 m²/m³	12.70	34.3	1,860	26.6	8.8	Denitrification, 1.2 l/kg, 40-mesh solids screen
Goldizen (1970)	Lobster Sea bream Lab. 0.5 m³	Filtration	Calcareous gravel (4% mg) 2.5 mm	0.10	20.0	15.0	70.0-500.0	14.0-100.0	Sea water, 99.9% recycle, aeration, UV, temp. control
Hilge (1980)	Channel catfish Pilot 60 m³	Activated sludge	—	—	—	—	7.2-18.9	—	24-26°C, 99.9% recycle
Jespersen & Hodal (1980)	R. trout fingerling 15.2 m³ Pilot	Biofilter sedimentation, hydroponic	Plastic 25 m²	0.7 / 0.7 / 11	4.6 / 4.6 / 72.4	20.6 / 20.6 / 1.3	1,520	70 / 70 / 1,100	16-18°C, 400 W fluorescent-light, cladophora, 3 g/m²/day DM produced
Lewis et al. (1980)	Channel catfish Lab. 3-8 m³	Biodisc sedimentation, hydroponic	45.6 m²	1.0 / 0.2 / 1.74	26.3 / 5.3 / 45.8	41.8 / 208.8 / 24.0	69	18.2 / 0.36 / 31.6	20-25°C, 95.5% recycle, non-continuous stocking
Kossman (1980)	Grass Carp Pilot 350 m³	Activated sludge, denitrifying filter, aeration tank	—	Total 263	75.1	1.33	45	34	25°C, lime-adjusted pH, 99.6% recycle, growth to 1 kg in 1 year, 5 kg in 2 years
	Prod'n 1500 m³ (planned)	Activated sludge sedimentation biofilter reservoir	—	Total 1,200 (3 x 400)	80.0	3.2	20	16	As above; aerated tanks total 2,500 m³ air/hour, waste pump/compresser heat to system
Otte & Rosenthal (1979)	Tilapia eel 5.5 m³	Trickle filter, settling, denitrification (act. sludge)	Plastic pack	1 / 1.9 / 1.06	18.8 / 34.5 / 19.3	168 / 88.4 / 158.5	24.3	4.4 / 8.4 / 4.6	8% salinity, ozonation, 22-26°C
Mayo (1980)	Salmonid 5500 m³	Downflow biofilter	Rock (1.2 m), crushed oyster (0.3 m)	2,600	47.3	31.9	63.8	30.1	90% recycle, 10-14°C, sand filter to ≤10μ, UV sterilisation

Table 6.30: continued

Reference	Type	Main Treatment	Media	Size (m³)	%	Flow (m³/m³/day)	Size/stock (l/kg) System	Treatment	Other
	4,600 m³ (planned)	Fluidised, upflow biofilter, clarifiers	0.8 m Quartz sand 1000 m²/m³	96	2.1	864	53.4	1.1	90-95% recycle, as above
				1,850	31.5	44.8	45.2	21.5	
	3,900 m³	Upflow biofilter	7 cm polythene beads 200 m²/m³	1,230	31.5	67.4		14.3	90% recycle, as above
Poxton et al. (1980)	Turbot 7 m³ Experimental	Flooded filter, integral	'Flocor' & gravel	0.5	50.0	57.6	250	125	
Van Toever & MacKay (1980)	Salmonids Pilot 2 m³	Biofilter, hydroponic, settling	Total	1.4	70.0	180	244	171	Now 100%, NO$_2$ problems, solar heated 7-14°C
Møller & Bjerk (1980)	Salmo salar production 200 m³	Settling		160	80.0	7.5	280	220	Heaters to 90 kW, 11°C, air 9,000 m³/hr, 1200 W UV
Rasmussen (1980)	Whitefish (Coregonus) 12.3 m³	Settling algae/daphnia, mussels		0.1	0.8	7.2	1,230	10	3-20°C, Cladephera sig. in NH$_3$ renewal, mussel growth not significant
				10	81.3	0.72		1,000	
				0.2	1.6	36		20	
Csávás & Váradi (1980)	Cyprinid Expm'tal 300 m³	Settling, biofiltration, aeration	3-5 mm (clinoptilolite)	56	18.7	34.3	stock weights not specified		
				72	24.0	26.7			
				29	8.7	66.2			
Chiba (1980)	Eel (typical) 200 m³	Submerged or trickle filter	4-6 cm stone, nets mats	50	25.0	48	64	28	2-4 kW aeration, 95-98% recycle
							128	68	
	Eel 54 m³	"	"	16.2	30.0	67.7	127	38.2	100% recycle (30 days).
	Carp 500 m³	"	"	42	8.4	206	100	8.4	99.5% recycle
	Ayu 114 m³	"	"	4.3	3.8	1,620	231.5	8.7	99.4% recycle
Mayo (1980)	5,475 m³	Settling, upflow biofilter	1.8 m Norton rings 8.9 cm	1,847	33.7	60.8	46.7	15.7	90% recycle, as above

Reference	Type	Main Treatment	Media	Size (m³)	%	Flow (m³/m³/day)	Size/stock (l/kg) System	Treatment	Other
Rosenthal (1980) review	9,380 m³	Downflow biofilter	Shell (0.3 m) rock (1.8 m)	5,210	55.5	31.7	109.3	60.7	90% recycle, as above, backwash 2/week
	Aquarium 35 m³	—	—	19	54.3	44.2	184.2	100.0	14°C, 99.9%
	Aquarium 250 m³	—	—	210	84.0	2.9	2,083	1,750	24-26°C, 96.2%
	Experimental— Lobster larvae 0.5 m³	—	—	0.4	80.4	23.1	1,389	1,111	22°C, 98.7%
	Experimental— Siganids 1.6 m³	—	—	0.95	59.4	—	246	146	22-24°C
	Experimental— Tilapia 2.0 m³	—	—	1.75	87.5	—	125	109.4	23°C
	Experimental— Tilapia Eel Dicentrarchus 8.0 m³	—	—	3.04	38.0	63.2	18.6	7.1	23-26°C, 95.8%
	Exp. FW Crayfish 12.2 m³	—	—	6.1	50.0	56.5	40.7	20.3	16°C, 99.8%
	Production Trout 160 m³	—	—	48	30.0	186.7	94	28	12°C
	Production Dicentrarchus 183 m³	—	—	43.2	23.6	5.6	6,655	1,570	18-21°C, 99%
Meske (1976)	Carp Pilot 60 m³	Act. sludge, sedimentation/ denitrification 'Ahrensburg' reservoir		29	48.3	9.9	40.5	19.3	23°C, 50% of flow through treatment
				17.3	28.8	16.6		11.5	
				6.8	11.3	42.4		4.5	
Berka et al. (1980) review	0.5 m³	Act. sludge, sed. tank		0.15	30.0	14.4	40	12.0	0.1 m³ denitrification added later
				0.22	44.0	9.9		17.6	

Table 6.30: continued

Reference	Type	Main Treatment	Media	Size (m³)	%	Flow (m³/m³/day)	Size/stock (l/kg) System	Treatment	Other
Nagel (1977)	Carp 2 m³	Act. sludge, nitrification, denitrification, sedimentation, aeration, plants		0.4 0.4 0.5 0.24					
Lu (1978)	Channel Catfish 4.1 m³	Trickle filter, settling tank, reservoir	Limestone	0.64 0.76 1.89	15.6 18.5 46.1				BOD 0.14 kg/m²/hr
Bohl (1977)	Eel 41 m³	Trickle filter	'Lavalit' slag	3.3	8.0	2.23			
Berka et al. (1980)	55 m³	Biodisc/O₂ settling		16 19	29.1 34.5	75.0 63.2	15.3	4.4 5.3	Oxygenated, 95-99% recycle, 12 tonnes/ annum
Honn & Chavin (1975)	Nurse Shark Lab. 2.27 m³	Filtration Algae	Marble/dolomite (2-5 mm, 10-15 mm) Enteromorphae		0.454 0.038	0.193 1.61	1,250	250 20.9	Seawater, 100% recycle, ozone treatment, 24°C
Cabrero	Catfish Pilot 32 m³	Polyculture, tilapia, algae, water hyacinth	—		16.0	—	1,392	696	Tilapia second crop, 100% recycle, aeration

System Design. Table 6.30 describes a number of working recycled water systems covering a range of species and operating duties.

There are insufficient working recycled systems to be able to identify specific design or layout types corresponding to a particular cultured species. As yet, feed rates, flow rates, water temperatures and husbandry procedures have not been optimised for particular species, and so process conditions have not yet been fully defined. Thus systems are generally designed on an ad hoc basis, and may not represent a fully developed design. It is instructive to note the wide variation in designed sizes and the range of treatments employed. There is unfortunately insufficient opportunity to examine the actual versus the designed performance of individual units, though the relatively low loading in many cases suggests that efficiencies are generally low.

As expected, on the basis of cost, stability and reliability, biological filters are most frequently employed, often with a layer of calcareous material for buffering. While many of the filters use rock or gravel, filters designed for higher loading frequently use plastic media. A number of systems do not incorporate a specific solids–removal stage, and hence nitrification efficiency will be lower.

As the systems described vary considerably in design, layout and duty, it is difficult to generalise design characteristics. However, with a few exceptions, a minimum treatment volume appears to be approximately 10 m^3/tonne, occupying 20-30 per cent of total volume.

The fact that recycle systems may entail pumped water and air, a number of connected containers, chemical dosing and specific process requirements implies a far greater degree of control than necessary in a simple gravity-fed or pond-type aquaculture system. It will normally be necessary to incorporate a number of level/flow operated alarms and possibly monitors for chemicals such as oxygen or chlorine. Any proposed system should be analysed carefully for consequences of simple events such as tank blockages and excess filter head loss, and intermittent-cycle systems should be analysed for operation sequences to ensure that capacities and flow rates of individual components are adequate.

The use of back-up facilities depends largely on the perception of risk by the operators and if reasonable reliability is available (e.g. public electricity supply, use-cycles in control switches), back-up or replacement schedules may be arranged accordingly. Note that if a recycled system is to be used for the early stages of a species not otherwise available, the cost of loss may have to be judged also against the lost production opportunity.

13. Husbandry and Production Management

The routine husbandry and stock control procedures employed in aquaculture must be considered carefully in recycled-system design. The effects of feeding and tank cleaning have been discussed earlier. Grading is likely to cause similar disturbances in stress and waste output, though connections between holding areas may be used to facilitate transfer, or even allow self-grading of stocks held.

Control and treatment of disease is of particular concern, as stocks may be isolated and treatments confined to the affected group. Care must also be taken to avoid contaminating biological treatments. The fate of disease organisms within recycled systems has not been adequately considered, though it is known that *Ichthyophthirius* can be particularly resistant, and unless a disinfection stage is incorporated, reinfection may be a problem. Populations of bacteria, such as *Aeromonas, Pseudomonas, Vibrio* and *Myxobacteria* spp., may occur in heavily loaded systems, and fungi such as *Saprolegnia* may also be present. The presence of untreated solid wastes appears to be particularly important. Both for treatment of affected stock and holding prior to introduction to a recycled system, a separate holding area would be desirable.

In terms of growth and stock management, recycled systems using temperature-controlled water may offer considerable advantages, particularly if sequential cropping is employed. Using the analysis described in Figure 6.24, it can be seen that production can reach 3-6 times normal capacity if optimal temperatures are employed, and so unit capital costs can be decreased. In practice, however, the increased number of holding units, limits of fry supply, and size distribution effects within the stock held may reduce this advantage. Berka (1980), however, refers to a production of 12.2 t of carp from a holding capacity of 3.6 t, and hence a P/C ratio of 3.4:1.

Figure 6.24: Multiple Cropping.

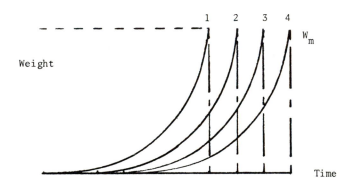

Capacity: sum of individual crop weights at specific time, $W = \Sigma W$.
Production: sum of market crops in 1 year, CW_m.
Production: capacity ratio: $(CW_m)/\Sigma W$.
Depending on growth rate and shape of growth curve, production:capacity ratio ranges from 1.5 (conventional farming systems) to a theoretical 8.0-8.5 (constant optimum temperature).

14. Economics of Recycling Systems

Recycled systems have been suggested in a number of instances for complete on-growing of cultured species, particularly where water shortages exist, or where a site is particularly near to market areas. By planned production it could also be possible to use heated recycled water to produce a number of crops throughout the year, and so to reduce the unit capital costs. Parker and Broussard (1977) have provided a bibliography of cost information on recycled systems.

In practice, there have been few completely successful recycle units for ongrowing, and most have been employed for the raising of relatively valuable (in terms of stock weights) early stages, e.g. of salmon, eels, carps, *Penaeus* and grouper. The small profit margins available with many cultured species, and the uncertainty of costs, risks and returns in recycled systems have contributed to this.

In an analysis of recycled system economics, Muir (1979) noted that even in heated systems with high production relative to capacity, the current cost breakdowns in trout culture indicated that energy and additional costs made recycled systems generally uneconomic. At current energy cost ranges (£0.2-£0.04/kWhr) operating costs were 20-40 per cent higher at 1 metre pumping head, and 44-69 per cent higher at 3 m pumping head. Alternatively, ratios of production to capacity would have to reach 1.5-2.0 for lowest recycle costs to reach highest open-flow costs. Unless specific advantages such as proximity to markets, reduced labour costs, or improved security could offset these higher costs, there was very little case for recycled systems for ongrowing. For species currently providing higher profitability, alternative lower-cost methods, and the higher fixed capital costs of recycled systems, tended to discourage their use.

Using heated, recycled water for early rearing, however, at 99 per cent recycle and 5°C average temperature rise, a weight fraction of 7.3-12.3 per cent of final crop weight at 3 metre pumping head, and 8.9-16 per cent at 1 metre pumping head could economically be produced at current energy costs. By comparison, however, using warm-water effluent, smolt production was found to be approximately 20 per cent cheaper than in conventional methods (Sumari and Westman, 1980). In such cases where a critical threshold of development occurs (i.e. an option of production at one-year or two-year age, as in the case of smolts), heated water, whether from waste heat or by recycling, may be particularly advantageous.

Macdonald *et al.* (1975) suggested that a recycled system should prove economic for salmonid smolt production. Mayo (1980) found that in trout hatchery design, costs in recycled systems were approximately 1.33 times those of open flow equivalents, though it was noted that if a choice exists between two small open flow systems on suitable sites and a single large recycled system, scale economies would favour the latter. However, the use of recycled systems to accelerate growth or smoltification in salmonids may not be currently feasible in terms of the ultimate quality of the fish produced (Wedemeyer, 1980).

15. Conclusions

Recycled systems have the potential to offer a number of important advantages for aquaculture production, including controlled environments, relative freedom from site and water supply constraints, control over disease agents and temperature-controlled growth rates and production. Economics of production, however, tend to limit their use to early-rearing stages, laboratory systems (e.g. to provide uniform conditions for experimental trials) and public display systems.

Current experience and economic developments suggest that recycled systems are unlikely to find acceptance for complete ongrowing, and it would be more appropriate to direct research towards improving effectiveness in fry and fingerling production. Areas of particular importance are: holding tank design related to solids removal with respect to biological filtration; the dynamics and control of recycle operation, particularly in terms of oxygen, ammonia, nitrite, carbon dioxide and pH changes; in terms of routine breeding and husbandry procedures, the use of oxidising agents such as ozone or chlorine as an adjunct to biological treatment; and the role and fate of pathogens within recycled systems. As fry and fingerlings will have to be physiologically adequate to grow-on successfully elsewhere, it is likely that environmental conditions should be of the highest standards, rather than those obtained by optimising the relationship between treatment cost and growth rate.

The information provided in this chapter and elsewhere should, however, be sufficient to enable the construction of working recycle systems to appropriate standards, and appropriate economic evaluation can be made on this basis. What is perhaps most critical, even with the package units currently being marketed or being conceived, is the understanding by the operator of the additional functions and dimensions to managing fish production in these systems, as successful, economical production is the only sure means to general acceptability.

References

Adams, C.E. (1973). Removing nitrogen from waste water. *Env. Sci. Tech., 7:* 696-701

Alabaster, J.S. (1974). The development of water quality criteria for marine fisheries. *Ocean Management, 2:* 101-15

Alabaster, J.S. and Lloyd, R. (1980). Water Quality Criteria for Freshwater Fish. Butterworth, London, 297 pp.

Alabaster, J.S., Shurben, D.G. and Knowles, G. (1979a). The effect of dissolved oxygen and salinity on the toxicity of ammonia to smolts of salmon *Salmo salar L. J. Fish Biol., 15:* 705-12

Alabaster, J.S., Shurben, D.G. and Mallett, M.J. (1979b). The survival of smolts of salmon (*Salmo salar*) at low concentrations of dissolved oxygen. *J. Fish Biol., 15:* 1-8

Albertini-Berhaut, J.C., Poizat, C. and Guerin, O. (1976). Mise au point d'un circuit ferme en eau de mer pour la survie en laboratoire des jeunes muges. *Aquaculture, 7 (4):* 347-55

Albertsson, J.G., McWhirter, I.R., Robinson, E.K. and Vahldieck, N.P. (1970). Investigation of the use of high-purity O_2 aeration in the conventional activated sludge process. Report, Union Carbide Corporation; summarised in *Water Pollution Abstracts, 121* (1971)

Albrecht, M.L. (1977). Bedeutung des sauerstoffs und schadigen durch sauerstoffmangel und kohlensaure uber-sattigung bei fischen. *Z. fur Binnenfischerei der DDR, 24:* 207-13

Alderson, R. (1979). The effect of ammonia on the growth of juvenile Dover Sole, *Solea solea* L. and turbot *Scopthalmus maximus* L. *Aquaculture, 17:* 291-309

Allen, K.O. (1974). Effects of stocking density and water exchange rate on growth and survival of Channel catfish *Ictalurus punctatus* (Rafinesque) in circular tanks. *Aquaculture, 4:* 29-39

Allen, P.G. and Johnston, W.E. (1976). Research direction and economic feasibility: an example of systems analysis for lobster aquaculture. *Aquaculture, 9:* 181-6

Al-Shamma, A. (1977). Economic and biological evaluation of possible uses of solar energy in trout culture. MSc dissertation, University of Stirling, Scotland, 95 pp.

Armstrong, D.A., Stephenson, M.F. and Knight, A.W. (1976). Acute toxicity of nitrite to larvae of the giant Malaysian prawn *Macrobrachium rosenbergii. Aquaculture, 9 (1):* 39-46

Atkinson, P., Swilley, E.L., Busch, A.W. and Williams, D.A. (1967). Kinetics of mass transfer, organism growth in a biological film reactor. *Trans. Inst. Chem. Engrs., 45:* T257

Ayles, G.B., Scott, K.R., Barica, J. and Lark, J.G.I. (1980). Combination of a solar collector with water recirculation units in a fish culture operation. Symposium on New Developments in the Utilisation of Heated Effluents and Recirculation Systems for Intensive Aquaculture. EIFAC, 11th Session, Stavanger, Norway, 28-30 May, EIFAC/80/Symp: E/32

Balarin, J.D. and Hatton, J.P. (1979). *Tilapia: A Guide to their Biology and Culture in Africa*. University of Stirling, Scotland, 174 pp.

Berener, R.A. (1968). Calcium carbonate concentrations formed by the decomposition of organic matter. *Science, 159:* 195-7

Berka, R., Kujal, B. and Lavicky, K. (1980). Recirculation systems in Eastern Europe. Symposium on New Developments in the Utilisation of Heated Effluents and Recirculation Systems for Intensive Aquaculture. EIFAC, 11th Session, Stavanger, Norway, 28-30 May, EIFAC/80/Symp: R/14.2

Birkbeck, A.E. and Walden, C.C. (1976). Enhancement of nitrification in biofilters for fish hatchery recycle systems. 43rd Ann. Meeting Pac. N.W. Polln. Control Assoc., Seattle

Bohl, M. (1977). Some initial aquaculture experiments in recirculating water systems. *Aquaculture, 11 (4):* 323-8

Boon, A.G. and Burgess, D.R. (1972). Effects of diurnal variations in flow of settled sewage on the performance of high rate activated sludge plants. *J. Inst. Wat. Poll. Cont., 71:* 493-7

Bower, C.E. and Bidwell, J.P. (1978). Ionisation of ammonia in seawater: effects of temperature, pH, and salinity. *J. Fish. Res. Bd Can., 35:* 1012-16

Boyd, C.E. (1970). Chemical analysis of some vascular aquatic plants. *Arch. Hydrobiol., 67:* 78-85

Braaten, B.R. (1978). Bioenergetics – a review of methodology. In: Halver, J. and Tiews, K. (eds), *Nutrition and Fishfeed Technology*. Proc FAO Conf., Hamburg, Heenemann Verlag, GmBH, pp. 461-505

Brett, J.R. (1970). Fish, the energy cost of living. In: McNeill, W.J. (ed.), *Marine Aquaculture*. Oregon State Univ. Press, Corvallis, Oregon, pp. 37-52

Brett, J.R. and Zala, C.A. (1975). Daily pattern of nitrogen excretion and oxygen consumption of sockeye salmon (*Oncorynchus nerka*), under controlled conditions. *J. Fish. Res. Bd Can., 32:* 2479-86

Brockway, D.R. (1950). Metabolic products and their effects. *Prog. Fish Cult., 12 (3):* 127-9

Bruce, A.M. (1970). Some factors affecting the efficiency of high-rate biological filters. In Jenkins, S.H. (ed.), *Advances in Water Pollution Research*. Proc. 5th Int. Wat. Pollut. Res., pp. 11-14

Bruce, A.M. and Boon, A.G. (1970). Aspects of high rate biological treatment of domestic and industrial waste waters. Inst. Wat. Polln Control, Public Works and Municipal Service Congress, paper 17

Bruce, A.M., Merkens, J.C. and Macmillan, S.C. (1970). Research and development in high rate biological filtration. *J. Inst. Publ. Hlth. Eng.:* 178-207

Bullock, G.L. and Stuckey, H.M. (1977). Ultraviolet treatment of water for destruction of five gram-negative bacteria pathogenic to fishes. *J. Fish Res. Bd Can., 34:* 1244-9

Burleson, G.R., Murray, T.M. and Pollard, M. (1975). Inactivation of viruses and bacteria by ozone, with and without sonication. *App. Micob., 29:* 340-4

Burrows, R.E. (1964). Effects of accumulated excretory products on hatchery-reared salmonids. *US Fishery and Wildlife Serv. Research Rept., 66:* 1-12

Burrows, R.E. and Chenoweth, H.H. (1970). The rectangular circulating rearing pond. *Prog. Fish Cult., 32 (2):* 67-80

Burrows, R.E. and Combs, B.D. (1968). Controlled environments for salmon propagation. *Prog. Fish Cult., 30 (3):* 123-36

Busch, C.D., Koan, J.L. and Allison, R. (1974). Aeration, water quality, and catfish production. *Trans. Am. Soc. Ag. Engr., 17 (3):* 443-5

Carmignani, G.M. and Bennett, J.P. (1977). Rapid start-up of a biological filter in a closed aquaculture system. *Aquaculture, 11 (1):* 85-8

Castro, W.E., Sielinski, P.B. and Sandifer, P.A. (1975). Performance characteristics of airlift pumps of short length and small diameter. Proc. 6th Ann. Meeting World

Mariculture Soc., Seattle, Washington, 27-31 January
Chapman, S.R., Chesness, J.L. and Mitchell, R.B. (1971). Design and operation of earthen raceways for channel catfish production. Joint Meeting S.E. Region Soil Cons. Soc./ S.E. Region Am. Soc. Ag. Engrs, Jacksonville, Fla, 31 January-3 February
Chave, K.E. (1965). Calcium carbonate; association with organic matter in surface seawater. *Science, 148:* 1723-4
Chesness, J.L. and Stevens, J.L. (1971). A model study of gravity flow aerators for catfish raceway systems. *Trans. Am. Soc. Agr. Engrs., 14 (6):* 1167-9
Chesness, J.L., Fussell, L.J. and Hill, T.K. (1973). Mechanical efficiency of a nozzle aerator. *Trans. Am. Soc. Agr. Engr., 16 (1):* 67-8
Chiba, K. (1980). Present state of recirculation and flow through systems and their problems in Japan. Symposium on New Developments in the Utilisation of Heated Effluents and Recirculation Systems for Intensive Aquaculture. EIFAC, 11th Session, Stavanger, Norway, 28-30 May, EIFAC/80/Symp. R16
Collins, M.T., Gratzek, J.B., Shotts, E.B., Dawe, D.L., Campbell, L.M. and Senn, D.R. (1975). Nitrification in aquatic recirculation systems. *J. Fish Res. Bd Can., 32:* 2025-31
Colt, J., Mitchell, S., Tchobanoglous, G. and Knight, A. (1979). The use and potential of aquatic species for wastewater treatment: Appendix B. The environmental requirements of fish. Publn. No. 65, State Water Resources Control Board, California
Colt, J. and Tchobanoglous, G. (1976). Evaluation of the short-term toxicity of nitrogenous compounds to channel catfish, *Ictalurus punctatus. Aquaculture, 8 (3):* 209-24
—— (1978). Chronic exposure of channel catfish, *Ictalurus punctatus*, to ammonia: effects on growth and survival. *Aquaculture, 15 (4):* 343-72
—— (1979). Design of aeration systems for aquaculture. Bioengineering Symposium, Culture Section, Am. Fish. Soc., Traverse City, Michigan, 15-19 October
Colt, J., Tchobanoglous, G. and Wang, B. (1975). The requirements and maintenance of environmental quality in the intensive culture of channel catfish. An annotated bibliography. Dept Civ. Eng., Univ. of Calif., Davis, USA, 119 pp.
Colberg, P.J. and Lingg, A.J. (1978). Effect of ozone on microbial fish pathogens, ammonia, nitrate, nitrite, and BOD in simulated reuse hatchery water. *J. Fish. Res. Bd Can., 35:* 1290-6
Conrad, J.F., Holt, R.A. and Kreps, T.D. (1975). Ozone disinfection of flowing water. *Prog. Fish. Cult., 37:* 134-6
Cooper, R.C., Oswald, W.J. and Bronson, J.C. (1965). Treatment of organic industrial wastes by lagooning. *Proc. 20th Ind. Waste Conf.,* Purdue Univ. Eng. Ext. Ser. 118, pp. 351-64
Crecelius, E.A. (1979). The measurement of oxidants in ozonised seawater and some biological reactions. *J. Fish. Res. Bd Can., 36:* 1006-8
Csavas, I. and Varadi, L. (1980). Design and operation of a large-scale experimental recycling system heated with geothermal energy at the Fish Culture Research Institute, Szarvas, Hungary. Symposium on New Developments in the Utilisation of Heated Effluents and Recirculation Systems for Intensive Aquaculture. EIFAC 11th Session, Stavanger, Norway, 28-30 May, EIFAC/80/Symp: E/16
Curds, C.R. and Cockburn, A. (1970). Protozoa in biological sewage treatment processes. *Wat. Res., 4:* 225-49
Davies, J.C. (1975). Minimal dissolved oxygen requirements of aquatic life with emphasis on Canadian species: a review. *J. Fish. Res. Bd Can., 32 (12):* 2295-332
Dawson, R.N. and Murphy, K.L. (1972). The temperature dependency of biological denitrification. *Wat. Res., 6:* 71-83
Degremont, R. (1974). *Water Treatment Handbook,* 3rd edn. Elliot, London
Doudoroff, P. and Katz, M. (1950). A critical review of the literature on the toxicity of industrial wastes and their components to fish. *Sew. and Ind. Wastes, 22:* 1432-58
Doudoroff, P. and Shumway, D.L. (1970). *Dissolved Oxygen Requirements of Freshwater Fishes.* FAO Fish. Tech. Paper 86, 291 pp.
Downing, A.L. (1966). Activated sludge design problems. *Proc. Biochem., 1:* 257
Downing, K.M. and Merkens, J.C. (1957). The influence of temperature on the survival of several species of fish in low tensions of dissolved oxygen. *Ann. Appl. Biol., 45:* 261-9
Downing, A.L., Painter, H.A. and Knowles, G. (1964). Nitrification in the activated sludge process. *J. Proc. Inst. Sew. Purif., 83:* 130-58

Duddles, G.A., Richardson, S.E. and Barth, E.E. (1974). Plastic medium trickling filters for biological nitrogen control. *J. Wat. Polln. Cont. Fed., 46:* 937-46

Eckenfelder, W.W. and Balakrishnan, S. (1968). Kinetics of biological nitrification and denitrification. Report, Center for Research in Water Resources, University of Texas

Eddy, F.B. and Morgan, R.I.G. (1969). Some effects of carbon dioxide on the blood of Rainbow trout (*S. gairdneri* Richardson). *J. Fish Biol., 1 (4):* 361-72

Eden, G.F., Truesdale, G.A. and Mann, H.T. (1966). Biological filtration using a plastic filter medium. *J. Proc. Inst. Sew. Purif., 31:* 562

Edwards, D.J. (1978). *Salmon and Trout Farming in Norway.* Fishing News Books, Surrey, 193 pp.

Edwardson, W. (1976). Energy demands of aquaculture, a worldwide survey. *Fish Farming Int., 3 (4):* 10-14

EIFAC (1969). Water quality criteria for European freshwater fish: water temperature and inland fisheries. *Wat. Res., 3:* 645-62

—— (1973). Water quality criteria for European freshwater fish, report on ammonia and inland fisheries. *Wat. Res., 7:* 1011-22

—— (1980). Symposium on New Developments in the Utilisation of Heated Effluents and Recirculation Systems for Intensive Aquaculture. 11th Session, Stavanger, Norway, 28-30 May

Elliott, J.M. (1969). Oxygen requirements in Chinook salmon. *Prog. Fish Cult., 31 (2):* 67-73

—— (1976). The energetics of feeding metabolism and growth of brown trout in relation to body weight, water temperature and ration size. *J. Anim. Ecol., 45:* 273-89

Ellis, J.E., Tackett, D.L. and Carter, R.R. (1978). Discharge of solids from fish ponds. *Prog. Fish. Cult., 40 (4):* 165-6

Fair, G.M., Geyer, J.C. and Okun, D.A. (1971). *Elements of Water Supply and Wastewater Disposal.* Wiley Interscience, New York, 752 pp.

Fickeisen, D.H. and Schneider, M.J. (1976). Gas Bubble Disease. Proc. Workshop, Richland, Wash., Oct. 8-9 1974, NTIS Conf. — 741033, US Dept of Commerce, Springfield, Va., 123 pp.

Flatow, R.E. (1980). High dosage UV water purification, an indispensible tool in recycling fish hatcheries and heated effluent aquaculture. Symposium on New Developments in the Utilisation of Heated Effluents and Recirculation Systems for Intensive Aquaculture. EIFAC, 11th Session, Stavanger, Norway, 28-30 May, EIFAC/80/Symp: E/44

Flis, J.R. (1968). Anatomico-histopathological changes induced in carp (*Cyprinus carpio*) by ammonia water. *Acta Hydrobiol., 10 (1/2):* 205-38

Forster, J.R.M. (1974). Studies on nitrification in marine biological filters. *Aquaculture, 4:* 387-97

Forster, J.R.M., Harman, J.P. and Smart, G.R. (1977). Water economy — its effect on trout farm production. *Fish Farming Int., 4 (1):* 10-13

Forster, J.R.M. and Smart, G.R. (1979). Water economy in aquaculture. In: Godfriaux, B.L., Eble, A.F., Farmanfarmaian, A., Guerra, C.R. and Stephen, C.A. *Power Plant Waste Heat Utilisation in Aquaculture.* Allanheld and Co., Montclair, New York

Fromm, P.O. and Gillette, J.R. (1968). Effect of ambient ammonia on blood ammonia and nitrogen excretion of Rainbow trout (*Salmo gairdneri*). *Comp. Biochem. Physiol., 26:* 887-96

Gerking, S.D. (1955). Influence of rate of feeding on body composition and protein metabolism of bluegill sunfish. *Physiol. Zool., 28:* 267-82

Gigger, R.P. and Speece, R.E. (1970). Treatment of fish hatchery effluent for recycle. *New Mexico State Univ. Tech. Rept No. 67,* 119 pp.

Goldizen, V.C. (1970). Management of closed system marine aquariums. *Helgolander Wiss. Meeresanters, 20:* 637-41

Hall, M.D. (1974). Design and operation of a fish barn. Paper given to Am. Soc. Agr. Engrs, Chicago, December

Hampson, B.L. (1976). Ammonia concentration in relation to ammonia toxicity during a rainbow trout rearing experiment in a closed freshwater-seawater system. *Aquaculture, 9:* 61-70

Harman, O.R. and Maurer, D. (1971). Environmental considerations for shellfish production. Paper for Am. Soc. Agr. Engrs., Washington, 26 pp.

Herbert, D.W.M. and Sherben, D.S. (1965). The susceptibility of salmonid fish to poisons under estuarine conditions – 11. Ammonium chloride. *Int. J. Air and Wat. Poll., 9:* 89-91

Haug, R.T. and McCarty, P.L. (1971). Nitrification with the submerged filter. Dept. Civ. Eng'g, Stanford, California, Tech. Rept 149

Hawkes, H.A. (1963). *The Ecology of Wastewater Treatment.* Pergamon, London

Harris, D. (1974). *Hydroponics.* David and Charles, London

Henderson-Arzapalo, A., Stickney, R.R. and Lewis, D.H. (1980). Immune hypersensitivity in intensively cultured Tilapia species. *Trans. Am. Fish. Soc., 109:* 244-7

Herald, E.S., Dempster, R.P., Walters, C. and Hunt, M.L. (1971). Filtration and ultraviolet sterilisation of seawater in large closed and semiclosed aquarium systems. *Bull. Just. Oceano. Num. Spec., 1B:* 49-61

Hess, J. (1979). Discussion paper on nitrifying filter modelling. Bioengineering symposium, Culture Section, Am. Fish. Soc., Traverse City, Michigan, 15-19 October

Heukelekian, H., Orford, H.E. and Manganelli, R. (1951). Factors affecting the quantity of sludge production in the activated sludge process. *Sew. Ind. Wastes, 23:* 9-45

Hilge, V. (1976). Biological and economic aspects of fish production in a closed warm-water system. FAO Tech. Conf. Aquaculture, Kyoto, Japan, E21, 6 pp.

——— (1980). Rearing of channel catfish (*Ictalurus punctatus* Raf.) in a closed warm-water system. Symposium on New Developments in the Utilisation of Heated Effluents and Recirculation Systems for Intensive Aquaculture. EIFAC, 11th Session, Stavanger, Norway, 28-30 May, EIFAC/80/Symp: E/5

Hill, T.K., Chesness, J.L. and Brown, E.E. (1973). A two-crop production system. *Trans. Am. Soc. Ag. Engrs. 16 (5):* 930-3

Hillaby, B.A. and Randall, D.J. (1979). Acute ammonia toxicity and ammonia excretion in Rainbow trout (*Salmo gairdnerii*). *J. Fish. Res. Bd Can., 36:* 621-9

Hoffmann, G.L. (1974). Disinfection of contaminated water by ultraviolet irradiation with emphasis on whirling disease (*Myxosoma cerebralis*) and its effect on fish. *Trans. Am. Fish. Soc., 103:* 541-50

Holmberg, L. (1978). Extract, in English, from article in *Ingeniøren, 27* (7 July), Denmark

——— (1980). Fish farm in a factory basement. *Fish Farming Intl* (September): 16-19

Honn, K. (1979). Ozonation as a critical component of closed marine system design. *Science and Engineering, 1:* 11-29

Honn, K. and Chavin, W. (1975). Prototype design for a closed marine system employing quaternary water processing. *Mar. Biol., 31:* 293-8

Huisman, E.A. (1974). Oplunalisering van de Groei bij de Karper(*Cyprinus carpio* L). PhD thesis, Agric. Univ. of Wageningen, Holland

Hunn, J.B. (1969). Chemical composition of rainbow trout urine following acute hypoxic stress. *Trans. Am. Fish. Soc., 98 (1):* 20-2

Itazawa, Y. (1971). An estimation of the minimum level of dissolved oxygen in the water required for normal life of fish. *Bull. Jap. Soc. Sci. Fish., 37 (4):* 273-6

Jeris, J.S. and Owens, R.W. (1975). Pilot-scale high biological denitrification. *J. Wat. Poll. Cont. Fed., 47:* 2043-57

Jespersen, T. and Hodal, J. (1980). Fingerling production in a recycled system. Symposium on New Developments in the Utilisation of Heated Effluents and Recirculation Systems for Intensive Aquaculture. EIFAC, 11th Session, Stavanger, Norway, 28-30 May, EIFAC/80/Symp: E/20

Joost, R.H. (1969). Systemisation in using the rotating biological surface (RBS) waste treatment process. *Proc. 24th Ind. Waste Conf.,* Purdue University, pp. 365-73

Jørgensen, S.F., Libor, O., Barkacs, K. and Luna, K. (1979). Equilibrium and capacity data of clinoptilolite. *Wat. Res., 13 (2):* 159-65

Katz, W.J. and Rohlich, G.A. (1956). A study of the equilibrium and kinetics of adsorption by activated sludge. In: McCabe, J. and Eckenfelder, W.W., Jr (eds), *Biological Treatment of Sewage and Industrial Wastes,* Vol. 1, Rheingold, New York, pp. 64-70

Kausche, H. (1973). The influence of spontaneous activity on the metabolic rate of starved and fed young carp (*Cyprinus carpio*). *Verh. Int. Wer. Limnol., 17:* 669-79

Kauschik, J. (1980). Influence of a rise in temperature on the nitrogen excretion of rainbow trout (*Salmo gairdnerii* R.). Symposium on New Developments on the Utilisation of Heated Effluents and Recirculation Systems for Intensive Aquaculture. EIFAC, 11th

Session, Stavanger, Norway, 28-30 May, EIFAC/80/Symp: E/6

Kawai, A., Yoshida, Y. and Kimata, M. (1965). Biochemical studies on the bacteria in the aquarium with a circulating system. *Bull. Jap. Soc. Sci. Fish., 31 (1):* 65-71

Kawamoto, N.Y. (1961). The influence of excretory substances of fishes on their growth. *Prog. Fish. Cult., 23 (2):* 70-5

Kerr, N.M. (1976). Farming marine flatfish using waste heat from seawater cooling. *Energy World, 4 (10):* 2-10
——— (1980). Design of equipment and selection of materials. Symposium on New Developments in the Utilisation of Heated Effluents and Recirculation Systems for Intensive Aquaculture. EIFAC, 11th Session, Stavanger, Norway, 28-30 May, EIFAC/80/Symp: R/5

Kimura, T., Yoshiruzu, M., Tajima, K., Ezura, Y., and Sakai, M. (1976). Disinfection of hatchery water supply by ultraviolet (UV) irradiation. Susceptibility of some fish pathogenic bacteria and microorganisms inhabiting pond waters. *Bull. Jap. Soc. Sci. Fish., 42:* 207-11

Kirk, W.L. (1974). The effects of hypoxia on certain blood and tissue electrolytes of channel catfish (*Ictalurus punctatus*) (Rafinesque). *Trans. Am. Fish. Soc., 103 (3):* 593-600

Knepp, G.L. and Arkin, G.F. (1973). Ammonia toxicity levels and nitrite tolerance of channel catfish. *Prog. Fish. Cult., 35 (4):* 221-4

Knowles, G., Downing, A.L. and Barrett, M.J. (1965). Determination of kinetic constants for nitrifying bacteria in mixed culture with the aid of an electronic computer. *J. Gen. Micr. 38:* 263-78

Konikoff, M. (1975). Toxicity of nitrite to channel catfish. *Prog. Fish. Cult., 37 (2):* 96-8

Kossmann, H. (1980). A warm-water recycling plant for production of grass carp (*Ctenopharyngodon idella*) in Sweden. Symposium on New Developments in the Utilisation of Heated Effluents and Recirculation Systems of Intensive Aquaculture. EIFAC, 11th Session, Stavanger, Norway, 28-30 May, EIFAC/80/Symp. E/71

Kujal, B. (1980). Pretreatment of water for intensive aquaculture by means of filters with inverted water passage. Symposium on New Developments in the Utilisation of Heated Effluents and Recirculation Systems for Intensive Aquaculture. EIFAC, 11th Session, Stavanger, Norway, 28-30 May, EIFAC/80/Symp: E/7

Larmoyeux, J.D. and Piper, R.G. (1973). Effects of water reuse on rainbow trout in hatcheries. *Prog. Fish Cult., 35 (1):* 2-8

Laudelout, H., Simionart, P.C. and Van Droogenbroeck, R. (1968). Free energy efficiency of *Nitrosomonas. Arch. Microbiol., 63 (3):* 256-77

Levenspiel, O. (1967). *Chemical Reaction Engineering.* Wiley, New York

Lewis, W.M. (1966). Odours and tastes in water derived from the River Severn. *Proc. Soc. Wat. Treat. and Exam., 15:* 30-5

Lewis, W.M. and Buynak, G.L. (1976). Evaluation of revolving plate type biofilter for use in recirculated fish production and holding units. *Trans. Am. Fish. Soc., 105 (6):* 704-8

Lewis, W.M., Yopp, J.H., Brandenburg, A.M. and Schnoor, K.D. (1980). On the maintenance of water quality for closed fish production systems by means of hydroponically grown vegetable crops. Symposium on New Developments in the Utilisation of Heated Effluents and Recirculation systems for Intensive Aquaculture. EIFAC, 11th Session, Stavanger, Norway, 28-30 May, EIFAC/80/Symp: E/57

Liao, I.C. (1969). Study on the feeding of 'Kuruma' prawn *Penaeus japonicus*. Bate. Tungkang Marine Laboratory, collected reprints, Vol. 1, 1969-71, pp. 17-24

Liao, P.B. (1970). Salmon hatchery wastewater treatment. *Wat. Sew. Wks, 117:* 439-43
——— (1971). Water requirements of salmonids. *Prog. Fish Cult., 33:* 210-15
——— (1980). Treatment units used in recirculation systems for intensive aquaculture. Symposium on New Developments in the Utilisation of Heated Effluents and Recirculation Systems for Intensive Aquaculture. EIFAC, 11th Session, Stavanger, Norway, 28-30 May, EIFAC/80/Symp: R/6

Liao, P.B. and Mayo, R.D. (1973). Salmonid hatchery water reuse systems. *Aquaculture:* 317-35
——— (1974). Intensified fish culture combining water reconditioning with pollution abatement. *Aquaculture, 3:* 61-85

Liao, P.B., Mayo, R.D. and Williams, W. (1972). A study for the development of a fish hatchery water treatment system. *Rept. U.S. Army Corps of Engineers,* Walla Walla, Wash.

Lloyd, R. and Herbert, D.W.M. (1960). The influence of carbon dioxide on the toxicity of unionised ammonia to rainbow trout (*Salmo gairdneri* Richardson). *Ann. Appl. Biol., 48 (2):* 399-404

Lloyd, R. and Orr, L.D. (1969). The diuretic response by rainbow trout to sublethal concentrations of ammonia, *Wat. Res., 3:* 335-44

Lu, J.D. (1978). A recirculating aquaculture system. *Bamidgeh, 30 (1):* 12-22

MacDonald, C.R., Meade, T.L. and Gates, G.M. (1975). A production cost analysis of a closed system culture of salmonids. Univ. Rhode Island NOAA Sea Grant, Mar. Tech. Rept, No. 41, 11 pp.

Mayer, F.L. and Kramer, R.H. (1973). Effects of hatchery water reuse on rainbow trout metabolism. *Prog. Fish Cult., 35 (1):* 9-11

Mayo, R.D. (1976). A technical and economic review of the use of reconditioned water in aquaculture. FAO Tech. Conf. Aquacult., Kyoto, Japan, R.30, 24 pp.

———— (1980). Recirculation systems in North America. Symposium on New Developments in the Utilisation of Heated Effluents and Recirculation Systems for Intensive Aquaculture. EIFAC, 11th Session, Stavanger, Norway, 28-30 May, EIFAC/80/Symp: E/76

McAuley, R. and Casselman, J. (1980). Temperature preference of fish as an index of the optimum temperature range for growth. Symposium on New Developments in the Use of Heated Effluents and Recirculation Systems for Intensive Aquaculture. EIFAC, 11th Session, Stavanger, Norway, 28-30 May, IEFAC/80/Symp: E/76

McGriff, E.C. (1972). Removal of nutrients and organics by activated algae. *Wat. Res., 6:* 1155-64

Meade, T.L. (1974). The technology of closed system culture of salmonids. Univ. Rhode Island NOAA Sea Grant, Mar. Tech. Rept, No. 30, 30 pp.

———— (1979). A combined physical and pulsed sand filter for solids removal from fish culture water. Bioengineering Symp, Culture Section, Am. Fish. Soc., Traverse City, Michigan, 15-19 October (Abstract only, paper not presented)

Meade, T.L. and Kenworthy, B.R. (1974). Denitrification in a water reuse system. *Proc. 5th Ann. Workshop,* World Mariculture Soc., Louisiana State Univ., Baton Rouge, pp. 33-42

Meske, C. (1973). *Aquakultur von Warmwasser Nutzfischen.* Verlag Fugen Ulmer, Stuttgart

———— (1976). Fish culture in a recirculation system with water treatment and activated sludge. FAO Tech. Conf. Aquacult., Kyoto, Japan, E.62, 7 pp.

Miklosz, J.C. (1971). Biological filtration. *The Marine Aquarist, 1 (5):* 22-9

Mitchell, R.E. and Kirby, A.M. (1976). Performance characteristics of pond aeration devices. *Proc. 7th Ann. Meeting,* World Mariculture Soc., Louisiana State Univ., Baton Rouge, pp. 561-81

Mitchell, R.E. and Lev, A.D. (1971). Economic comparison of U-tube aeration with other methods for aerating wastewater. Am. Inst. Chem. Engrs, *Chem. Eng. Prog. Symp. Ser. 67,* pp. 558-65

Møller, D. and Bjerk, O. (1980). Smolt production in a recirculation system in Northern Norway. Symposium on New Developments in the Utilisation of Heated Effluents and Recirculation Systems for Intensive Aquaculture. EIFAC, 11th Session, Stavanger, Norway, 28-30 May, EIFAC/80/Symp: E/52

Muir, J.F. (1975). Waste recycling systems in fish farming. *Fish Farming Int., 2 (2):* 14-15, 48

———— (1976). How filters improve water quality for fish farmers. *Fish Farming Int., 3 (3):* 35-8

———— (1978). Aspects of water treatment and reuse in intensive fish culture. PhD thesis, University of Strathclyde, Glasgow, 451 pp.

———— (1979). Management and cost implications in water reuse systems. Bioengineering Symp, Culture Section, Am. Fish. Soc., Traverse City, Michigan, 15-19 October

Mukherjee, S. and Bhattachanya, S. (1974). Effect of some industrial pollutants on fish brain cholinesterase activity. *Environ. Physiol. Biochem., 4:* 226-31

Munro, A.L.S. (1978). The aquatic environment. In: Roberts, R.J. (ed.), *Fish Pathology,* Balliere Tindall, London

Murphy, J.P. and Lipper, R.I. (1970). BOD production of channel catfish. *Prog. Fish Cult.*, *32 (4):* 195-8

Murray, K.R., Poxton, M.G., Linfoot, B.T. and Watret, D.W. (1980). The design and performance of low pressure air lift pumps in a closed marine recirculation system. Symposium on New Developments in the Utilisation of Heated Effluents and Recirculation Systems for Intensive Aquaculture. EIFAC, 11th Session, Stavanger, Norway, 28-30 May, EIFAC/80/Symp: E/24

Nagel, L. (1977). Combined production of fish and plants in recirculating water. *Aquaculture, 10 (1):* 17-24

Ogden, C.G., Gibbs, J.W. and Gameson, A.L.H. (1959). Some factors affecting the aeration of flowing saline water. *Wat. Waste Treat. J.,* 7: 392-6

Olsen, K.R. and Fromm, P.O. (1971). Excretion of urea by two teleosts exposed to different concentrations of ambient ammonia. *Comp. Biochem. Physiol., 40:* 999-1007

Otte, G., Hilge, V. and Rosenthal, H. (1977). Effect of ozone on yellow substances accumulating in a recycling system for fish culture. ICES, CM, 1977/E 27

Otte, G. and Rosenthal, H. (1979). Management of closed brackish-water system for high-density fish culture by biological and chemical water treatment. *Aquaculture, 18:* 169-81

Painter, H.A. (1970). A review of the literature on the inorganic nitrogen metabolism in micro-organisms. *Wat. Res., 4:* 393-450

Paloheimo, J.E. and Dickie, L.M. (1966). Food and growth of fishes. *J. Fish. Res. Bd Can., 23:* 133-9

Parisot, T.J. (1967). A closed recirculated seawater system. *Prog. Fish Cult., 29 (7):* 133-9

Parker, N.C. and Broussard, M.C. (1977). Selected bibliography of water reuse systems for aquaculture. Texas Agric. Exp. Stat., 33 pp.

Parker, N.C. and Simco, B.A. (1974). Evaluation of recirculating systems for the culture of channel catfish. *Proc. 27th Ann. Conf. S.E. Fish and Game Comm.,* Hot Springs, Ark., pp. 474-87

Parker, S.S. (1972). Biological pretreatment at Strensham. *Wat. Treat. Exam., 21:* 315

Perrone, S.J. and Meade, T.L. (1977). Protective effect of chloride on nitrite toxicity to Coho salmon (*Onchorynchus kisutch*). *J. Fish. Res. Bd Can., 34:* 486-92

Perry, R.H. and Chilton, C.H. (1973). *The Chemical Engineers Handbook*, 5th edn. McGraw-Hill, New York

Petit, J. (1980). Utilisation de l'oxygene pur en pisciculture. Symposium on New Developments in the Utilisation of Heated Effluents and Recirculation Systems for Intensive Aquaculture. EIFAC, 11th Session, Stavanger, Norway, 28-30 May, EIFAC/80/Symp: E/54

Pettigrew, T., Henderson, E.B., Saunders, R.L. and Sochasky, J.B. (1978). A review of water reconditioning reuse technology for fish culture with a selected bibliography. Fish. and Marine Service Tech. Rep., No. 801, Biol. Stat., New Brunswick, 19 pp.

Pfuderer, P., Williams, P. and Francis, A.A. (1974). Partial purification of the crowding factor from *Carassius auratus* and *Cyprinus carpio. J. Exp. Zool., 187:* 375-82

Pike, E.S. and Curds, C.R. (1971). Microbial ecology of the activated sludge process. In: Sykes, G. and Skinner, F.A. (eds), *Microbial Aspects of Pollution.* Academic Press, New York

Poston, H.A. (1978). Neuroendocrine mediation of photoperiod and other environmental influences on physiological responses in salmonids, a review. Tech. Paper 96, US Fish and Wildlife Service, USDI

Poxton, M.G., Murray, K.R., Linfoot, B.T. and Pooley, A.B.W. (1980). The design and performance of biological filters in an experimental facility. Symposium on New Developments in the Utilisation of Heated Effluents and Recirculation Systems for Intensive Aquaculture. EIFAC, 11th Session, Stavanger, Norway, 28-30 May, EIFAC/80/Symp: E/10

Pugh, N.J. (1945). The treatment of doubtful wastes for public water supply. *Trans. Inst. Water Engrs, 50:* 80

Rapaport, A., Sarig, S. and Marek, M. (1976). Results of tests of varied aeration systems on the oxygen regime in the Ginossar experimental ponds, and the growth of the fish there in 1975. *Bamidgeh, 28 (3):* 35-49

Rasmussen, K. (1980). Culture of whitefish (*Coregonus* spp). in recirculated water with reuse of dissolved nutrients. Symposium on New Developments in the Utilisation of Heated Effluents and Recirculation Systems for Intensive Aquaculture. EIFAC, 11th Session, Stavanger, Norway, 28-30 May, EIFAC/80/Symp: E/19

Risa, S. and Skjervold, H. (1975). Water reuse system for smolt production. *Aquaculture, 6:* 191-5

Robinette, H.R. (1976). Effects of selected sublethal levels of ammonia on the growth of channel catfish (*Ictalurus punctatus*). *Prog. Fish Cult., 38 (1):* 26-9

Robinson-Wilson, E.F. and Seim, W.K. (1975). The lethal and sublethal effects of a zirconium process effluent on juvenile salmonids. *Wat. Res. Bull., 11:* 975-86

Rosenthal, H. (1980a). Ozonation and sterilisation. Symposium on New Developments in the Utilisation of Heated Effluents and Recirculation Systems for Intensive Aquaculture. EIFAC, 11th Session, Stavanger, Norway, 28-30 May, EIFAC/80/Symp: R/8

——— (1980b). Recirculation systems in Western Europe. Symposium as above, EIFAC/80/Symp: R/13

Rosenthal, H., Andjus, R. and Kruner, G. (1980). Daily variations of water quality parameters under intensive culture conditions in a recycling system. Symposium as above, EIFAC/80/Symp: E/59

Rosenthal, H., Kruner, G. and Otte, G. (1978). Effects of ozone treatment on recirculating water in a closed fish culture system. ICES CM/1978/F. 9, pp. 1-16

Russo, R.C., Smith, C.E. and Thurston, R.V. (1974). The acute toxicity of nitrite to rainbow trout. *J. Fish. Res. Bd Can., 31:* 1653-5

Rychly, J. and Marina, B.A. (1977). The ammonia excretion of trout during a 24-hour period. *Aquaculture, 11:* 173-8

Saeki, A. (1958). Studies on fish culture in the aquarium of a closed circulating system, its fundamental theory and standard plan. *Bull. Jap. Soc. Sci. Fish., 23 (1):* 884-95

Seegert, G.L. and Brooks, A.S. (1978). Dechlorination of water for fish culture: comparison of the activated carbon, sulfite reduction, and photochemical methods. *J. Fish. Res. Bd Can., 35:* 88-92

Serfling, S.A., Van Olst, J.C. and Ford, R.F. (1974). A recirculating culture system for larvae of the American lobster, *Homarus americanus. Aquaculture, 3 (3):* 303-9

Shell, G. and Cassady, T. (1973) Selecting mechanical aerators. *Ind. Wat. Eng., 10 (4):* 21-5

Shepherd, C.J. (1973). Studies on the biological and economic factors involved in fish culture, with special reference to Scotland. PhD thesis, University of Stirling, Scotland

Shirahata, S. (1964). Problems of water quality in food trout production. *Bull. Fac. Fish. Nagasaki Univ., 17:* 68-82

Short, C.S. (1973). Removal of ammonia from river waters. *Wat. Res. Assn Tech. Rept, 101:* 42 pp.

Smart, G.R. (1976). The effect of ammonia exposure on the gill structure of the rainbow trout (*Salmo gairdnerii* R.). *J. Fish Biol., 8 (6):* 474-5

Smart, G.R., Knox, D., Harrison, J.G., Ralph, J.A., Richards, R.H. and Cowey, C.B. (1978). Nephrocalcinosis in rainbow trout: the effect of exposure to elevated carbon dioxide concentrations. *J. Fish Diseases, 2:* 279-89

Smith, C.E. and Piper, R.G. (1975). Lesions associated with chronic exposure to ammonia. In: Ribelin, W.E. and Migaki, G. (eds), *Pathology of Fishes*. University of Wisconsin Press, Madison, pp. 497-514

Smith, C.E. and Russo, R.C. (1975). Nitrite-induced methemoglobinaemia in rainbow trout. *Prog. Fish Cult., 37 (3):* 150-2

Smith, C.E. and Williams, W.G. (1974). Experimental nitrite toxicity in rainbow trout and chinook salmon. *Trans. Am. Fish. Soc., 103:* 389-90

Smith, J.M., Masse, A.M., Fiege, W.E.A. and Kamphake, L.J. (1972). Nitrogen removal from municipal waste water by columnar denitrification. *Env. Sci. Tech., 6 (3):* 260-7

Sousa, R.J. and Meade, T.L. (1977). Influence of ammonia on the oxygen delivery system of coho salmon haemoglobin. *Comp. Biochem. Physiol., 58A:* 23-8

Sowerbutts, B.J. and Forster, J.R.M. (1980). Gas exchange and reoxygenation. Symposium on New Developments in the Utilisation of Heated Effluents and Recirculation Systems for Intensive Aquaculture. EIFAC, 11th Session, Stavanger, Norway, 28-30 May, EIFAC/80/Symp: R/7

Speece, R.E. (1973). Trout metabolism characteristics and the rational design of nitrification facilities for water reuse in hatcheries. *Trans. Am. Fish. Soc., 102:* 323-3

Speece, R.E., Adams, J.L. and Woolridge, C.B. (1969). U-tube aeration operating characteristics. *J. Sanit. Eng. Div. Proc. Am. Soc. Civil Engrs, 95 (SA3):* 563-74

Speece, R.E. and Orosco, R. (1970). Design of U-tube aeration systems. *J. Sanit. Eng. Div. Proc. Am. Soc. Civil Engrs, 96 (SA4):* 715-25

Spotte, S.H. (1970). *Fish and Invertebrate Culture: Water Management in Closed Systems.* Wiley, New York, 145 pp.

——— (1979). *Seawater Aquariums, the Captive Environment.* Wiley, New York, 413 pp.

Srna, R. and Baggaley, A. (1975). The kinetic response of perturbed marine nitrification systems. *J. Wat. Poll. Cont. Fed., 47 (3):* 472

St Amant, P.P. and McCarty, P.L. (1969). Treatment of high-nitrate waters. *J. Am. Wat. Wks Assn 61 (12):* 659-62

Steels, I.H. (1974). Design basis for the rotating disc process. *Effluent Wat. Treat. J., 14 (8):* 431-45

Stewart, N.E., Shumway, D.L. and Doudoroff, P. (1967). Influence of oxygen concentration on the growth of juvenile largemouth bass. *J. Fish. Res. Bd Can., 24 (3):* 475-94

Strand, J.A., Cummins, J.T. and Vaughan, B.E. (1969). A fast-flow sealed disc filter system for marine aquaria. *Limnol. and Ocean., 14:* 444-8

Sumari, O. and Westman, K. (1980). Biological and economic aspects in the use of heated water for salmon smolt production as compared with traditional rearing. Symposium on New Developments in the Utilisation of Heated Effluents and Recirculation Systems for Intensive Aquaculture. EIFAC, 11th Session, Stavanger, Norway, 28-30 May, EIFAC/80/Symp: E/33

Thomas, H.A. and Archibald, R.S. (1952). Longitudinal mixing measured by radioactive tracers. *Trans. Am. Soc. Civil Engrs, 117:* 839

Tischler, L.F. and Eckenfelder, W.W. (1969). Linear substrate removal in the activated sludge process. *Wat. Res., 2:* 427-47

Tomasso, J.R., Simco, B.A. and Davis, K.B. (1979). Chloride inhibition of nitrite-induced methemoglobinaemia in channel catfish (*Ictalurus punctatus*). *J. Fish. Res. Bd Can., 36:* 1141-4

Tomlinson, T.G., Loveless, J.E. and Sear, L.G. (1962). Effect of treatment in percolating filters on the number of bacteria in sewage in relation to the size and composition of filtering medium. *J. Hyg. Camb., 60:* 365

Van Toever, W. and MacKay, K.T. (1980). A modular recirculating hatchery and rearing system for salmonids utilising ecological design principles. Symposium on New Developments in the Utilisation of Heated Effluents and Recirculation Systems for Intensive Aquaculture. EIFAC, 11th Session, Stavanger, Norway, 28-30 May, EIFAC/80/Symp: E/58

Waggott, A. and Bayley, R.W. (1972). The use of activated carbon for improving the quality of polished sewage effluent. *J. Inst. Wat. Poll. Cont., 71 (4):* 417-25

Warren, C.E., Doudoroff, P. and Shumway, D.L. (1973). Developments of dissolved oxygen criteria for freshwater fish. Ecological Research Series, EPA-R3-73-019, US Environmental Protection Agency, Wash. DC, 121 pp.

Warrer-Hansen, I. (1979). Fish farming and pollution problems – methods to reduce discharged matters. SMBA/HIDB, Fish Farmers Meeting, Oban, Scotland, 22-3 Feb., 17 pp.

Wedemeyer, G.A. (1976). Physiological response of juvenile Coho Salmon (*Oncorynchus kisutch*) and Rainbow trout (*Salmo gairdneri*) to handling and crowding stress in intensive culture. *J. Fish. Res. Bd Can., 33 (12):* 2699-702

——— (1980). The physiological response of fishes to the stress of intensive aquaculture in recirculation systems. Symposium on New Developments in the Utilisation of Heated Effluents and Recirculation Systems for Intensive Aquaculture. EIFAC, 11th Session, Stavanger, Norway, 28-30 May, EIFAC/80/Symp: R/9

Wedemeyer, G.A. and Nelson, N.C. (1977). Survival of two bacterial fish pathogens (*Aeromonas salmonicida* and the Enteric Redmouth Bacterium) in ozonated, chlorinated, and untreated waters. *J. Fish. Res. Bd Can., 34:* 429-32

Wedemeyer, G.A. and Yasutake, W.T. (1978). Prevention and treatment of nitrite toxicity in juvenile steelhead trout (*Salmo gairdnerii*). *J. Fish. Res. Bd Can., 35:* 822-7

Wedemeyer, G.A., Meyer, F.P. and Smith, L. (1976). Environmental stress and fish diseases.

In: Sniesko, S.F. and Axelrod, H.R. (eds), *Diseases of Fishes*, Book 5. TFH Publications, New Jersey

Wedemeyer, G.A., Nelson, N.C. and Smith, C.A. (1978). Survival of the salmonid viruses infectious haemopoietic necrosis (IHNV), and infectious pancreatic necrosis (IPNV), in ozonated, chlorinated, and untreated waters. *J. Fish. Res. Bd Can., 35:* 875-9

Wedemeyer, G.A., Nelson, N.C. and Yasutake, W.T. (1979). Physiological and biochemical aspects of ozone toxicity to rainbow trout (*Salmo gairdnerii* R.). *J. Fish. Res. Bd Can., 36:* 605-14

Westin, D.T. (1974). Nitrate and nitrite toxicity to salmonid fishes. *Prog. Fish Cult., 36 (2):* 86-9

Wheatland, A.B., Bairet, M. and Bruce, A. (1959). Some observations on denitrification in rivers and estuaries. *J. Inst. Sew. Purif., 49:* 149-59

Wheaton, F.W. (1977). *Aquaculture Engineering.* Wiley, New York, 708 pp.

Wickins, J.F. (1976). The tolerance of warm-water prawns to recirculated water. *Aquaculture, 9:* 19-37

—— (1980). Water quality requirements for intensive animal aquaculture — a review. Symposium on New Developments in the Utilisation of Heated Effluents and Recirculation Systems for Intensive Aquaculture. EIFAC, 11th Session, Stavanger, Norway, 28-30 May, EIFAC/80/Symp: R/2

Wijler, J. and Delwiche, C.C. (1954). Investigations on the denitrifying process in soil, *Plant and Soil, 5 (2):* 155-69

Wild, H.E., Sawyer, C.N. and McMahon, T.C. (1971). Factors affecting nitrification kinetics. *J. Wat. Poll. Cont. Fed., 43 (9):* 1845-54

Willoughby, H., Larsen, H.N. and Bowen, J.T. (1972). The pollutional effects of fish hatcheries. *Am. Fish and Trout News. 17 (3):* 6-7, 20

Winberg, C.G. (1956). Rate of metabolism and food requirement of fishes. *Fish. Res. Bd Can., Trans. Ser.,* No. 194

Wohlfarth, G. and Lahman, M. (1971). Preliminary experiments in growing carp in recirculating water, *Bamidgeh, 23 (4):* 103-16

Wood, J.D. (1958). Nitrogen excretion in some marine teleosts. *Can. J. Biochem. Physiol. 36:* 1237-42

Wuhrmann, K. (1963). Effects of oxygen tension on biochemical reactions in sewage purification plants. In: Eckenfelder, W.W. and McCabe, J. (eds), *Advances in Biological Waste Treatment.* Pergamon, Oxford, pp. 27-38

Yu, M. and Perlmutter, A. (1970). Growth inhibiting factors in the Zebra-fish, *Brachydanio rerio*, and the Blue gourami, *Trichogaster trichopterus. Growth, 34:* 153-75

CONTRIBUTORS

John D. Balarin, BSc, MSc, is a consultant with Baobab Farms, Mombasa, Kenya.

René D. Haller, BAgron, is the managing director of Baobab Farms, Mombasa, Kenya.

Kim Jauncey, BSc, PhD, is an ODA Research Lecturer in the Institute of Aquaculture, University of Stirling, Scotland.

Donald J. Macintosh, BSc, PhD, is an ODA Research Lecturer in the Institute of Aquaculture, University of Stirling.

James F. Muir, BSc, PhD, is Lecturer in Aquaculture Engineering, Institute of Aquaculture, University of Stirling.

Ronald J. Roberts, BVMS, PhD, MRCPath, MRCVS, FIBiol., FRSE, is Professor and Director of the Institute of Aquaculture, University of Stirling.

Kok Leong Wee, BSc, MSc, is a Research Student, Institute of Aquaculture, University of Stirling.

John F. Wickins, BSc, MIBiol., is Senior Scientific Officer at the MAFF Fisheries Experimental Station, Conwy, Gwynedd, Wales.

INDEX

Acetes 26, 60

airlift: configurations 422; pumps 421-3

algae: *Chaetomorpha* and *Enteromorpha* 34; digestibility in carp feeds 231; in mangroves 14, 17

alpha-tocopherol (vitamin E) 243

amino acids, in carp feeds 218-34; qualitative requirements 218-19; quantitative requirements 219-22, availability 220-2; supplementation 230

ammonia: air stripping of 420; and hydraulic flow-rate 390; and hydraulic loading 391; dissociation in water 377; limitations 363-4; oxidation, amount and rate of 156, 416; removal from river wastes 383; total nitrogen 158

anchovy, Peruvian 229

anorexia, in carp fed riboflavin-poor diets 239, 243

ants, tailor (*Oecophylla* spp.) 23

Aphanomyces astaci (Schikora) 123

aquaculture: energy use in 420-4; husbandry and production management 433-4; organisms of in mangrove swamps 35-74; pumping energy 420-1; technology in snakehead farming 209-10

Artemia (brine shrimp) 95, 253

ascorbic acid (vitamin C) 242-3

Astacus astacus (Noble crayfish) 123, 124, 128

Australian marron (*Cherax tenuimanus*) 101, 125, 146, 148

Austropotamobius pallipes 124, 125, 126, 128, 163

Avicennia 14, 24, 25, 31, 37

'Avomarin' enzyme preparation 249

Bacillus, in denitrification 405

backwashing 413; and filter nitrification 393-4

bacteria, nitrifying 149, 157

Baobab Farm, Kenya, and tilapia rearing 311

bed systems, fluidised 404

'biodisc', 'biodrum' and 'biosedimentation' 401; systems 402, 403-4

biotin 242

Bohr effect, and oxygen dissociation curves 281

Boleophthalmus (mudskipper) 26, 47-8

bream, Red Sea 246

Brighton Marina, use of for containing hatchery reared shellfish 122

brine shrimp (*Artemia salina*) 95, 253

Britain, *Palaemon* and *Pandalus* spp. in 116

cage culture, in tilapia 320-30; *see also under* tilapia

calcium 246-7

Cancer pagurus see crab

Candida lipolytica 253

carbohydrate 237-8; dietary protein sparing by 227-9

carbonaceous oxidation, in water reuse 404-7

carbonate system, and pH control in water reuse 417-20; activated carbon treatment 419-20; ion exchange treatment 419

carbon dioxide, and *S. mossambicus* 282-4; limitations 364

carp (*Cyprinus carpio* L.): algae digestibility in feeds 231; anorexia due to riboflavin-poor diets 239, 243; carbohydrates in 237-8, glucose oxidation in 238; cystine, in feeds 230; effect of temperature on growth of mirror 249-51; feed additives 249; growth responses of 252; Halver's test diet and 219; larval nutrition 253-4; lipids in 234-7; minerals in 244-9, availability of dietary phosphorus to 247; proteins and amino acids in feeds 218-34, digestibility of sources by 232, 233; ration size of feed in 251-3; vitamins in 238-44, summary of 245; water quality requirements 280

Caulerpa seaweed 73

cellulose activity, in digestive tracts of estuarine fish 238

Chaetomorpha algae 34

Chanidae *see* milkfish

Channa punctatus (snakehead) lethal effects of DDVP on 191

channel catfish 228, 229, 237, 250; water quality requirements for 280

Channidae *see* snakeheads

chemical oxidation processes, in water reuse 416-17

Cherax tenuimanus (Australian marron) 101, 125, 146, 148

chinook salmon *(Oncorhynchus tschawytscha*) 218, 223, 239, 240; dextrin in diets 228

chlorine, and sterilisation 416

choline 242

cichlids (*Tilapia* or *Sarotherodon*) 46-7, 271; pearlspot 47; *S. mossambicus*

448

DATE DUE

F